Topics in
Current Physics

17

# Topics in Current Physics

Founded by Helmut K. V. Lotsch

1 **Beam-Foil Spectroscopy**
Editor: S. Bashkin

2 **Modern Three-Hadron Physics**
Editor: A. W. Thomas

3 **Dynamics of Solids and Liquids by Neutron Scattering**
Editors: S. W. Lovesey and T. Springer

4 **Electron Spectroscopy for Surface Analysis**
Editor: H. Ibach

5 **Structure and Collisions of Ions and Atoms**
Editor: I. A. Sellin

6 **Neutron Diffraction**
Editor: H. Dachs

7 **Monte Carlo Methods** in Statistical Physics
Editor: K. Binder

8 **Ocean Acoustics**
Editor: J. A. DeSanto

9 **Inverse Source Problems** in Optics
Editor: H. P. Baltes

10 **Synchrotron Radiation** Techniques and Applications
Editor: C. Kunz

11 **Raman Spectroscopy** of Gases and Liquids
Editor: A. Weber

12 **Positrons in Solids**
Editor: P. Hautojärvi

13 **Computer Processing of Electron Microscope Images**
Editor: P. W. Hawkes

14 **Excitons** Editor: K. Cho

15 Physics of **Superionic Conductors**
Editor: M. B. Salamon

16 **Aerosol Microphysics I** Particle Interactions
Editor: W. H. Marlow

17 **Solitons**
Editors: R. K. Bullough, P. J. Caudrey

18 **Magnetic Electron Lens Properties**
Editor: P. W. Hawkes

19 Theory of **Chemisorption**
Editor: J. R. Smith

20 **Inverse Scattering Problems** in Optics
Editor: H. P. Baltes

21 **Coherent Nonlinear Optics** Recent Advances
Editors: M. S. Feld and V. S. Letokhov

22 **Electromagnetic Theory of Gratings**
Editor: R. Petit

# Solitons

Edited by
R. K. Bullough and P. J. Caudrey

With Contributions by
R. K. Bullough F. Calogero P. J. Caudrey
A. Degasperis L. D. Faddeev H. M. Gibbs
R. Hirota G. L. Lamb, Jr. A. H. Luther
D. W. McLaughlin A. C. Newell S. P. Novikov
M. Toda M. Wadati V. E. Zakharov

With 20 Figures

Springer-Verlag Berlin Heidelberg New York 1980

Professor Dr. Robin K. Bullough
Dr. Philip J. Caudrey

Department of Mathematics, University of Manchester,
Institute of Science and Technology,
Manchester M60 1QD, UK



ISBN 3-540-09962-X Springer-Verlag Berlin Heidelberg New York
ISBN 0-387-09962-X Springer-Verlag New York Heidelberg Berlin

Library of Congress Cataloging in Publication Data. Main entry under title: Solitons. (Topics in current physics ; v. 17), Bibliography: p. Includes index.
1. Solitons. I. Bullough, Robert Keith, 1929–. II. Caudrey, P. J., 1943–. III. Series.
QC174.26.W28S63 530.1′24 80-12790



Offset printing and bookbinding: Konrad Triltsch, Graphischer Betrieb, Würzburg.
2153/3130-543210

"This is a most beautiful and extraordinary phenomenon: the first day I saw it was the happiest day of my life. Nobody had ever had the good fortune to see it before or, at all events, to know what it meant. It is now known as the solitary wave of translation. No one before had fancied a solitary wave as a possible thing. When I described this to Sir John Herschel, he said 'It is merely half of a common wave that has been cut off'. But it is not so, because the common waves go partly above and partly below the surface level; and not only that but its shape is different. Instead of being half a wave it is clearly a whole wave, with this difference, that the whole wave is not above and below the surface alternately but always above it. So much for what a heap of water does: it will not stay where it is but travels to a distance."

John Scott Russell: The Modern System of Naval Architecture, Vol. I (Day & Son, London 1865) p. 208

# Preface

Until the beginning of this decade the number of significant exactly soluble problems in physics was limited to a very few: the classical or quantised harmonic oscillator, the linearised many-body problem, the quantised hydrogen atom, Newton's solution of the planetary orbit problem, Onsager's solution of the two-dimensional Ising problem, almost exhaust the list. Now the situation is quite different. We have a large number of exactly soluble nonlinear systems of physical significance and the number of these is growing steadily. Recent examples include a limited solution of Einstein's field equations[1], an apparently exact solution of the quantised sine-Gordon equation $u_{xx} - u_{tt} = \sin(u)$ which establishes connections with the Ising models[2], and a solution of the equations of motion of a rigid body in g dimensions[3].

This book is concerned with problems such as these. But more specifically it is concerned with solitons. These mathematical objects are exact, analytical, solutions of nonlinear wave or evolution equations like the sine-Gordon equation (s-G) or the Korteweg-de Vries equation (KdV) $u_t + 6uu_x + u_{xxx} = 0$. The discovery by Gardner, Greene, Kruskal and Miura in 1967 that there was an analytical method for solving the initial value problem for this particular equation, and the subsequent discovery that similar methods could be applied to the s-G and other nonlinear equations, have created a revolution in nonlinear physics which has changed our views and understanding of many physical problems. There is no sign of this revolution subsiding and what it has already achieved seems to be of permanent value.

For the uninitiated reader it is worth remarking that the applications of the s-G, for example, already span fields as various as crystal dislocation theory, [4]the theory of Josephson junctions,[4] a theory of spin excitations in liquid $^3$He below 2.6 mK,[5] the theory of ns or shorter resonant optical pulses,[4] the charge density wave theory of one-dimensional organic conductors like TTF-TCNQ[6], and model

1 See [1.207-209].
2 See Chap.12. Reference [12.17] shows the solution *is* exact.
3 This solution is referred to in [1.187] and described in Matveev's *Abelian Functions* [1.119]. For g = 3 it is the celebrated Sonja Kowalewski (1888) elliptic function solution.
4 See Chaps.1 and 2.
5 See Chap.3.
6 See [3.21]. Another beautiful application of the s-G to a problem in the solid state is that made by Mikeska [J. Phys. C *11*, L29 (1978)] to the one-dimensional ferromagnetic $CsNiF_3$. For experiments and more theory see Kjems and Steiner [Phys. Rev. Lett. *31*, 1137 (1978)] and much subsequent literature which has appeared in 1979.

field theories.[7] The KdV emerged as long ago as 1895 in the theory of gravity waves in shallow water, but it has appeared again in lattice theory, plasma theory, and magnetohydrodynamics. The list of applications of these and related nonlinear equations in physics is now very extensive[8] and there seems no doubt of their physical importance.

This book on solitons was conceived in 1975 when Dr. H. Lotsch of Springer-Verlag suggested to the two of us that a collection of articles on the present state of this exciting field of applicable mathematics would be a useful addition to their Topics in Current Physics series. This suggestion was motivated in part by the obvious potential of the field for further physical applications and in part by the need, already apparent, for a comprehensive exposition of the mathematical techniques demanded of anybody working in it. At that time (1975) the number of groups active in soliton research as we then conceived it was very small and these groups were well defined. It was plain to all of us in the field that the subject was an important one; but it was still possible at that time clearly to identify two or three groups in each of the different areas of the world, USA, Europe, Russia, and Japan, where the exciting new results of the integrable (that is soluble) nonlinear evolution equations had only recently been achieved.

Accordingly the two Editors welcomed Lotsch's suggestion and immediately realised that a properly balanced book had to combine representative work from each of the different national groups. Of course, even then it was not possible to ask every significant worker in the field to contribute and we tried to find contributors each of whom could represent their several immediate colleagues. In Japan our choice fell naturally on Toda, Hirota, and Wadati - although even in 1975 it was becoming apparent that these outstanding workers were already representative of many other persons whose interest in solitons they had themselves helped to create. The same was perhaps even more true of Faddeev, Zakharov, and Novikov in the Soviet Union where barriers of language had prevented us from wholly grasping the achievement of this country in soliton research. We did, however, fully comprehend the exceptional achievement of these three workers in the field and were delighted when they agreed to contribute to this book.

For the USA we did not at that time have the temerity to approach Martin Kruskal; and the Princeton group which, through its work on the KdV had done so much for the whole subject, had apparently broken up. It was obvious that at least the three 'schools' at Potsdam, Courant Institute, and Arizona-Wisconsin had to be represented however. As senior partner in the AKNS inverse scattering scheme Newell delighted us by agreeing to contribute an article. Peter Lax and co-workers at the Courant Institute were sympathetic to providing an article but in the event, and to our regret, found themselves unable under pressure of constraints to do so. Lamb and

7 See references Sect.1.2.
8 See, e.g. [1.22,23].

McLaughlin, however, were able to represent the Arizona-Wisconsin 'school' and we were particularly pleased that they could.

This left us with the problem of Europe where, apart from the work coming out of Manchester, little seemed to be happening. This was an illusion, however: the work of Calogero in Rome burst on the two Editors at the Arizona meeting in January, 1976, and it was obvious he had to be a party to the book. Then in the summer of 1976 we discovered the amazing sequence of transformations used by Luther to find the spectrum of the quantised s-G. It seemed essential to have something of this work, heralding as it does significant advances in statistical mechanics, solid-state physics, and particle physics and the study of the connections between these disciplines, inside this Topics in Current Physics volume. In fact by the summer of 1976 it was becoming plain that one ought to have specific contributions on plasma physics and particle physics as well as all the other topics and that the strict mathematics of the theory also needed further articles. It was then that the horrifying truth emerged: this book on solitons could be unbounded in size and certainly unending in the writing!

Events themselves then took a turn when the Editors found themselves enmeshed in unexpected administrative and other duties which prevented them from giving their full attention (and full attention was certainly needed) to the problem of producing this volume. The contributors were exemplary, all of them, or almost all of them, providing publishable forms of their articles either on time or thereabouts. It is therefore with some embarrassment that the Editors find themselves going to press so late. The situation does not do proper credit to the originality and creativity of many of the articles as these had been written in 1976; and we take this opportunity to offer a profound apology to our long suffering colleagues who had produced them then, and a further apology for subsequently asking for the revisions which would bring these articles up to date in 1979.

Nevertheless, and in some defence of ourselves, it is worth indicating the size of the problem we found ourselves facing: in 1975 the foundations of classical soliton theory had been laid; the inverse scattering method for solving the KdV found by Kruskal and colleagues in 1967 had been applied by Zakharov and Shabat to the nonlinear Schrödinger equation, Wadati and Toda had extended explicit solutions of the KdV, and Wadati had gone on to solve the modified KdV $u_t + 6u^2u_x + u_{xxx} = 0$, whilst AKNS had found the generalised Zakharov-Shabat $2 \times 2$ scattering problem. In 1975 also, results on periodic boundary conditions for the KdV were appearing from the Courant Institute whilst Novikov's first work on this problem had appeared a little earlier in the Russian literature. There had also been a number of successful physical applications of soliton theory — perhaps most noticeably to nonlinear optics by Lamb and by the Manchester group. Further the review article by *Scott*, *Chu*, and *McLaughlin*[9] had been available for nearly two years — just about the time required to shift the viewpoint of a whole physics community.

9 Ref.[1.4]

In fact,however, the particle physicists were finding their own way to their own ideas on solitons through the work of Dashen, Hasslacher, and Neveu in particular, but more than substantially helped by Fadeev from the 'classical' soliton[10] school. Present ideas on solitons in particle physics coincide most notably with the classical solitons[10], which constitute the principal study of this book only in the s-G and some related systems like the nonlinear $\sigma$ models. Nevertheless this does not detract from the fact that classical soliton theory has profoundly influenced particle physics in the last four years whilst the many developments in this field have also influenced classical soliton theory.

A considerable number of physical applications of solitons have been made since 1975 also — particularly in the solid state where work on charge density wave theories of linear conductors by the Cornell group and in statistical mechanics as exemplified by Luther's work must be mentioned. Ideas on solitary waves, in plasmas, in solitonlike theories of Langmuir turbulence[11], and in other nonlinear physical contexts, have also had their effect. Strictly mathematical advances associated with soliton theory have also been numerous since 1975: the complete integrability of the infinite dimensional systems with soliton solutions had been understood following the remarkable paper by Zakharov and Faddeev on the KdV in 1970; but applications of algebraic geometry to the discovery of soliton solutions,[12] the discovery of the connections with Jacobian varieties and integrable dynamical systems[13], the jet-bundle formulation of Bäcklund transformations, and the application of Cartan's theory of differential forms in the 'prolongation structures' of Wahlquist and Estabrook,[14] are all much more recent. The consequence of all this activity has been that the book could have been finished in 1975; if only it could have been, whilst it has proved impossible, or almost impossible, to complete it since.

We have completed this book in fact by doing three things: we have lowered the range of the material presented in detail in the contributory articles and confined that material with two exceptions to the classical solitons, the inverse scattering methods for finding them, and the Hamiltonian structures associated with them; for those chapters very much on the frontiers of current understanding in 1976 we have requested additions from their authors which bring them thoroughly up to date for April 1979; and we have ourselves written as the first chapter in this volume a long introductory article which attempts to cover, at least through references, the whole of the present range of this remarkable subject.

The book which has resulted is *not* an elementary one; in some respects this fact reflects the nature of the subject; but it also reflects our belief in a present need to provide a first substantial 'source' book on the subject of solitons and

---

10 A working definition of this 'classical' soliton appears in Chap.1, p.6
11 See [1.193]
12 See Sect.1.5.
13 See Sect.1.7.
14 Refer to Sects.1.3 and 1.7.

inverse methods for those persons who either already work on nonlinear problems or who wish to do so. For these persons an understanding of the present capabilities of analytical soliton theory is essential. Nevertheless we have attempted to provide in our long introductory article a smooth transition from elementary physico-mathematical ideas, rendered, so we hope, of additional interest through historical anecdote, on towards the quite complicated mathematical apparatus now surrounding the integrable systems.

We recognise that the book which has emerged is large; but a moment's thought about what has been left out soon convinces the reader that a second volume is already needed. With our own experience behind us we promise nothing in this connection: soliton theory is already in a Tristram Shandy situation[15]. We hope what is here will nevertheless stimulate the research worker still further to find a new field of mathematics or a new field of physics or other science where solitons are relevant or where instanton[16] or other new solutions throw up totally new ideas. We hope also that the beginner in the subject will be able to find from the articles here presented what soliton theory is all about: he will need to work hard to do so; but we believe that what he will achieve thereby will brings its own real reward.

We are grateful to our colleagues who have contributed to this volume for their skill, knowledge and continuing patience. We are grateful to Springer-Verlag for the speed and accuracy with which they have produced this volume once they had received the manuscripts. We are grateful to Pamela Queeley for the typing of it and to our several colleagues who have provided translations of the Russian articles. Finally we are grateful to Dr. H. Lotsch of Springer-Verlag whose patience under recurrent delays has been a wonder to behold!

Manchester
May, 1980

*Robin Bullough*
*Philip Caudrey*

---

15 Tristram Shandy employed two years chronicling the first two days of his life. Would he ever complete his biography? For the answer see Chap.V of Bertrand Russell's "Mysticism and Logic".

16 See Sect.1.7.

# Contents

# List of Contributors

Bullough, Robert K.

Department of Mathematics, University of Manchester,
Institute of Science and Technology
Manchester M60 1QD, United Kingdom

Calogero, Francesco

Istituto di Fisica, Università di Roma; and Istituto Nazionale
di Fisica Nucleare, Sezione di Roma, I-00185 Roma, Italy

Caudrey, Philip J.

Department of Mathematics, University of Manchester,
Institute of Science and Technology
Manchester M60 1QD, United Kingdom

Degasperis, Antonio

Istituto di Fisica, Università di Roma; and Istituto Nazionale
di Fisica Nucleare, Sezione di Roma, I-00185 Roma, Italy

Faddeev, Ludwig D.

Academy of Sciences USSR, Steklov Mathematical Institute, Fontanka 27
Leningrad D-11, USSR

Gibbs, H.M.

Bell Laboratories, Murray Hill, NJ 07974, USA

Hirota, Ryogo

Department of Applied Mathematics, Faculty of Engineering,
Hiroshima University, Hiroshima 730, Japan

Lamb, George Lawrence

Department of Mathematics, University of Arizona, Tucson, AZ 85721, USA

Luther, Allan H.

NORDITA, Blegdamsvej 17, DK-2100 Copenhagen Ø, Denmark

McLaughlin, David Warren

Department of Mathematics, University of Arizona, Tucson, AZ 85721, USA

Newell, Alan Clive

Department of Mathematics, Clarkson College of Technology,
Potsdam, NY 13676, USA

Novikov, Sergey Petrowitz

L.D. Landau Institute for Theoretical Physics, Academy of Sciences,
V-334 Moskau, USSR

Toda, Morizaku

Institute for Applied Mathematics, Faculty of Engineering,
Yokohama National University, Yokohama 240, Japan

Wadati, Miki

Institute for Optical Research, Kyoiku University, 3-22-17 Hyakunintyo
Shinjuku-ku, Tokyo, Japan

Zakharov, V.E.

L.D. Landau Institute for Theoretical Physics, Academy of Sciences of the USSR,
V-334 Moskau, USSR

# 1. The Soliton and Its History

R. K. Bullough and P. J. Caudrey

We review the history of the soliton, since the first recorded observation of the 'great solitary wave' by Russell in 1834, as a means of developing the mathematical properties of a large class of solvable nonlinear evolution equations. This class embraces, amongst others, the Korteweg-de Vries, sine-Gordon, and nonlinear Schrödinger equations. Solitary waves, solitons, Bäcklund transformations, conserved quantities and integrable evolution equations as completely integrable Hamiltonian systems are all introduced this way and form a basis for the more detailed discussions which follow in the remaining chapters. The differential geometry of one large class of nonlinear evolution equations is described. Some connections with nonlinear field theories and with solvable many-body problems are established. The short biography of John Scott Russell which forms much of the first section is continued as an appendix at the back of this volume.

## The Soliton and Its History

### 1.1 Russell's Discovery of the 'Great Solitary Wave'

The inverse scattering method for solving nonlinear evolution equations of the type $u_t = K[u]$ where $K[u]$ is some nonlinear functional of $u(x,t)$ is now some twelve years old. Its discovery is due to KRUSKAL and his co-workers [1.1] and was first reported in 1967. These authors showed how to solve the Korteweg-de Vries (KdV) equation

$$u_t = 6uu_x - u_{xxx} \tag{1.1}$$

this way and in particular how to find all its soliton solutions. The properties of the soliton had been discovered and the object defined in the paper of ZABUSKY and KRUSKAL [1.2] in 1965. If current research activity is a proper measure of the significance of these two discoveries they represent the most significant advance in the theory of nonlinear waves since the work of Riemann (1826-1866) and Cauchy (1789-1857) on characteristics, and in the theory of partial differential equations since the work of Fourier (1758-1830) on linear equations. Moreover, the capacity the inverse scattering method has given us to obtain exact analytical solutions of what are now a considerable number of nonlinear equations of physical significance has sparked a revolution in the approach to nonlinear physics. One consequence is that nonlinear problems are under scrutiny in areas where until fairly recently only linear theories were capable of yielding digestible results. In this first chapter we adumbrate those steps in the history of the soliton which have seemed to us important either to the mathematical development of the subject or to the application of that mathematics to problems of physical interest.

An attractive feature of soliton theory is the intimate way in which physics and mathematics interact within it. The subject surely begins with the physical observation [1.3] of what we now identify as the single soliton solution of the KdV equation in the month of August 1834 by John Scott Russell. His own report of that observation is already much quoted [1.4]. We repeat it here because it serves to illustrate the fascination the soliton immediately exercised over RUSSELL. And his striking description may serve in part to explain the interest this same object has aroused in mathematicians and physicists some one hundred and forty years later.

Russell (1808-1882) was a supreme example of the Victorian entrepreneur. He was precocious: he attended all three of the Scottish Universities, St. Andrews, Edinburgh and Glasgow, before graduating from the last at the age of sixteen [1.5]. Working in the Department of Natural Philosophy at Edinburgh during 1832-1833 he was asked to investigate the capacity of the Union Canal which from Edinburgh, joins with the Forth-Clyde canal and thence to the two coasts of Scotland, for its economical use by steamships. We believe that is was in the course of investigations to that end he reported as follows [1.3]:

"I was observing the motion of a boat which was rapidly drawn along a narrow channel by a pair of horses when the boat suddenly stopped – not so the mass of water in the channel which it had put in motion: it accumulated round the prow of the vessel in a state of violent agitation, then suddenly leaving it behind, rolled forward with great velocity, assuming the form of a large solitary elevation, a rounded smooth and well defined heap of water, which continued its course along the channel apparently without change of form or diminution of speed. I followed it on horseback, and overtook it still rolling on at a rate of some eight or nine miles an hour, preserving its orginal figure some thirty feet long and a foot to a foot and a half in height. Its height gradually diminished and after a chase of one or two miles I lost it in the windings of the channel. Such, in the month of August 1834, was my first chance interview with the singular and beautiful phenomenon which I have called the Wave of Translation, a name which it now generally bears: which I have since found to be an important element in almost every case of fluid resistance, and ascertained to be of the type of that great moving elevation of the sea, which, with the regularity of a planet, ascends our rivers and rolls along our shores.

To study minutely this phenomenon with a view to determining accurately its nature and laws, I have adopted other more convenient modes of producing it than that which I just described, and have employed various methods of observation. A description of these will probably assist me in conveying just conception of the nature of this wave.

Genesis of the Wave of the First Order .........."

As the last lines of the quotation suggest RUSSELL was then categorising fluid wave motions and conducting experiments in water to observe them. He distinguished four orders, I, II, III and IV, and two types, solitary and gregarious: the gregarious waves included oscillatory water waves and wave groups, order II, as well as capillary surface waves, order III. His wave of translation was of order I and solitary; and he later called it 'the Great Solitary Wave'. His wave of order II he called corpuscular, was solitary, and was actually a sound wave for the particular reasons noted below.[1]

The idea of the solitary wave, at least, has remained with us and we use this now to mean any, typically bell-shaped, plane wave pulse which translates in one space direction without changing its shape (it is a wave of permanent profile or permanent type [1.6]). Any bell-shaped function $u(x - Vt)$ is a solitary wave translating with speed $V$ along $x$. The solitary wave solution of the KdV in the form (1.1) is

$$u = -2\eta^2 \text{sech}^2[\eta(x - 4\eta^2 t)] \tag{1.2}$$

with speed $4\eta^2$. It is negative solely because of the choice of sign for $u$ implicit in (1.1). It is easy to see that by scaling $u$ so that $u \to -u/6$ an alternative form of the KdV is

$$u_t + uu_x + u_{xxx} = 0 \tag{1.3}$$

with the solitary wave solution

$$u = 12\eta^2 \text{sech}^2[\eta(x - 4\eta^2 t)] \tag{1.4}$$

$$= 3\alpha^2 \text{sech}^2[\tfrac{1}{2}\alpha(x - \alpha^2 t)] \tag{1.5}$$

in which $\alpha \equiv 2\eta$. Note the connection between the speeds $4\eta^2$ or $\alpha^2$ and the amplitudes $12\eta^2$ or $3\alpha^2$. Evidently bigger pulses travel faster. Such a connection occurs

---

[1] See the Appendix.

only in nonlinear systems.[2]

The KdV described any weakly nonlinear, weakly dispersive plane wave system. In the solitary wave the nonlinearity $uu_x$ is just balanced by the dispersion $u_{xxx}$. The equation arises naturally in the form

$$u_t + \frac{3}{2}\frac{c_s}{h} uu_\xi = \frac{1}{2} c_s h^2 \left(\frac{\gamma}{\rho g h^2} - \frac{1}{3}\right) u_{\xi\xi\xi} \tag{1.6}$$

in the theory of gravity waves in water of finite depth [1.7]. The independent variable $\xi$ is $x - c_s t$ and the sound speed (linearised velocity) is $c_s = \sqrt{gh}$:h is the depth of undisturbed water, $\gamma$ is the surface tension of the water and $\rho$ its density: the solitary wave solution is entirely positive for $\gamma/\rho g h^2 < 1/3$ and is entirely negative for $\gamma/\rho g h^2 > 1/3$. For $\gamma/\rho g h^2 \sim 1/3$ a higher order derivative $u_{\xi\xi\xi\xi\xi}$ is needed to prevent the breakdown of the theory.

Equation (1.6) will govern waves in canals of sufficiently wide and uniform cross section and its positive solution is that observed by RUSSELL in 1834. These facts were not known to RUSSELL. But he found empirically [1.3], by the series of well executed experiments mentioned in the quotation, the important relation between the speed c of the solitary wave solution of (1.6) and its maximum height k above the free water surface

$$c^2 = g(h + k) \quad . \tag{1.7}$$

He also disagreed over this formula with AIRY who found theoretically a very different result [1.3,8]. Actually providing $\gamma$ is set identically zero the speed c of the solitary wave solution of (1.6) is given by $c = c_s(1 + k/2h)$. And the result $c = c_s(1 + k/h)^{\frac{1}{2}} = \sqrt{g}(h + k)^{\frac{1}{2}}$ is found only for the solitary wave solutions of the Boussinesq equation [1.9]

$$u_{tt} = c_s^2 \frac{\partial^2}{\partial x^2}\left(u + \frac{3}{2}\frac{u^2}{h} + \frac{h^2}{3} u_{xx}\right) \quad . \tag{1.8}$$

This admits the solitary wave solutions $u = k\,\mathrm{sech}^2[(3k/h^2)^{\frac{1}{2}}(x \pm ct)]$ travelling either along the positive or the negative x directions. The Boussinesq equation scales to the KdV and a single direction of propagation by $\xi = x - c_s t$ and $\tau = \varepsilon t$ and neglect of terms $O(\varepsilon^2)$.

RUSSELL later based some remarkable and perhaps over-imaginative speculations on his formula (1.7). The interested reader will find reference to these and to

---

[2]The KdV equations (1.1,3) are invariant under the *scale transformation* $u(x,t) \to \eta^2 u(\eta x, \eta^3 t)$. Since $12\,\mathrm{sech}^2(x - 4t)$ is a solution of (1.3), the 1-parameter solution (1.4) follows.

the further professional career of Russell himself in the Appendix. The solitary wave gained the attention of de Boussinesq, Rayleigh and others and remained under active discussion until the paper of KORTEWEG and DE VRIES [1.7] of 1895. The still fascinating interplay of comment and ideas which surrounded the solitary wave at this time is reviewed, following the remarks on Russell himself, in the Appendix (see also [1.10-20]). If suffices to say here that Russell's solitary wave was well conceived and has provided the essential foundation on which the present theory of solitons could be based. The concept of the soliton itself is introduced in Sect.1.2.

## 1.2 Definition of a Soliton: N-Soliton Solutions of Nonlinear Evolution Equations

The later history of the soliton is really that of three nonlinear evolution equations, namely the KdV equations (1.1, 4 or 6), the nonlinear Schrödinger equation (NLS)

$$iu_t + u_{xx} + 2u|u|^2 = 0 \tag{1.9}$$

and the sine-Gordon equation (s-G)

$$u_{xx} - u_{tt} = \sin u \quad . \tag{1.10}$$

The s-G takes the form

$$u_{xt} = \sin u \tag{1.11a}$$

in light cone coordinates $\frac{1}{2}(x + t) \rightarrow x$, $\frac{1}{2}(x - t) \rightarrow t$. This becomes the evolution equation

$$u_t = \int_{-\infty}^{x} \sin[u(x',t)]dx' \tag{1.11b}$$

under the boundary conditions $u(x,t) \rightarrow u_0 \equiv 0 \pmod{2\pi}$, $u_x$, $u_{xx}$, etc $\rightarrow 0$, $|x| \rightarrow \infty$, assumed, for (1.11a); it is in any case an evolution equation in $v = u_x$ which is the usual point of view adopted for solving this equation. The s-G occurred first in physics in an application to crystal dislocation theory [1.21]. It describes the propagation of spins, notional or real, in various physical systems [1.22,23]. These include the propagation of fluxons in Josephson junctions [1.23-25] and the propagation of resonant short optical pulses [1.23,24,26,27]. Some applications are discussed by LAMB and McLAUGHLIN in Chap.2 as well as in the later one by ourselves (Chap.3).

The NLS is exceptional amongst the three equations in that $u(x,t)$ is a complex rather than a real field. It governs the evolution of any weakly nonlinear strongly dispersive almost monochromatic wave and in particular governs the evolution of *deep* water waves [1.28]. The KdV governs the weakly nonlinear weakly dispersive regime in shallow water, and it is a fourth equation, the equation of a simple wave

$$u_t + uu_x = 0 \tag{1.12}$$

which governs the strongly nonlinear nondispersive regime. As we shall see this equation also has played a role in the development of the subject, but unlike the KdV, NLS and s-G it does not have soliton solutions. Let us quickly say what we mean by these since they are the subject of the whole of this volume.

The solitary wave solution (1.5) of the KdV (1.3) is the unique solution of the form $u(x-Vt)$ satisfying the boundary conditions $u$, $u_x$, $u_{xx}$, etc., $\to 0$ as $|x| \to \infty$. ZABUSKY and KRUSKAL [1.2] were concerned with periodic boundary conditions. Nevertheless they in effect made the remarkable discovery that arbitrary initial data $u(x,0)$ evolved for $t \to \infty$ into a train of solitary waves each of the form (1.5), asymptotically well spaced and travelling with different speeds. This breakup behaviour was actually known to RUSSELL [see e.g., Ref.1.3, Fig.6 in Plate 47]. ZABUSKY and KRUSKAL highlighted the collision property of solitons — namely that two pulses like (1.5) at $x = -\infty$ for $t \to -\infty$ could collide in the finite x region but would emerge from the collision unchanged in shape and speed. There is a shift of position, the phase shift, in the expressions (1.5) as $t \to +\infty$. The shift is usually small, but not always [1.29]. *We shall therefore take as a working definition of a soliton that it is a solitary wave which preserves its shape and speed in a collision with another such solitary wave.*

The condition on the solitary wave is restrictive and nontrivial. For example the "$\phi$-four equations" take their name from the Hamiltonian density $\mathscr{H} = \frac{1}{2}\phi_t^2 + \frac{1}{2}\phi_x^2 \pm (\frac{1}{2}\phi^2 - \frac{1}{4}\phi^4)$. The energy is bounded below in the case of the negative sign and this case has been used as a model field theory [1.30,31]. The equations of motion in terms of a dependent variable $u \equiv \phi$ are

$$u_{xx} - u_{tt} = \pm (u - u^3) \quad . \tag{1.13}$$

These have the solitary wave solutions

$$u = \pm \operatorname{sech} \frac{1}{\sqrt{2}} \frac{(x-Vt)}{\sqrt{1-V^2}} \tag{1.14}$$

[+ve sign in (1.13)] and the "kink-like" solutions of permanent profile

$$u = \pm \tanh \frac{1}{\sqrt{2}} \frac{x-VT}{\sqrt{1-V^2}} \tag{1.15}$$

[-ve sign in (1.13)]. Neither the solutions (1.14) satisfying $u \to 0$, $|x| \to \infty$, nor the kink solutions (1.15) satisfying $u \to \pm 1$, $x \to -\infty$ and $u \to \mp 1$, $x \to +\infty$ have the simple collision properties of solitons as we have now defined them. These solutions can bump, lock or annihilate each other [1.32]; but in addition they always emit some oscillatory disturbance or 'radiation' in the course of a collision [1.32]. On the other hand the solitary wave or kink solution of quite a large number of equations in two dimensions do have the soliton collision property. Amongst these are the KdV, NLS and s-G equations.

Notice that a soliton like (1.5) with parameter $\alpha = \alpha_2 > \alpha_1$ will overtake a second soliton with parameter $\alpha_1$. It will appear to pass through the second soliton so that asymptotically the two solitons appear to have changed places. It is also so that a set of N KdV solitons with parameters $\alpha_N > \alpha_{N-1} > \ldots > \alpha_1$ ordered in the sequence N, N-1, ..., 1 for $t \to -\infty$ appear to become the set in natural order 1,2, ..., N for $t \to +\infty$. This interpretation is subjective since the pulses may also be considered to maintain their orders and to simply transfer energy, momentum (and amplitude) in the collision process. Such particle-like behaviour was the origin of the name "soliton" [1.2].

The second interpretation can be maintained, but not so easily, in the collision of a breather solution of the s-G with a kink solution of the s-G. The analytical forms of these two individual solutions are exhibited in (1.29) and (1.23), respectively. The kink plus breather solution itself involves three of the parameters $a_i$ and can be found from (1.22). A first impression from it is that the breather passes through the kink with no more than the appropriate phase shift. The key points about these different but equally tenable interpretations, however, are that I) the collisions are clean, that is no additional disturbance such as 'radiation' is created by the collision process; and II) the solutions can be obtained analytically throughout the whole motion by, for example, the inverse scattering method.

To fix ideas we give the N-soliton solution of the KdV equation (1.3). The boundary conditions are $u$, $u_x$, $u_{xx}$, etc, $\to 0$, $|x| \to \infty$ and the N-soliton solution is

$$u = 12(\ln F)_{xx}$$

$$F = \det F_{nm}, \quad F_{nm} = \delta_{mn} + \frac{2(\alpha_m\alpha_n)^{\frac{1}{2}}}{\alpha_m + \alpha_n} f_n \tag{1.16}$$

$$f_n = \exp[-\alpha_n(x - x_n) + \alpha_n^3 t]$$

in which $\alpha_n$ and $x_n$ are free real parameters. As far as we are aware the solution was first given in this form by HIROTA [1.33] who used his direct method (see Chap.5) and not the inverse scattering method. HIROTA worked on (1.1) with a consequently different scaling.

Notice that in the case N = 2 as a particular example

$$F = 1 + f_1 + f_2 + \left(\frac{\alpha_2-\alpha_1}{\alpha_2+\alpha_1}\right)^2 f_1 f_2 \quad . \tag{1.17}$$

From (1.16) one finds that for $f_1 \approx 1$, $f_2 << 1$

$$u = \frac{12\alpha_1^2 f_1}{(1+f_1^2)} = 3\alpha_1^2 \operatorname{sech}^2\left[\frac{\alpha_1}{2}(x - x_1 - \alpha_1^2 t)\right] \tag{1.18}$$

whilst for $f_1 \approx 1$, $f_2 >> 1$

$$u = \frac{12\alpha_1^2 \bar{f}_1}{(1+\bar{f}_1^2)} \quad , \quad \bar{f}_1 \equiv \left(\frac{\alpha_2-\alpha_1}{\alpha_2+\alpha_1}\right)^2 f_1$$

and $u = 3\alpha_1^2 \operatorname{sech}^2\left[\frac{\alpha_1}{2}(x - \bar{x}_1 - \alpha_1^2 t)\right]$,

$$\text{with } \bar{x}_1 = x_1 - \frac{1}{\alpha_1} \ln\left(\frac{\alpha_2+\alpha_1}{\alpha_2-\alpha_1}\right)^2 \quad . \tag{1.19}$$

Thus for $f_1 \approx 1$, u takes on the solitary wave form (1.5) whether $f_2$ is very small or very large; but in the latter case it undergoes a shift of argument (the phase shift) $+\ln[(\alpha_2 + \alpha_1)/(\alpha_2 - \alpha_1)]^2$ with respect to the former case.

We suppose $\alpha_2 > \alpha_1 > 0$ and $x_2 << x_1 < 0$. Then at t = 0, the larger soliton with parameter $\alpha_2$ is substantially to the left of the smaller with parameter $\alpha_2$. Indeed on $x = x_1 + \alpha_1^2 t$ for $t \approx 0$, $f_1 = 1$ and $f_2 << 1$, whilst on $x = \bar{x}_1 + \alpha_1^2 t$ for $t \to \infty$, $\bar{f}_1 = 1$, $f_1 \approx 1$, whilst $f_2 >> 1$. One can see that as t increases from zero the larger soliton passes through the smaller inducing the shift of argument $\ln[(\alpha_2 + \alpha_1)/(\alpha_2 - \alpha_1)]^2$. Similarly on $x = \bar{x}_2 + \alpha_2^2 t$ for $t \approx 0$, $f_1 \approx 1$, $f_1 >> 1$ whilst, on $x = x_2 + \alpha_2^2 t$, $f_2 = 1$ and $f_1 << 1$, and one can see that in passing through the smaller soliton the larger soliton receives the shift of argument $-\ln[(\alpha_2 + \alpha_1)/(\alpha_2 - \alpha_1)]^2$. If one analyses the N-soliton solution in a similar way one finds that the $i^{th}$ soliton receives the shift of argument

$$\delta_i = \sum_{j \neq i} \delta_{ij} \ , \quad \delta_{ij} = \operatorname{sgn}\{i-j\} \ln\left(\frac{\alpha_i+\alpha_j}{\alpha_i-\alpha_j}\right)^2 \quad . \tag{1.20}$$

Thus the shift $\delta_i$ is made up of pairwise shifts only whilst $\sum_{i=1}^{N} \delta_i = 0$, and total phase shift is conserved. The pairwise shift property is connected [1.34] with the conserved densities of the KdV. These densities are described in Sect.1.4. The asymptotic behaviour of the KdV was described by GARDNER et al. [1.1], and by

GIBBON and EILBECK [1.35]. The argument adopted here is substantially due to WHITHAM [1.36].

The point now is that the three very different looking equations KdV (1.3), NLS (1.9) and s-G (1.10) or (1.11) all have N-soliton solutions whilst the $\phi$-four equations (1.13) do not. The 'simple wave' (1.12) also does not but this equation does not have a solitary wave solution, although it does actually have an infinity of conserved densities which are in involution [1.37]. The significance of this for soliton solutions is treated in Sect.1.4.

Before going further we think it may be helpful to the reader if we now interpolate a remark on the inverse scattering method of solving nonlinear evolution equations $u_t = K[u]$. The method and its history are described in Sects.1.4-6. For those who have not heard of the method before, its essential idea is this: because no direct method of finding $u(x,t)$ from initial data $u(x,0)$ is usually available one associates a scattering problem with the NLEE such that $u(x,t)$ is the scattering potential; for example (and see Sect.1.4) for the KdV the scattering problem is the Schrödinger eigenvalue problem $[-\partial^2/\partial x^2 + u(x,t)]\psi = \pm k^2\psi$. The initial data $u(x,0)$ are mapped via the scattering problem onto so-called scattering data. It now turns out that the evolution of this scattering data can be computed from the NLEE. From the scattering data at time t the final step is to compute the potential $u(x,t)$ corresponding to the scattering data. This can be done via a linear method—the so called 'inverse' problem. Thus $u(x,t)$ is determined! The KdV, s-G and NLS equations can all be solved by this method. The simplest scattering problem for the s-G and NLS equations is not a Schrödinger eigenvalue problem, however.

The symmetry of the scattering problem has some consequences to the forms of the soliton solutions. For example the single soliton solution of the NLS (1.9) was found by ZAKHAROV and SHABAT through a $2\times 2$ scattering problem [1.35]. It takes the form

$$u = \frac{2\eta\exp[-4i(\xi^2-\eta^2)t-2i\xi x+i\delta]}{\cosh[2\eta(t-t_0)+8\eta\xi t]} \tag{1.21}$$

in which $\xi$, $\eta$, $\delta$ and $x_0$ are all free parameters. It is an 'envelope soliton' in which a sech envelope modulates the harmonic 'carrier wave' of the numerator. In this soliton the connection between speed and amplitude exhibited in the KdV soliton (1.2) is lost, but the inverse method of solution [1.38] provides a ready explanation. It is the bound states of the scattering problems which describe solitons; in the case of the NLS the scattering problem admits bound state eigenvalues $\zeta$ which are strictly complex, $\zeta = \xi + i\eta(\eta > 0)$ and $\xi$ and $\eta$ then prove to be the parameters in (1.21). In contrast the bound state eigenvalues of the Schrödinger scattering problem lie solely on the imaginary axis, $\zeta = i\eta(\eta > 0)$. Thus the parameter $\eta$ determines the amplitude and $\xi$ the speed of the envelope soliton solution (1.21) of the NLS; the parameter $\eta$ determines both the speed *and* the amplitude of the soliton solution (1.2) of the KdV.

In the case of the s-G several new features emerge: the N-soliton solution of the s-G equation (1.9) in covariant form is [1.39]

$$\cos u(x,t) = 1 - 2\left(\frac{\partial^2}{\partial x^2} - \frac{\partial^2}{\partial t^2}\right) \ln F$$

$$F = \det M_{ij}$$

$$M_{ij} = 2(a_i + a_j)^{-1} \cosh[\tfrac{1}{2}(\theta_i + \theta_j)]$$

$$\theta_i = \gamma_i(x - V_i t - x_i)$$

$$a_i^2 = (1 - V_i)(1 + V_i)^{-1}\,, \quad \gamma_1^2 = (1 - V_i^2)^{-1} \tag{1.22}$$

where $a_i$ and $\gamma_i$ have the same sign and in which the $a_i$ and $x_i$ are free parameters. Notice first of all that the single soliton is the "kink"

$$u = 4 \tan^{-1} \exp\left[\pm \frac{(x-x_1-V_1t)}{(1-V_1^2)^{\frac{1}{2}}}\right] \tag{1.23}$$

The kink satisfies $u \to 0$, $x \to -\infty$, $u \to 2\pi$, $x \to +\infty$ (the $2\pi$ kink) and the positive sign is taken in (1.23). But there is also the corresponding antikink from the negative sign satisfying $u \to 2\pi$, $x \to -\infty$ and $u \to 0$, $x \to +\infty$ (the $-2\pi$ antikink). In contrast with the kinks and antikink (1.15) of the $\phi$-four equation the kinks and antikinks of the s-G collide amongst themselves or each other as solitons. Notice in any case that

$$u_x = \pm\frac{2}{(1-V_1^2)^{\frac{1}{2}}} \operatorname{sech}\left[\frac{x-x_1-V_1t}{(1-V_1^2)^{\frac{1}{2}}}\right] \tag{1.24}$$

is of (bell-shaped) solitary wave character so that $u_x$ is made up of actual solitons. The boundary conditions on the s-G (1.10) are therefore $u \to 0 \pmod{2\pi}$, $u_x$, $u_{xx}$, etc $\to 0$ as $|x| \to \infty$, as was assumed for (1.11).

The $4\pi$-kink solution is the 2-soliton solution

$$u = 4 \tan^{-1} \frac{\sin \tfrac{1}{2}(\theta_1+\theta_2)}{(a_{12})^{\frac{1}{2}} \cosh \tfrac{1}{2}(\theta_1+\theta_2)}\,.$$

$$a_{12} = (a_1 - a_2)^2(a_1 + a_2)^{-2} \tag{1.25}$$

Asymptotically [1.39]

$$u(x,t) \equiv 4 \tan^{-1} \exp(\theta_1 + \delta_1^\pm) + 4 \tan^{-1} \exp(\theta_2 + \delta_2^\pm)$$

as $t \to \pm\infty$, and

$$\delta_1^\pm = -\delta_2^\pm = \pm \frac{1}{2} \ln a_{12} \quad . \tag{1.26}$$

In an N-soliton collision (the $2N\pi$-kink solution)

$$\delta_i^\pm = \pm \frac{1}{2} \sum_{j=1}^{i-1} \ln a_{ji} \mp \sum_{j=i+1} \ln a_{ij}$$

$$= \pm \frac{1}{2} \sum_{j=1, j\neq i}^{N} \mathrm{sgn}(i - j) \ln a_{ij} \tag{1.27a}$$

and

$$\sum_{i=1}^{N} (\delta_i^+ - \delta_i^-) = 0 \quad . \tag{1.27b}$$

Secondly notice that by taking the arbitrary constants in complex conjugate pairs so that

$$a_1 = a_2^* = a_R + i\, a_I = ae^{i\mu}$$

$$x_1 = x_2^* = x_R + i\, x_I \tag{1.28}$$

one obtains the "breather solutions"

$$u(x,t) = 4 \tan^{-1} r \sin\theta_I \operatorname{sech}\theta_R \tag{1.29a}$$

with

$$\theta_R = \frac{\cos\mu}{(1-V^2)^{\frac{1}{2}}}(x - Vt) - \frac{1}{2} a_R x_R \quad , \quad \theta_I = \frac{\sin\mu}{(1-V^2)^{\frac{1}{2}}}(t - Vx) - \frac{1}{2} a_I x_I$$

$$r = a_R a_I^{-1} = \cot\mu \quad , \quad \text{and} \quad V = (1 - a^2)(1 + a^2)^{-1} \quad . \tag{1.29b}$$

The breather solution is called a "$0\pi$ pulse" in nonlinear optics since it takes u from zero to zero as x goes from $-\infty$ to $+\infty$. Similarly the kink becomes the $2\pi$ pulse. The pulse terminology arises because (see Chap.2) it is the electric field envelope $\varepsilon \propto (u_t + u_x)$ which characterizes the observable disturbance (it is the intensity $\varepsilon^2$ which is actually observed). In the theory described in Chap.2 $\varepsilon$ has 'pulse area' $2\pi$ when it is associated with the $2\pi$-kink; it has area zero, or $0\pi$, if it is associated with the breather (1.28).

In an arbitrary reference frame the energy of the breather is easily computed to be [1.39]

$$16\gamma_R[r^2(1 + r^2)^{-1}]^{\frac{1}{2}} = 16\gamma_R \cos\mu \qquad (1.30)$$

in which $\gamma_R^2 = (1 - V^2)^{-1}$. The breather therefore has an apparent rest mass 16 cos$\mu$. The corresponding energy of the $2\pi$ kink (1.23) is $8\gamma_1$ and that of the $4\pi$ kink (1.25) is $8\gamma_1 + 8\gamma_2$. The rest mass of the kink is therefore 8. Notice that the breather has an internal oscillation of frequency $(1 - V^2)^{-\frac{1}{2}} \sin\mu$. In the rest frame $V = 0$, $a = 1$ and $u(x,t)$ is the standing wave

$$u(x,t) = 4 \tan^{-1}\{\cot\mu \sin[(\sin\mu)(t - \tfrac{1}{2} x_I)]\mathrm{sech}[(\cos\mu)x - \tfrac{1}{2} x_R]\} \qquad (1.31)$$

which oscillates with internal frequency sin$\mu$. It is this internal degree of freedom which takes an interesting discrete spectrum when the s-G is quantised [1.40-42]. COLEMAN [1.43] gave a simple argument for this spectrum. The quantisation of the s-G is treated from the very different point of view of related models in statistical mechanics in Chap.12.

As we have indicated breathers, kinks and antikinks all act like solitons in collision. The s-G can be solved by a $2\times 2$ inverse scattering scheme [1.44,45]. The NLS and KdV can be also [1.38,44,45]. For those equations solvable by the $2\times 2$ inverse scattering schemes due to ZAKHAROV and SHABAT [1.38], and ABLOWITZ, KAUP, NEWELL, and SEGUR (AKNS) [1.44,45] there can be no more complicated solitons than the sech (or sech$^2$ for the KdV) solitons, the kinks, and the breathers.

Notice that two KdV solitons or two s-G kinks must travel at different speeds [look at the expressions (1.16) and e.g., (1.25)]: thus a disturbance containing them must break up. On the other hand any number of solitons of the NLS can travel at the same speed; and any number of breathers of the s-G can travel at the same speed which may also be that of a single kink or antikink. This is because the speeds of the breathers of the s-G are determined from its scattering problem by the moduli of pairs of complex eigenvalues $(\zeta, -\zeta^*)$; likewise [compare (1.9)] the speed of an NLS soliton is determined by Re$\{\zeta\}$. There can of course be any number of distinct complex eigenvalues $\zeta$ with given $|\zeta|$ or Re$\{\zeta\}$.

If readers wish to know more here and now about the way the behaviour of solitons can be quickly determined by looking at the scattering problems associated with them we refer them to the Chaps.2 and 6, for example. We also refer them forward to the latter part of the present article, notably to (1.100,101) and the further discussion there. It is already easy to see, however, that the KdV cannot have any breather solutions: its bound state eigenvalues lie on the imaginary axis, $\zeta = i\eta (\eta > 0)$ and cannot arise in pairs. Notice also that for more general scattering problems the behaviour of the soliton solutions can become more complicated.

ZAKHAROV and MANKOV [1.46], and KAUP [1.47] solved the 3-wave interaction (we quote here only the decay type problem [1.22,46,47])

$$u_{1,t} + v_1 u_{1,x} = iqu_2 u_3^*$$

$$u_{2,t} + v_2 u_{2,x} = iqu_1 u_3$$

$$u_{3,t} + v_3 u_{3,x} = iqu_1 u_2 \qquad (1.32)$$

with different degrees of completeness in 1973 and 1976. The inverse scattering solution involves a $3 \times 3$ scattering problem. The system (1.32) is nondispersive but has soliton solutions: the three wave envelopes $u_i$ are complex, $q$ is a coupling constant, and the $v_i$ are constant real velocities. Soliton number is not conserved in the scattering process [1.46,47]. In Chap.7 ZAKHAROV discusses the solution of the N-wave interaction which exhibits still more unusual behaviour, whilst in Chap.9 CALAGERO and DEGASPERIS describe 'boomerons' and 'trappons' which are simple solitons with bizarre trajectories.

We have used these various remarks to introduce some elementary ideas surrounding the concept of the soliton. In the rest of this chapter we use the history of the subject to introduce the reader to some mathematical and physical ideas of rather greater depth.

## 1.3 Bäcklund Transformations and Conserved Densities

Russell's work influenced DE BOUSSINESQ [1.9], RAYLEIGH [1.16] and KORTEWEG and DE VRIES [1.7]. It was physically based and stimulated no significant mathematics before the work of ZABUSKY and KRUSKAL [1.2] and GARDNER et al. [1.1]. On the other hand the s-G first arose in a mathematical context, namely in differential geometry in the study of pseudo-spherical surfaces — surfaces of constant negative curvature [1.27]. Recent work by LUND [1.48] exemplifies and extends these earlier studies and we refer to this again in Sect.1.6. We also refer there to some still more recent work on the geometry of solitons.

We note, however, that before the emergence of the s-G around 1883 LIOUVILLE [1.49] had given the general solution of the not unrelated nonlinear Klein-Gordon equation

$$u_{xt} = \exp(mu) \qquad (1.33)$$

($m$ = parameter) in 1853. We have recently used a Bäcklund transformation to regain, from the general solution $u = f(x) + g(t)$ of $u_{xt} = 0$, Liouville's general solution of (1.32) in the form [1.50] (and see LAMB [1.51])

$$\exp(mu) = \exp[m(f - g)]\left[k \int^{x} \exp(mf)dx' + \frac{1}{2} k^{-1} m \int^{t} \exp(-mg)dt\right]^{-2} \tag{1.34}$$

in which k is a parameter. This is a convenient point to introduce the Bäcklund transformations therefore and we do this now.

The Bäcklund transformation (BT) which yields the solution (1.34) of (1.33) is [1.50]

$$\begin{aligned} u'_x &= u_x + \frac{2k}{m} \exp\left[\frac{1}{2} m(u + u')\right] \\ u'_t &= - u_t - k^{-1} \exp\left[\frac{1}{2} m(u - u')\right] \quad . \end{aligned} \tag{1.35}$$

This is a transformation which connects a solution u of one equation with a solution u' of a second equation and is typical of a BT. Notice that the *integrability condition* $u'_{xt} = u'_{tx}$ yields both $u_{xt} = \exp(mu)$ and $u'_{xt} = 0$ which are therefore the two equations connected by the BT. From (1.35) it is easy to see, if $u' = f(x) + g(t)$ that

$$d\left\{\exp\left[-\frac{1}{2} m(u - f + g)\right]\right\} = k \exp(mf)dx + \frac{m}{2k} \exp(-mg)dt \quad . \tag{1.36}$$

From this (1.34) easily follows.

Both Liouville's equation (1.33) and the s-G (1.11a) in light cone co-ordinates are examples of the generalised Klein-Gordon equation

$$u_{xt} = F(u) \quad . \tag{1.37}$$

However, there are no soliton solutions of (1.33) since neither u = 0 nor u = constant is a solution.[3] Nonlinear Klein-Gordon equations like (1.37) can have both Bäcklund transformations (BTs) and auto-Bäcklund transformations (aBTs). Whilst the BTs connect solutions of two different equations, the aBTs connect solutions of the same equation. Equation (1.37) has an aBT if and only if $\ddot{F}(u) + \alpha^2 F(u) = 0$ for some $\alpha$ including $\alpha = 0$ [1.50,54] ($\ddot{F}$ denotes second derivative). A particular case of this, the aBT with free real parameter k

$$\begin{aligned} u'_x &= u_x + 2k \sin \frac{1}{2} (u + u') \\ u'_t &= - u_t - 2k^{-1} \sin \frac{1}{2} (u - u') \end{aligned} \tag{1.38}$$

---

[3] Incidentally we now see the origin of the name 'sine-Gordon' for (1.10) or (1.11): in detail RUBINSTEIN [1.52] ascribes the name to Kruskal; COLEMAN [1.43] indicates otherwise, and KRUSKAL himself is 'not quite sure' [1.53]. Whichever way the name is certainly here to stay!

connecting two solutions $u$ and $u'$ of the s-G in light cone coordinates (1.11a) was known to BÄCKLUND before 1883 [1.55]. In this example the integrability condition $u'_{xt} = u'_{tx}$ implies both $u_{xt} = \sin(u)$ and $u'_{xt} = \sin(u')$. That a similar aBT exists for the KdV was realised much more recently [1.51,56]. It is

$$u'_x = -u_x - k + (u' - u)^2$$

$$u'_t = -u_t + 4[k^2 + ku_x - k(u' - u)^2 + u_x^2 + u_x(u' - u)^2 + u_{xx}(u' - u)] \quad . \quad (1.39)$$

The integrability condition $u'_{xt} = u'_{tx}$ in this case is

$$(u' - u)(u_{xxx} + u_t - 6u_x^2) = 0 \quad . \quad (1.40)$$

The identity transformation $u' = u$ is trivial. The remaining bracket yields the KdV (1.1) (with factor 12 not 6) by differentiation with respect to $x$ and setting $u_x \to u$. The function $u'$ satisfies

$$u'_{xxx} + u'_t - 6u'^2_x = 0 \quad (1.41)$$

so the transformation is an aBT.

There is an intimate connection between aBTs, solitons and inverse scattering methods, although the example of Liouville's equation (which also has an aBT [1.50,54]) shows that one does not necessarily imply the other. Bäcklund transformations were discussed and extended by a number of authors in the period 1900-1920 [1.57]. CLAIRIN [1.57] in particular may have found the 'superposition principle' for the aBT (1.34) described by LAMB and McLAUGHLIN in Chap.2, although EISENHART [1.57] quotes Bianchi as the source of it. FORSYTH [1.57] offers the finding of the aBT (1.34) for $u_{xt} = \sin u$ as a problem (!) and refers to Bianchi and Darboux for the origin of it. Certainly WAHLQUIST and ESTABROOK [1.56] first found the superposition principle ('theorem of permutability') for (1.39). CHEN [1.58] showed how the existence of aBTs could be connected with the $2 \times 2$ inverse scattering scheme of AKNS [1.44].

That the BT may be a linearising transformation emerged first in the work of HOPF [1.59] and COLE [1.60] in 1950. These authors separately linearised the Burgers equation

$$u_t + uu_x - \nu u_{xx} = 0 \quad (\nu > 0) \quad . \quad (1.42)$$

The BT

$$u'_x = -(2\nu)^{-1} uu'$$

$$u'_t = -(4\nu)^{-1}(2\nu u_x - u^2)u' \quad (1.43)$$

is not an auto-BT: it maps the solution u of (1.42) into a solution u' of the heat equation $u'_t - \nu u'_{xx} = 0$. The transformation $u = -2\nu(\ln u')_x$ is the Hopf-Cole transformation: it linearises the Burgers equation as the heat equation and the Burgers equation can be exactly solved this way.

The BT

$$u'_x = -2u - u'^2$$

$$u'_t = 8u^2 + 4uu'^2 + 2u_{xx} - 4u_x u' \tag{1.44}$$

was found by WAHLQUIST and ESTABROOK [1.61] in 1975. It maps between

$$u_{xxx} + u_t + 12uu_x = 0 \tag{1.45}$$

and the modified Korteweg-de Vries equation (with solution u')

$$u_{xxx} + u_t - 6u^2u_x = 0 \quad . \tag{1.46}$$

The transformation between $u'_x$, u and u' in (1.44) is MIURA's transformation [1.62]. This proved fundamental in the discovery of the inverse scattering method as we show below.

There are also sets of 'conserved densities' associated with nonlinear evolution equations which have soliton solutions; these also are fundamental and have played a fundamental role in the history of the subject. These densities are often 'polynomial conserved densities' (p.c.ds): such densities are polynomials in $u, u_x, u_{xx}, \ldots$, etc. For example the density $u_x^2/2$ satisfies

$$\left(\frac{1}{2}u_x^2\right)_t + \left[-\int^u F(u')du'\right]_x = 0 \tag{1.47}$$

if u satisfies (1.37). From this and $u \to 0$, $|x| \to \infty$ it follows that $\int_{-\infty}^{\infty}(u_x^2/2)dx$ is a constant of the motion. The existence of a 'conservation law' like (1.47) therefore defines $u_x^2/2$ as a locally conserved (polynomial) density which is globally conserved if the boundary conditions are right. Equation (1.37) does not have soliton solutions for arbitrary F(u). But in all cases when it does, and indeed in all examples associated with solitons known to the authors with the single exception of the equations concerning 'inhomogeneously broadened optical pulse propagation' [1.63,64] (see also Sect.1.4 and Chaps.2,6), the locally conserved quantities are also globally conserved.

For solitons it seems, however, that an infinite set of such conserved densities is required. As we shall discover equations with soliton solutions prove to be examples of *infinite dimensional* completely integrable Hamiltonian systems. A finite dimensional completely integrable Hamiltonian system with 2n degrees of freedom has n constants of the motion and these are in 'involution'. A necessary condition that an infinite dimensional Hamiltonian system be completely integrable is that there is an infinite number of constants of the motion. Sufficient conditions are that there is precisely the right number (the 'right infinity') and that these are in 'involution'. Involution means that these constants satisfy a bracket condition, the Poisson bracket condition. We refer to this briefly again in Sect.1.5. For further details, however, the reader is referred to Chaps.2,6,11, among others.

The conserved densities of the KdV equation were the key to the discovery of the inverse scattering method [1.1]. We describe their role in the history of the subject in Sect.1.4.

As for the Klein-Gordon equations (1.37) KRUSKAL some five years later [1.65] made the useful remark that (1.37) always had the one *polynomial* conserved density $u_x^2/2$ (a polynomial in $u_x$) but that a necessary and sufficient condition for a second polynomial density involving $u_{xx}$ was that $\ddot{F}(u) + \alpha^2 F(u) = 0$ for some $\alpha$. It then proves [1.65] that there is an infinity of polynomial conserved densities, (p.c.ds). KRUSKAL [1.65] also used the BT

$$u'_x = k^{-1} \sin(u' - u)$$

$$u'_t = u_t + k \sin u' \tag{1.48}$$

which maps a solution u of $u_{xt} = \sin u$ to a solution u' of

$$u'_{xt} = [1 - k^2 u_x'^2]^{\frac{1}{2}} \sin u' \tag{1.49}$$

to obtain an infinity of p.c.ds for the sine-Gordon equation. We know about the origins of this BT only that it appears in Kruskal's work and was also found independently but later by ourselves [1.50]. We now show how the infinite set of p.c.ds for the s-G can be obtained from this BT [1.50,65,66].

We write

$$u' = \sum_{n=0}^{\infty} f_n(u) k^n$$

and find from (1.48) that

$$u' = u + ku_x + k^2 u_{xx} + k^3\left(u_{xxx} + \frac{1}{3!} u_x^3\right) + k^4(u_{xxxx} + u_x^2 u_{xx}) + \dots \quad . \tag{1.50}$$

A conservation law for (1.49) is

$$[(1 - k^2 u_x'^2)^{\frac{1}{2}}]_t - k^2(\cos u')_x = 0 \quad . \tag{1.51}$$

If (1.50) is inserted into (1.51) we get, after dropping constants and constant factors, the expression (1.52) for the conserved density in (1.51)

$$u_x^2 + k(2u_x u_{xx}) + k^2(u_{xx}^2 + 2u_x u_{xxx} + \frac{1}{4}u_x^4)$$
$$+ k^3(2u_x u_{xxxx} + 2u_{xx}u_{xxx} + 2u_x^3 u_{xx})$$
$$+ k^4(u_{xxx}^2 + 2u_{xx}u_{xxxx} + 2u_x u_{xxxxx} + \frac{1}{8}u_x^6$$
$$+ 3u_{xxx}u_x^3 + \frac{13}{2}u_x^2 u_{xx}^2) + \dots \quad . \tag{1.52}$$

Since k is arbitrary, the quantities $u_x^2$, $u_{xx}^2 + 2u_x u_{xxx} + \frac{1}{4}u_x^4$, ... are conserved separately. The perfect differentials $2u_x u_{xx}$, ... which arise at odd powers of k are trivially conserved since they may be moved from the density T to the flux X in any conservation law

$$T_t + X_x = 0 \tag{1.53}$$

of the form (1.47). Modulo certain perfect differentials, it follows from (1.52) that a nontrivial set of conserved densities for the s-G is therefore [1.66]

$$T^2 = \frac{1}{2}u_x^2 \ , \quad T^4 = \frac{1}{2}u_{xx}^2 - \frac{1}{8}u_x^4 \ , \quad T^6 = \frac{1}{2}u_{xxx}^2 - \frac{5}{4}u_{xx}^2 u_x^2 + \frac{1}{16}u_x^6 \ ,$$
$$T^8 = \frac{1}{2}u_{xxxx}^2 - \frac{7}{4}u_{xxx}^2 u_x^2 + \frac{7}{8}u_{xx}^4 + \frac{35}{16}u_{xx}^2 u_x^4 + \frac{5}{128}u_x^8 \ ,$$
$$T^{10} = \dots \quad . \tag{1.54}$$

Notice that the densities are polynomials in $u_x$, $u_{xx}$, etc., that is they are p.c.ds, and that every term $\Pi_j\, u_j^{a_j}$ in $T^r$ (with $u_j$ the $j^{th}$ partial derivative) has the same rank r defined by $r = \sum_j ja_j$. It is proved [1.66] that there is an infinite set of such nontrivial polynomials $T^r$ for each even integer r. The corresponding infinite set of polynomial conserved densities for the KdV, the first of its kind discovered [1.67], is obtained from the BT (1.44) instead of (1.47).

The *polynomial* character of the conserved densities (1.54) and of those for the KdV [1.67] is associated with the scattering problems which solve these equations. The set (1.54) is an example of the more general set derivable from the $2\times 2$ AKNS scattering problem [1.42,68]. The conserved densities of the s-G (1.10) in Lorentz covariant form are not polynomials [1.66]. Notice in any case that at least one half of the conserved densities of the s-G in light cone coordinates are not polynomial. For in (1.51) the flux $k^2 \cos u'$ is from (1.50)

$$k^2 \cos u' = k^2 \cos u[1 - \frac{1}{2}k^2u_x^2 - k^3u_x u_{xx} - k^4u_x(u_{xxx} + \frac{3}{4}u_x^3) + \dots$$
$$- k^2 \sin u(ku_x + k^2u_{xx} + k^3u_{xxx} + k^4[u_{xxxx} + u_x^2 u_{xx}$$

$$- 3u_x(u_{xxx} + \frac{1}{3!} u_x^3)] + \ldots \quad . \tag{1.55}$$

Then

$$X^2 = -(1 - \cos u), \quad X^3 = -u_x \sin u \ , \quad X^4 = -(\tfrac{1}{2} u_x^2 \cos u + u_{xx} \sin u)$$

$$X^5 = -( u_x u_{xx} \cos u + u_{xxx} \sin u) \ , \quad X^6 = -\Big[u_x\Big(u_{xxx} + \frac{3}{4!} u_x^3\Big)\cos u$$

$$+ \Big(u_{xxxx} + u_x^2 u_{xx} - u_x u_{xxx} - \frac{1}{2} u_x^4\Big)\sin u\Big] \ , \ \ldots \tag{1.56}$$

and

$$X^m = f_m \cos u + g_m \sin u \tag{1.57}$$

where $f_m$ and $g_m$ are polynomials of rank m-2 in $u_x$, $u_{xx}$, etc. The densities $T^m$ and fluxes $X^m$ then satisfy a conservation law like (1.53) for each m after correcting for the perfect differentials omitted from (1.53).

Since the s-G (1.10a) in light cone coordinates is invariant under interchange of x and t it follows that

$$\frac{\partial}{\partial t} (\cos u - 1) + \frac{\partial}{\partial x} (\tfrac{1}{2} u_t^2) = 0$$

$$\frac{\partial}{\partial t} (- \tfrac{1}{2} u_t^2 \cos u) + \frac{\partial}{\partial x} (\tfrac{1}{2} u_{tt}^2 - \tfrac{1}{8} u_t^4) = 0 \tag{1.58}$$

. . . .

and that, in general, every flux $X^{2n}$ yields a conserved density by replacing $\partial/\partial x$ everywhere in this expression by $\partial/\partial t$. There is therefore a second infinite set of conserved densities $T^{2n}$

$$T^2 = (\cos u - 1)$$

$$T^{2n} = f_{2n} \cos u + g_{2n} \sin u \ , \quad n \geq 2 \tag{1.59}$$

where $f_{2n}$ and $g_{2n}$ are polynomials in $u_t$, $u_{tt}$, etc. of rank (in t) 2n-2. It is easy to show [1.42,69] (see also Chap.11) that the density $T^2$ in (1.59) serves as a *Hamiltonian* density for the s-G (1.11a) in light cone coordinates. It seems, however that the constants of the motion $\int_{-\infty}^{\infty} T^{2n}\, dx$ for $n \geq 2$ are not in involution.

Fortunately there is yet a third infinite set of conserved densities for the s-G (1.11a) which provide constants of the motion which are in involution. Their existence is connected with the dispersion relation $\omega = k^{-1}$ and its pole at k = 0. How this comes about is described by NEWELL in Chap.6. Details for the s-G are also given in [1.42,69], for example, where it is indicated how, after appropriate scaling, the set (1.53) acts as momentum densities and the third infinite set acts as energy densities for an infinite hierarchy of generalised s-G equations each of which is solvable by the inverse method and each of which is a completely integrable infinite dimensional Hamiltonian system. This infinite hierarchy is analogous to the infinite set of KdV equations first found by LAX [1.70] in 1968. Lax's set of KdV equations is described more precisely in Sect.1.4. Fadeev handles the momentum and energy densities of the s-G in Chap.11. Members of the infinite set of conserved densities which provide constants of the motion in involution are calculated at the end of Sect.1.6, (1.153).

The members of the infinite set of completely integrable s-G equations are easily characterised in terms of scattering data [1.69]. Except for the s-G itself the equations of motion are complicated and seem to involve nonlocal operators.[4] It follows that the s-G is the unique equation invariant under the infinitesimal Lie transformation $x \to (1-\varepsilon)x$, $t \to (1+\varepsilon)t$ and the finite *scale transformations* $x \to a^{-1}x$, $t \to a\,t$. It is then easy to see that, in laboratory coordinates, it is the unique equation which is Lorentz covariant. From the theorem that the nonlinear Klein-Gordon equation (1.37) has the p.c.ds $\frac{1}{2}u_x^2$ of rank 2 but then has a second

---

[4]The Hamiltonian densitiy for the next member of the infinite s-G sequence is calculated by the methods of [1.42,69] to be

$$\mathcal{H}_2 = -\frac{1}{2}u_x \exp\left[-\int_{-\infty}^{x}\left(\int_{-\infty}^{x'} \sin u\, dx''\right)\frac{\partial}{\partial x'}\left(-\tan\frac{1}{2}u\right)dx'\right]$$

$$\times\left\{\int_{-\infty}^{x}\left[-\frac{1}{2}\left(\int_{-\infty}^{x'} \sin u\, dx''\right)\sec^2\frac{1}{2}u + \frac{1}{4}\left(\int_{-\infty}^{x'} \sin u\, dx''\right)^2 \sec^4\frac{1}{2}u\right]\right.$$

$$\left.\times \exp\left[\int_{-\infty}^{x'}\left(\int_{-\infty}^{x''} \sin u\, dx'''\right)\frac{\partial}{\partial x''}\left(-\tan\frac{1}{2}u\right)dx''\right]dx'\right\}$$

$\mathcal{H}_2$ is actually trivial and $\mathcal{H}_3$ is the next nontrivial density. We know [1.42,69] that the integrable s-G equations have 1-soliton solutions $4\tan^{-1}\exp[kx+2\Omega(\frac{1}{2}ik)t]$ where $\Omega(\zeta) = -\frac{i}{2}\omega(-2\zeta)$ and $\omega(k)$ is the linearised dispersion relation of the equation. For the s-G itself $\omega(k) = k^{-1} = 2\Omega(\frac{1}{2}ik)$. For the next member $\omega(k) \propto k^{-3}$ [1.69]. Thus the linearised equation is $u_{xxxt} = Au$ ($A$ = constant), or $u_{xt} = A\int_{-\infty}^{x}\int_{-\infty}^{x'} u(x'',t)dx''\,dx'$. In [1.42] we implied (see footnote 5 there) that the equation of motion derived from the conserved density $-\frac{1}{2}u_t^2\cos u$ in (1.58), namely $u_{xt} = \frac{1}{2}u_t^2\sin u - u_{tt}\cos u$ was the next number of the integrable s-Gs. But this equation does not have solutions $4\tan^{-1}\exp[kx + 2\Omega(\frac{1}{2}ik)t]$ whilst the linearised dispersion relation of this equation is $\omega = k$. This remark also corrects the impression given in [1.195 ]referenced below that the conserved density $-\frac{1}{2}u_t^2\cos u$ yields the next integrable s-G. The reader is referred to the end of Sect.1.6, where the integrable s-G's are looked at again.

of rank 4 involving $u_{xx}$ if and only if $\ddot{F}(u) + \alpha^2 F(u) = 0$ [1.65,66] and then has an infinity [1.66], it follows that $F(u) = Ae^{\alpha u} + Be^{-\alpha u}$. From the requirements that u and F(u) are real and $u = \delta$(= constant) is a solution of (1.37) it follows that equation will have a second polynomial density of rank 4 if and only if

$$F(u) = A \sinh(u - \delta) \tag{1.60a}$$

or

$$F(u) = B \sin(u - \delta) \tag{1.60b}$$

and in both cases there is an infinity of conserved densities.

The sinh-Gordon equation $u_{xt} = \sinh(u - \delta)$ has no bounded solitary wave or permanent profile solution defined on the whole of the x axis (the solution is $u = 4 \tanh^{-1}\theta$, $\theta = kx + k^{-1}t$, defined for $\theta \leqq \theta_0 < 0$). It is related to the s-G by $u - \delta \to i(u - \delta)$ whence $u_{xt} = \sin u$. It is easy to see, by referring to Sect.1.4 and the discussion of the AKNS inverse scattering method there following (1.100, 101), that the scattering problem (1.100) has a Hermitean matrix $\hat{L}$ and therefore no bound states. There are therefore no multisoliton solutions of the sinh-Gordon equation. The situation surrounding the simple wave (1.12) is similar.

One result therefore is that the existence of an infinite set of conserved densities is not sufficient for multisoliton solutions. The sinh-Gordon equation is one example which demonstrates this; the equation of a simple wave (1.12) is another (a third is the *linear* Klein-Gordon equation [1.71]). The main result here, however, is that we have shown that the s-G is the unique nonlinear Klein-Gordon equation with two successive p.c.ds of ranks 2 and 4, with a finite energy permanent profile solution, and with soliton solutions. KULISH [1.72] also deduced that the S-G is the unique equation with finite energy solutions, a second conserved density, and hence an infinity of densities. That it is the unique equation in a single field with an infinity of conserved densities apparently determines the S-matrix of the *quantised* s-G completely [1.71]. The s-G is thus a very special and remarkable equation. Luther, in Chap.12, indicates from his different point of view that the quantised s-G is a very special system.

The sinh-Gordon equation $u_{xt} = \sinh u$ yields the s-G (1.11a) by $u \to iu$. We now investigate briefly the situation when the independent variables x,t are rotated in the complex plane. Consider in particular the s-G (1.11a) when $x \to \frac{1}{2}(x + iy)$ $t \to \frac{1}{2}(x - iy)$. Evidently

$$u_{xx} + u_{yy} = \sin u \tag{1.61}$$

with solutions u in 2-dimensional Euclidean space. This equation has some interest in statistical mechanics in connection with 2-dimensional vortex models [1.73,74]. It has the aBT [compare (1.37)]

$$u'_x = -iu_y + k\sin\frac{1}{2}(u'+u) + k^{-1}\sin\frac{1}{2}(u'-u)$$

$$-iu'_y = u_x + k\sin\frac{1}{2}(u'+u) - k^{-1}\sin\frac{1}{2}(u'-u) \quad . \tag{1.62}$$

The integrability conditions $u'_{xy} = u'_{yx}$ and $u_{xy} = u_{yx}$ yield

$$u'_{xx} + u'_{yy} - \sin u' = u_{xx} + u_{yy} - \sin u = 0 \quad . \tag{1.63}$$

For $u = 0$ a solution of (1.62) is the 'kink'

$$u' = 4\tan^{-1}\exp[(x - Vy)/(1 + V^2)^{\frac{1}{2}}] \quad . \tag{1.64}$$

Providing we choose $|k| = 1$ at each stage, real multikink solutions can be built up by repeated use of this aBT though the intermediate stages are not all real. The two-kink solution is

$$u(x,y) = 4\tan^{-1} r\cosh\Theta_I \operatorname{sech}\Theta_R \tag{1.65a}$$

with $r = \cot\mu$ and

$$\Theta_R = \frac{\cos\mu}{(1+V^2)^{\frac{1}{2}}}(x - Vy) \ , \quad \Theta_I = \frac{\sin\mu}{(1+V^2)^{\frac{1}{2}}}(y + Vx) \quad . \tag{1.65b}$$

As far as we are aware (1.61) has not yet been studied in much greater depth.[5] In principle other solutions like (1.65a) can be found by methods analogous to those developed from (1.66a) below.

In Sect.1.3 we have noted the connection between nonlinear evolution equations with soliton solutions, infinities of conserved densities, and Hamiltonian systems. We refer to this connection again in Sect.1.4 and it is treated rather thoroughly in later chapters. We have also seen a connection between BTs and infinite sets of conserved densities. More recently the close connection between BTs, nonlinear

---

[5] McCARTHY [1.215] lists the obvious BTs

$$\left.\begin{aligned} u'_x + u_y &= \sin u' \cosh u \\ u'_y - u_x &= \cos u' \sinh u \end{aligned}\right\} \quad \text{between} \quad \left\{\begin{aligned} u_{xx} + u_{yy} &= \sinh u \cosh u \\ u'_{xx} + u'_{yy} &= \sin u' \cos u' \end{aligned}\right. \quad .$$

$$\left.\begin{aligned} u'_x + u_y &= \cosh u' \cos u \\ u'_y + u_x &= \sinh u' \sin u \end{aligned}\right\} \quad \text{between} \quad \left\{\begin{aligned} u_{xx} + u_{yy} &= \sin u \cos u \\ u'_{xx} + u'_{yy} &= \sinh u' \cosh u' \end{aligned}\right.$$

and

$$\left.\begin{aligned} u'_x + u_y &= \sin u' \, e^u \\ u'_y - u_x &= \cos u' \, e^u \end{aligned}\right\} \quad \text{between} \quad \left\{\begin{aligned} u_{xx} + u_{yy} &= e^{2u} \\ u'_{xx} + u'_{yy} &= 0 \end{aligned}\right. \quad .$$

evolution equations and solitons has been still further exemplified in the work of WAHLQUIST and ESTABROOK on prolongation structures [1.61,75,76]. PIRANI and co-workers [1.77] have recently (1978) given a jet bundle formulation of BTs which provides a rational framework for many of these results. We refer the reader to the published literature for further study of this important development of the mathematics of the subject. On a different track CALOGERO [1.78] was able to illuminate the connection between BTs and the inverse scattering method in the course of extending the applications of that method (the unifying idea is that both evolution equations and their BTs can have a simple structure in terms of scattering data despite their complexity in x space). This work is described in Chap.9. We refer the reader also to [1.79,80] for other recent work on BTs whilst still other work connected with BTs and prolongation structures is referenced in Sect.1.5. Hirota treats BTs from his own particular and unconventional standpoint in Chap.5.

We finish Sect.1.3 with some further historical remarks on the s-G (1.11a) and note some of its recent applications in field theory. Lorentz covariance of the s-G (1.10) means Lie invariance of the s-G (1.11a) and under the more general change of coordinates

$$\xi = a\,t + a^{-1}x \quad , \qquad \eta = a\,t - a^{-1}x$$

it is Lorentz covariant. If we set

$$u = 4\tan^{-1}[F(\xi)/G(\eta)] \tag{1.66a}$$

then

$$F'^2 = -\,kF^4 + mF^2 + n$$

$$G'^2 = kG^4 + (m-1)G^2 - n \tag{1.66b}$$

in which k,m,n are arbitrary constants and prime denotes differentiation with respect to argument [for solutions of (1.61) $G'^2$ changes sign]. There is some hint here of the direct methods for solving the s-G for its N-soliton solution found by HIROTA [1.81] and CAUDREY et al. [1.82] in 1972-1973. These are mentioned again in Sect.1.5. However, the various pseudo-spherical surfaces corresponding to the solutions of (1.64) (called surfaces of Enneper of constant curvature) were catalogued by STEURWALD [1.83] in 1936 and contain the $2\pi$ kink, and $4\pi$ kink and the breather. These points are made by LAMB [1.27] to whom the reader is referred for much other rewarding historical comment.[6] (Footnote see page 24.)

Independently SKYRME [1.84,85] discovered these three solutions of the s-G whilst constructing the first soluble nonlinear model field theory in 1959-1962. Investigation of the phase shifts shows that kinks (antikinks) repel each other but a kink and an antikink attract each other. SKYRME assigned 'topological charges' +1 and -1 to the kink and antikink. As we have explained the kink (antikink) interpolates between 0 and $2\pi$ ($2\pi$ and 0). The charges are the jumps $u(\infty,t) - u(-\infty,t)$ scaled against $2\pi$ and are constants of the motion stabilised by the boundary conditions. Topological quantum numbers have this character [1.43]. The breather therefore has no charge: it is its own antiparticle and was assigned the role of a meson. As far as we known SKYRME actually guessed his $4\pi$ kink and breather solutions after fitting analytical expressions to numerical work.

An enormous activity in comparable model field theories has developed only since 1974 [1.39-41,43,86,87]. One remarkable result is the connection between the *quantised* s-G and the massive Thirring model found by COLEMAN [1.43,88] in 1975. The connection is described by Luther in Chap.12. Connections between the conserved densities of the quantised s-G and the massive Thirring model have been established by BERG et al. [1.89], and KULISH and NISSIMOV [1.90,91]. The latter show [1.90,91] that quantisation leads only to slight modifications of the classical conservation laws. MIKHAILOV [1.92] established that the classical massive Thirring model was soluble by an inverse scattering method and had an infinity of conservation laws. The classical results for the s-G are of course essentially those in (1.54,59). An important result then is that the existence of an infinite set of conserved densities for a nonlinear evolution equation seems to imply that its S-matrix is factorisable [1.72,93-95]. In consequence the S-matrices for both the quantised s-G and massive Thirring models have been calculated exactly [1.71,93,95] and are necessarily identical. S-matrices for other field theoretical models in 1 + 1 dimensions have also now been calculated [1.95,96].

## 1.4 Other Physical Problems and the Discovery of the Inverse Method

In the period 1960-1967 a number of physical problems influenced the developement of soliton theory. McCALL and HAHN [1.29,97] reported the theory of self-induced transparency (SIT) and experimental results in support of it. They had found an analytical sech pulse solution for an electric field envelope traversing a resonant 2-level atomic medium after first finding it numerically. They also discovered the break up of an optical pulse into a train of such sech pulses and realised the

[6] Pseudo-spherical surfaces are surfaces of constant negative Gaussian curvature. An early example (with curvature $-a^{-2}$) was Beltrami's surface of revolution obtained by spinning the tractrix $y = \sqrt{a^2 - x^2} - a \ln[(a + \sqrt{a^2 - x^2})/x]$ about the y axis. Amongst the solutions of nonlinear evolution equations it is not only those of the s-G which represent surfaces of constant negative curvature. We show at the end of Sect.1.7 that at least the solutions of the KdV, modified KdV, NLS and indeed all AKNS systems have this property.

connection of these with the $2\pi$ kinks of the s-G. The N-kink solution (1.22) of the s-G was, however, found only in 1972-1973 [1.81,82,98]. The theory of SIT is treated by LAMB and McLAUGHLIN in Chap.2. Early work on SIT is due to ARECCHI and BONIFACIO [1.99].

In 1964 CHIAO et al. [1.100] found the NLS equation in the study of optical self-focussing and optical filamentation. And they found its sech solution (1.21) for the single filament transverse amplitude profile. A simple argument for a neutral dielectric is the following [1.22,23]. For isotropic atoms the scalar atomic dipole $P(x,t)$ induced by a scalar field $E(\underset{\sim}{x},t)$ is

$$P(\underset{\sim}{x},t) = \alpha E(\underset{\sim}{x},t) + \alpha_{NL} E(\underset{\sim}{x},t)^3 + \dots \tag{1.67}$$

in which $\alpha$ and $\alpha_{NL}$ are the linear and nonlinear polarisabilities. Maxwell's equation is of course linear and is

$$\nabla^2 E - c^{-2}\partial^2 E/\partial t^2 = 4\pi n c^{-2}\partial^2 P/\partial t^2 \tag{1.68}$$

in which n is the smoothed atomic number density. If we look for complex envelope solutions modulating carrier waves for E, namely,

$$E(\underset{\sim}{x},t) = \varepsilon(x,y,z,t)\exp[i(\omega t - kz)] + \text{c.c.} \tag{1.69}$$

and we impose the linear dispersion relation $\omega^2 = c^2k^2 - 4\pi n\omega^2\alpha$ we find

$$\varepsilon_{xx} + \varepsilon_{yy} + \omega^2 c^{-2} 12\pi n\alpha_{NL}|\varepsilon|^2\varepsilon + 2ic^{-2}[\omega\varepsilon_t(1 + 4\pi n\alpha) - c^2 k\varepsilon_z] = 0 \quad . \tag{1.70}$$

In particular for steady state solutions after scaling

$$\varepsilon_{xx} + \varepsilon_{yy} + 2|\varepsilon|^2\varepsilon - i\varepsilon_z = 0 \quad . \tag{1.71}$$

This is the two-dimensional NLS in (x,y) and a 'time' (z). The equation is apparently unstable [1.101] — solutions in two space dimensions grow to infinity in a finite time — but in one space dimension (x) it reduces to (1.9). CHIAO et al. reached this equation via a nonlinear intensity dependent dielectric constant. LAMB and McLAUGHLIN (Chap.2) reach it in a still different context in nonlinear optics. GINZBURG and PITAEVSKII [1.102], PITAEVSKII [1.103], and GROSS [1.104] found the (steady state) Gross-Pitaevskii equation [1.105]

$$a + b|u|^2 - \frac{h^2}{2m}\nabla^2 u = 0 \tag{1.72}$$

in the study of superconductors. BENJAMIN and FEIR [1.106] found the instability in deep water waves which bears their names: this instability is due to the bunching of harmonic waves into envelope solitons governed by the NLS. BENNEY and NEWELL [1.107] studied the NLS in this context. Extensions of the NLS equation to

$$u_{xx} + 2[1 - \exp(-|u|^2)]u + iu_t = 0 \tag{1.73}$$

in the theory of plasma cavitons, photon bubbles, and the interaction of radiation with plasmas are described [1.22,23]. The connection with the equations of Langmuir turbulence due to ZAKHAROV [1.108] is made in [1.64].

The NLS was solved by ZAKHAROV and SHABAT [1.38] in 1971 and subsequently [1.109] in 1973 by the first example of a $2 \times 2$ inverse scattering scheme. These superb papers are remarkable in the number of analytical results and new methods they contain: not least amongst these is a derivation in [1.38] of the infinite set of conserved densities for the NLS by a method which can be applied to all AKNS systems [1.44,45] with a $2 \times 2$ scattering problem [1.42,69]. The principal significance of this paper, however, is that it showed that the inverse scattering method [1.1] for solving the KdV equation was not unique to that equation. We think as an introduction to the inverse scattering method, it will be helpful to the reader if we describe the discovery of the inverse method for the KdV as we have heard it from KRUSKAL [1.110] and we do this next.

KRUSKAL in the period 1955-1965 was concerned with the recurrences exhibited by one-dimensional anharmonic lattices—the Fermi-Pasta-Ulam (FPU) problem [1.12]. This led ZABUSKY and KRUSKAL [1.2] to study the equation obtained for a cubic nonlinearity in the continuum limit

$$y_{tt} = y_{xx}(1 + \varepsilon y_x) \tag{1.74}$$

where the parameter $\varepsilon$ scales the nonlinearity $\sim 1/10$. By looking at the Riemann invariants of the linear problem $y_{tt} = y_{xx}$, namely

$$u = \frac{1}{2}(y_x + y_t) = O(1) \quad , \quad v = \frac{1}{2}(y_x - y_t) = O(\varepsilon)$$

they reduced (1.74) to the simple wave

$$u_\tau + uu_x = 0 \qquad [\tau \equiv 4\varepsilon(t - x)] \quad . \tag{1.75}$$

Characteristics of this equation are $dx/d\tau = u =$ constant. Intersection of these suggests that the equation shocks. It may be integrated by the hodograph transformation to yield the solution $u = u(x - ut)$ which shocks (the speed $u$ is large where the disturbance $u$ is large). An element of dispersion was next added by returning to the discrete lattice model: the result was

$$y_{tt} = y_{xx}(1 + \varepsilon y_x) + \frac{\Delta x^2}{12} y_{xxxx} \quad , \tag{1.76}$$

where $\Delta x$ was the lattice spacing and from this

$$u_\tau + uu_x + \delta^2 u_{xxx} = 0 \quad , \tag{1.77}$$

where $\delta^2 \equiv \Delta x^2/12\varepsilon$. For $\Delta x \sim 1/64$, $\varepsilon \sim 1/10$, $\delta \approx 0.022$.

Numerical solutions of (1.77) showed [1.2] that, under periodic boundary conditions, harmonic initial data developed a smooth jump approaching the simple wave shock for the chosen small value of $\delta^2$. Thereafter the profile broke up into spikes the collision properties of which caused ZABUSKY and KRUSKAL [1.2] to call them solitons. In order to characterise the solution across the jump KRUSKAL and co-workers then looked for conservation laws. Equation (1.77) is already in conservation form with conserved density $T^1 = u$. However, $T^2 = u^2$ and $T^3 = 2u^3 - u_x^2$ are also conserved densities. MIURA et al. [1.67] found 10 polynomial conserved densities for the KdV in the form (1.3) and proved that there was an infinity of nontrivial $T^r$. Each term $\Pi u_j^{a_j}$ in $T^r$ (where again $u_j$ is the derivative of order j) has in this case rank $r = \sum_j (1 + \frac{1}{2} j)a_j$: $T^{10} = u^{10}/10 - 36u^7u_1^2 \ldots + 419904\, u_8^2/12155$ has rank 10 and contains 32 terms [1.67].

FPU also looked at a lattice with quartic rather than cubic anharmonicity. This leads to $y_{tt} = y_{xx}(1 + \varepsilon y_x^2)$ and thence to the modified KdV

$$v_t + v^2 v_x + v_{xxx} = 0 \quad . \tag{1.78}$$

This also has conservation laws. MIURA [1.62] found the transformation $u = v^2 + \sqrt{-6}v_x$ connecting these. The $\sqrt{-6}$ is innocuous: the MIURA transformation [1.62]

$$u = v^2 + v_x$$

is the BT (1.44) scaled in this case to connect the KdV and modified KdV equations in the forms

$$u_t - 6uu_x + u_{xxx} = 0 \tag{1.79a}$$

$$v_t - 6v^2 v_x + v_{xxx} = 0 \quad . \tag{1.79b}$$

The Riccati transformation $v = \partial(\ln\psi)/\partial x$ linearises $v^2 + v_x$ and $u = \psi_{xx}/\psi$. The Galilean transformation

$$x' = x - Vt \quad , \quad t' \to t \quad , \quad u \to u + V$$

leaves $u_t - uu_x + u_{xxx} = 0$ invariant. Hence $u \to u + \lambda = \psi_{xx}/\psi$ and the Schrödinger eigenvalue problem

$$- \psi_{xx} + u\psi = -\lambda\psi \tag{1.80}$$

has emerged. Furthermore $u = \psi_{xx}/\psi - \lambda$ into the KdV (1.79a) yields [1.1]

a) $\lambda_t = 0$ for the bound states $(\lambda > 0)$ even though u depends on t;

$$\text{b) } \psi_t + \psi_{xxx} - 3(u - \lambda)\psi_x = C\psi + D\psi \int^x \frac{dx}{\psi^2} \tag{1.81}$$

and [1.1] D vanishes for $\lambda > 0$, whilst C vanishes through the normalisation of the bound state eigenfunctions. From the property $u \to 0$ as $|x| \to \infty$ the eigenfunction $\psi_n$ belonging to $\lambda_n > 0$ takes the form

$$\psi_n \sim c_n(t) \exp(-k_n x) \quad \text{for} \quad x \to \infty \tag{1.82}$$

$(k_n = \lambda_n^{1/2})$ and from (1.81) $c_n(t) = c_n(0) \exp 4k_n^3 t$.

For $\lambda = -k^2 < 0$, a solution of (1.80) for large $|x|$ is a linear combination of $\exp \pm ikx$. If

$$\psi \sim \exp(-ikx) + b \exp(ikx) \qquad x \to \infty \tag{1.83a}$$

$$\psi \sim a \exp(-ikx) \qquad x \to -\infty \quad , \tag{1.83b}$$

a(k,t) is a transmission coefficient and b(k,t) a reflection coefficient and $|a|^2 + |b|^2 = 1$. Since the spectrum for $\lambda < 0$ is continuous, $\lambda$ can be chosen so that $\lambda_t = 0$ here also. Equations (1.83a,b) in (1.80) yield $D = 0$, $C = 4ik^3$ and two equations which easily give

$$a(k,t) = a(k,0)$$

$$b(k,t) = b(k,0) \exp(8ik^3 t) \quad . \tag{1.84}$$

The key point now is that the *scattering data* a, b, $c_n$ and $\lambda_n$ are sufficient to allow reconstruction of u at any time t. Let K(x,y) for $y \geqq x$ be the solution of the Gel'fand-Levitan equation [1.111,112]

$$K(x,y) + B(x + y) + \int_x^\infty K(x,z)B(y + z)dz = 0 \tag{1.85a}$$

with

$$B(\xi) \equiv \frac{1}{2\pi} \int_{-\infty}^{\infty} b(k) \exp(ik\xi)dk + \sum_n c_n^2 \exp(k_n)\xi \quad . \tag{1.85b}$$

Then

$$u(x,t) = 2\frac{\partial}{\partial x}K(x,x) \quad . \qquad (1.85c)$$

The procedure for solving the KdV equation (1.79a), which is the KdV equation (1.1), is therefore

I) map the initial data $u(x,0)$ onto scattering data $S = \{b(k,0),\ c_n(0)$ and $\lambda_n(n = 1, \ldots, N)\}$;

II) compute the time evolution of the scattering data as indicated above;

III) solve the Gel'fand-Levitan equation (1.85a) and calculate $u(x,t)$ $(t > 0)$.

In particular, in the reflectionless case, $b(k,0) = 0$ and the solution is an N-soliton solution corresponding to that displayed in (1.16) (but with the different scaling). This remarkable discovery has been the source of all of soliton theory with the single exception of the direct methods by HIROTA [1.33,81] (see also Chap.5) and to a lesser extent by CAUDREY [1.82]. The precise connection of these methods with the inverse scattering method is still to be established although of course both methods find the same N-soliton solutions.

We should pause now and explain some terminology: procedure I) is now sometimes called the direct scattering (or spectral) transform. Procedure III) is sometimes called the inverse scattering (or spectral) transform. The total procedure is 'the inverse scattering method' or 'the method of the inverse spectral transform'. We use mostly 'inverse scattering method' or 'inverse method' in this book. It is treated from different points of view in at least five of the contributions to this volume. We believe that the measure of overlap thus created will be a source of help rather than irritation to the reader. Section 1.4 is intended as a first exposure to the method which is developed at length by NEWELL in Chap.6 and, differently, by ZAKHAROV in Chap.7.

## 1.5 Operator Pair Formulation of Nonlinear Evolution Equations

LAX [1.70] stimulated an important development and added mathematical understanding to the inverse method. Given the relatively prime differential operators $u \to \hat{L}_u \equiv (-\partial^2/\partial x^2 + u)$, $\hat{B}_u \equiv \partial^{2n+1}/\partial x^{2n+1}$ + lower degree ($\hat{B}$ is skew symmetric $\hat{B}_u^+ = -\hat{B}$) such that there is a one parameter family of unitary operators $\hat{U}$ satisfying $\hat{U}_t = \hat{B}\hat{U}$; and $\hat{L}_u$ is unitary equivalent under $\hat{U}$, i.e., $\hat{U}^{-1}(t)\hat{L}_u(t)\hat{U}(t)$ is independent of t; then, as LAX notes,

I) the eigenvalues $\lambda_u$ of $\hat{L}_u$ are integrals of the motion, i.e., $\lambda_{u,t} = 0$. [The proof is trivial: $\hat{L}_u(0) = \hat{U}(t)\hat{L}_u\hat{U}(t)$, $\hat{L}_u(0)\psi(\hat{0}) = \lambda_u\psi(0)$. Then $\hat{L}_u(t)\psi(t) = \lambda_u\ (t)$ with $\psi(t) = \hat{U}(t)\psi(0)$, i.e., $\hat{U}(t)$ is the evolution operator.]

II) $\frac{\partial}{\partial t}\hat{U}^{-1}(t)\hat{L}_u(t)\hat{U}(t) = 0$ implies that

$$-\hat{U}^{-1}\hat{U}_t\hat{U}^{-1}\hat{L}_u\hat{U} + \hat{U}^{-1}\hat{L}_{u,t}\hat{U} + \hat{U}^{-1}\hat{L}_u\hat{U}_t = 0 \quad .$$

Then if $\hat{U}_t = \hat{B}\hat{U}$ we have $-\hat{U}^{-1}\hat{B}\hat{L}_u\hat{U} + \hat{U}^{-1}\hat{L}_{u,t}\hat{U} + \hat{U}^{-1}\hat{L}\hat{B}\hat{U} = 0$ , and

$$\hat{L}_{u,t} = [\hat{B},\hat{L}_u] \quad . \tag{1.86a}$$

III) Since $\hat{L}_{u,t} = u_t$, the operator equation (1.86a) is an evolution equation $u_t = K[u]$ in which $K[u]$ is a functional of u. LAX [1.70] proved that initial data $u(x,0)$ determine the solution of such nonlinear evolution equations uniquely. By trying for the skew symmetric operator $\hat{B} = \partial^3 + b\partial + \partial b(\partial \equiv \partial/\partial x)$ LAX also found that with $b = -3/4\ u$ (our scaling is different from that used by LAX [1.70]) he regained from (1.86a) the KdV equation. Further the method generalised to yield an infinite hierarchy of KdV equations of degrees 3 (the KdV), 5, 7, etc. ZAKHAROV and FADEEV [1.113], and LAX [1.114] subsequently indicated the symplectic structure behind this result which we now know to be a property of all the 'integrable systems' like the KdV, modified KdV, NLS and s-G equations. The representation of nonlinear evolution equations in the 'Lax pair' form

$$\hat{L}_t = [\hat{B},\hat{L}] \tag{1.86b}$$

remains the most powerful technique for developing analytical solutions of such equations. Notice that such a representation guarantees the constancy of the eigenspectrum. For if $\hat{B}$ is skew the solution of $U_t = \hat{B}\hat{U}$ is unitary and the argument to (1.86a) in II) can be reversed. $\hat{L}$ is then unitary equivalent under U and the eigenvalues of $\hat{L}$ are constants of the motion by I).

It is remarkable that Lax's operators $\hat{B}$ for the hierarchy of KdV equations can be found by defining the square root $\hat{R}$ of the operator $-\hat{L}_u$ by $\hat{R}^2 \equiv -\hat{L}_u$ and $\hat{R} = \partial + c_0 + c_1\partial^{-1} + c_2\partial^{-2} + \ldots$ . Then $-\hat{L}_u = \partial^2 - u$ and $\hat{R}^2 = -\hat{L}_u$ is $\partial^2 + 2c_0\partial + (c_{0,x} + c_0^2 + 2c_1) + (c_{1,x} + 2c_2 + 2c_0c_1)\partial^{-1} + \ldots$, so that $c_0 = 0$, $c_1 = -u/2$ $c_2 = u_x/4$. Evidently the principal part of $\hat{R}$ (that excluding inverse powers of $\partial$) is simply $\partial$ and

$$[\hat{R},\hat{L}_u] = \hat{L}_{u,t} \quad \text{is}$$

$$u_x = u_t \quad . \tag{1.87}$$

However, the principal part of $\hat{R}^{3/2} = (-\hat{L}_u)\hat{R}$ is

$$\partial^3 - \partial u - \frac{1}{2}u\partial + \frac{1}{4}u_x = \partial^3 - \frac{3}{4}\partial u - \frac{3}{4}u\partial \tag{1.88}$$

and this is Lax's operator for the KdV. Other fractional powers $\hat{R}^{5/2}$, etc., generate Lax's KdV hierarchy. This hierarchy is apparently not unique [1.115].[7] This fact has been partly understood in terms of prolongation structures [1.116]. The essential point seems to be that if $\hat{L}$ is of degree 2 it is the scalar Schrödinger operator $\hat{L}_u$ and so generates Lax's hierarchy of KdV equations. But if $\hat{L}$ is of degree 3 as the work on prolongation structures suggests a different hierarchy of equations is generated. This interesting problem needs further study. It raises interesting questions concerning the hierarchies of other integrable systems.

GEL'FAND and DICKII [1.117] and other workers now referenced by MANIN [1.118] have provided a basis for the argument to (1.88). They showed how to solve the KdV for its N-soliton solutions on the real line $-\infty<x<\infty$ by reducing the problem by algebraic-geometric methods to a system of linear algebraic equations. Analogous but perhaps more difficult methods were introduced by MATVEEV and ITS [1.119,120] and by DUBROVIN and NOVIKOV and their collaborators [1.119,120] in order to solve the KdV under periodic boundary conditions. The present form of the method is described by Novikov in Chap.10. LAX [1.114] gave a dense set of solutions of the KdV equation under periodic boundary conditions in 1974 and this stimulated work by McKEAN and VAN MOERBEKE [1.121].

Following the work of LAX [1.70] two remarkable papers created much of soliton theory as we now know it. These are the paper by ZAKHAROV and SHABAT [1.38] solving the NLS equation and the paper by ZAKHAROV and FADEEV [1.113] indicating the symplectic structure of the KdV equation. The authors of the latter paper showed that the KdV was the first known example of an infinite dimensional completely integrable Hamiltonian system. They noted that the KdV had an infinity of first integrals $I_n[u] = \int_{-\infty}^{\infty} P_u(u,u_x,\ldots)dx$ (the $P_u$ are the conserved densities $T^n$ of Sect.1.4). They took the Hamiltonian form of the KdV equation

$$u_t = \frac{\partial}{\partial x}\frac{\partial I_3[u]}{\partial u(x)} \tag{1.89}$$

quoted by LAX [1.70] as due to Gardner. They demonstrated that the infinity of constants $I_n$ was precisely the right number by constructing the symplectic form

$$\Omega(\delta_1 u,\delta_2 u) = \int_{-\infty}^{\infty} dx \int_{-\infty}^{x} dy[\delta_1 u(x)\delta_2 u(y) - \delta_1 u(y)\delta_2 u(x)] \quad , \tag{1.90}$$

showing this was closed under the transformation to scattering data (the variations $\delta_1 u$ and $\delta_2 u$ must be expressed in terms of corresponding variations in scattering data), demonstrating thereby that the inverse and direct scattering transform are canonical transformations from and to action-angle variables, constructing

---

[7] See also [1.172]. The BT reported there is also of interest.

the Hamiltonian in terms of these variables, showing that this Hamiltonian depended only on the action variables which were constants of the motion, and explicitly integrating the equations of motion in these variables. This beautiful analysis is the source of much of the material now appearing in this volume. In particular the study of soliton equations as completely integrable Hamiltonian systems constitutes the theme of Chap.11 (by Fadeev) and much of that of Chap.6 (by Newell), whilst in Chap.2 Lamb and McLaughlin introduce again the canonical Hamiltonian structure of nonlinear evolution equations with soliton solutions in a fashion useful as an introduction to the later chapters. There is therefore no point in pursuing this elegant topic further here.

We should, however mention for historical reasons the important paper by GARDNER [1.122] of 1971 which contains the published results relating to (1.89,90) GARDNER defined the Poisson bracket of two functionals $K_1$ and $K_2$ of u by

$$\{K_1,K_2\} = \int \frac{\delta K_1}{\delta u}\frac{\partial}{\partial x}\left(\frac{\delta K_2}{\delta u}\right) dx \quad . \tag{1.91}$$

He proved that the constants $I_n[u]$ associated with the densities $T^n$ are in involution, i.e., $\{I_n, I_m\} = 0$. We will trace elsewhere [1.123] how these ideas relate to the Frobenius integrability theorem, to Cartan's theory of exterior differential forms, and to the prolongation structures of WAHLQUIST and ESTABROOK [1.61,75,76].

From the $2\times 2$ scattering problems introduced by ZAKHAROV and SHABAT in 1971 and 1973 [1.38,109] to solve the NLS there is a natural extension found by AKNS [1.44,45] in 1973. This step was taken only after other work had emerged but it is not clear to us how far this work influenced AKNS.

## 1.6 Discovery of Some Other N-Soliton Solutions: The AKNS-Zakharov-Shabat $2\times 2$ Scattering Scheme and Its Geometry

In 1971 HIROTA [1.33] found the N-soliton solution of the KdV by the first application of his direct method. In 1972 WADATI and TODA [1.124] used the inverse method described by GARDNER et al. [1.1] to obtain explicitly the N-soliton solution of the KdV. WADATI [1.125] then solved the modified KdV (1.78) by a $2\times 2$ scattering scheme. G.L. LAMB was puzzling over possible connections between the KdV and s-G at this time—an idea he transmitted to us in Manchester. In 1972 he used the conservation laws of the s-G to describe pulse break-up [1.126]. In 1973 [1.127,128] he reported an inverse scattering solution of the SIT equations (see Chap.2).

In Manchester we were also studying ultra-short optical pulses and had earlier [1.129] noticed the collision properties of the solitary wave sech solutions of the Maxwell-Bloch (MB) equations. We are here using the nomenclatire SIT and MB equations of [1.130]. The SIT equations are called the Maxwell-Bloch equations by Lamb and McLaughlin in Chap.2. We now know [1.131] that the MB equations as defined in [1.130] are not an integrable system and do not have *exact* soliton solutions as we defined these in Sect.1.2. In 1972 GIBBON and EILBECK [1.132] conjectured an N-soliton solution of the sharp line resonant SIT equations: these equations are equivalent to the sine-Gordon equation and relate to the reduced Maxwell-Bloch (RMB) equations [1.64,133]. The RMB equations are

$$E_x + E_t = \alpha s$$

$$r_t = -\mu s$$

$$s_t = \mu r + E_u$$

$$u_t = -Es \qquad (1.92)$$

solved by CAUDREY et al. [1.82] by the Hopf-Cole type substitution

$$E = 2g/f \quad , \quad r = 2h/f \quad , \quad E^2 = \frac{4\partial^2}{\partial t^2} \ln f \qquad (1.93)$$

which yields

$$g^2 = ff_{tt} - f_t^2 \qquad (1.94)$$

directly. The RMB equations then yield two further homogeneous equations in f, g, and h. The solution now follows [1.82] by assuming it to be of the form

$$f = \det M_{ij}$$

$$M_{ij} = \frac{2}{E_i+E_j} \cosh\left[\frac{1}{2} (\theta_i + \theta_j)\right]$$

$$\theta_j = \omega_j t - k_j x + \delta_j \quad . \qquad (1.95)$$

The similarity with the solution (1.22) of the s-G in Lorentz covariant form is plain. Notice in any case that for $\mu = 0$ the equations (1.92) are solved by $u = \cos\phi$, $s = -\sin\phi$ with $\phi_t = E$ and $r = 0$ providing

$$\phi_{xt} + \phi_{tt} = -\alpha \sin\phi \quad . \qquad (1.96)$$

This yields the s-G (1.10) by $\xi = \sqrt{\alpha}(t - 2x)$, $\eta = \sqrt{\alpha}t$ and $\xi \to x$, $\eta \to t$. Thus the solution (1.95) necessarily includes the solution of the s-G (1.10).

The solution (1.95) was reported in brief early in 1973 and then at greater length [1.82]. HIROTA [1.81] independently published his solution of the s-G, also found by direct methods, in late 1972. AKNS published their inverse scattering solution of the s-G in 1973 [1.98] and subsequently gave their generalisation [1.44,45] of the ZAKHAROV and SHABAT inverse scattering scheme [1.38,109] later in the same year. The RMB equations (1.92) were solved by this method late in 1973 [1.133]. The application of the RMB equations to the theory of SIT is described in [1.130], for example. Its applications and canonical structure are treated at greater length in [1.64].

With hindsight a natural route to the AKNS inverse scattering scheme for the s-G (given by AKNS in their paper [1.45] is from the aBT (1.38). Set $\Gamma = \tan (u + u')/4$. Then

$$\Gamma_{,x} = \frac{1}{2} u_x (1 + \Gamma^2) + k\Gamma$$

$$\Gamma_{,t} = k^{-1}\Gamma\cos u - (2k)^{-1}(1 - \Gamma^2) \sin u \quad . \tag{1.97}$$

The Riccati transformation $\Gamma = v_2 v_1^{-1}$ now yields

$$\begin{pmatrix} v_{1,x} \\ v_{2,x} \end{pmatrix} = \begin{pmatrix} -\frac{1}{2}k & -\frac{1}{2}u_x \\ +\frac{1}{2}u_x & +\frac{1}{2}k \end{pmatrix} \begin{pmatrix} v_1 \\ v_2 \end{pmatrix} \tag{1.98}$$

$$\begin{pmatrix} v_{1,t} \\ v_{2,t} \end{pmatrix} = \frac{-1}{2k} \begin{pmatrix} \cos u & \sin u \\ \sin u & -\cos u \end{pmatrix} \begin{pmatrix} v_1 \\ v_2 \end{pmatrix} \quad . \tag{1.99}$$

This is the combination of scattering problem and time evolution used by AKNS to solve the s-G (1.11a) by an inverse method if k is interpreted as the eigenvalue $k \equiv 2i\zeta$. Their subsequent generalisation [1.39,40] was to the scattering and time evolution equations

$$\hat{L}v = \zeta v \quad , \quad \hat{L} = \begin{bmatrix} i\partial/\partial x & -iq \\ ir & -\partial/\partial x \end{bmatrix} \tag{1.100}$$

$$\hat{A}v = v_t \quad , \quad \hat{A} = \begin{bmatrix} A & B \\ C & -A \end{bmatrix} \tag{1.101}$$

where A, B, C, q and r are chosen so that, under $\zeta_t = 0$, $\hat{L}_t = [\hat{A},\hat{L}]$ is the required *pair* of evolution equations $q_t = K_1[q,r]$, $r_t = K_2[q,r]$ (in which $K_i[q,r]$ is a functional of the two potentials q,r). Equations (1.101) and (1.100) reduce to (1.99) and (1.98) if $k = 2i\zeta$, $r = -q = -u_x/2$ and $A = (i/4\zeta)\cos u$, $B = C = (i/4\zeta)\sin u$. Under the symmetry $r = -q$ the bound state eigenvalues of the non-Hermitean operator (1.100) lie on the imaginary axis $\zeta = i\eta (\eta > 0)$ or in pairs $(\zeta, -\zeta^*)$ in the upper half $\zeta$ plane. In the case of the s-G these are associated with the $2\pi$-kink and breather solutions, respectively. (Notice as we have already foreshadowed that if $q = r^*$, as is the case for the sinh-Gordon equation $u_{xt} = \sinh u$, for example, the operator $\hat{L}$ is Hermitean and only real eigenvalues are possible. There are no bound states and no solitons or breathers. The sinh-Gordon equation has an aBT [1.50,54] but it does not have soliton solutions even though $u = 0$ is a solution.)

We shall not discuss the AKNS generalisation as such further here since Newell develops the whole theory in Chap.6. It is worth remarking, however, that, although the operators $\hat{A}$ and $\hat{L}$ satisfy a Lax pair equation of the type (1.86b), $\hat{L}$ is not Hermitean in general as we have seen, whilst $\hat{A}$, not necessarily skew is not now a differential operator and depends on the eigenvalue $\zeta$. Thus if I) $[\hat{A},\hat{L}] = \hat{L}_t$, II) $\hat{L}v = \zeta v$, III) $\zeta_t = 0$, then $\hat{L}(Av - v_t) = \zeta(Av - v_t)$ and $Av - v_t$ is an eigenfunction belonging to $\zeta$, that is $\hat{A}v - v_t = \lambda(\zeta)v$, or $\lambda(\zeta) = 0$. It is easy to see that $\zeta^{-1}$ in $\hat{A}$ acts as an integral operator by calculating $[\hat{L},\hat{A}]$ from (1.98,99). One finds[8]

$$[\hat{L},\hat{A}] = -\frac{\sin u}{2\zeta}\begin{bmatrix} 0 & 1 \\ 1 & 0 \end{bmatrix}\begin{bmatrix} \partial/\partial x & \frac{1}{2}u_x \\ \frac{1}{2}u_x & -\partial/\partial x \end{bmatrix}$$

$$= i\,\frac{\sin u}{2}\begin{bmatrix} 0 & 1 \\ 1 & 0 \end{bmatrix} = i\begin{bmatrix} 0 & \frac{1}{2}u_{xt} \\ \frac{1}{2}u_{xt} & 0 \end{bmatrix} = \hat{L}_t \quad . \tag{1.102}$$

On the other hand (I) and (II) and $\hat{A}v = v_t$ implies $\zeta_t = 0$ and this deformation is isospectral.

TAKTADJAN and FADEEV [1.134] found a true Lax pair inverse scattering scheme for the covariant s-G equation (1.10) in 1974. Their pair of operators are differential operators: they are $4 \times 4$ matrix operators. The singular character of these oper-

---

[8] AKNS used the integrability condition $v_{xt} = v_{tx}$ to derive (in component form) $[\hat{A},\hat{L}]v = \hat{L}_t v$ (1). Plainly the Lax form $\hat{L}_t = [\hat{A},\hat{L}]$ (2) follows if the form (1) is true for all v and the form (2) is independent of $\zeta$. The argument for (1.102) appears to raise questions concerning the eigenvector v belonging to $\zeta = 0$. On the other hand, if $\hat{A}$ is polynomial in $\zeta$, $\zeta$ acts as a differential operator and so does $\hat{A}$.

ators causes some difficulties [1.135]. The connection between this inverse scattering scheme and that of (1.98) with (1.99) is established by POHLMEYER, cf. [Ref.1.141, p.320]. For the s-G incorporating a mass m in the form

$$u_{xx} - u_{tt} = m^2 \sin u \tag{1.103}$$

TAKHTADJAN and FADEEV [1.134] were able to find a remarkable expression for the Hamiltonian. In terms of a dimensionless coupling constant $\gamma_0$ this was

$$H = \sum_{i=1}^{L} (64m^2\gamma_0^{-2} + p_i^2)^{\frac{1}{2}} + \sum_{j=1}^{M} (256m^2\gamma_0^{-2} \sin^2\theta_j + \hat{p}_j^2)^{\frac{1}{2}}$$

$$+ \int_{-\infty}^{\infty} P(\xi)[m^2 + p^2(\xi)]^{\frac{1}{2}} d\xi \quad . \tag{1.104}$$

This beautiful result is rederived by Fadeev in Chap.11. The route to it appeals to the fact that the inverse (and direct) scattering transforms a canonical transformations. We have subsequently found the comparable expression for the s-G (1.11a) in light cone coordinates and derived (1.105) from it [1.42] using the canonical structure derived from the $2 \times 2$ scattering problem (1.98). Fadeev carries out a comparable calculation for the s-G (1.11a) in Chap.11.

Notice that the Hamiltonian (1.104) is again of action-angle type. Its semiclassical quantisation is easy [1.41,42] since the coordinates conjugate to the momenta $p_i$ and $\hat{p}_j$ can be found from the symplectic form as Fadeev indicates. The result agrees exactly with that found by DASHEN, HASSLACHER, and NEVEU (DHN) [1.40] by WKB methods up to a renormalisation of the coupling constant $\gamma_0$. In particular the (discrete) breather spectrum takes the form

$$M_n = \frac{16m}{\gamma_0} \sin \frac{n\gamma_0}{16} \quad , \quad n = 1, \ldots, N \tag{1.105}$$

where N is the largest integer $< 8\pi\gamma_0^{-1}$. KOREPIN and FADEEV [1.41] examined the first corrections to $\gamma_0$ and found these agreed with those reported by DHN [1.40]. The remarkable result (1.105) also agrees with the bound state eigenvalues of the S-matrix for the quantised s-G [1.71,95].

Notice that each 'particle', soliton, breather or mode $\xi$, whether quantised or not, contributes to (1.104) as would a free relativistic particle of appropriate mass: these masses are $8m\gamma_0^{-1}$ for the kink, $16m\gamma_0^{-1} \sin\theta$ for the breather, and m for each mode $\xi$. These masses agree with those computed near to equation (1.30) earlier but now depend on the coupling constant $\gamma_0$ as well as on the mass of the field m (the coupling constant $\gamma_0$ has nothing to do with the Lorentz factors $\gamma_1$, $\gamma_2$, etc.). It is possible to express the Hamiltonian of the covariant s-G in the particular form (1.104) because of the linearised dispersion relation $\omega^2 = m^2 + k^2$ of (1.103). Similarly, the Hamiltonian for the modified KdV equation (1.78) depends on a dynamical law in which the energy is cubic in the momentum [1.42]; this has less physical appeal than (1.104). It is not clear whether co-ordinates can always be found so that the solitons and breathers satisfy the energy momentum relationship of the linearised equation: a counter example seems to be the RMB equations [1.64].

LEE [1.136] has recently stressed the role played by the coupling constant $\gamma_0$ in nonlinear field theories. Consider the Lagrangian density in a field v(x,t)

$$\mathscr{L} = \frac{1}{2} v_t^2 - \frac{1}{2} v_x^2 - \frac{m^2}{\beta^2} \mathscr{V}(\beta^2 v^2) \quad . \tag{1.106}$$

Then: I) since we suppose

$$\frac{m^2}{\beta^2} \mathscr{V}(\beta^2 v^2) = \frac{1}{2} m^2 v^2 + O(\beta^2 v^4) + O(\beta^4 v^6) + \dots \tag{1.107}$$

and the only solutions with argument (x - Vt) are harmonic waves in linear theory, no solitary waves can be constructed from these dispersive linearised waves and the solitary wave solutions of (1.106) are singular in $\beta$.

II) This singularity is a simple pole. For set $\gamma_0 \equiv \beta^2$, and $\beta v \equiv u$. Then

$$\mathscr{L} = \gamma_0^{-1}\left[\frac{1}{2} u_t^2 - \frac{1}{2} u_x^2 - m^2 \mathscr{V}(u^2)\right] \tag{1.108}$$

and $\gamma_0$ does not figure in the equations of motion for u (compute the Lagrangian equations of motion).

III) Since the quantum mechanical action is $\hbar^{-1}$ times the classical action, $\gamma_0$ plays the role of $\hbar$ in any quantised theory. The rest energy $E_{classical}$ for the solitary wave is $\propto \gamma_0^{-1}$ and

$$E_{quantal} = E_{classical} + O(\gamma_0^0) + O(\gamma_0^1) + \dots \quad . \tag{1.109}$$

IV) $\gamma_0$ plays the role of $\hbar$ since the canonical momentum from (1.106) is $\pi(x,t) \equiv \delta\mathscr{L}/\delta u_t = \gamma_0^{-1} u_t$, the Poisson bracket is

$$\{u(x,t), \quad \pi(y,t)\} = \delta(x - y) \tag{1.110a}$$

and the canonical quantisation would be through the commutator

$\hbar = h/2\pi$ (normalised Planck's constant)

$$[u(x,t),\; u_t(y,t)] = i\gamma_0\delta(x-y) \quad . \tag{1.110b}$$

Notice that I) and II) mean that solitary waves cannot be found by perturbation theory about linear theory. Lee's results apply in 4-dimensional Minkowski space [(3 + 1) dimensions] as well as in (1 + 1) dimensions and are not restricted to the solitons as these were defined in Sect.1.1. In (1 + 1) dimensions the solitary wave mass M is from (1.108) just

$$M = \frac{m}{\gamma_0}\int_{u_1}^{u_2} \sqrt{2V(u^2)}\,du$$

where $u_1$ and $u_2$ are successive zeros of $V(u^2)$. For the $2\pi$-kink solution of the s-G, $V(u^2) = (1 - \cos u)$ and $M = 8m\gamma_0^{-1}$ in agreement with (1.104).

The free particle result (1.104) offers the s-G as a model system in classical statistical mechanics as well as in quantised field theories. Note, however, that the precise form (1.104) depends on the initial data: the number of kinks L, of breathers M, and the occupation of momentum states $P(\xi)$ must all be calculated from the initial data $u(x,0)$, $u_t(x,0)$ for the s-G (1.10). $P(\xi)$ arises first of all as the canonical momentum associated with the mode labelled by $\xi$ and we find we should keep it as such (see [1.221]). Note that the modes labelled by $\xi$ do not contribute to the quantised sine-Gordon theory. Their role is apparently taken over by the breather contributions. We refer the reader to the articles in [1.137] where some applications of soliton theory to solid-state theory are described. A study by BISHOP of the statistical mechanics of the s-G is reported there. Luther, in Chap.12, establishes connections between the spin-$\frac{1}{2}$ x y z model of statistical mechanics [1.138], the massive Luttinger model and by implication his backward scattering model [1.139], the massive Thirring model, and the *quantised* s-G. These connections allow him to compute the breather spectrum (1.105) 'exactly'.

It is remarkable that the calculation of the S-matrix which assumes (I) an infinity of conservation laws and therefore factorisation into a product of 2-body S-matrices (II) crossing symmetry (III) unitarity (IV) certain analytical behaviour, yields the spectrum (1.105) for the bound states. The equation of motion for the s-G is not used [1.71,95]. The result appears to rely on the proposition that the existence of 2 polynomial conserved densities $T^2$ and $T^4$ of a nonlinear Klein-Gordon equation (1.37) means that it is the sine-Gordon (sinh-Gordon) equation as noted in Sect.1.3 near (1.60). However, the argument is developed in terms of the massive Thirring model [1.95] and the existence of a coupling region with no bound states (a property of the MT ≡ s-G models, U(1) symmetry of the 2-particle S-matrix and a minimal condition (absence of redundant poles and zeros in the physical sheet for the transmission amplitudes) are imposed conditions. Nor is it proved that the resultant unique S-matrix is the S-matrix of the MT ≡ s-G models.

The argument which obtains a scattering problem like (1.100) and a time evolution like (1.101) from an aBT like (1.38) has been generalised by POHLMEYER [1.86] to the classical nonlinear $\sigma$ models in 1+1 dimensions. These are Lorentz covariant equations of the form

$$\underset{\sim}{q}_{tt} - \underset{\sim}{q}_{xx} - \lambda \underset{\sim}{q} = 0 \tag{1.112a}$$

with constraint

$$q^2 = 1 \quad . \tag{1.112b}$$

The constraint determines the Lagrangian multiplier $\lambda(x,t) = -(\underset{\sim}{q}_t^2 - \underset{\sim}{q}_x^2)$. We suppose $\underset{\sim}{q} = (q_1, \ldots, q_n)$ consists of n scalar fields. The n equations with constraint (1.112b) are invariant under the action of the group $O_n$. POHLMEYER constructed an aBT for (1.112) and derived from it an infinite set of conserved densities (notice that in Sect.1.2 we used a BT, not an aBT to find the conserved densities; but WADATI [1.140] has shown how to use aBTs to find conserved densities). For $n \leq 6$ POHLMEYER found the linear eigenvalue problem corresponding to (1.100) and the isospectral time evolution corresponding to (1.101). Note that in a remarkable analysis POHLMEYER accomplishes a linearisation of system (1.112) for arbitrary n with the help of Clifford algebras and Levin's linearisation of matrix Riccati equations. Equations (1.112) in light cone coordinates are invariant under local scale transformations [1.86,141, pp.339-355] and there are no soliton solutions. However, by breaking scale invariance soliton solutions can arise. In particular, for n = 3 ($O_3$ invariant model) the system includes the sine-Gordon theory which has finite energy soliton (kink) solutions.

In light cone coordinates $\xi = (t + x)/2$, $\eta = (t - x)/2$ the equation for n = 3 is

$$\frac{\partial^2 \underset{\sim}{q}}{\partial\eta\partial\xi} + \left(\frac{\partial \underset{\sim}{q}}{\partial\eta} \cdot \frac{\partial \underset{\sim}{q}}{\partial\xi}\right) \underset{\sim}{q} = 0$$

$$\underset{\sim}{q} = (q_1, q_2, q_3) \quad . \tag{1.113}$$

Energy and momentum conservation imply that

$$\frac{\partial}{\partial\xi} \underset{\sim}{q}_\eta^2 = 0 = \frac{\partial}{\partial\eta} \underset{\sim}{q}_\xi^2 \tag{1.114}$$

and scale invariance is broken by distorting each light cone variable in such a way that $\underset{\sim}{q}_\eta^2$ and $\underset{\sim}{q}_\xi^2$ are normalised

$$q_\eta^2 = q_\xi^2 = 1 \quad . \tag{1.115}$$

Introduce now the local frame $q$, $q_\xi$, $q_\eta$: $q$ is orthogonal to $q_\xi$ and $q_\eta$ but $q_\xi$ and $q_\eta$ form the nontrivial angle

$$q_\eta \cdot q_\xi = \cos u \quad . \tag{1.116}$$

Equation (1.113) with normalisation (1.115) and definition (1.116) yields

$$\frac{\partial^2 u}{\partial\eta\partial\xi} = -\sin u \tag{1.117}$$

[the minus sign from the choice $\eta = (t - x)/2$ not $(x - t)/2$].

POHLMEYER [1.86,141, pp.339-355] linearised the system (1.112) (i.e., found spectral transforms) for the cases $n \leq 6$ and for $n = 4$ by breaking scale invariance as at (1.115), found equations of motion in two fields $\alpha$, $\beta$

$$\alpha_{\xi\eta} + \sin\alpha + \frac{\cot^2 \frac{1}{2}\alpha}{\sin\alpha} \beta_\xi \beta_\eta = 0$$

$$\beta_{\zeta\eta} = \frac{\beta_\xi \alpha_\eta + \alpha_\xi \beta_\eta}{\sin\alpha} \quad . \tag{1.118}$$

These reduce to the s-G (1.117) for $\beta \equiv 0$. POHLMEYER [1.86] noted that equations (1.118) are the conditions for embedding a 2-dimensional surface in a three-dimensional sphere which is itself embedded in a four-dimensional Euclidean space. LUND [1.48] independently reached the same equations of motion by starting from this point of view. Since his argument provides a geometrical basis for spectral transforms in the case he considers we say a little more about it here. The sketched remarks we give follow LUND's review [1.142] and for the details the reader is referred there.

Consider the problem of embedding the n-dimensional surface $V_n$ in the n + 1 dimensional Euclidean space E

$$E = x^i \quad (i = 1,2, \ldots, n + 1)$$

$$V_n = y^\mu \quad (\mu = 1,2, \ldots, n) \quad . \tag{1.119}$$

$V_n$ has metric $ds^2 = g_{\mu\nu} dy^\mu dy^\nu$. The embedding is isometric (preserves scalar products) if $V_n$ can be defined through

$$x^i = x^i \, (y^1, \ldots, y^n) \tag{1.120}$$

such that $g_{\mu\nu} = (\partial x^i/\partial y^\mu)(\partial x^i/\partial y^\nu)$. Vectors $X_\mu$ in the tangent space of $V_n$, and the surface normal $X_{n+1}$, define a basis in E, $X_\mu \equiv \partial \underset{\sim}{x}/\partial y^\mu$, $X_{n+1}$. The expression of the vectors $\partial X_\mu/\partial y^\nu$, $\partial X_{n+1}/\partial y^\nu$ in terms of this basis is the (linear) system of Gauss-Weingarten equations [1.143]. The integrability conditions $\partial^2 X/\partial y^\nu \partial y^\sigma = \partial^2 X/\partial y^\sigma \partial y^\nu$ (for all pairs $y^\sigma$, $y^\nu$) lead to a nonlinear system, the Gauss-Codazzi equations [1.144].

For a 2-dimensional surface of particular metric in 3-dimensional Euclidean space LUND [1.48,142] found Gauss-Codazzi equations which take a form equivalent to (1.118)[9]

$$\frac{\partial}{\partial \xi}(\cot^2\theta\lambda_\eta) + \frac{\partial}{\partial \eta}(\cot^2\theta\lambda_\xi) = 0$$

$$\theta_{\xi\eta} - \frac{1}{2}\sin 2\theta + \frac{\cos\theta}{\sin^3\theta}\lambda_\xi\lambda_\eta = 0 \tag{1.121}$$

[evidently $\theta = \frac{1}{2}\alpha$, $\lambda = 2\beta$ compared with (1.118) whilst $\eta = \frac{1}{2}(x - t)$]. For the Gauss-Weingarten system

$$\begin{pmatrix} v_{1,\xi} \\ v_{2,\eta} \end{pmatrix} = \begin{pmatrix} -i\zeta + ip & q \\ -q^* & i\zeta - ip \end{pmatrix} \begin{pmatrix} v_1 \\ v_2 \end{pmatrix}$$

$$\begin{pmatrix} v_{1,\eta} \\ v_{2,\eta} \end{pmatrix} = i \begin{pmatrix} r & s \\ s & -r \end{pmatrix} \begin{pmatrix} v_1 \\ v_2 \end{pmatrix} \tag{1.122}$$

in which $p = (\cos 2\theta/2\sin^2\theta)\lambda_\xi$, $q = -\theta_\xi + i\cot\lambda_\xi$, $r = (1/4\zeta)\cos 2\theta - (\sin^2\theta)^{-1}\lambda_\eta$ $s = (1/4\zeta)\sin 2\theta$. An eigenvalue has been introduced by appealing to the Lie invariance (≡ Lorentz invariance — compare the introduction of this eigenvalue with the eigenvalue $\lambda$ introduced in (1.80) by Galilean invariance, of (1.121). The conditions (1.122) are sufficient and *necessary* for the integration of (1.121). The scattering problem and the time evolution (1.122) are, however, a generalised AKNS scheme comparing with (1.100,101). As far as we known results similar to those of LUND and POHLMEYER have also been obtained by TAKHTADJAN [1.144].

LUND's arguments [1.142] give a nice geometrical basis to the inverse scattering scheme for the s-G (1.10) or (1.11) and its generalisation to two fields $\theta,\lambda$ which is (1.121). They indicate the way in which the s-G arose in differential geometry in the first place [1.42] as we noted in Sect.1.2. LAMB [1.145], and also LAKSHMANAN [1.146], in different ways have also given a geometric basis to the inverse scattering method whilst a very recent (1979) analysis by SASAKI

---

[9] A second condition is required. See the explanatory 'Note in Proof' at the end of the book.

[1.147]has shown that *all* AKNS systems represent surfaces of constant negative curvature. The argument is concise and elegant so we present it here.

The AKNS scattering and time evolution equations (1.100 and 101) are equivalent to a completely integrable Pfaffian system expressible in the forms

$$dv - \Omega v = 0 \ , \quad \mathrm{Tr}\{\Omega\} = 0 \quad , \tag{1.123}$$

in which $\Omega$ is $2 \times 2$ matrix of 1-forms. The integrability condition is

$$\begin{aligned} 0 = d^2 v &= d\Omega v - \Omega_\wedge dv \\ &= (d\Omega - \Omega_\wedge \Omega) v \end{aligned} \tag{1.124}$$

so the nonlinear evolution equations $\hat{L}_{,t} = [\hat{A}, \hat{L}]$ are equivalent to the vanishing of the 2-forms

$$\Theta \equiv d\Omega - \Omega_\wedge \Omega \ , \quad \text{i.e.,}\ \Theta = 0 \quad . \tag{1.125}$$

For differential n forms we are using the 'wedge' notation "$\wedge$" of the exterior differential calculus. An excellent book for reference is that by FLANDERS [1.148].

It is easy to see that $\Omega$ is not unique: it is form invariant under the 'gauge transformation'

$$\begin{aligned} v &\to v' = Bv \\ \Omega &\to \Omega' = dBB^{-1} + B\Omega B^{-1} \\ \Theta &\to \Theta' = B\Theta B^{-1} \end{aligned} \tag{1.126}$$

in which B is an arbitrary $2 \times 2$ matrix of determinant unity.

Compare (1.125) with a result derivable for pseudo-spherical surfaces: one introduces (in a 3-space) a local orthonormal basis on the tangent plane $T_p$ at each point P of the surface: this basis is spanned by two vectors $\underline{e}_1$, $\underline{e}_2$ (say) and one can always choose

$$\underset{\sim}{e}_i \cdot \underset{\sim}{e}_j = \delta_{ij} \quad . \tag{1.127}$$

The "structure equations" are (see FLANDERS [1.148])

$$dP = \sigma^1 \underline{e}_1 + \sigma^2 \underline{e}_2 \tag{1.128a}$$

$$d\underline{e}_1 = \omega \underline{e}_2 \tag{1.128b}$$

$$de_2 = -\omega e_1 \tag{1.128c}$$

in which $\sigma^1$ and $\sigma^2$ are 1-forms; $\omega$ is the 'connection' 1-form. The first equation (1.128a) states that P is confined to the surface; the remaining two equations follow from (1.127).

The integrability condition in this case is $d^2P = 0$ from which follows

$$d\sigma^1 = \omega\sigma^2$$

$$d\sigma^2 = -\omega\sigma^1 \quad . \tag{1.129}$$

The Gaussian curvature K is defined by [1.128]

$$d\omega = -K\sigma^1{\wedge}\sigma^2 \quad . \tag{1.130}$$

For K = -1 precisely the choice of $\Omega$

$$\Omega \equiv \begin{pmatrix} -\frac{1}{2}\sigma^2 & \frac{1}{2}(\omega + \sigma^1) \\ \frac{1}{2}(-\omega + \sigma^1) & \frac{1}{2}\sigma^2 \end{pmatrix}$$

(for example) satisfies (1.125) exactly.

It is plain now that every AKNS system can be represented as a surface of constant negative curvature with respect to a particular metric.[10] The metric is given by

$$ds^2 = (\sigma^1)^2 + (\sigma^2)^2 \quad . \tag{1.131}$$

It is known [1.149] that surfaces of constant curvature have a maximum number of isometries [transformations mapping the pseudo-spherical surfaces into themselves preserving distance (i.e., scalar products)]. For the 2-surface of constant negative curvature the isometry groups are SL(2,R) and SU(1,1). This 'explains', for example, the emergence of the Lie algebras of these groups in the prolongation structures of the AKNS systems [1.150].

From this point of view the aBT is a gauge transformation (1.126) from one pseudo-spherical surface to another and can be given abstract form for the general AKNS system. The conserved quantities which we now derive explicitly here are exactly the conserved densities of the AKNS system found by expansion about $\infty$ in the $\zeta$ plane [1.69] but may be still more complete. To fix ideas we derive the conserved densities for the sine-Gordon equation and show how these can be complete in this case.

---

[10] A second condition is required. See the explanatory 'Note in Proof' at the end of the book.

From $\theta = 0$ we can always define two completely integrable Pfaffian systems

$$0 = \gamma_1 \equiv d\Gamma_1 - \omega_3 - 2\Gamma_1\omega_1 - \Gamma_1^2\omega_2 \tag{1.132a}$$

$$0 = \gamma_2 \equiv d\Gamma_3 - \omega_2 - 2\Gamma_2\omega_1 + \Gamma_2^2\omega_2 \tag{1.132b}$$

where

$$\Omega = \begin{pmatrix} \omega_1 & \omega_2 \\ \omega_3 & -\omega_1 \end{pmatrix} \tag{1.132c}$$

and the $\Gamma_i$ are two unknown functions. Then from $\theta = 0$ (1.125),

$$d\gamma_1 = 2\gamma_1(\omega_1 + \Gamma_1\omega_2)$$

$$d\gamma_2 = -2\gamma_2(\omega_1 - \Gamma_2\omega_3) \tag{1.133a}$$

whilst from the closures $d^2\gamma_1 = d^2\gamma_2 = 0$ it follows that the 1-forms

$$\delta_1 \equiv \omega_1 + \Gamma_1\omega_2 \tag{1.133b}$$

$$\delta_2 \equiv \omega_1 - \Gamma_2\omega_3 \tag{1.133c}$$

are closed, i.e.,

$$d\delta_1 = d\delta_2 = 0 \quad . \tag{1.134}$$

These are the conservation laws. But because each depends on $\zeta$ each represents an infinite set of apparently nontrivial conservation laws.

From (1.98,99) one finds for the s-G that a choice is

$$\omega_1 = -\frac{1}{2}\,k dx - \frac{1}{2k}\cos u\, dt$$

$$\omega_2 = \frac{1}{2}\,u_x dx - \frac{1}{2k}\sin u\, dt$$

$$\omega_3 = -\frac{1}{2}\,u_x dx - \frac{1}{2k}\sin u\, dt \tag{1.135}$$

in which k is a real parameter. From this one easily finds for example that the Pfaffian (1.131b) is exactly equivalent to (1.97). [The Pfaffian (1.131a) corresponds to choosing $\Gamma_1 = v_1/v_2$.] Whilst the conserved densities (1.54) are relatively easy to find this way, it is convenient to find a route which also isolates the set (1.59). To do this (as SASAKI pointed out to us) one can choose the form for $\Omega$

$$\begin{bmatrix} \frac{1}{4}(u_x dx - u_t dt) & \frac{1}{2}\eta\, e^{iu/2\eta}\, dx + \frac{1}{2\eta} e^{iu/2}\, dt \\ \frac{1}{2}\eta\, e^{iu/2}\, dx + \frac{1}{2} e^{-iu/2\eta}\, dt & \frac{-i}{4}(u_x dx - u_t dt) \end{bmatrix} \tag{1.136}$$

The identification is

$$\Omega = \begin{bmatrix} \frac{i}{2}\omega & \frac{1}{2}(\sigma^1 - i\sigma^2) \\ \frac{1}{2}(\sigma^1 + i\sigma^2) & -\frac{i}{2}\omega \end{bmatrix} \tag{1.137}$$

with

$$\sigma^1 = \cos\frac{u}{2}(dx + dt)$$

$$\sigma^2 = \sin\frac{u}{2}(dx - dt)$$

$$\omega = \frac{1}{2}(u_x dx - u_t dt) \tag{1.138}$$

followed by the scale transformation

$$x \to \eta x \quad , \quad t \to \eta^{-1} t \quad .$$

After this the metric is

$$ds^2 = \eta^2 (dx)^2 + 2\cos u\, dx\, dt + \eta^{-2}(dt)^2 \tag{1.139}$$

and is the metric of (1.135) (when $\eta \equiv k$).

One finds now that the conserved quantities are complex conjugates $\delta_1^* = \delta_2$ and we need consider only one of these. The Riccati equation is

$$0 = d\Gamma - \omega_3 + 2\Gamma\omega_1 + \Gamma^2\omega_2 \tag{1.140}$$

so

$$\Gamma_t - \frac{1}{2\eta} e^{-iu/2} - \frac{i}{2} u_t + \frac{1}{2\eta} e^{iu/2} = 0 \tag{1.141a}$$

$$\Gamma_x - \frac{1}{2}\eta\, e^{iu/2} + \frac{1}{2} u_x \Gamma + \frac{1}{2}\eta\, e^{-iu/2}\Gamma^2 = 0 \quad . \tag{1.141b}$$

These two equations are invariant under

$$t \leftrightarrow x \ , \quad u \leftrightarrow -u \ , \quad \eta \leftrightarrow \eta^{-1} \quad . \tag{1.142}$$

They can be solved by two alternative expansions

$$\Gamma = \sum_{k=0}^{\infty} \eta^{-k}\gamma_k \tag{1.143a}$$

for (1.141b) and

$$\Gamma = \sum_{k=0}^{\infty} \eta^{k}\tilde{\gamma}_k \tag{1.143b}$$

for (1.141a). And, because of the symmetry, if $\gamma_k = f_k(u, u_x, \ldots)$, $\tilde{\gamma}_k = f_k(u, u_t, \ldots)$. The simplest recursion formulae emerge if we choose

$$\gamma_0 = -e^{iu/2}$$
$$\gamma_1 = -i\, e^{iu/2}\, u_x \tag{1.144}$$

and define $\xi_k$ by

$$\xi_k = e^{-iu/2}\gamma_k \quad . \tag{1.145}$$

Then one finds from (1.141b) that

$$\xi_k = (\xi_{k-1})_x + \frac{1}{2}\sum_{\ell=2}^{k-2} \xi_{k-\ell}\, \xi_\ell \qquad (k \geqq 3)$$
$$\xi_0 = -1 \ , \quad \xi_1 = -iu_x \ , \quad \xi_2 = M \equiv \frac{1}{2} u_x^2 - iu_{xx} \tag{1.146}$$

From

$$\delta = \omega_1 + \Gamma\omega_2 = \omega_1 + \left(\sum_{k=0}^{\infty} \eta^{-k}\, e^{iu/2}\xi_k\right)\omega_2 \ ,$$
$$\delta = -\frac{1}{2}\eta^1 dx - \frac{1}{4}\eta^0 du$$
$$+ \frac{1}{2}\sum_{k=1}^{\infty} (\xi_{k+1}\, dx + e^{iu}\xi_{k-1} dt)\eta^{-k}$$

and the $\mathrm{Re}\{\xi_{k+1}\}$ are real conserved densities. One finds $\xi_0 = -1$, $\xi_1 = -iu_x$, $\xi_2 = M$, $\xi_3 = M_x$, $\xi_4 = M_{xx} + \frac{1}{2}M^2$, $\xi_5 = M_{xxx} + (M^2)_x$, ... .

Comparison with (1.54) shows that modulo the perfect derivatives

$$\mathrm{Re}\{\xi_2\} = T^2 \ , \quad \mathrm{Re}\{\xi_4\} = -T^4 \ , \ \ldots \quad . \tag{1.147}$$

Note that precisely the 'canonical' forms $T^2$ and $T^4$ arise this way and that up to this rank the additional perfect derivatives are obtained explicitly as such.

The transcendental conserved densities (1.59) now follow from (1.143a), (1.141a) and the definition

$$\tilde{\xi}_k = \tilde{\gamma}_k\, e^{-iu/2} \quad . \tag{1.148}$$

Because of (1.42)

$$\tilde{\xi}_k = \xi_k(u \to -u,\ u_x \to -u_t,\ \ldots) \tag{1.149}$$

and comparison with (1.46) shows that the densities (1.59) arise as $\mathrm{Re}\{\exp(-iu)\tilde{\xi}_{k-1}\}$.

One can also switch the expansions (1.143) and apply (1.143b) to (1.141b). The significant complex densities arise in the sequence

$$\mathscr{T}^2 = e^{-iu}, \quad \mathscr{T}^3 = ie^{-iu}\int_{-\infty}^{x} \sin u\, dx' \ , \quad \mathscr{T}^4 = -ie^{-iu}\int_{-\infty}^{x} e^{-iu}\int_{-\infty}^{x'} \sin u\, dx''dx' \ ,$$

$$\mathscr{T}^5 = ie^{-iu}\int_{-\infty}^{x} e^{-iu}\int_{-\infty}^{x'} e^{-iu}\int_{-\infty}^{x''} \sin u\, dx'''dx''dx' + \tfrac{1}{2}e^{-iu}\int_{-\infty}^{x} e^{-iu}\left(\int_{-\infty}^{x'} \sin u\, dx''\right)^2 dx,$$

$$\mathscr{T}^6 = \ldots \quad . \tag{1.150}$$

One easily checks for the s-G that

$$\begin{aligned} &\mathscr{T}^2{}_{,t} - (\mathscr{T}^4 e^{iu})_x = 0 \\ &\mathscr{T}^3{}_{,t} - (\mathscr{T}^3 e^{iu})_x = 0 \end{aligned} \tag{1.151}$$

and one finds

$$\mathscr{T}^n{}_{,t} - (\mathscr{T}^{n+2} e^{iu})_x = 0 \tag{1.152}$$

so the fluxes are $\mathscr{T}^{n+2} e^{iu}$.

The relevant densities appear to be

$$\mathrm{Re}\{\mathscr{T}^2\} = \cos u \ , \quad \mathrm{Re}\{\mathscr{T}^4\} = -\sin u \int_{-\infty}^{x} \cos u \int_{-\infty}^{x'} \sin u\, dx''dx'$$

$$- \cos u \int_{-\infty}^{x} \sin u \int_{-\infty}^{x'} \sin u\, dx''dx' \quad ,$$

$$\mathrm{Re}\{\mathscr{T}^6\} = \ldots \quad . \tag{1.153}$$

These are nonlocal and, viewed as Hamiltonian densities, certainly yield equations of motion with kink solutions and the correct linearised dispersion relations (see [1.42,69] and footnote 4). There is the complication that the integrated Hamiltonian densities, the Hamiltonians, must be finite. On these and other details the reader is referred to the $5^{th}$ and $6^{th}$ papers referenced in [1.147]. A few other geometrical results surrounding the AKNS systems are also mentioned there and in the other papers cited. Notice *for the s-G* that $\mathscr{T}^4 = -(u_{tt} \sin u + \frac{1}{2} u_t^2 \cos u)$ which becomes $\frac{1}{2} u_t^2 \cos u$ on adding a perfect t derivative. Evidently for the s-G the sequence (1.153) regenerates (1.59) as it should.

## 1.7 Further Progress on Inverse Scattering Methods

It is convenient to close this history of the soliton essentially at the point in 1973 where the Zakharov-Shabat-AKNS inverse scattering scheme had emerged. This is not because the subject has no further history. Indeed, in connection with particular points as they have arisen, we have already referred to advances in the theory as recent as 1979 — at the end of Sect.1.6 for example. But as far as inverse scattering methods go, the articles which follow in this volume each develop the theory of solitons from a starting point roughly equivalent to the state of knowledge in 1973. This state was well reviewed in [1.4].

We can now see that the appearance of this article by SCOTT et al. [1.4] marks a watershed in the history of the subject: in the period 1965-1973 the soliton was discovered, the physically applicable equations KdV, modified KdV, NLS, s-G were being understood and scattering methods were devised for them: since 1973 the mathematical structures of soliton systems are being vastly extended and so is the range of their physical applications. In Sect.1.7 we simply cite a few more important results obtained since 1973. We cannot hope to reference all the advances in the subject. Some of these advances are described in other chapters. Where not we hope we give enough references to enable the reader to find his own way through the particular literature.

Steps of importance since mid 1973 seem to us to include WAHLQUIST's and ESTABROOK's discovery of the aBT for the KdV [1.56] reported in December 1973; ZAKHAROV's and MANAKOV's solution of the 3-wave interaction [1.46] in August 1973 and KAUP's solution [1.47] in 1974; TAKTADJAN's and FADEEV's result (1.104) [1.134] for the s-G in 1974; McLAUGHLIN's published results [1.151] on the canonical structure of a number of nonlinear wave equations including the s-G in 1974; the solution of the Toda lattice by FLASCHKA [1.152] and MANAKOV [1.153] (Toda describes the history of his lattice and its solution in Chap.4); the remarkable work of NOVIKOV, MATVEEV and their collaborators [1.119,120] on the periodic KdV and the first paper by Novikov published in 1974 (this analysis is developed further in Chap.10); FLASCHKA's and NEWELL's article [1.154] on the canonical structure of the 'integrable' nonlinear evolution equations in 1975 (now further extended in Chap.6); LAX's solution [1.114] of the KdV under periodic boundary conditions in 1975; WADATI's extension [1.155] of scattering problems to bigger matrices in 1974 and described further in Chap.8; CALOGERO's use of generalised Wronskian techniques in 1975 [1.78,156] and his extension of the scattering theory to more 'time-like' independent variables (noted by Newell in Chap.6); and CALOGERO's and DEGASPERIS's extension of scattering theory to bigger matrices in 1976 leading to the boomeron solutions of [1.157] as described in Chap.9.

In 1974 there also appeared the remarkable paper on the inverse method by ZAKHAROV and SHABAT [1.158]: the ideas of this paper form the germ of the comprehensive and creative article presented by ZAKHAROV in Chap.7 which starts from a study of the representation of nonlinear evolution equations as commuting pairs of differential operators as in the Lax pair differential operator formulation described in Sect.1.4.

In 1975 and 1976 the papers [1.61] and [1.75] on the theory of prolongation structures by WAHLQUIST and ESTABROOK appeared: this subject is being actively developed at the present time, it is not represented in this volume, and we refer the reader to the articles by ESTABROOK [1.76], DODD and GIBBON [1.150], MORRIS [1.159-161], DODD [1.162], CORONES and collaborators [1.163], as well as the work of PIRANI et al. [1.77] cited in Sect.1.3. In our view the geometrical methods of Sect.1.6 form a good starting point for deriving prolongation structures and seeing their significance to, for example, the conservation laws and Bäcklund transformations. SASAKI [1.147, third and fourth references] derives the prolongation structure of the AKNS inverse scattering scheme from the 'gauge field' formulation (1.123) this way and finds similar structures for larger scattering schemes.

A rather different class of problems has emerged from a discovery by MOSER and we say rather more about this here. The discovery depended on a result MOSER found in 1975. He showed [1.164], how to use the idea of the Lax pair of matrix operators in the fashion pioneered by FLASCHKA [1.152] and MANAKOV [1.153] to solve the one-dimensional many-body problem with Hamiltonian

$$H = \frac{1}{2}\sum_j y_j^2 + g\sum_j \sum_{j<k} (x_j - x_k)^{-2}\ , \qquad g > 0 \tag{1.154}$$

($y$ = momentum, $\partial H/\partial y_j = \dot{x}_j$, $\dot{y}_j = -\partial H/\partial x_j$; $1 \leq j \leq N$). The equations of motion take the Lax pair form (1.86b) where $\hat{L}$ and $\hat{B}$ are now $N \times N$ *matrices* given by

$$L_{jk} = \delta_{jk} v_j + i\sqrt{g}(1 - \delta_{jk})(x_j - x_k)^{-1}$$

$$B_{jk} = -i\sqrt{g}\gamma_{jk} \sum_{\substack{\ell=1 \\ \ell \neq j}}^{N} (x_j - x_\ell)^{-2} + i\sqrt{g}(1 - \delta_{jk})(x_j - x_k)^{-2} \quad . \tag{1.155}$$

$\hat{L}$ is Hermitean and $\hat{B}$ is skew; it is easily checked in any case that since $(L^+)_t = (L_t)^+$ an alternative Lax pair representation is always

$$\hat{L}_t = \left[\frac{1}{2}(\hat{B} - \hat{B}^+), \hat{L}\right] \tag{1.156}$$

in which the matrix $(\hat{B} - \hat{B}^+)/2$ is certainly skew. Just as for (1.86b) it now follows that the eigenvalues of $\hat{L}$ are constants of the motion (which are also in involution). So are the symmetric functions $I_n$

$$\det[L_{jk} - \lambda\delta_{jk}] = \lambda^N + \sum_{n=1}^{N} \lambda^{N-n} I_n \quad ; \tag{1.157a}$$

and so are the traced powers

$$F_n \equiv \mathrm{Tr}(\hat{L}^n/n) \quad . \tag{1.157b}$$

The system is a completely integrable Hamiltonian system with 2N degrees of freedom.

The quantum mechanical problem (1.154) was solved by CALOGERO in 1971 [1.165]. He conjectured that the classical problem was also solvable and has subsequently shown [1.166] (and see [1.72]) that the soluble classical many-body problems with pair potentials $V(x_j - x_k)$ and a Lax pair representation with an ansatz generalising (1.155) must have $V(x)$ the Weirstrass elliptic function $\mathscr{P}(x)$ : $x^{-2}$, $\sinh^{-2}x$, $\sin^{-2}x$ are particular cases of this function. Moser's discovery was a remarkable connection between the classical N-body problem (1.154) and the *rational* solutions of the KdV satisfying $u \to 0$, $x \to \infty$, about which we have said nothing so far.

The soliton solution (1.2) of the KdV equation has double poles in the complex x plane: KRUSKAL [1.65] was the first to study the motion of these poles. The double pole character may motivate the study of rational solutions of the KdV in the form

$$u(x,t) = 2\sum_{j=1}^{N} [x - x_j(t)]^{-2} r_j(t) \quad . \tag{1.158}$$

By direct substitution into the KdV equation taken in the form

$$u_t - 3uu_x + \frac{1}{2}u_{xxx} = 0 \tag{1.159}$$

one finds that the ansatz (1.158) is a rational solution of this equation if

$r_i(t) = 1$ and

$$\dot{x}_j = 6 \sum_{k \neq j} (x_j - x_k)^{-2} \tag{1.160a}$$

$$\sum_{k \neq j} (x_j - x_k)^{-3} = 0 \; ; \quad 1 \leqq j \leqq N \quad . \tag{1.160b}$$

The argument is facilitated by using the addition theorem for elliptic functions $\phi(x) = x^{-2}$

$$\phi(x)\phi'(y) + \phi'(x)\phi(y) = \phi(x-y)[\phi'(y) + \phi'(x)] - \phi'(x-y)[\phi(y) - \phi(x)] \quad . \tag{1.161}$$

The remarkable fact is that this restricted set of equations (1.160) for the evolution of the coordinates $x_j$ [the equations of motion (1.160a) are restricted to the manifold satisfying equations (1.160b)] is also that of a restricted Hamiltonian flow associated with (1.154) — namely, with the $F_r$ defined as in (1.157b), $F_2 \equiv H$ given by (1.154), whilst the flow with Hamiltonian $F_3$ restricted by the condition grad $F_2 = 0$ yields, for $g = 6$, precisely the set (1.160) since

$$F_3 \equiv \frac{1}{3} \sum y_j^3 + g \sum_{j \neq i} y_i(x_i - x_j)^{-2} \quad . \tag{1.162}$$

Because the $F_n$ are in involution the flows with these Hamiltonians commute with each other (see [1.114]): then if grad $F_k = 0$ is imposed for any k at any t this restriction is maintained throughout the motion (the $F_n$ flow restricted by grad $F_k = 0$ is an invariant flow [1.167]). It remains to show that the set (1.160) has a solution: it does so if first the $x_j$ lie in the complex x plane and second $N = n(n+1)/2$ for some $n = 1,2, \ldots$ . For example, if $N = 1$, $u = 2(x-c)^{-1}$; if $N = 3$, the roots $x_i$ of (1.160a) are proportional to the cube roots of unity [1.168] and $u(x,t) = 6x^2(x^3 - 2t + c)^{-1}$. Notice that [1.168], u is a solution (1.158) (with $r_j = 1$) if and only if $u(x,t) = 2\partial^2 \ln P(x,t)/\partial x^2$ where $P(x,t) = \Pi_{j=1}^N [x - x_j(t)]$ and the $x_j$ satisfy (1.157a). This formula mirrors the N-soliton formulae for the real line [see (1.16,22)] and the $\Theta$-function formula for the solutions of the KdV with periodic boundary conditions derived by Novikov in Chap.10. Matveev established the connection with Hirota's N-soliton formula in the lectures [1.119]. (In the 1-dimensional case the formula may actually be due first of all to Baker — see the note after [1.219] in the references.)

A similar connection between a soluble many-body problem and an integrable evolution equation has been found for rational solutions (1.160a) of the Boussinesq equation [1.167]: the many-body problem is still (1.154) but the restricted flow is the $F_2$ flow with grad$(F_3 - F_1) = 0$ [1.167]. The KdV equation (1.159) also admits elliptic function solutions

$$u(x,t) = \sum_{j=1} \mathscr{P}\{2^{-\frac{1}{2}}[x - x_j(t)]\} \tag{1.163}$$

if and only if [1.169]

$$\dot{x}_j = 3 \sum_{k \neq j} \mathscr{P}[2^{-\frac{1}{2}}(x_j - x_k)] \tag{1.164a}$$

$$\sum_{k \neq j} \mathscr{P}'[2^{-\frac{1}{2}}(x_j - x_k)] = 0 \quad . \tag{1.164b}$$

The corresponding many-body problem is

$$H' = \frac{1}{2} \sum y_j^2 + 6 \sum_{j \neq i} \mathscr{P}(x_i - x_j) \tag{1.165}$$

with invariants $J_2' \equiv H'$, $J_3'$, ... . The flow is the $J_3'$ flow restricted by grad $J_2' = 0$. Likewise the flows of the higher order KdV equations in Lax's hierarchy seem to be the $J_n'$ flows restricted by grad $J_2' = 0$. The higher order KdV equations have rational solutions (1.158) associated with the restricted $J_n$ flows of (1.154) [1.169].

The fundamental character of (1.154) for rational solutions is presumably connected with the fact that the Burgers equation $u_t - 2cuu_x - cuu_x = 0$ [compare (1.42) and set $u \to -\frac{1}{2}\nu u$, $t \to c^{-1}\nu t$] has rational solutions

$$u(x,t) = \sum_{j=1}^{N} [x - x_j(t)]^{-1} r_j(t) \tag{1.166}$$

if and only if [1.169] $r_j(t) = 1$ and

$$\dot{x}_j = -2c \sum_{k \neq j} (x_j - x_k)^{-1} \quad , \quad 1 < j < N \quad . \tag{1.167}$$

This system of equations implies [1.169]

$$\ddot{x}_j = -8c^2 \sum_{k \neq j} (x_j - x_k)^{-3} \tag{1.168}$$

which (for $c = 1/2$) is the *unrestricted* $J_2 \equiv H$ flow of (1.154). Since [see (1.43)] the Burgers equation is integrable through the Hopf-Cole transformation, this presumably explains the complete integrability of (1.154). Nevertheless the Burgers equation does not have soliton solutions, is not Hamiltonian, and does not have an infinite set of conserved quantities in involution. It does have a Lax pair representation (in which, however, $\hat{L} \neq \hat{L}^+$ and $\hat{B} \neq \hat{B}^+$) and an infinite hierarchy of equations [1.169] (as well as a prolongation structure [1.150]) and these facts seem sufficient for its equivalence to (1.167).

Every restricted flow (1.164) is a flow of the Hamiltonian [1.169]

$$H = \frac{1}{2}\sum y_j^2 + \frac{1}{8}\sum_i \sum_{i\neq j} \mathscr{P}^2(x_i - x_j) \tag{1.169}$$

but this fact does not seem to detract from the fundamental character of (1.165) and (1.154). Recently CHEN et al. [1.170] have actually found the soliton solutions of the apparently integrable BENJAMIN-ONO equation [1.171] from the motion of the poles (1.154) (the solitons are the rational solutions in this case). It seems probable that similar connections between rational solutions and many-body problems can be established for all the integrable equations: partial results for the modified KdV and Kadomtsev-Petviashvili equations [(1.78) and (1.172) below] are given in [1.169], and partial results for the alternative KdV equations [1.115,169] are given in [1.169,170], and for the Toda lattice in [1.173], for example, but we know of no other particular results in this connection at this time. Nor do we fully understand the role played by the particular N-body problem (1.154) in most of the results obtained so far. We refer the reader to the paper by CALOGERO [1.166] for a survey of many results, to that by AIRAULT et al. [1.167] for further analysis concerning the KdV and Boussinesq equations, and to that by CHOODNOVSKY and CHOODNOVSKY [1.169] for an interesting alternative analysis. Addition theorems like (1.161) are associated with Abelian (Jacobian) varieties. We refer the reader to connected work on Jacobian varieties described in [1.167] and further described in McKEAN's lecture notes [1.174] and in [1.175].

In a very different direction the applicable side of soliton theory has been vastly extended by the development since 1976 of singular perturbation theories for nonintegrable equations which are in some sense 'close' to the integrable systems. We refer the reader to pioneering papers by KAUP and NEWELL [1.176-178], and to applications by NEWELL [1.179], MASON [1.180], and KITCHENSIDE et al. [1.181] to the 'double sine-Gordon' equation

$$u_{xx} - u_{tt} = \pm\left(\sin u + \frac{1}{2}\varepsilon \sin\frac{1}{2}u\right) \tag{1.170}$$

especially (physical applications of this double s-G are described in our Chap.3). We refer also to the perturbation theories of SCOTT and McLAUGHLIN [1.25,182] and KARPMAN [1.183], as well as to a recent paper by NEWELL [1.184]. Newell devotes some portion of Chap.6 to his form of singular perturbation theory.

Recent exact results on the integrable systems include the solution of the equations of the one-dimensional Heisenberg ferromagnet in continuum approximation

$$\frac{\partial \underset{\sim}{s}}{\partial t} = \underset{\sim}{s} \times \frac{\partial^2 \underset{\sim}{s}}{\partial x^2} \tag{1.171}$$

[in which s(x,t) is a continuous spin density] found by TAKHTADJAN [1.185] and LAKSHMANAN [1.186]; and the solution of the equations of a rigid body by MANAKOV [1.187], as well as the solutions of the many-body problems [1.166-170] already referred to, and some extensions [1.188]. We refer the reader also to a number of proceedings of conferences [1.137,189-194]: these indicate some of the most recent topics associated with soliton theory, the mathematical results, and the present range of physical applications. For a simple look at this very wide range of physical applications we refer also to [1.22,23].

We have said virtually nothing in this article about solitons in more than one space dimension (one space and one time dimension: the two Euclidean space dimensions looked at in Sect.1.3 are of course an aspect of this). The restrictions of the Hobart-Derrick theorem [1.195-197] on solutions of the generalised Klein-Gordon equation in more than one space dimension are well known [1.22,43]. For other work on more than one space dimension we cite, for example, [1.101,158,161,198,199] and especially Chap.7. We mention also Calogero's extension of the AKNS scheme to more "time-like" space variables (see Chaps.6,9 and [1.69]). Explicit examples of plane wave solitons in more than one space dimension have been given (see, e.g., [1.29,199-201]); in particular the Kadomtsev-Petviashvili equations

$$3\beta^2 u_{yy} + (u_t + u_{xxx} + 6uu_x)_x = 0 \tag{1.172}$$

($\beta^2 = \pm 1$) have multisoliton plane wave solutions [1.29,118,158]. Recently MANAKOV et al. [1.202] found, by a suitable limit of the multisoliton plane wave solution, nonsingular strictly localised soliton solutions of the KP equations (1.172) in the case $\beta^2 = -1$. The solutions are localised but singular for $\beta^2 = +1$. In both cases these solitons are rational functions; but a connection with rational solutions associated with many-body problems is not established; nor is it clear whether these localised solitons represent a special case or something more general. The KP equations have been solved by algebraic-geometrical methods [1.118]. ATIYAH [1.203] has (or is in principle able to find) by comparable algebraic-geometrical methods all the 'instanton' solutions of classical self-dual non-Abelian gauge theories: these are solutions in four Euclidean dimensions. To our understanding the connection between instantons and solitons is not fully established. Nevertheless BELAVIN and ZAKHAROV [1.204] have already shown how to find the instanton solutions by an extended inverse scattering method. This method is applicable to an arbitrary number of dimensions.[11]

---

[11]McCARTHY [1.214,215] finds BT's in $2^n$ dimensions ($n \geq 1$) and, in particular, reports one for the 4-Euclidean-dimensional self-dual $\overline{SU}_2$ gauge fields considered by YANG [1.216] and N-soliton solutions for it.

This latest achievement in the history of the soliton and of the inverse scattering method is essentially the most recent fundamental result we are able to quote in this article. As a help to those readers whose interests lie more in the application of soliton theory to their own branches of physics we end by quoting the following further references here: for solitons in plasma physics Chap.2 and [1.22,189,193] and the references therein; for laser physics Chaps.2,3 and [1.23,27,64,193,205] and references; for solid state physics Chap.12 and the articles of [1.137] and its references; for particle physics [1.43,71,86-91,93,95, 206]; for a lead to the quantisation of the sine-Gordon equation Chap.11 and for the quantised s-G and massive Thirring models Chap.12; for statistical mechanics Chap.12 and [1.137] and references. One should also mention the beginnings of the application of solitons to problems in cosmology. A Bäcklund transformation for the Ernst equation of general relativity $(\mathrm{Re}\{E\})\nabla^2 E = (\nabla E)^2$ for axially symmetric stationary vacuum fields has been found [1.207]: it had been established that the equation had an infinity of potentials [1.208]. Solitonlike solutions are reported by BELINSKY and ZAKHAROV [1.209].

For those interested in solitonlike objects rather than in specific integrable systems, and for those interested in computational work on almost integrable systems we refer also to the review article by MAKHANKOV [1.210]. Results include some for nonintegrable systems in three space dimensions with spherical symmetry (the "pulsons" [1.210]). Further numerical work on pulsons is reported by EILBECK [1.211] and CHRISTIANSEN [1.212] (who incidentally reports also a Bäcklund transformation for the sine-Gordon equation in three space dimensions: this aBT was found independently by LEIBBRANDT [1.213]. In contrast CALOGERO and DEGASPERIS [1.217] recently reported the *exact* integration of a Korteweg-de Vries equation with cylindrical symmetry.

This ends our survey of the history of the soliton and the inverse scattering method and the interaction of mathematics and physics which has produced it. The following chapters all provide further connections between soliton theory and physical problems. But perhaps their main purpose is to make readily accessible the most generally useful parts of the mathematics of the *analytical* methods presently available for solving nonlinear evolution equations. Many of these equations have direct physical significance and the body of knowledge they represent will surely play the role in nonlinear physics in the future that the conventional linear 'equations of mathematical physics' have played in physics these last 150 years.

*Acknowledgments*

The authors are grateful to Dr. David Cambell for a critical reading of the manuscript and to Professor Francesco Calogero for reading and commenting on parts of it. Most of this article was written at the Aspen Center for Physics in August 1978, and that part which was not was written in Copenhagen. One of us (R.K.B.) is grateful to the Aspen Center and to Nordita, Copenhagen, for the hospitality which made the writing of this article possible.

References


1.1 C.S. Gardner, J.M. Greene, M.D. Kruskal, R.M. Miura: Phys. Rev. Lett. *19*, 1095 (1967)
1.2 N. Zabusky, M.D. Kruskal: Phys. Rev. Lett. *15*, 240 (1965)
1.3 J.S. Russell: "Report on Waves", British Association Reports (1844)
1.4 A.C. Scott, F.Y.F. Chu, D.W. McLaughlin: Proc. IEEE *61*, 1443 (1973)
1.5 G.S. Emmerson: "John Scott Russell. A Great Victorian Engineer and Naval Architect" (John Murray: London, 1977); Encyclopedia Britannica, 9th edn., p.66. Only the latter quotes Edinburgh as well as St. Andrews and Glasgow. The former gives Russell's age on leaving the University of Glasgow as 17
1.6 Sir Horace Lamb: *Hydrodynamics* 6th ed. (Cambridge University Press, 1952) pp.417-420, 423-429. For the solitary wave see pp.423-426. For a quotation from Russell on the generation of surface waves on water by the wind see pp.630-631
1.7 D.J. Korteweg, G. de Vries: "On the change of Form of Long Waves Advancing in a Rectangular Canal, and on a New Type of Long Stationary Waves". The London, Edinburgh, and Dublin Philosophical Magazine and Journal of Science Series 5, *39*, No.241, June 1895, pp.422-443 [Phil. Mag. *39*, 422 (1895)]
1.8 Airy: "Tides and Waves" Encycl. Metrop. (1845) (reference from [1.7])
1.9 J. de Boussinesq: "Théorie de l'intumescence liquide appeleé on de solitaire ou de translation se propageant dans un canal rectangulaire", Comptes Rendus *72*, 755-759 (1871); J. Math. Pures et Appliqueés *2*, 55 (1872)
1.10 J.D. Forbes: *Travels through the Alps*, new ed. revised and annotated by W.A.B. Coolidge (Adam and Charles Black, London 1900)
1.11 J. Tyndall: *The Glaciers of the Alps* (John Murray, London 1860)
1.12 E. Fermi, J.R. Pasta, S.M. Ulam: *Studies of Nonlinear Problems*, Vol.1, Los Alamos Rpt. LA-1940 (May 1955); *Collected Works of E. Fermi*, Vol.2 (Univ. of Chicago Press, 1965) pp.978-988
1.13 E.A. Jackson: "Nonlinearity and Irreversibility in Lattice Dynamics", Rocky Mountain J. Math. *8*, 127 (1978)
M. Toda: "Solitons and heat conduction", Phys. Scr. *20*, 424 (1979)
1.14 J.S. Russell: *The Modern System of Naval Architecture* (London, 1865). Of course there are previous works, e.g., J.W. Griffiths *A treatise on Marine and Naval Architecture* (new ed., London 1857)
1.15 J.S. Russell: "The Wave of Translation in the Oceans of Water, Air and Ether" (Trubner: London, 1895) This is called a new edition. We have referenced an earlier edition in [1.17] below as "Teubner London 1878" but this may be incorrect. The preface to the 1895 edition rather indicates that the material had not been published in book form before
1.16 Lord Rayleigh (J.W. Strutt): "On Waves", Philos. Mag. *1*, 257 (1976). The solitary wave is discussed on pp.262-267
1.17 C.G. Stokes: "On the theory of Oscillating Waves", Trans. Camb. Phil. Soc. Vol.viii (1847)
1.18 St. Venant: Comptes Rendus, Vol.ci (1885) (reference taken from [1.7])
1.19 J. McCowan: "On the solitary wave", Philos. Mag. *32*, 45 (1891)
1.20 Bazin: Mém des Savants étrangers, tom. xix (reference taken from [1.19])
1.21 J. Frenkel, T. Kontorova: "On the theory of plastic deformation and twinning", J. Phys. (USSR) *1*, 137 (1939)
1.22 R.K. Bullough: "Solitons", Phys. Bull., February 1978, pp.78-82
1.23 R.K. Bullough: "Solitons", in *Interaction of Radiation and Condensed Matter*, Vol.1, IAEA-SMR-20/51 (International Atomic Energy Agency, Vienna 1977) pp.381-469
1.24 A. Barone, F. Eposito, C.J. Magee, A.C. Scott: "Theory and Applications of the sine-Gordon Equation", Riv. Nuovo Cimento *1*, 227-267 (1971)
1.25 D.W. McLaughlin, A.C. Scott: Appl. Phys. Lett. *30*, 545 (1977)
1.26 S.L. McCall, E.L. Hahn: Phys. Rev. *183*, 457 (1969)
1.27 G.L. Lamb: Rev. Mod. Phys. *43*, 99 (1971)
1.28 H.C. Yuen, B.M. Lake: Phys. Fluids *18*, 956 (1975)

1.29 J.D. Gibbon, N.C. Freeman, R.S. Johnson: Phys. Lett. *65*A, 380 (1978) also D. Anker, N.C. Freeman: Proc. Roy. Soc. (London) A*360*, 529 (1978) J. Miles: J. Fluid Mech. *79*, 171 (1977)
1.30 J. Goldstone, R. Jackiw: Phys. Rev. D *11*, 1486 (1975)
1.31 R.F. Dashen, B. Hasslacher, A. Neveu: Phys. Rev. D *10* (1974)
1.32 S. Aubry: J. Chem. Phys. *24*, 291 (1976) B.S. Getmanov: JETP Lett. *24*, 291 (1976) Results of numerical investigations communicated to us by M.J. Ablowitz, M.D. Kruskal; and remarks by A.R. Bishop and J.C. Eilbeck, J.D. Gibbon. The 'lock' is into an oscillating breatherlike bound kink-antikink pair, and arises only in the collision of a kink [+ve sign in (1.15)] with an antikink [-ve sign in (1.15)]. J.C. Eilbeck has a film showing how this arises
1.33 R. Hirota: Phys. Rev. Lett. *27*, 1192 (1971)
1.34 P.P. Kulish: Reported at "Symposium on Nonlinear Evolution Equations Solvable by the Inverse Spectral Transform", (Accademia dei Lincei, Rome, June, 1977). The published form "Factorisation of Scattering Characteristics and Integrals of Motion" in *Nonlinear Evolution Equations Solvable by the Spectral Transform*, ed. by F. Calogero (Pitman, London 1978) pp.252-257 does not cover this material)
1.35 J.D. Gibbon, J.C. Eilbeck: J. Phys. A: Proc. Phys. Soc. Lond. *5*, L132 (1972)
1.36 G.B. Whitham: *Linear and Nonlinear Waves* (Wiley, New York 1974) Chap.17
1.37 R.K. Bullough, R.K. Dodd: "Solitons in Mathematics: Brief History", in *Solitons and Condensed Matter Physics*, ed. by A.R. Bishop, T. Schneider, Springer Series in Solid State Sciences, Vol.8 (Springer, Berlin, Heidelberg, New York 1978); "Solitons in Physics: Basic Concepts" and "Solitons in Mathematics", in *Structural Stability in Physics*, ed. by W. Güttinger, H. Eikemeier, Springer Series in Synergetics, Vol.4 (Springer, Berlin, Heidelberg, New York 1979) pp.219-232 and 233-253
1.38 V.E. Zahkharov, A.B. Shabat: Zh. Eksp. Teor. Fiz. *61*, 118 (1971) [English transl.: Soviet Phys. JETP *34*, 62 (1972)]
1.39 P.J. Caudrey, J.C. Eilbeck, J.D. Gibbon: Nuovo Cimento *25*, 497 (1975)
1.40 R.F. Dashen, B. Hasslacher, A. Neveu: Phys. Rev. D *11*, 3424 (1975)
1.41 V.E. Korepin, L.D. Fadeev: Teor. Mat. Fiz. *25*, 147 (1975)
1.42 R.K. Bullough, R.K. Dodd: In "Synergetics. A Workshop", Proc. Intl. Workshop on Synergetics at Schloss Elmau, Bavaria, May 1977, Springer Series in Synergetics, Vol.2, ed. by H. Haken (Springer, Berlin, Heidelberg, New York 1977) pp.92-119. Also see [1.37]
1.43 S. Coleman: "Classical Lumps and their Quantum Descendants", Lectures at the 1975 Intern. School of Subnuclear Physics "Ettore Majorana" (1975)
1.44 M.J. Ablowitz, D.J. Kaup, A.C. Newell, H. Segur: Phys. Rev. Lett. *31*, 125 (1973)
1.45 M.J. Ablowitz, D.J. Kaup, A.C. Newell, H. Segur: Stud. Appl. Math. *53*, 249 (1974)
1.46 V.E. Zahkharov, S.V. Manakov: JETP Lett. *18*, 243 (1973)
1.47 D.J. Kaup: Studies in Appl. Math. *55*, 9 (1976); Rocky Mountain J. Math. *8* (1,2), 283 (1978)
1.48 F. Lund: Phys. Rev. Lett. *38*, 1175 (1977)
1.49 J. Liouville: "Sur l'equation aux différences partielles $d^2 \log\lambda / dudv \pm \lambda/2a^2 = 0$". J. Math. Pures et Appliqueés (Paris) *18* (1) 71-72 (1853)
1.50 R.K. Dodd, R.K. Bullough: Proc. Roy. Soc. (London) A*351*, 499 (1976)
1.51 G.L. Lamb: In *Bäcklund Transformations*, Lecture Notes in Mathematics, Vol. 515, ed. by R.M. Miura (Springer, Berlin, Heidelberg, New York 1976) pp.69-79
1.52 J. Rubinstein: J. Math. Phys. *11*, 258 (1970)
1.53 M.D. Kruskal: Private communication
1.54 D.W. McLaughlin, A.C. Scott: J. Math. Phys. *14*, 1817 (1973)
1.55 A.V. Bäcklund: "Einiges über Curven und Flächentransformationen", Lund Universitets Arsskrift *10* (1875); "Om Ytor med konstant negativ Krökning", Lund Universitets Arsskrift *19* (1883)
1.56 H.D. Wahlquist, F.B. Estabrook: Phys. Rev. Lett. *31*, 1386 (1973)

1.57 M.J. Clairin: Ann. Scient. Éc. Norm. Sup., 3[e]sér. Suppl. *19*, S1-S63; "Sur quelques equations aux derivées partielles du second ordre", Ann. Toulouse *2*, Sér. *5*, 437-458
A.R. Forsyth: *Theory of Differential Equations*, Vol.6, ch.21 (Dover, New York 1959)
P.P. Eisenhart: *A treatise on the differential geometry of curves and surfaces*, ch.8 (Dover, New York 1960)
E. Goursat: Ann. de Toulouse 3[e]Sér. *10*, 65 (1918)
A. Seger, H. Donth, A. Kochendoerfer: Z. Physik *134*, 173 (1953)
1.58 H. Chen: Phys. Rev. Lett. *33*, 925 (1974)
1.59 E. Hopf: "The partial differential equation $u_t + uu_x = \mu u_{xx}$", Comm. Pure Appl. Math. *3*, 201-230 (1950)
1.60 J.D. Cole: "On a quasilinear parabolic equation occurring in aerodynamics", Q. Appl. Math. *9*, 225-236 (1950)
1.61 H.D. Wahlquist, F.B. Estabrook: J. Math. Phys. *16*, 1 (1975)
1.62 R.M. Miura: J. Math. Phys. *9*, 1202 (1968)
1.63 M.J. Ablowitz, D.J. Kaup, A.C. Newell: J. Math. Phys. *15*, 1852 (1974)
1.64 R.K. Bullough, P.W. Kitchenside, P.M. Jack, R. Saunders: "Solitons in laser physics", Phys. Scr. *20*, 364 (1979)
1.65 M.D. Kruskal: "The Korteweg de Vries equation and related evolution equations", Lect. appl. Maths. *15*, ed. by A.C. Newell (AMS: Providence, Rhode Island 1974) pp.61-83
1.66 R.K. Dodd, R.K. Bullough: Proc. Roy. Soc. (London) A*352*, 481 (1977)
1.67 R.M. Miura, C.S. Gardner, M.D. Kruskal: J. Math. Phys. *9*, 1204 (1968)
1.68 J. Satsuma: Prog. Theor. Phys. *52*, 1396L (1974)
1.69 R.K. Dodd, R.K. Bullough: "The Generalised Marchenko Equation and the Canonical Structure of the AKNS-ZS Inverse Method", Phys. Scr. *20*, 514 (1979)
1.70 P.D. Lax: Comm. Pure Appl. Math. *21*, 467 (1968)
1.71 See, e.g., B. Schroer: Lectures at the Soliton Workshop, Università di Salerno, Italy, July 1977
1.72 P.P. Kulish: Teor. Mat. Fiz. *26*, 198 (1976) [Theoretical and Math. Phys. *26*, 132 (1976)]
1.73 J.V. José, L.P. Kadanoff, S. Kirkpatrick, D.R. Nelson: Phys. Rev. B*16*, 1217 (1977)
D.R. Nelson, J.M. Kosterlitz: Phys. Rev. Lett. *39*, 1201 (1977)
D.R. Nelson: Phys. Reports *49*, 255 (1979). These authors are concerned with the Coulomb gas and its relation to an X-Y model. Similar techniques can be used to relate to the Euclidean sine-Gordon equation
1.74 J.M. Kosterlitz, D.J. Thouless: J. Phys. C *6*, 1181 (1973). A connection between the vortex model of these authors and a quantised field theory relating to Luther's work described in Sect.11 is: A Luther, D.J. Scalapino: Phys. Rev. B*16*, 1153 (1977)
1.75 F.B. Estabrook, H.D. Wahlquist: J. Math. Phys. *17*, 1292 (1976)
1.76 F.B. Estabrook, H.D. Wahlquist: "Prolongation Structures, Connection Theory and Bäcklund Transformations", in *Evolution Equations Solvable by the Spectral Transform*, ed. by F. Calogero (Pitman, London 1978) pp.64-83
1.77 F.A.E. Pirani, D.C. Robinson, W.F. Shadwick: "Local jet bundle formulation of Bäcklund transformations (to be published 1978). Also see R. Herman 'Geometric Theory of Nonlinear Differential Equations, Bäcklund Transformations and Solitons Parts A and B, Math. Sci. Press (53 Jordan Road, Brookline, MA 02146, USA)
1.78 F. Calogero: Lett. al Nuovo Cimento *14*, 537 (1975)
F. Calogero, A. Degasperis: Lett. al Nuovo Cimento *16*, 181 (1976)
1.79 R.K. Dodd, R.K. Bullough: Phys. Lett. *62*A, 70 (1977)
1.80 R.K. Dodd: J. Phys. A: Math. Gen. *11*, 81 (1978)
1.81 R. Hirota: J. Phys. Soc. Jpn. *33*, 1459 (1972)
1.82 P.J. Caudrey, J.D. Gibbon, J.C. Eilbeck, R.K. Bullough: Phys. Rev. Lett. *30*, 237 (1973)
P.J. Caudrey, J.C. Eilbeck, J.D. Gibbon: J. Inst. Maths. Applics. *14*, 375 (1974)
1.83 R. Steurwald: Abh. Bayer. Akad. Wiss. (München) *40*, 1 (1936)

1.84 T.H.R. Skyrme: Proc. Roy. Soc. (London) A *247*, 260 (1958)
1.85 T.H.R. Skyrme: Proc. Roy. Soc. (London) A *262*, 237 (1961)
J.K. Perring, T.H.R. Skyrme: Nucl. Phys. *31*, 550 (1962)
1.86 K. Pohlmeyer: Commun. Math. Phys. *46*, 207 (1976)
1.87 R. Jackiw: Rev. Mod. Phys. *49*, 681 (1977)
1.88 S. Coleman: Phys. Rev. D*11*, 2088 (1975)
1.89 B. Berg, M. Karowski, H.J. Thun: Phys. Lett. *64*B, 286 (1976)
1.90 P.P. Kulish, E.R. Nissimov: Pis'ma Zk. Eksp. Teor. Fiz. *24* (4), 247-250 (1976)
1.91 E.R. Nissimov: Bulg. J. Phys. IV *2*, 113 (1977)
1.92 A.V. Mikhailov: JETP Letts. *23*, 356 (1976)
1.93 A.B. Zamolodchikov: Comm. Math. Phys. *55*, 183 (1977); JETP Lett. *25*, 468 (1977); also ITEP (Moscow) Preprint (ITEP 12) (1977)
A. and A.L. Zamolodchikov: JETP Lett. *26*, 457 (1977)
1.94 D. Iagolnitzer: Phys. Rev. D *18*, 1275 (1978)
1.95 M. Karowski, H.J. Thun, T.T. Truong, P.H. Weisz: Phys. Lett. *67*B, 321 (1977)
M. Karowski, H.J. Thun: Nuclear Phys. B *130*, 295-308 (1977)
M. Karowski: Phys. Rpts. *49*, 229 (1979)
1.96 R. Shankar, E. Witten: Phys. Rev. D *17*, 2134 (1978); 'The S-matrix of the kinks of the $(\bar{\psi}\psi)^2$ Model' Preprint (1977)
B. Berg, M. Karowski, P. Weisz: 'Construction of Green Functions from an Exact S-matrix' (Preprint FUB-HEP June 1978). These authors construct the S-matrix of the Ising model in the scaling limit
1.97 S.L. McCall, E.L. Hahn: Phys. Rev. Lett. *18*, 908 (1967)
1.98 M.J. Ablowitz, D.J. Kaup, A.C. Newell, H. Segur: Phys. Rev. Lett. *30*, 1262 (1973)
1.99 F.T. Arecchi, R. Bonifacio: IEEE *QE-1*, 169 (1965)
1.100 R.Y. Chiao, E. Garmire, C.H. Townes: Phys. Rev. Lett. *13*, 479 (1964); also P.L. Kelley: Phys. Rev. Lett. *15*, 1005 (1965)
1.101 H. Cornille: J. Math. Phys. *20*, 199 (1979)
For remarks on the instability see also A.C. Newell: "Bifurcation and Nonlinear Focussing", in *Proc. Pattern Formation by Dynamic Systems and Pattern Recognition*, ed. by H. Haken, Springer Series in Synergetics, Vol.5 (Springer, Berlin, Heidelberg, New York 1979)
1.102 V.L. Ginzburg, L.P. Pitaevskii: Zh. Eks. Teor. Fiz. (USSR) *34*, 1240 (1958)
1.103 L.P. Pitaevskii: Zh. Eks. Teor. Fiz. (USSR) *35*, 408 (1958) [English transl.: Sov. Phys. JETP *8*, 282 (1958)]
1.104 E.P. Gross: Ann. Phys. *4*, 57 (1958)
1.105 W. Yourgrau, S. Mandelstam: *Variational Principles in Dynamics and Quantum Theory*, 3rd ed., ed. by L. Mittag, M.J. Stephen, W. Yourgrau (Pitman, London 1968) Sect.13, p.158
1.106 T.B. Benjamin, J.E. Feir: J. Fluid Mech. *27*, 417 (1966)
1.107 D.J. Benney, A.C. Newell: J. Math. Phys. *46*, 133 (1967)
1.108 V.E. Zakharov: Zh. Eksp. Teor. Fiz. (USSR) *62*, 1745 (1972) [English transl.: Sov. Phys. JETP *35*, 908 (1972)]
1.109 V.E. Zakharov, A.B. Shabat: Zh. Eksp. Teor. Fiz. (USSR) *64*, 1627 (1973) [English transl.: Sov. Phys. JETP *37*, 823 (1973)]
1.110 M.D. Kruskal: "The birth of the soliton", paper presented to the Symposium on Nonlinear Evolution Equations Solvable by the Inverse Spectral Transform", Accademia dei Lincei, Rome, June 1977. Published in *Nonlinear evolution equations solvable by the spectral transform*, ed. by F. Calogero (Pitman, London 1978) pp.1-8. Lecture to the NATO Advanced Study Institute on *Nonlinear Problems in Physics and Mathematics*, Istanbul, August, 1977, ed. by A.O. Barut (D. Reidel, Dordrecht, Holland 1978)
1.111 I.M. Gel'fand, B.M. Levitan: Izv. Akad. Nauk. SSSR, Ser. Mat. *15*, 309 (1951) Translated in American Math. Soc. Translations (AMS: Providence, Rhode Island, 1955) Ser.2, Vol.1, p.253. The extension to $-\infty < x < \infty$ is due to Marchenko. Also see K. Chadan and P.C. Sabatier: "Inverse Problems in Quantum Scattering Theory", in Texts and Monographs in Physics (Springer, Berlin, Heidelberg, New York 1977)

1.112 I. Kay: Courant Institute of Mathematical Sciences, New York University, Report No. EM-74, 1955
I. Kay, H.E. Moses: Nuovo Cimento *3*, 276 (1956)
1.113 V.E. Zakharov, L.D. Fadeev: Funkt. Anal. Prilozh. *5*, 18 (1971); Funct. Anal. Appl. *5*, 280 (1971)
1.114 P.D. Lax: Commun. Pure Appl. Math. *28*, 141 (1975)
1.115 P.J. Caudrey, R.K. Dodd, J.D. Gibbon: Proc. Roy. Soc. (London) A *351*, 407 (1976)
1.116 R.K. Dodd, J.D. Gibbon: Proc. Roy. Soc. (London) A *358*, 287 (1977)
1.117 I.M. Gel'fand, L. Dikii: Uspeki mat. nauk *30*, 67 (1975); Russian Math. Surveys *30*, 77 (1975); Funkt. Anal. Prilozh. *10*, 18 (1976)
1.118 Yu.I. Manin: Itogi Nauki Tekn.*11*, 5 (1978)
1.119 V.B. Matveev: Usp. Mat. Nauk *6*, 201-203 (1975)
A.P. Its, V.B. Matveev: Funkt. Anal. Ego Prilozh. *9*, 65 (1976); Theor. Math. Phys. *23*: 1, 56-67 (1975). A complete list of references appears in D.A. Dubrovin, V.B. Matveev, S.P. Novikov: Usp. Mat. Nauk *1*, 55-136 (1976); and in V.B. Matveev:'Abelian functions and solitons', Lectures at the University of Wroclaw, Preprint No.373, Inst. Theoretical Physics, University of Wroclaw, Wroclaw, Cybulskiego 36, Poland, June 1976. The NLS and s-G under periodic boundary conditions are also treated in these lectures and the references given there
1.120 S.P. Novikov: Funkt. Anal. Prilozh. *8*:3, 54 (1974)
B.A. Dubrovin, S.P. Novikov: JETP *67*, 2131 (1974)
S.P. Novikov, B.A. Dubrovin, I.M. Krichever: Dokl. Akad. Nauk. SSR *229*:1 (1976). Also see the references in Matveev's 'Abelian functions' [1.119] and in Novikov's Chap.10
1.121 H.P. McKean, P. van Moerbeke: Inventiones Math. *30*, 217-274 (1975)
1.122 C.S. Gardner: J. Math. Phys. *12*, 1548 (1971)
1.123 R.K. Dodd, R.K. Bullough: "Integrability of Nonlinear Evolution Equations: Prolongation and Solitons" (to be published)
1.124 M. Wadati, M. Toda: J. Phys. Soc. Jpn. *32*, 1403 (1972)
1.125 M. Wadati: J. Phys. Soc. Jpn. *32*, 1681 (1972)
1.126 D.D. Schnack, G.L. Lamb, Jr.: In *Proceedings of Third Rochester Conference on Coherence and Quantum Optics* (Plenum Press, New York 1973) p.23
1.127 G.L. Lamb, Jr.: Phys. Rev. Lett. *31*, 196 (1973)
1.128 G.L. Lamb, Jr.: Phys. Rev. A*9*, 422 (1974)
1.129 R.K. Bullough, F. Ahmad: Phys. Rev. Lett. *27*, 330 (1971)
J.C. Eilbeck, R.K. Bullough: J. Phys. A: Math. Gen. *5*, 820 (1972)
1.130 J.C. Eilbeck, P.J. Caudrey, J.D. Gibbon, R.K. Bullough: J. Phys. A: Math. Gen. *6*, 1337 (1973)
1.131 P.J. Caudrey, J.C. Eilbeck: Phys. Lett. *62*A, 65 (1977)
1.132 J.D. Gibbon, J.C. Eilbeck: J. Phys. A: Proc. Phys. Soc. Lond. *5*, L122 (1972)
1.133 J.D. Gibbon, P.J. Caudrey, R.K. Bullough, J.C. Eilbeck: Lett. Nuovo Cimento *8*, 775 (1973)
1.134 L.A. Takhtadjan, L.D. Fadeev: Teor. Mat. Fiz. *21*, 160 (1974)
1.135 V.E. Zakharov, L.A. Takhtadjan, L.D. Fadeev: Dokl. Akad. Nauk. SSSR *219*, 1334 (1974) [English transl.: Sov. Phys. Doklady *19*, 824 (1975)]
1.136 T.D. Lee: "Nontopological Solitons and Applications to Hadrons", Phys. Scr. *20*, 440 (1979)
1.137 Symposium on "Nonlinear (soliton) structure and dynamics in condensed matter", Oxford, June, 1978, Proceedings, in *Solitons and Condensed Matter Physics*, Springer Series in Solid State Sciences, Vol.8, ed. by A.R. Bishop, T. Schneider (Springer, Berlin, Heidelberg, New York 1978)
1.138 R.J. Baxter: Phys. Rev. Lett. *26*, 832 (1971)
J.D. Johnson, S. Krinsky, B.M. McCoy: Phys. Rev. A *8*, 2526 (1973)
1.139 A. Luther, V.J. Emery: Phys. Rev. Lett. *33*, 589 (1974)
1.140 M. Wadati, H. Sanuki, K. Konno: Prog. Theor. Phys. *53*, 419-436 (1975)
1.141 K. Pohlmeyer: "Interactions via Quadratic Constraints", in *New Developments in Quantum Field Theory and Statistical Mechanics*, ed. by M. Lévy, P. Mitter (Plenum Press, New York 1977)

1.142 F. Lund: *Proc. NATO Advanced Study Institute on 'Nonlinear Problems in Physics and Mathematics'*, Istanbul, August 1977, ed. by A.O. Barut (D. Reidel, Dordrecht, Holland 1978)
1.143 L. Bianchi: *Lezioni di Geometria Diferenziale* (Spoerri, Pisa 1896)
1.144 Results reported by P.P. Kulish at 'Symposium on Nonlinear Evolution Equations Solvable by the Inverse Spectral Transform' Accademia dei Lincei, Rome, June 1977. The published paper, 'Factorisation of scattering characteristics and integrals of motion' in *Nonlinear evolution equations solvable by the spectral transform*, ed. by F. Calogero (Pitman, London 1978) pp.252-257, does not cover this material
1.145 G.L. Lamb, Jr.: J. Math. Phys. *18*, 1654 (1977)
1.146 M. Lakshmanan: Phys. Lett. *64*A, 353 (1978)
1.147 R. Sasaki: Phys. Lett. *71*A, 390 (1979); Nucl Phys. B*154*, 343 (1979); Phys. Lett. *73*A, 77 (1979); "Geometric approach to soliton equations", Niels Bohr Institute preprint NBI-HE-79-31
R. Sasaki, R.K. Bullough: "Geometric theory of local and nonlocal conservation laws for the sine-Gordon equation", Niels Bohr Institute preprint NBI-HE-79-32
R. Sasaki, R.K. Bullough: "Geometry of the AKNS-ZS inverse scattering scheme", in Proc. Intl. Meeting "Nonlinear Evolution Equations and Dynamical Systems", Lecce, Italy, June 20-23, 1979. A Springer Lecture Notes in Physics Volume, M. Boiti, F. Pempinelli, G. Soliani (eds.) (Springer, Berlin, Heidelberg, New York 1980)
Also, M. Crampin, P.J. McCarthy: Lett. Math. Phys. *2*, 303 (1978) and reference 2 there
M. Crampin, F.A.E. Pirani, D.C. Robinson: Lett. Math. Phys. *2*, 15 (1977)
R. Herman in [1.77] and in his subsequent volumes on Interdisciplinary Mathematics
1.148 H. Flanders: *Differential Forms* (Academic Press, New York 1963)
1.149 S. Weinberg: *Gravitation and Cosmology* (Wiley, New York 1972)
1.150 R.K. Dodd, J.D. Gibbon: Proc. Roy. Soc. (London) A*359*, 411 (1978)
1.151 D.W. McLaughlin: J. Math. Phys. *16*, 96 (1975)
1.152 H. Flaschka: Phys. Rev. B*9*, 1924 (1974); Progr. Theor. Phys. *51*, 703 (1974)
1.153 S.V. Manakov: Zh. Eksp. Teor. Fiz. *67*, 543 (1974) [English transl.: Sov. Phys. JETP *40*, 269 (1974)]
1.154 H. Flaschka, A.C. Newell: "Integrable systems of non-linear evolution equations", in *Dynamical Systems, Theory and Applications*, Lecture Notes in Physics, Vol.38, ed. by J. Moser (Springer, Berlin, Heidelberg, New York 1975)
1.155 M. Wadati, T. Kamijo: Progr. Theor. Phys. *52*, 397 (1974)
1.156 F. Calogero: Lett. Nuovo Cimento *14*, 443 (1975)
1.157 F. Calogero, A. Degasperis: Lett. Nuovo Cimento *16*, 425-433 (1976)
1.158 V.E. Zakharov, A.B. Shabat: Funkt. Anal. Prilozh. *8*, 43 (1974)
1.159 H.C. Morris: Int. J. Theor. Phys. *16*, 227 (1977)
1.160 H.C. Morris: J. Math. Phys. *18*, 530; *18*, 533 (1977)
1.161 H.C. Morris: J. Math. Phys. *17*, 1870 (1976); *18*, 285 (1977)
1.162 R.K. Dodd: "Prolongation Structure Techniques for the New Hierarchy of Korteweg-de Vries Equations", in *Proc. NATO Advanced Study Institute on Nonlinear Problems in Physics and Mathematics*, Istanbul, August 1977, ed. by A.O. Barut, (D. Reidel, Dordrecht, Holland 1978)
1.163 J. Corones: J. Math. Phys. *18*, 163 (1977); *19*, 2431 (1978)
J. Corones, B.L. Markovski, V.A. Rizov: J. Math. Phys. *18*, 2207 (1977)
1.164 J. Moser: Adv. Math. *16*, 197 (1975)
1.165 F. Calogero: J. Math. Phys. *12*, 419 (1971)
1.166 F. Calogero: "Motion of poles and zeros of special solutions of nonlinear and linear partial differential equations and related "solvable" many body problems", in *Proc. NATO Advanced Study Institute on Nonlinear Problems in Physics and Mathematics*, Istanbul, August 1977, ed. by A.O. Barut (D. Reidel, Dordrecht, Holland 1978)
1.167 H. Airault, H.P. McKean, J. Moser: Commun. Pure Appl. Math. *30*, 95 (1977)
1.168 H. Airault:"Poles of Nonlinear Evolution Equations and Integrable Dynamical Systems", in *Nonlinear Evolution Equations Solvable by the Spectral Transform*, ed. by F Calogero (Pitman, London 1978) pp.244-251

1.169 D.V. Choodnovsky, G.D. Choodnovsky: Nuovo Cimento *40*B, 339 (1977)
1.170 H.H. Chen, Y.C. Lee, N.R. Pereira: "Algebraic internal wave solitons and the integrable Calogero-Moser-Sutherland N-body problem". Preprint (1978)
1.171 T.B. Benjamin: J. Fluid Mech. *25*, 241 (1966); *29*, 559 (1967)
H. Ono: J. Phys. Soc. Jpn. *39*, 1082 (1975); also see
R.I. Joseph: J. Math. Phys. *18*, 2251 (1977)
1.172 J. Satsuma, D.J. Kaup: J. Phys. Soc. Jpn. *43*, 692 (1977)
1.173 J.D. Gibbon: "Poles of the Toda lattice", in *Solitons and Condensed Matter Physics*, Springer Series in Solid State Sciences, Vol.8, ed. by A.R. Bishop, T. Schneider (Springer, Berlin, Heidelberg, New York 1978) pp.44-47
1.174 H.P. McKean: Lecture Notes at Calgary (1978)
1.175 E. Trubowitz: Commun. Pure Appl. Math. *30*, 321 (1977)
1.176 D.J. Kaup: SIAM J. Appl. Math. *31*, 121 (1976)
1.177 D.J. Kaup, A.C. Newell: Proc. Roy. Soc. (London) A*361*, 413 (1978)
1.178 A.C. Newell: "Soliton Perturbations and Nonlinear Focussing" in *Solitons and Condensed Matter Physics*, Springer Series in Solid State Sciences, Vol.8, ed. by A.R. Bishop, T. Schneider (Springer, Berlin, Heidelberg, New York 1978) pp.52-67
1.179 A.C. Newell: J. Math. Phys. *18*, 922 (1977)
1.180 A.L. Mason: "Perturbation Theory for the Double sine-Gordon Equation", in *Proc. NATO Advanced Study Institute, Nonlinear Equations in Physics and Mathematics*, Istanbul, August 1977, ed. by A.O. Barut (D. Reidel, Dordrecht, Holland 1978)
1.181 P.W. Kitchenside, A.L. Mason, R.K. Bullough, P.J. Caudrey: "Perturbation Theory of the Double sine-Gordon Equation", in *Solitons and Condensed Matter Physics*, Springer Series in Solid State Sciences, Vol.8, ed. by A.R. Bishop, T. Schneider (Springer, Berlin, Heidelberg, New York 1978) pp.48-51
1.182 D.W. McLaughlin, A.C. Scott: "Soliton Perturbation Theory", in *Nonlinear Evolution Equations Solvable by the Spectral Transform*, ed. by F. Calogero (Pitman, London 1978) pp.225-243; Phys. Rev. A *18*, 1632 (1978)
1.183 V.I. Karpman: "Soliton Perturbation in the Presence of Perturbation", Phys. Scr. *20*, 462 (1979)
1.184 A.C. Newell: "Near integrable systems, nonlinear tunnelling and solitons in slowly changing media", in *Nonlinear Evolution Equations Solvable by the Spectral Transform*, ed. by F. Calogero (Pitman, London 1978) pp.127-179
1.185 L.A. Takhtadjan: Phys. Lett. *64*A, 235 (1977)
1.186 M. Lakshmanan: Phys. Lett. *61*A, 53 (1977)
1.187 S.V. Manakov: Reported at the Warsaw (Jadwisin) Meeting, September, 1977
1.188 M.A. Olshanetsky, A.M. Perelomov: Inventiones Math. *37*, 93 (1976)
1.189 *Proceedings of the Conference on Solitons and Nonlinear Waves* (University of Arizona, Tucson, January 1976) ed. by H. Flaschka, D.W. McLaughlin. Published in Rocky Mount. J. Math. *8*, Nos.1 and 2, Winter and Spring (1978)
1.190 R. Miura (ed.): *Bäcklund Transformations*, Lecture Notes in Mathematics, Vol.*515* (Springer, Berlin, Heidelberg, New York 1976)
1.191 F. Calogero (ed.): *Proc. Symp. Nonlinear Evolution Equations Solvable by the Inverse Spectral Transform*, Accademia dei Lincei, Rome, 1977 (Pitman, London 1978)
1.192 A.O. Barut (ed.): *Proceedings of the NATO Advanced Study Institute on Non-linear Equations in Physics and Mathematics*, Istanbul, August 1977 (D. Reidel, Dordrecht, Holland 1978)
1.193 Chalmers Symp. on "Solitons and their Applications in Science and Technology", Göteborg, June 1978. Articles published in Phys. Scr. *20* (1979)
1.194 K. Lonngren, A. Scott (eds.): *Solitons in Action* (Academic Press, New York 1979)
1.195 R.K. Bullough: "Solitons in Physics, in Ref. [1.192] pp.99-141
1.196 R. Hobart: Proc. Phys. Soc. (London) *82*, 201 (1963)
1.197 G.H. Derrick: J. Math. Phys. *5*, 1252 (1964)
G.H. Derrick, Wan Kay-Kong: J. Math. Phys. *9*, 232 (1968)
1.198 M.J. Ablowitz, R. Haberman: Phys. Rev. Lett. *35*, 1185 (1975); J. Math. Phys. *16*, 2301 (1975)
1.199 H. Cornille: J. Phys. A*11*, 1509 (1978)

1.200 R. Hirota: J. Phys. Soc. Jpn. *33*, 1459 (1972)
J.D. Gibbon, G. Zambotti: Nuovo Cimento *28*B, 1 (1975)
1.201 J.D. Gibbon, N.C. Freeman, A. Davey: J. Phys. A: Math. Gen. *11*, L93 (1978)
1.202 S.V. Manakov, V.E. Zakharov, L.A. Bordag, A.R. Its, V.B. Matveev: Phys. Lett. *63*A, 205 (1977)
1.203 A.A. Belavin, A.M. Polyakov, A.S. Schwartz, Yu.S. Tyupkin: Phys. Lett. *59*B, 85 (1975)
M.F. Atiyah, R.S. Ward: Comm. Math. Phys. *55*, 117 (1977)
M.F. Atiyah, N.J. Hitchin, V.G. Drinfeld, Yu.I. Manin: Phys. Lett. *65*A, 185 (1978)
E. Corrigan, D.B. Fairlie, R.G. Yates, P. Goddard: Comm. Math. Phys. *58*, 233 (1978)
1.204 A.A. Belavin, V.E. Zakharov: Pis'ma Zh. Eksp. Teor. Fiz. (USSR) *25*, 603 (1977) [English transl.: Sov. Phys. JETP Lett. *25*, 567 (1977)]
1.205 D.J. Kaup: Phys. Rev. A*16*, 704 (1977)
1.206 P. Goddard, J. Nuyts, D. Olive: "Gauge Theories and Magnetic Charge", CERN Preprint (1978) (Ref. TH. 2255-CERN)
1.207 B.K. Harrison: Phys. Rev. Lett. *41*, 1197 (1978)
1.208 W. Kinnersley, D.M. Chitre: J. Math. Phys. *18*, 1539 (1977)
1.209 V.A. Belinsky, V.E. Zakharov: Ref.15 in [1.207]. We have been unable to consult this reference
1.210 V.G. Makhanov: Phys. Rep. *35*C, 1 (1978)
1.211 J.C. Eilbeck: "Numerical Studies of Solitons", in *Solitons and Condensed Matter Physics*, Springer Series in Solid State Sciences, Vol.8, ed. by A.R. Bishop, T. Schneider (Springer, Berlin, Heidelberg, New York 1978) pp.28-43
1.212 P.L. Christiansen, O.H. Olsen: Phys. Scr. *20*, 531 (1979)
1.213 H. Leibbrandt: Phys. Rev. Lett. *41*, 435 (1978)
1.214 P.J. McCarthy: Lett. Math. Phys. *2*, 493 (1978)
1.215 P.J. McCarthy: Lett. Math. Phys. *2*, 167 (1977)
1.216 C.N. Yang: Phys. Rev. Lett. *36*, 1377 (1977)
1.217 F. Calogero, A. Degasperis: "Novel class of nonlinear evolution equations solvable by spectral transforms technique, including the so called cylindrical KdV equation", in *Solitons and Condensed Matter Physics*, Springer Series in Solid State Sciences, Vol.8, ed. by A.R. Bishop, T. Schneider (Springer, Berlin, Heidelberg, New York 1978) pp.68-70

A valuable additional reference is:

1.218 L.D. Faddeev, V.E. Korepin: Phys. Rep. *42*C, 1 (1978)

Two recent papers on solitons in statistical mechanics are:

1.219 J.F. Currie, S. Sarker, A.R. Bishop, S.E. Trullinger: Phys. Rev. A*20*, 2213 (1979)
1.220 A.R. Bishop, J.F. Currie, S.E. Trullinger: Adv. in Phys. In press (1980)

Also, in connection with the work of Lax [1.70,114], Matveev, Its, Dubrovin, Novikov, Krichever [1.119,120], and the references and work of Manin [1.118] as described in Sect.1.5 (as well as in connection with the 'systems of commuting differential operators' approach of Zakharov in this volume), Drs. D.V. and G.D. Chudnovsky have drawn our attention to the prescient papers on ordinary differential equations by:

1.221 J.L. Burchnall, T.W. Chaundy: Proc. Lond. Math. Soc. *21*, 420 (1922), Proc. Roy. Soc. (Lond.) *118*, 557 (1928) and a note by H.F. Baker, Proc. Roy. Soc. (Lond.) *118*, 584 (1928).

Apparently Baker first realised the significance of $\theta$-functions in this connection. Mumford (in 1978) realised the relevance of this early work to the much more recent work on soliton theory which had developed independently. Connections and further extensions are established by

1.222 D.V. Chudnovsky: 'Riemann monodromy problem, isomonodromy deformation equations, and completely integrable systems", in*Bifurcation Phenoma in Mathematical Physics and Related Problems*, Proc. NATO Advanced Study Institute, Cargèse, Corsica, June 24 - July 8, 1979, ed. by D. Bessis and C. Bardos (Reidel, Dordrecht, Holland). To be published early 1980

1.223 R.K. Bullough: "Solitons: Inverse Scattering Theory and Its Applications", in *Bifurcation Phenomena in Mathematical Physics and Related Problems*, Proc. NATO Adv. Study Inst., Cargèse, Corsica, June 24 - July 8, 1979, ed. by D. Bessis, C. Bardos (Reidel, Dordrecht, Holland 1980) to be published early 1980

In connection with the papers [1.207, 208, 209] and the Einstein field equations we should also cite

1.224 D. Maison: Phys. Rev. Lett. *41*, 521 (1978) essentially deriving a Lax pair for the Ernst equations, and work reported by B.K. Harrison at the 'Study Group on Solitons, Partial Differential Equations, and Spectral Methods' ICTP, Trieste (July, 1979)

In connection with the footnote p.26 and the NLS with negative selfinteraction (as well as BTs more generally) we cite

1.225 V.S. Gerdzhikov, P.P. Kulish: Teor. Mat. Fiz. *39*, 69 (1979)

An application of soliton theory to meterology is made in

1.226 J.D. Gibbon, I.N. James, I.M. Moroz: Phys. Scripta *20*, 402 (1979) and subsequent papers

1.227 D.J. Kaup: "The Estabrook-Wahlquist method with examples of application", Physica D. To be published (1980)

# 2. Aspects of Soliton Physics

G. L. Lamb, Jr. and D. W. McLaughlin

With 5 Figures

Solitons are nonlinear pulses which result from a balance between nonlinearity and dispersion. In this article examples are used to illustrate some connections between physical interacting systems and the solitons which such systems support. In different instances the interaction of an electric field with an assemblage of two-level atoms is shown to yield the solitons of self-induced transparency and the solitons of a nonlinear Schroedinger equation. One case of an electromagnetic wave interacting with a plasma is discussed which yields the solitons of the Korteweg-de Vries equation. The interaction of a two-level atomic system with an electric field is used to introduce and to motivate the inverse scattering transform. Finally, connections between this transform, higher conservation laws, and Hamiltonian dynamics are discussed.

## 2.1 Historical Remarks and Summary

Research on solitons up to the present time may be divided, like all Gaul, into three parts. A number of summaries of the work associated with these three phases are currently available. In this introduction we shall reference sources where extensive discussions of the three eras can be found.

The first era of soliton research belongs to the last century and was lucidly recorded during that time in the writings of RUSSELL [2.1] and KORTEWEG and DE VRIES [2.2]. A recent review is given by MIURA [2.3]. The hydrodynamic soliton discovered by Russell is a localized water wave travelling at constant speed. It represents an equilibrium between two competing effects. On the one hand, the nonlinearities inherent in the hydrodynamic equations lead to steepening and breaking of waves. On the other hand, the linear description (which is appropriate for small amplitude disturbances) yields waves which experience dispersion. Such dispersion causes a localized disturbance to spread. At amplitudes which are slightly nonlinear, the competition between the two effects of steepening and spreading can intuitively be expected to lead to a balance that gives rise to a steady-state pulse. This is indeed the case, and the resulting steady pulse is the soliton. Such balances have been discussed by KADOMTSEV and KARPMAN [2.4]. A similar interplay between the two competing effects will be considered quantitatively in Sect.2.2.2 where the case of ion-acoustic waves in plasma physics is treated in some detail.

Solitons exhibit far more remarkable behaviour than mere steady state propagation. In their interaction they are found to pass through one another in nearly elastic fashion [2.5], such interaction of solitons was observed by RUSSELL [2.1]. Unlike the steady-state pulse, the details of this nonlinear interaction do not appear to be amenable to any simple intuitive physical interpretation.

Also during this first period of soliton history certain mathematical transformations of nonlinear partial differential equations were being developed, primarily by BÄCKLUND [2.6-8]. These transformations would play a role in the subsequent confluence of various analytical techniques which are now employed to solve those equations which exhibit soliton behaviour.

The second era, which began in the 1940's with the researches of FERMI, PASTA and ULAM (FPU) [2.9] on a problem in lattice dynamics, saw the development of various numerical and transformation techniques. The FPU work represents one of the first successful attempts to use the newly developed computing machines to analyze problems that present a complexity slightly beyond the analytical capabilities of one's time. The importance of computing machines during this second era of soliton history cannot be overestimated. Indeed, the discovery of soliton behaviour took place repeatedly during the analysis of numerical simulations of various physical processes. These developments have been extensively described in the soliton literature, for example [2.10-12].

The latter advances in soliton research were parallelled by independent experimental and theoretical research on the propagation of coherent light pulses in two-level atomic media by McCALL and HAHN [2.13]. Again it was the examination of computer output that showed the development of a steady-state pulse, as well as the decomposition of a pulse into a sequence of smaller ones. In this field soliton behaviour became known as self-induced transparency (SIT) [2.14]. At this time it was recognized [2.15] that the equations arising in a simple model of optical pulse propagation [2.16] had been analyzed by Bäcklund transformation methods [2.17]. Analytical expressions describing decomposition into pure N-soliton shapes were thus provided [2.18]. Simultaneously, formulas for multi-soliton solutions of the Korteweg-de Vries equation were developed by use of inverse scattering techniques. This method led to a solution to the full initial value problem for the Korteweg-de Vries equation [2.19]. In 1971, ZAKHAROV and SHABAT [2.20] generalized the inverse scattering method as it was described by LAX [2.21] and thus provided a basis for a subsequent synthesis of the various equations that were known to exhibit soliton behaviour [2.22-24].

The inverse scattering method converts the nonlinear equations of soliton theory into linear integral equations. Some research is presently directed toward asymptotic solution of these integral equations [2.25-28]. Another exciting area of current interest is the extension of inverse spectral methods to include the study of nonlinear equations with periodic boundary conditions [2.29-31]. Completion of the periodic boundary value problem should provide a culmination to one of the major studies in the second era of soliton history.

The third era has just begun to emerge with extensions to more than one spatial dimension and with the application of soliton equations to areas of science such as nonlinear aspects of solid-state physics and quantum field theory [2.32,33]. These subjects are developing rapidly at the present time, and it seems premature to assess them here. The following sections are concerned solely with certain aspects of the second era of soliton development; and at that, only a few of its many features will be singled out for consideration.

In the numerous references just cited there are many surveys which discuss the general properties of solitons for nonlinear, dispersive wave equations. Another general reference is [2.8]. In this article, rather than repeat these general properties, we shall use examples to illustrate some connections between physical *interacting systems* and the solitons which such systems support. We shall consider several specific coupled systems which, in the absence of coupling, are linear. In Sect.2.2.1 we discuss some examples of the interaction of an electric field with an assemblage of two level atoms. In different instances this interaction will be shown to yield the solitons of self-induced transparency and the solitons of a nonlinear Schroedinger equation. In Sect.2.2.2 we discuss one case of an electromagnetic wave interacting with a plasma, a case which yields the solitons of the Korteweg-de Vries equation. In Sect.2.3 we use the interaction of a two level atomic system with an electric field to introduce and to motivate the inverse scattering transform. Finally, we discuss connections between this transform and the "higher conservation laws" which, as conserved quantities, played a useful role in the early development of the Korteweg-de Vries equation and of the equation of self-induced transparency.

## 2.2 Model Interacting Systems

The balance of dispersion with weak nonlinearity is a general physical process, one which arises in a wide variety of diverse physical applications [2.5]. The physical situations which yield these balances fall into a few categories, each of which can be loosely characterized by rather general physical principles [2.10,34]. Thus, the dynamics of diverse physical systems can be described by a relatively short list of equations. Although the classifications which define these categories may appear rather vague, they cover most of the nonlinear, dispersive, conservative systems which are discussed in the current literature.

To illustrate typical categories, we consider a conservative, dispersive weakly nonlinear wave which consists of a rapidly oscillating carrier with slowly varying amplitude and phase. These slow modulations can be described by some *nonlinear Schroedinger* equation provided I) the carrier is very oscillatory, II) there are no resonance effects in the system, and III) the response to the nonlinearity is over space and time distances which are long compared with the wavelength and frequency of the carrier. In the presence of resonance effects, the modeling is more difficult, for example, if the resonance is between the carrier and the medium supporting the wave, some dynamics of the medium must be included in the equations. Frequently the microscopic medium is described quantum mechanically, a description which yields some form of the coupled *Maxwell-Schroedinger equations* such as the Maxwell-Bloch equations. Finally, if there is more than one carrier wave, an appropriate *N-wave generalization* to one of the classes must be adopted [2.35,36].

In contrast with a rapidly oscillating carrier, another common starting point is the conservation laws of fluid mechanics. In this case, the circumstances of weak nonlinearity balancing weak dispersion yield a *Korteweg-de Vries equation*.

In order to illustrate the manner in which these various categories arise, we shall consider several very specific models of interacting systems and apply approximations which reduce the "exact systems" to members of the different categories. Such reduction procedures provide us with a concrete feeling about the type of approximations which characterize the various classes. In Sect.2.2.1, we describe interactions of an electric field with a two-level medium, and consider one case which is near resonance and another which is far from resonance. As one example from fluid mechanics, we describe a special case of the interaction of an electric field with the ions of a plasma.

### 2.2.1 The Interaction of an Electromagnetic Field with a Two-Level Medium

*General Framework*

Consider a highly oscillatory electromagnetic wave propagating through some medium. A very simple idealized model which describes such propagation is the interaction of a classical electric field with an assemblage of two-level quantum oscillators. The dynamics of each oscillator is described by Schroedinger's equation, which for two-level atoms [2.37-39] takes the form

$$i\hbar\partial_t\begin{pmatrix}\alpha\\ \beta\end{pmatrix}=\begin{pmatrix}\frac{1}{2}\hbar\omega & -\underline{q}\cdot\underline{E}\\ -\underline{q}\cdot\underline{E} & -\frac{1}{2}\hbar\omega\end{pmatrix}\begin{pmatrix}\alpha\\ \beta\end{pmatrix} \quad . \tag{2.1}$$

---

$\hbar = h/2\pi$ (normalized Planck's constant)

Here $\alpha$ and $\beta$ denote the probability amplitudes for the upper and lower states, respectively. These two states are separated by energy $\hbar$. Each oscillator is coupled to the electromagnetic wave $\underline{E}$ through dipole coupling, with the parameter $\underline{q}$ representing the dipole matrix element.

The $\underline{E}$ field is essentially a plane wave oriented in the z direction and propagating in the x direction. It is governed by Maxwell's equations in the form

$$(\partial_{tt} - c^2\partial_{xx})\underline{E} = -4\pi\partial_{tt}\underline{\mathscr{P}} \quad . \tag{2.2}$$

Here $\underline{\mathscr{P}}$ denotes the macroscopic polarization induced in the medium by the action of the $\underline{E}$ field on the two-level oscillators.

Consider an oscillator located near the plane $x = X$, and being driven by the field $\underline{E}(X,t)$ according to (2.1). This interaction polarizes the oscillator by creating a microscopic polarization $\underline{P} \equiv \underline{q}(\alpha\beta^* + \alpha^*\beta)$. The macroscopic polarization $\underline{\mathscr{P}}$ is given in terms of $\underline{P}$ by

$$\underline{\mathscr{P}}(x,t) = n_0 \quad \langle\underline{P}(x,t)\rangle \ , \quad \underline{P} = \underline{q}(\alpha\beta^* + \alpha^*\beta) \quad , \tag{2.3}$$

where $n_0$ is the density of the resonant atomic systems and $\langle.\rangle$ represents an averaging process which is to be considered as a map which converts microscopic dipoles near x into a macroscopic polarization at x. The precise specification of $\langle.\rangle$ depends upon the physical situation at hand. Later in this section we will consider three different choices for $\langle.\rangle$.

Equations (2.1) and (2.2) together with definition (2.3) constitute a closed system of equations to be solved for $(\alpha,\beta,\underline{E})$. Our primary concern is to calculate the modulation in amplitude and phase for an "almost plane wave" $\underline{E}$ due to its interaction with the two-level medium.

Since it is the polarization $\underline{P} \equiv \underline{q}(\alpha\beta^* + \alpha^*\beta)$ which drives the $\underline{E}$ field, it seems convenient to introduce new variables which explicitly contain the combination $(\alpha\beta^* + \alpha^*\beta)$

$$T \equiv \alpha\alpha^* + \beta\beta^* \qquad N \equiv \alpha\alpha^* - \beta\beta^*$$

$$P_+ \equiv \alpha\beta^* + \alpha^*\beta \ ; \qquad P_- \equiv i(\alpha\beta^* - \alpha^*\beta) \quad . \tag{2.4}$$

In terms of these variables the interacting system takes the form

$$\partial_t T = 0$$

$$\partial_t N = - \frac{2}{\hbar}(\underline{q} \cdot \underline{E})P_-$$

$$\partial_t P_+ = -\omega P_-$$

$$\partial_t P_- = \omega P_+ + \frac{2}{\hbar}(\underline{q} \cdot \underline{E})P_-$$

$$(\partial_{tt}P_- - c^2\partial_{xx})\underline{E} = 4\pi n_0 \quad \langle\omega^2 P_+\underline{q} + \omega(\underline{q} \cdot \underline{E})N\underline{q}\rangle \tag{2.5}$$

where we have used

$$\partial_{tt}P_+ = -\omega^2 P_+ - \frac{2\omega}{\hbar}(\underline{q} \cdot \underline{E})N$$

which follows from (2.5). This "exact" interacting system is rich enough to model many physical situations and is called the coupled *Maxwell-Bloch system*.

First, we consider the case when the $\underline{E}$ field and the medium are nearly at resonance. That is, the carrier frequency of the $\underline{E}$ field ($\omega_c$) is chosen to be near the oscillator frequency $\omega$. Further, we assume that this frequency is quite large, so that in Maxwell's equation we need only keep the resonance term $\omega^2 P_+$ and may neglect the other term $(\omega(\underline{q} \cdot \underline{E})N)/\hbar$.

Next, we select a micro to macro map <.>. Physically each quantum oscillator could not be driven exactly at resonance because each oscillator has some motion of its own center of mass. This motion introduces a Doppler frequency shift in the laboratory frame and cause the collection of oscillators to have an essentially continuous distribution of transition frequencies. We assume that this distribution of frequencies is peaked at the carrier frequency $\omega_c$, and we model its spread through a probability density function $g(\Delta\omega)$, $\omega = \omega_c + \Delta\omega$. In terms of this density g, the micro to macro map <.> is defined by

$$\langle f\rangle \equiv \int_{-\infty}^{\infty} f(\Delta\omega)g(\Delta\omega)d(\Delta\omega) \quad . \tag{2.6}$$

With the prescription for <.> and this nearly resonant choice of carrier frequency $\omega_c$, we seek an approximate solution of the Maxwell-Bloch equations (2.5). First, notice that in the absence of induced polarization the medium is essentially a vacuum in which case the carrier frequency $\omega_c$ and wave number $k_c$ would satisfy $\omega_c = ck_c$. In this model all dispersion is introduced through the interaction which yields an induced polarization. We assume this polarization is weak (as it is in most applications). This assumption leads us to seek a solution of the Maxwell-Bloch equations in the form

$$E = \frac{\hbar}{q}\,\mathcal{E}(x,t)\cos[k_c x - \omega_c t + \phi(x,t)]$$

$$N = \mathcal{N}(t;x,\Delta\omega)$$

$$P_+ = \mathcal{Q}(t;x,\Delta\omega)\cos[k_c x - \omega_c t + \phi(x,t)] + P(t;x,\Delta\omega)\sin[k_c x - \omega_c t + \phi(x,t)] \quad (2.7)$$

where $\omega_c = ck_c$ and where the amplitudes $(E,N,P,\mathcal{Q})$ and the phase $\phi$ vary slowly on the $\omega_c t$ and $k_c x$ scales. Equations which govern the slow modulations may be obtained by inserting ansatz (2.7) into the Maxwell-Bloch system, while neglecting all second derivatives of the envelopes as well as any higher harmonics. The result yields an interacting system for the Maxwell-Bloch equations in the slowly varying envelope and phase approximation

$$\frac{\partial E}{\partial t} + c\frac{\partial E}{\partial x} = c\alpha'\langle P(\Delta\omega,x,t)\rangle$$

$$E\left(\frac{\partial \phi}{\partial t} + c\frac{\partial \phi}{\partial x}\right) = c\alpha'\langle \mathcal{Q}(\Delta\omega,x,t)\rangle$$

$$\frac{\partial P}{\partial t} = EN + \left(\Delta\omega + \frac{\partial}{\partial t}\phi\right)\mathcal{Q}$$

$$\frac{\partial N}{\partial t} = -EP$$

$$\frac{\partial \mathcal{Q}}{\partial t} = -\left(\Delta\omega + \frac{\partial}{\partial t}\phi\right)P \quad (2.8)$$

Here $\alpha' = 2\pi n_0 \omega_c q^2/\hbar c$ where $n_0$ is the density of the resonant atomic systems. In the next section we discuss some of the physical information contained in these equations.

Case I. *Self-Induced Transparency*

As noted in the introduction, one of the early observations of soliton behaviour was in the propagation of coherent optical pulses [2.13,15]. Such resonant interactions between light and matter lead to novel propagation effects that are now known to be characteristic of solitons. Perhaps the most spectacular of these effects is that of self-induced transparency. In this effect, the leading edge of the pulse is used to invert an atomic population while the trailing edge returns the population to its initial state by means of stimulated emission. The process is realizable if it takes place in a time that is short compared to the phase memory time of the resonant atomic systems (i.e., if the pulses are ultrashort), and also if the pulse has sufficient intensity to effect a population inversion. When conditions for the process are met, it is found that a steady-state pulse profile is established and that this pulse then propagates without attenuation at a velocity that may be as much as three orders of magnitudes less than the phase velocity of light in the medium. Pulses with intensity below the threshold required for this process merely attenuate in the usual manner. Within the theoretical framework that has been used to describe this effect, it has been shown that the above-mentioned steady-state propagation takes place after the profile of the electric field has evolved to the soliton wave form.

Equations (2.8) provide a theoretical basis for the analysis of self-induced transparency. One may make satisfactory contact with many experimental results by considering a version of the equations in which phase variation $\phi$ is ignored. In this case, a steady-state solution is found to be [2.13]

$$E = \frac{2}{\tau_p} \operatorname{sech}[(t - x/v)/\tau_p] \quad , \tag{2.9}$$

where the constant $\tau_p$ is related to the pulse half-width and the velocity v is defined by

$$\frac{1}{v} \equiv \frac{1}{c} + \alpha' \tau_p^2 \int_{-\infty}^{\infty} \frac{g(\Delta\omega)}{1+(\tau_p\Delta\omega)^2} d(\Delta\omega) \quad . \tag{2.10}$$

As noted above, early experiments established that this pulse velocity v may be two or three orders of magnitude lower than the phase velocity. Numerical values indicative of these experimental results [2.40-42] are $\omega_c \sim 10^{15}\ s^{-1}$, $n_0 \sim 10^{12} cm^{-3}$, $q \sim 6 \times 10^{-18}$, $\tau_p \sim 7 \times 10^{-9}$ s. Then $\alpha' c \sim 2 \times 10^{20}\ s^{-2}$ and from (2.10)

$$\frac{c}{v} \sim 3000 \quad . \tag{2.11}$$

Hence, in this experiment the pulse velocity is actually three orders of magnitude less than the phase velocity c.

The "area" $\theta$ under the pulse profile given in (2.9) is readily shown by direct integration to be $2\pi$, i.e.,

$$\theta(x) = \int_{-\infty}^{\infty} E(x,t)dt = 2\pi \quad . \tag{2.12}$$

As a result, it has become customary to refer to this pulse as a $2\pi$ pulse. Equation (2.9) for the $2\pi$ pulse is the simplest example of the optical soliton. Computer solutions for the evolution of pulses with initial area between $\pi$ and $3\pi$ were found to evolve into a $2\pi$ pulse of this soliton shape. Pulses smaller than $\pi$ decayed to zero area (as would be expected of a pulse propagating in an attenuator). On the other hand, pulses larger than $3\pi$ (but less than $5\pi$) were found to decompose into two $2\pi$ pulses. The relative heights of the two pulses were dependent upon the shape of the initial pulse. Pulses with initial area between $5\pi$ and $7\pi$ decomposed into three pulses, etc. Such pulse decomposition is characteristic of soliton behaviour.

In the present problem certain aspects of pulse decomposition are readily systematized by employing the so-called area theorem. The equation governing the area $\theta(x)$ is readily obtained by integrating (2.8) over all times. One finds

$$\frac{d\theta}{dx} = \pm \frac{\alpha}{2} \sin\theta \quad , \tag{2.13}$$

where

$$\alpha = 2\pi g(0)\alpha' \quad , \tag{2.14}$$

and the upper sign is used if the atomic population is initially inverted (an amplifier), while the lower is used if the population is initially in the lower level (an attenuator). The constant $\alpha'$ is defined below (2.8).

Equation (2.13) is known as the area theorem [2.13,14,38]. It is a key to understanding some of the effects which occur in the propagation of ultrashort pulses. First, one may see the physical significance of $\alpha$ from the linearized version of (2.13) in which sin is replaced by $\theta$. The field is then seen to amplify or decay in the characteristic length $\alpha^{-1}$.

Next, consider the exact solution of (2.13) which satisfies $\theta = \theta_0$ at $x = x_0$,

$$\tan \frac{\theta}{2} = \tan \frac{\theta_0}{2} \exp\left[\pm \frac{\alpha}{2} (x - x_0)\right] \quad . \tag{2.15}$$

Its behaviour is depicted schematically by Fig.2.1. Since (2.13) contains a choice of signs, it is actually two distinct differential equations. The two solutions are obtained from Fig.2.1 by reading the diagram from the right to left for the plus sign (amplifier) and from left to right for the minus sign (attenuator). Hence one sees that an infinitesimal area will grow to $\pi$ in an amplifier while any area less than $\pi$ will evolve to zero in an attenuator. The second result not only permits the well-known decay of a pulse as it propagates in an attenuator, but also allows for the evolution into a nonvanishing zero-$\pi$ pulse, i.e., one in which the total area under the pulse envelope is zero but the area under the pulse ($\sim E^2$) is not zero. (This is possible if the positive portions of a pulse envelope are equal in area to the negative portions. Physically, the regions of positive and negative envelope are merely regions in which there is a relative difference of $180^\circ$ in the phase of the carrier wave.) In an attenuator, initial pulse areas between $\pi$ and $3\pi$ will evolve into a steady-state $2\pi$ pulse of self-induced transparency. On the other hand, the $2\pi$ pulse is unstable in an amplifier and will evolve into either a $\pi$ or $3\pi$ pulse. We emphasize that Fig.2.1, which summarizes the content of the area theorem, refers only to the total area of a pulse and gives no information about the exact pulse shape. For example, the area theorem does not predict the possible breakup of a pulse into several pulses with the same total area.

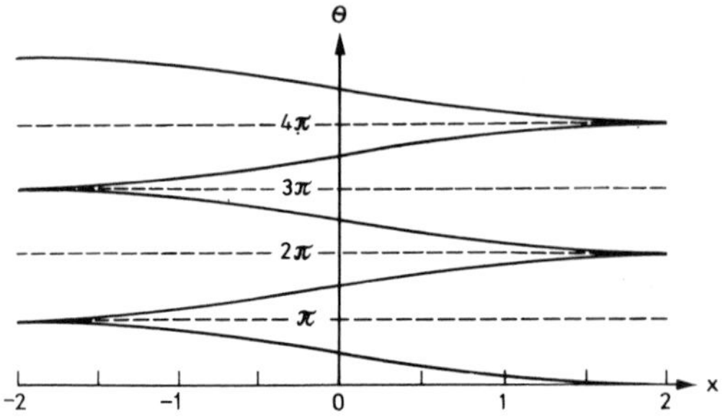


Fig.2.1. Area Theorem — graph of (2.15)

Analytical expressions for the various $2N\pi$ pulses have been obtained by a variety of methods. Results for $4\pi$ and zero $\pi$ pulses were first obtained by using a technique developed by Bargmann to construct phase equivalent potentials in quantum scattering theory. Since the Bargmann potential provides a class of particularly simple solutions to the integral equations of inverse scattering theory, it was inevitable that these results would be recovered in this manner as well. Various applications of the inverse method have since been developed for this purpose [2.18,43]. A discussion of inverse methods will be given in Sect.2.3.

Case II. *Sine-Gordon (Resonant Without Broadening)*

While (2.8) provides a fairly complete description of coherent optical pulse propagation in an inhomogeneously broadened medium, the essential nonlinearities of the propagation process may be exposed in an extremely simple manner if Doppler broadening as well as phase variation are ignored [i.e., one assumes $g(\Delta\omega) = \delta(\Delta\omega)$ and sets $\phi = 0$]. In this limit, one obtained

$$\frac{\partial E}{\partial \xi} = P \tag{2.16}$$

$$\frac{\partial P}{\partial \tau} = EN \tag{2.17}$$

$$\frac{\partial N}{\partial \tau} = -EP \quad , \tag{2.18}$$

where $\tau = t - x/c$ and $\xi = \alpha' x$.

These equations possess the first integral $P^2 + N^2 = 1$, the constant of integration being chosen from the initial condition $N(-\xi,-\infty) = 1$. As in (2.13), the upper sign refers to an amplifier, the lower sign to an attenuator. In terms of the parametric representation,

$$P = \pm \sin\sigma$$
$$N = \pm \cos\sigma \tag{2.19}$$

one finds that

$$E = \frac{\partial\sigma}{\partial\tau} \tag{2.20}$$

and

$$\frac{\partial\sigma}{\partial\xi\partial\tau} = \pm \sin\sigma \quad . \tag{2.21}$$

Hence, in the limit of no Doppler broadening, the coupled Maxwell-Bloch equations may be reduced to a single nonlinear partial differential equation that has come to be known as the *sine-Gordon* equation. This equation arose in early research in differential geometry and many particular solutions of it are known. Before taking up such considerations, we first consider its steady-state solution. Although this solution gives no indication of the effects which take place during interactions among solitons, it is a solution of prime concern in experimental considerations of coherent pulse propagation. A solution depending only on the combination $(a\tau - \xi/a)$ is seen to satisfy

$$\sigma'' = \sin\sigma \quad , \tag{2.22}$$

where prime indicates differentiation with respect to $a\tau - \xi/a$ and only the lower sign in (2.21) has been retained. (The steady-state pulse is stable only in the attenuator.) Although other solutions of (2.22) could be considered, the single soliton

$$\sigma = 4 \tan^{-1} \exp(a\tau - \xi/a) \tag{2.23}$$

is the solution of most importance in the pulse propagation problems. Since $E = \sigma_\tau$ one finds

$$E(\xi,\tau) = 2a \operatorname{sech}(a\tau - \xi/a) \quad . \tag{2.24}$$

The argument of the sech function may be written as $a(t - x/v)$ where

$$\frac{1}{v} = \frac{1}{c} - \frac{\alpha'}{a^2} \tag{2.25}$$

which agrees with (2.10) in the limit of no inhomogeneous broadening. At this point we note that the area under this pulse is

$$\int_{-\infty}^{\infty} d\tau E(\xi,\tau) = 2\pi \tag{2.26}$$

so that in this model the sech pulse is again a $2\pi$ pulse. (Equivalently, one notes that $\theta$ turns through an angle of $2\pi$ as the pulse passes a given point in space.)

This hyperbolic secant pulse profile is typical of the results obtained in soliton propagation. However, the method by which it was obtained here gives no clue to the more distinctive features of soliton propagation. These features manifest themselves in the interaction among such pulses as they propagate through each other.

In order to discuss pulse interaction, one requires a more elaborate technique for obtaining solutions of the sine-Gordon equation (2.21). Much consideration has been given to the construction of such solutions in recent years. Here we shall consider only the oldest method, namely the use of Bäcklund transformations. Equation (2.21) arose in classical differential geometry [2.44], and this method for obtaining particular solutions was developed before the turn of the century.

The Bäcklund transformation may be interpreted as a mapping of one solution of a given equation (here the sine-Gordon equation) into another solution of the same equation. The transformation is given by the equations

$$\frac{\partial}{\partial\tau}\left(\frac{\sigma_1-\sigma_0}{2}\right) = -\, a \sin\left(\frac{\sigma_1+\sigma_0}{2}\right) \tag{2.27}$$

$$\frac{\partial}{\partial\xi}\left(\frac{\sigma_1+\sigma_0}{2}\right) = \frac{1}{a} \sin\left(\frac{\sigma_1-\sigma_0}{2}\right) \quad . \tag{2.28}$$

These equations may be derived by using a technique devised by CLAIRIN [2.45] as shown in [2.46].

One may easily show that both $\sigma_0$ and $\sigma_1$ satisfy the sine-Gordon equation. Hence, from a given solution $\sigma_0$, one may obtain a new solution $\sigma_1$. The transformation may be used repeatedly to generate a solution $\sigma_2$ from $\sigma_1$, etc. The usefulness of the Bäcklund technique is enhanced by the fact that it generates explicitly an infinite sequence of multisoliton solutions.

The first step in this process of generating solutions is known as the theorem of permutability which is the relation

$$\tan\frac{\sigma_3-\sigma_0}{4} = \frac{a_1+a_2}{a_1-a_2} \tan\frac{\sigma_1-\sigma_2}{4} \quad . \tag{2.29}$$

Here $\sigma_1$ and $\sigma_2$ are particular solutions generated from $\sigma_0$ by (2.27,28) with $a_1$ and $a_2$ the associated transformation constants. The new solution is $\sigma_3$. For example, when $\sigma_0$ is chosen to be zero, $\sigma_1$ and $\sigma_2$ are of the single soliton wave forms of (2.23), and (2.29) yields

$$\sigma_3 = 4 \tan^{-1} \left[\left(\frac{a_1+a_2}{a_1-a_2}\right) \frac{\sinh \frac{1}{2}(v_1-v_2)}{\cosh \frac{1}{2}(v_1+v_2)}\right] \tag{2.30}$$

in which

$$v_i = a_i\tau - \xi/a_i \ , \quad i = 1,2 \quad . \tag{2.31}$$

The associated electric field may be put in the form

$$E = A \frac{a_1 \text{ sech } X - a_2 \text{ sech } Y}{1-B(\tanh X \tanh Y - \text{sech } X \text{ sech } Y)} \tag{2.32}$$

where

$$A = \frac{a_1^2-a_2^2}{a_1^2+a_2^2} \tag{2.33a}$$

$$B = \frac{2a_1a_2}{a_1^2+a_2^2} \tag{2.33b}$$

$$X = a_1(t - x/v_1) \tag{2.33c}$$

$$Y = a_2(t - x/v_2) \quad . \tag{2.33d}$$

The Bäcklund transformation technique has been used to describe the interaction of three pulses [2.18] and also six pulses [2.47].

*A System Which is Far From Resonance (Nonlinear Schroedinger Equation)*

After considering an electric field nearly at resonance with the two-level oscillators, we turn to a case where the field and the oscillators are far from resonance. Typically, the physical situation could begin with a medium which consists of a collection of long thin cigar shaped molecules. Initially, these molecules are oriented randomly due to "thermal agitation". If each molecule is polarizable, an intense electric field of high frequency will 1) induce a polarization in each molecule and 2) orient the microscopic dipoles just created. Although thermal agitation competes with this orientation process, the net result is still an induced macroscopic polarization of the medium. Such "Kerr effects" [2.40-42,48], cause the field envelope to self focus, self modulate, etc.

A collection of two-level oscillators can constitute a crude model of such a medium, provided each oscillator is labeled by a unit vector $\hat{\underline{p}}$ (which describes the orientation of this "cigar shaped molecule") in addition to its position label

x. Furthermore, we assume that for an oscillator oriented in the $\hat{\underline{p}}$ direction, the induced microscopic dipole moment is entirely in the $\hat{\underline{p}}$ direction. With these provisions the Maxwell-Bloch system (2.5) still describes the dynamics for $N(t;x,\hat{\underline{p}})$, $P_{\pm}(t;x,\hat{\underline{p}})$, and $E(x,t)$.

In the next few paragraphs we obtain an approximation to the Maxwell-Bloch system which is appropriate to the physical situation at hand. The first step in the approximation scheme is the calculation of a nonresonant response of a two level "molecule" to a rapidly oscillating electric field. This response induces microscopic dipoles, and the next step is the definition of an appropriate map <.> which converts these induced diples into a macroscopic polarization. Finally we calculate the modulation in the electric field amplitude, a modulation which is caused by this induced polarization. In this derivation we make no attempt to keep track of all the constants; rather, at each step we incorporate them into a constant G.

First, we assume that, under the averaging process, any components of the microscopic dipoles which are transverse to the applied field will average to zero. That is, we assume that the induced *macroscopic* polarization will be collinear with the applied field, so $\underline{E}$, initially in the $\hat{\underline{z}}$ direction, will remain so. Next, for simplicity we set the population inversion N at -1, and reduce the Maxwell-Bloch equations to the simpler form

$$\partial_{tt}P_{+} = -\omega^{2}P_{+} + \frac{2\omega q}{\hbar}(\cos\theta)E \tag{2.34a}$$

$$(\partial_{tt} - c^{2}\partial_{xx})E = G\langle q\omega^{2}(\cos\theta)P_{+} - \omega q^{2}(\cos^{2}\theta)E\rangle \quad , \tag{2.34b}$$

where $\underline{E} = E\hat{\underline{z}}$, $q = |\underline{q}|$, and $\cos\theta = \hat{\underline{q}} \cdot \hat{\underline{z}}$. System (2.34) is equivalent to a collection of classical harmonic oscillators interacting with an electric field, a model which forms the basis for the Lorentz theory of dispersion. (A more accurate treatment of the population inversion N would merely alter the final coefficient of nonlinearity. Recall that in the case of self-induced transparency, the resonance forced rather careful consideration of the dynamics of the medium. Here, since the system is far from resonance, the precise model of the medium is far less important. For qualitative behaviour, linear oscillators will suffice.)

We consider an E field of the form

$$E = E_{c}\cos(k_{c}x - \bar{\omega}_{c}t) + E_{s}\sin(k_{c}x - \bar{\omega}_{c}t) \quad , \tag{2.35}$$

where the amplitudes ($E_c$ and $E_s$) are slowly varying functions (on the $k_c x$ and $\bar{\omega}_c t$ scales of the carrier wave). Here $\bar{\omega}_c = \bar{\omega}(k_c)$ is the dispersion relation of the entire linearized system and will be computed later. In order to compute the induced dipole $P_+$, we treat the slowly varying amplitudes $E_c$ and $E_s$ as constants in

(2.34) and, since the E field is not at resonance with the medium, find

$$P_{+} \simeq \frac{2\omega q \cos\theta}{\hbar(\omega^{2}-\bar{\omega}_{c}^{2})} E \quad . \tag{2.36}$$

Inserting this approximation into (2.34b) yields an equation for E

$$(\partial_{tt} - c^{2}\partial_{xx})E = G\langle(\cos^{2}\theta)E\rangle \quad . \tag{2.37}$$

Next, we specify the micro to macro map <.> which in this model is the source of the nonlinearity. Recall that the collection of dipoles, which have been induced by the electric field, will also be oriented by the field. Thermal agitation will compete with the orientation process. We assume thermal equilibrium has been reached and adopt the techniques of GUSTAFSON et al. [2.49] to compute the average polarization. The energy u for one of these induced dipoles in an E field is given by

$$u = -\underline{P}_{+} \cdot \underline{E} = G \cos^{2}\theta E^{2} \quad .$$

If one considers situations in which the orientational process occurs on a slow time scale (compared with the $\bar{\omega}_{c}t$ scale), it is appropriate to average u over one period T of the fast oscillation,

$$\bar{u} = \frac{1}{T}\int_{t}^{t+T} dt' u(x,t') = \frac{1}{2} G \cos^{2}\theta (E_{c}^{2} + E_{s}^{2}) \quad .$$

Using $\bar{u}$, we define the map <.> through the Boltzmann distribution function

$$\langle f(\cdot)\rangle = \frac{\int_{-1}^{1} f(\theta)\exp[-\bar{u}(\theta)]d(\cos\theta)}{\int_{-1}^{1} \exp[-\bar{u}(\theta)]d(\cos\theta)} \quad . \tag{2.38}$$

This definition of the map <.> yields the macroscopic field equation

$$(\partial_{tt} - c^{2}\partial_{xx})E = G\langle\cos^{2}(\cdot)\rangle E \quad , \tag{2.39}$$

where

$$\langle\cos^{2}(\cdot)\rangle = \frac{\int_{-1}^{1} \cos^{2}\theta[\exp - \frac{G}{2}\cos^{2}\theta(E_{c}^{2}+E_{s}^{2})]d(\cos\theta)}{\int_{-1}^{1} \exp[-\frac{G}{2}\cos^{2}\theta(E_{c}^{2}+E_{s}^{2})]d(\cos\theta)} \quad .$$

For moderate field amplitudes, one can expand this exponential and place $\langle\cos^2(\cdot)\rangle$ in the form

$$\langle\cos^2(\cdot)\rangle \simeq \frac{1}{3} + G(E_c^2 + E_s^2) \quad .$$

It remains to compute the dispersion relation $\bar{\omega}$ and to find an equation which governs the modulations of the envelopes $E_c$ and $E_s$. In complex notation the field equation (2.39) takes the form

$$[(\partial_{tt} - c^2\partial_{xx}) + G_1 - G_2(FF^*)]F = 0 \quad , \tag{2.40}$$

where $F = F\exp[i(k_c x - \bar{\omega}_c t)]$ and $F = (E_s + iE_c)$. From (2.40) one sees immediately that the dispersion relation, complete with nonlinear correction, is given by

$$\bar{\omega}^2(k) = c^2k^2 + G_1(k) - [G_2(k)]FF^* \quad . \tag{2.41}$$

The final step in the approximation scheme is to reduce (2.40) to an equation which describes modulations in $F$. For this reduction we assume the nonlinearity is weak. If it were missing altogether ($G_2 = 0$), the linear field $F = F\exp[i(k_c x - \bar{\omega}_c t)]$ could be represented as the Fourier integral.

$$F(x,t) = F(x,t)e^{i[k_c x - \bar{\omega}_c t]} = \int dk\, e^{i[kx - \bar{\omega}(k)t]}\tilde{F}(k) \quad ,$$

or equivalently, after changing variables of integration from k to $\kappa = k - k_c$,

$$F(x,t) = \int d\kappa \exp(i\{\kappa x - [\bar{\omega}(k_c + \kappa) - \bar{\omega}_c]t\})\tilde{F}(k_c + \kappa) \quad .$$

Since the amplitude $F(x,t)$ is a slowly varying function of x and t, $\tilde{F}(k_c + \kappa)$ is sharply peaked about $\kappa = 0$

$$F(x,t) \simeq \int d\kappa \exp\left\{i\left[\kappa x - \left(\bar{\omega}_c'\kappa + \frac{\bar{\omega}_c''}{2}\kappa^2 + \ldots\right)t\right]\right\}F(k_c + \kappa) \quad .$$

From this formula it is apparent that the envelope F must satisfy

$$\left[i(\partial_t + \bar{\omega}_c'\partial_x) + \frac{\bar{\omega}_c''}{2}\partial_{xx} + \ldots\right]F = 0 \quad . \tag{2.42}$$

Equation (2.42) is actually a general result [2.4]. *Provided resonances are absent, the amplitude modulations of any linear dispersive wave, which consist of a rapidly oscillating carrier with a slowly varying envelope, are governed by a linear Schroedinger equation.* Since our system is weakly nonlinear, the nonlinear correction to the dispersion relation (2.41) yields

$$i(\partial_t + \bar{\omega}_c'\partial_x) + \frac{\bar{\omega}_c''}{2}\partial_{xx} F + G(FF^*)F = 0 \quad . \tag{2.43}$$

It must be emphasized that the nonlinear Schroedinger equation has far wider applicability than this model indicates. It applies to any weakly nonlinear wave which satisfies the conditions summarized in the last paragraph. Because of its generality, it arises in diverse physical situations such as propagation of optical pulses [2.50], waves in water [2.51] and waves in plasmas [2.52]. It has current technological importance in that it models self-focussing effects in laser pulses [2.50].

In order to get some feeling for the solitons of the Schroedinger equation, consider a traveling wave solution of the form

$$F(x,t) = f(x - vt)\exp[i(Kx - \Omega t)] \quad .$$

Inserting this form of $F$ into the nonlinear Schroedinger equation (2.43) yields

$$f'' = -\frac{\partial}{\partial f}\left[\frac{G}{2\bar{\omega}_c''} f^4 + \frac{(\Omega - K\bar{\omega}_c' - K^2\bar{\omega}_c''/2}{\bar{\omega}_c''}\right] f^2 \tag{2.44}$$

provided the pulse velocity v satisfies

$$v = \bar{\omega}_c' + K\bar{\omega}_c'' \quad . \tag{2.45}$$

If $(G/\bar{\omega}_c'') > 0$, a solution of (2.44) is given by

$$F = F_s \operatorname{sech}\left(\frac{t - x/v}{\tau}\right) \exp[i(\kappa x - \Omega t)] \quad ,$$

with v given by (2.45), $\Omega = K\bar{\omega}_c' + K^2\bar{\omega}_c''/2 - GF_s^2/2$, and $F_s^2 = \bar{\omega}_c''/Gv^2\tau^2$. In this case the field envelope assumes a sech shape with amplitude $F_s$ and half-width $\tau$. This pulse represents an intense, localized field propagating at constant velocity v, with its carrier frequency and wavelength shifted by $\Omega$ and K, respectively.

On the other hand, if $(G/\bar{\omega}_c'') < 0$, the amplitude $F$ takes the form

$$F = F_s \tanh\left(\frac{t - x/v}{\tau}\right) \exp[i(Kx - \Omega t)] \quad ,$$

with v given by (2.45), $\Omega = K\bar{\omega}_c' + K^2\bar{\omega}_c''/2 - GF_s^2$, and $F_s^2 = -\bar{\omega}_c''/Gv^2\tau^2$. In this case the field is expelled from a region, and a dark spot of half-width $\tau$ is formed which travels at a constant speed v [2.53].

In both cases [$G/\bar{\omega}_c'' > 0$ and $G/\bar{\omega}_c'' < 0$] this equation also has multiple soliton solutions. Both cases have been solved by the inverse method [2.23,54].

The nonlinear Schroedinger equation is very important and it is difficult to restrict oneself to a few remarks about it. We only intend to indicate how the equation can arise from a coupled system in which the interaction introduces both the dispersion and the nonlinearity. In the next section we consider an example of a different type of interacting system, that of an electric field interacting with a fluid of ions.

### 2.2.2 Solitons in Plasma Physics

Solitons have arisen in a number of areas of plasma physics. Foremost of these are hydromagnetic waves [2.55-58] and ion acoustic waves [2.55,59-61]. As in the more familiar example of shallow water waves studied originally by Kortweg-de Vries, the soliton in plasma physics arises from a balance between self-steepening effects associated with nonlinearities and the spreading out of a disturbance by dispersion.

Due to space limitations, only the ion-acoustic [2.59] wave will be treated in detail here. The situation envisaged is one in which the ions execute low frequency oscillations near the ion plasma frequency. The electrons follow the ion motion and preserve an approximate charge neutrality. The governing equations are [2.62].

$$\frac{\partial n_i}{\partial t} + \nabla \cdot n_i \underline{v}_i = 0 \qquad \text{(conservation of ions)} \qquad (2.46)$$

where $n_i$ and $\underline{v}_i$ are the ion density and average velocity, respectively.

$$n_e m_e \frac{d\underline{v}_e}{dt} = - n_e e \quad - \nabla p_e \qquad \text{(conservation of electron momentum)} \qquad (2.47)$$

$$n_i m_i \frac{d\underline{v}_i}{dt} = n_i Z\underline{E} - \nabla p_i \qquad \text{(conservation of ion momentum)} \qquad (2.48)$$

where Ze is the ion charge, $m_i$ and $m_e$ are ion and electron masses, respectively, $p_i$ and $p_e$ are pressure terms and $\underline{E}$ is the electric field set up by the lack of complete charge neutrality, i.e.,

$$\nabla \cdot \underline{E} = 4\pi e(Zn_i - n_e) \quad . \qquad (2.49)$$

Collisions have been neglected in these expressions, and we shall consider the case in which ion temperatures are much less than electron temperatures (actually

$T_i = 0$) and neglect the pressure term $p_i$. The current is defined by

$$\underline{j} = e(Zn_i\underline{v}_i - n_e\underline{v}_e) \quad , \tag{2.50}$$

and if one neglects $(m_e/m_i)(n_iZ^2/n_e)$ compared to unity, one finds that

$$\frac{m_e}{e^2n_e}\frac{d\underline{j}}{dt} = \underline{E} + \frac{1}{en_e}\nabla p_e \quad . \tag{2.51}$$

Attention will be confined to situations in which the plasma carries no current, i.e., $j = 0$. Also, the customary isothermal equation of state is used for the electron pressure, i.e., $p_e = n_ekT_e$.

Considering spatial variation in only one dimension and introducing the dimensionless variables

$$x' = x/L \ , \quad t' = t/T \ , \quad v' = vT/L \ , \quad E' = eLE/kT_e \ , \quad n_i' = n_i/n_0$$

where $n_0$ is the average ion density, $L^2 = kT_e/4\pi n_0e^2$ and $T^{-2} = 4\pi n_0e^2Z/m$, one finds that (2.46,48,51 and 49) respectively take the form

$$\frac{\partial n_i'}{\partial t'} + \frac{\partial n_i'v_i'}{\partial x'} = 0 \tag{2.52}$$

$$\frac{\partial v_i'}{\partial t'} + v_i'\frac{\partial v_i'}{\partial x'} = E' \tag{2.53}$$

$$E' + \frac{1}{n_e'}\frac{\partial n_e'}{\partial x'} = 0 \tag{2.54}$$

$$\frac{\partial E'}{\partial x'} = Zn_i' - n_e' \quad . \tag{2.55}$$

Linearization of these equations about $n_i' = n_e' = 1$ and $v_i' = E' = 0$ plus the assumption that all quantities vary as $kx' - \omega t'$ leads to the dispersion relation (Z has been set at 1)

$$V^2 \equiv \frac{\omega^2}{k^2} = 1 - \omega^2 \quad . \tag{2.56}$$

If, on the other hand, complete charge neutrality had been assumed so that $Zn_i' = n_e'$, then one would obtain the dispersionless result $V = \omega/k = 1$. At the same time, complete charge neutrality in the full nonlinear equations (2.52-55) would have led to (2.52) plus

$$\frac{\partial v_i'}{\partial t} + v_i' \frac{\partial v_i'}{\partial x'} = - \frac{1}{n_i'} \frac{\partial n_i'}{\partial x'} \quad . \tag{2.57}$$

It is known that (2.52,57) lead to continued steepening until a shock wave develops. Thus it is the lack of complete charge neutrality that introduces dispersion and provides the balancing effect that prevents shock formation and leads to a steady-state profile.

For weak dispersion (2.56) yields

$$k = \omega + \frac{1}{2} \omega^3 \tag{2.58}$$

and $kx' - \omega t' = (x' - t') + \frac{1}{2} \omega^3 x'$. Setting

$$\xi = \omega(x' - t') \ , \qquad \eta = \omega^3 x' \quad , \tag{2.59}$$

(2.52-55) take the form with ($n_i' = n$, $v_i' = u$)

$$- \frac{\partial n}{\partial \xi} + \frac{\partial (nu)}{\partial \xi} + \omega^2 \frac{\partial nu}{\partial \eta} = 0 \tag{2.60}$$

$$- \omega \frac{\partial u}{\partial \xi} + \omega u \left( \frac{\partial u}{\partial \xi} + \omega^2 \frac{\partial u}{\partial \eta} \right) = E \tag{2.61}$$

$$\omega \frac{\partial n_e'}{\partial \xi} + \omega^3 \frac{\partial n_e'}{\partial \eta} = -n_e' E \tag{2.62}$$

$$\omega \frac{\partial E}{\partial \xi} + \omega^3 \frac{\partial E}{\partial \eta} = n - n_e' \quad . \tag{2.63}$$

A perturbation solution of these equations is now developed. One finds that a consistent expansion procedure is

$$n = 1 + \omega^2 n^{(1)} + \omega^4 n^{(2)} + \ldots \tag{2.64}$$

$$u = \omega^2 u^{(1)} + \omega^4 u^{(2)} + \ldots \tag{2.65}$$

$$n_e = 1 + \omega^2 n_e^{(1)} + \omega^4 n_e^{(2)} + \ldots \tag{2.66}$$

$$E = \omega^3 E^{(1)} + \omega^5 E^{(2)} + \ldots \quad . \tag{2.67}$$

Substitution of these expansions into (2.60-63) yields the first-order equations

$$\frac{\partial n^{(1)}}{\partial \xi} - \frac{\partial u^{(1)}}{\partial \xi} = 0 \ , \quad \frac{\partial u^{(1)}}{\partial \xi} = \frac{\partial n_e^{(1)}}{\partial \xi} = -E^{(1)} \ , \quad n^{(1)} = n_e^{(1)} \quad . \tag{2.68}$$

The second-order ones are

$$\frac{\partial u^{(1)}}{\partial \eta} + \frac{\partial}{\partial \xi} \ n^{(1)}u^{(1)} + u^{(2)} \ - \frac{\partial n^{(2)}}{\partial \xi} = 0 \tag{2.69}$$

$$\frac{\partial u^{(2)}}{\partial \xi} - u^{(1)} \frac{\partial u^{(1)}}{\partial \xi} = -E^{(2)} \tag{2.70}$$

$$\frac{\partial n_e^{(2)}}{\partial \xi} + \frac{\partial n_e^{(1)}}{\partial \eta} = -n_e^{(1)}E^{(1)} - E^{(2)} \tag{2.71}$$

$$\frac{\partial E^{(1)}}{\partial \xi} = n^{(2)} - n_e^{(2)} \quad . \tag{2.72}$$

These equations reduce to a Korteweg-de Vries equation for $u^{(1)}$

$$\frac{\partial u^{(1)}}{\partial \eta} + u^{(1)} \frac{\partial u^{(1)}}{\partial \xi} + \frac{1}{2} \frac{\partial^3 u^{(1)}}{\partial \xi^3} = 0 \quad . \tag{2.73}$$

According to (2.68) a similar equation holds for $n^{(1)}$ and $n_e^{(1)}$.

The simplest solution is the steady-state solution which is the KdV soliton $u^{(1)} = f(x' - u_0 t') = g(u_0\xi - a\eta)$ where $a = (u_0 - 1)/\omega^2$; then (2.73) becomes the ordinary differential equation

$$- ag' + u_0 gg' + \frac{1}{2} u_0^3 g''' = 0 \quad , \tag{2.74}$$

and integration yields

$$g = \frac{3a}{u_0} \operatorname{sech}^2 \left[ \sqrt{\frac{a}{2u_0^3}} \ (u_0\xi - a\eta) \right] \quad . \tag{2.75}$$

For the density $\rho$ this becomes

$$\rho = n_0 \omega^2 n^{(1)} = \delta n \operatorname{sech}^2[(x - vt)/D] \tag{2.76}$$

where

$$v = u_0 \sqrt{\frac{kTe}{m_i}} \tag{2.77}$$

$$u_0 = 1 + \frac{1}{3}\frac{\delta n}{n_0} \tag{2.78}$$

and

$$D^2 = k_D^2\left(\frac{2u_D^3}{u_0-1}\right) = k_D^2\,\frac{6n_0}{\delta n} \quad . \tag{2.79}$$

One sees that the width of the soliton (~ D) decreases with increasing amplitude $\delta n$. Also the solitary wave velocity v increases with increasing $\delta n$. Experimental verification of these results has been obtained. For values of $n_0 \sim 10^9 - 10^{10}\,\mathrm{cm}^{-3}$, $T_e = 3$ev and $T_i < 0.2$ eV, IKEZI [2.63] has obtained confirmation of the relationship between pulse velocity and pulse width. In his experiments $\delta n/n_0 \approx 0.2$. Pulse breakup was also observed and the number of pulses is in agreement with theoretical predictions obtained by solution of the initial value problem for the Korteweg-de Vries equation.

## 2.3 Inverse Spectral Transformation and Motion Invariants

Much of this volume is devoted to discussions of the inverse scattering transform as it applies to nonlinear systems. Here we introduce and motivate this spectral transform through the physics of self-induced transparency. A glance over the rest of the volume shows that the spectral method is far more general than this particular application in optics; nevertheless, we believe many of the features of the inverse method seem less mysterious when described in a physical context.

### 2.3.1 Physical Introduction to the Inverse Scattering Method

The direct problem of self-induced transparency is to find the solution $[E(t,x), A(t,x;\xi), B(t,x;\zeta)]$ of the coupled system

$$\partial_x E = -\langle 2AB^*\rangle \; ; \quad i\partial_t\begin{pmatrix}A\\B\end{pmatrix} = \frac{1}{2}\begin{pmatrix}2\zeta & iE\\ -iE^* & -2\zeta\end{pmatrix}\begin{pmatrix}A\\B\end{pmatrix} \tag{2.80a,b}$$

subject to the prescribed boundary conditions

$$E(t,x)\Big|_{x=0} = E_0(t) \to 0 \quad \text{as} \quad t \to \pm\infty \; ; \quad \begin{pmatrix}A(t,x,\zeta)\\B(t,x,\zeta)\end{pmatrix} \simeq \begin{pmatrix}0\\e^{i\zeta t}\end{pmatrix} \quad \text{as } t \to -\infty \; . \tag{2.81a,b}$$

Here $<f(\cdot)> = \int_{-\infty}^{\infty} f(\zeta)g(\zeta)d\zeta$. Equations (2.80) are a complex form of the system (2.8). $E$ denotes the slowly varying, complex envelope of the electric field. Recall that this field has been tuned to resonate with the medium which is described as an ensemble of two-level quantum oscillators. A and B denote slowly varying amplitudes for the upper and lower quantum levels, respectively. They determine the polarization $\lambda$ and population inversion N by the formulas

$$\lambda(t,x;\zeta) = -2A(t,x;\zeta)B^*(t,x,\zeta)N(t,x,\zeta) = |A(t,x;\zeta)|^2 - |B(t,x;\zeta)|^2 \quad . \qquad (2.82)$$

The parameter $\zeta$ measures the difference between the carrier frequency and the natural frequency of the oscillator. At perfect resonance $\zeta = 0$. We remark that this approximation of the two level dynamics (A,B) may be obtained from the "exact two-level description $(\alpha,\beta)$ of (2.1)" by approximations which are analogous to those which were employed in Sect.2.2.1 to reduce the Bloch equations (2.5-8). This reduction is discussed in some detail in [2.64].

Turning to the boundary conditions, notice that the $E$ field is prescribed at $x = 0$, while in the "distant past" all quantum states are in the lower energy level. Such an ensemble of oscillators describes a medium which will attenuate the field. (As already discussed in Sect.2.2.1, in an amplifying medium, initially more oscillators must be in the upper state than the lower, thus enabling the field to extract energy from the medium and amplify.)

For the moment, fix the position $x = x_0 > 0$ and consider the electric field envelope $E$ at a position $x_0$ as a function of time t. The field $E$ is a localized pulse; hence, for large negative times, the field is negligible at $x_0$ since the pulse has not reached $x_0$ as yet. For such negative times, the two-level oscillators near $x_0$ remain in their lower states, and their polarization $\lambda$ remains at zero. As time increases, the field $E$ reaches $x_0$ and, via interaction (2.80b), excites the upper levels of the quantum oscillators near $x_0$. Later, at large positive times, the field at $x_0$ returns to zero since the local pulse $E$ has passed, and the quantum oscillators near $x_0$ are left in a state which is a mixture of the two levels[1]

$$\begin{bmatrix} A(t,x_0;\zeta) \\ \\ B(t,x_0;\zeta) \end{bmatrix} \simeq \begin{bmatrix} -\bar{b}(\zeta - x_0)e^{-i\zeta t} \\ \\ \bar{a}(\zeta;x_0)e^{i\zeta t} \end{bmatrix} \quad \text{as} \quad t \to +\infty \quad . \qquad (2.83)$$

---

[1]This notation, while somewhat unnatural here, has been selected to conform with [2.22,24,25,43,65].

Here the coefficients ($\bar{a}$ and $\bar{b}$) are determined by the specific details of the pulse $E$; however, since $|A|^2 + |B|^2$ is constant with respect to t (conservation of probability), it must equal its initial value 1, which forces the coefficients $\bar{a}$ and $\bar{b}$ to satisfy the constraint

$$|\bar{a}(\zeta;x_0)|^2 + |\bar{b}(\zeta;x_0)|^2 = 1 \quad . \tag{2.84}$$

Notice that after the pulse has passed $x_0$, the polarization $\lambda = 2i\overline{ba^*\exp(-2i\zeta t)}$ oscillates in t; the medium is left "ringing". This ringing is an excited state for the medium. Since the entire system (field and medium) conserves energy, this gain in medium energy must result from a loss of energy in the field. Indeed, the conservation law

$$\partial_t\langle N(t,x)\rangle + \frac{1}{2}\,\partial_x|E(t,x)|^2 = 0 \quad ,$$

which follows immediately from (2.80,82), shows this balance. Here $|E(t,x)|^2$ represents the density of field energy while the population inversion N represents the medium energy above the ground state.

In the last paragraph we have described the generic situation; however, one could envision certain special pulses which, after passing the point $x_0$, return the medium entirely to its ground state. For such special pulses, $b(\zeta,x_0) = 0$ for all (real) values of the detuning parameter $\zeta$. It follows from the constant of motion (2.84) that $|a(\zeta,x_0)| = 1$ for such pulses, and that the only lasting effect on the medium (at $x_0$) is a phase shift in the lower state. The medium would not be left ringing, and no energy would be transferred from the field to the medium. If this occurred for all locations $x_0$, the medium would be transparent to these special pulse shapes, and the resonance would induce a transparent medium. This remarkable situation does occur, and forms the "sech" pulses discussed in Sect. 2.2.1; furthermore, as we shall see, one does not have to establish these special pulses externally The interacting system itself ultimately decomposes the field into two components, one of which loses energy to the medium and leaves the oscillators ringing, while the other component shapes itself into the special soliton shapes which yield distortionless propagation. This is actually the generic situation as is best seen by solving interacting system (2.80) by the mathematical technique of the inverse method.

The inverse method goes much farther than the observations of the last two paragraphs towards exploiting the dynamics of the two-level oscillators. The idea is to use the medium at $x = x_0$, at time t after the pulse has passed $x_0$, as a measuring device to determine the field $E$ at $x_0$ for all time t. Clearly the pulse $E$ determines the scattering coefficients $\bar{a}$ and $\bar{b}$, but can the measurement of the medium's response to the pulse (that is the measurement of the scattering coefficients at $x_0$, $\bar{a}(\zeta,x_0)$ and $\bar{b}(\zeta,x_0)$, for all values of the detuning parameter $\zeta$) determine the pulse $E$? It turns out that certain information at x, known as the scattering data $\sum(x)$, determines $\sum(t,x)$ for all time t. Moreover, $\sum(x)$ can be computed explicitly from $\sum(x_0)$, the scattering data at $x_0$. Thus, measurement of $\sum(x_0)$ for all x, and therefore $E(t,x)$ for all x and t. But let us be more specific.

First, we rewrite the two-level dynamics as an eigenvalue problem,

$$\left[i\begin{pmatrix}1 & 0\\ 0 & -1\end{pmatrix}\partial_t - \frac{i}{2}\begin{pmatrix}0 & E(t,x_0)\\ E^*(t,x_0) & 0\end{pmatrix}\right]\begin{pmatrix}A\\ B\end{pmatrix} = \zeta\begin{pmatrix}A\\ B\end{pmatrix} \quad ,$$

where $x_0$ is fixed and $E$ is treated as a known function of t (as it is at $x_0 = 0$). Even though this eigenvalue problem is not self-adjoint, its spectral properties are known. Its spectrum consists of the entire real $\zeta$ axis as continuous spectrum together with a finite number of bound states in the upper half $\zeta$ plane, $\mathrm{Im}\{\zeta\} \geqq 0$. The "potential" $E$ determines the following spectral data $\sum$,

$$\sum \quad \{R(\xi) \;\; (\text{for all real } \zeta) \; ; \;\; \zeta_j \text{ and } \bar{c}_j \;\; [\text{for all } j\varepsilon(1,2,\ldots,N)]\} \quad .$$

Here $R(\zeta) \equiv \bar{b}(\zeta)/[\bar{a}(\zeta)]^*$, N denotes the number of bound states $\zeta_j$ in the upper half $\zeta$ plane, and $\bar{c}_j$ denotes a natural normalization of the bound state eigenfunctions. (If $E$ has compact support in t, $\bar{c}_j = d\{\bar{b}(\zeta)/[\bar{a}(\zeta)]^*\}/d|_{\zeta=\zeta_j}$.

A key step in the inverse method is the realization that the map $E \to \sum$ is invertible; that is, a knowledge of $\sum$ uniquely determines $E$ for all time. This realization permits consideration of the now familiar diagram of the inverse spectral transform (Fig.2.2). In this diagram, direct integration is depicted as paths represented by path ④, while the inverse method of integration is depicted as paths ① → ② → ③. Path ① maps the prescribed field $E(t,x = 0)$ into spectral data $\sum(x = 0)$. This step is analogous to (and as difficult as) the direct scattering problem of quantum mechanics. Along path ② one uses the evolution of $E$ in x as given by (2.80a) to find the x evolution of the spectral data $\sum$. One of the miracles of inverse scattering is that the evolution of $\sum$ in x is trivial and that $\sum(x)$ can be found explicitly. This step ② is extensively and clearly described throughout the literature and other articles in this volume. For this SIT system, we merely summarize the result which is derived in considerable detail in [2.43]

$$\sum(x) = \left\{R(\zeta,x) = R(\zeta,x = 0)\exp\left[-\frac{i}{2}x\int_{\Gamma_u}\frac{g(\zeta')d\zeta'}{\zeta-\zeta'}\right];\quad \zeta_j(x) = \zeta_j(x = 0)\;;\right.$$

$$\left.\bar{c}_j(x) = \bar{c}_j(x = 0)\exp\left(-\frac{i}{2}x\int_{\Gamma_u}\frac{g(\zeta')d\zeta'}{\zeta_j-\zeta'}\right)\right\} \quad . \tag{2.85}$$

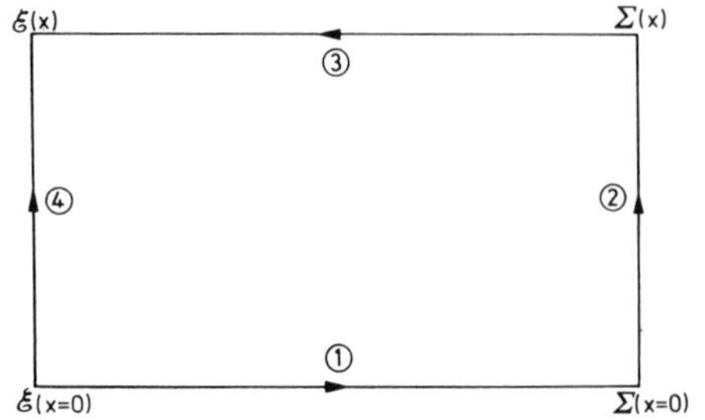


Fig.2.2. Schematic of the inverse spectral transform .

Here $\Gamma_u$ is a contour in the $\zeta$ plane running along the real axis from $-\infty$ to $+\infty$, indented under the point $\zeta' = \zeta$. Notice in particular that the eigenvalues $\{\zeta_j\}$ are constant in x, while the x evolution of R and $\bar{c}_j$ depends in a simple manner on the broadening $g(\cdot)$.

The last step ③ is to construct $E(\cdot,x)$ from $\sum(x)$ and is similar to the "inverse scattering problem" of quantum mechanics. In the appendix we present a derivation of the method of inversion which we feel is shorter, clearer, and easier to follow than the derivations which are generally available in the literature. Here we merely state the final steps.

First, one constructs the kernel $\hat{R}(t+y)$ from the scattering data $\sum$,

$$\hat{R}(t+y) \equiv \frac{1}{2\pi}\int_{\infty}^{\infty} R(\zeta)\exp[-i\zeta(t+y)]d\zeta - i\sum_{j=1}^{N} \bar{c}_j \exp[-i\zeta_j(t+y)] \quad , \qquad (2.86)$$

where we have suppressed the x dependence. Using $\hat{R}(t+y)$ as the known kernel, one then solves the Marchenko integral equation for

$$K = \begin{pmatrix} K_{22}^* & K_{12} \\ -K_{12}^* & K_{22} \end{pmatrix} \quad .$$

$$\begin{pmatrix} 0 \\ 0 \end{pmatrix} = \begin{pmatrix} R(t+y) \\ 0 \end{pmatrix} + K(t,y)\begin{pmatrix} 0 \\ 1 \end{pmatrix} + \int_{-\infty}^{t} k(t,y')\begin{pmatrix} R(y'+y) \\ 0 \end{pmatrix} dy' \ , \quad t > y \quad . \qquad (2.87)$$

Finally, one calculates $E(t,\cdot)$ by the formula

$$E(t,\cdot) = 4K_{12}(t,t) \quad . \qquad (2.88)$$

These last three formulas, together with the x evolution of the scattering data as given by (2.85) complete our description of paths ② and ③. In the next section we use these formulas to analyze the evolution of the $E$ field from its initial profile to its propagating wave form.

### 2.3.2 Physical Information from the Inverse Method

If the broadening $g(\cdot)$ is known, the scattering data $\sum(x = 0)$ determine $\sum(x)$ as formula (2.85) shows. This fact leads us to consider the medium at $x = 0$ as a measuring device which measures $\sum(x = 0)$. From this measured data, we can construct $E(t\cdot x)$ by using (2.85) and the Marchenko equations.

First, consider a special pulse shape which returns all the oscillators in the measuring device at $x = 0$ to their ground state once the pulse has passed. For such a special pulse, $R(\zeta, x = 0)$ vanishes for all real $\zeta$, and the kernel $\hat{R}(t + y, x = 0)$ reduces to a finite sum over the discrete spectrum. If this reduction to a finite sum occurs at $x = 0$, (2.85) shows that $\hat{R}$ remains a finite sum for all $x$. In this situation, the Marchenko equation reduces to a finite dimensional, linear, algebraic system [2.5,11,66] whose solution yields the "N-soliton formulas". (Here N denotes the number of eigenvalues and is the size of the algebraic system.) These N solitons are the distortionless pulses.

The single soliton waveform results when there is exactly one bound state, $\zeta_1 = \xi + i\eta$. In this instance, the waveform is given by [2.67,68]

$$E(x,t) = 4\eta \exp[-i(\phi_0 + \omega_1 x - 2\xi t)]\mathrm{sech}[\theta_0 + \omega_2 x - 2\eta t] \quad . \tag{2.89}$$

Here $\omega_1 + i\omega_2 = -\frac{1}{2}\int_{-\infty}^{\infty}(\zeta_1 - \zeta')^{-1}g(\zeta')d\zeta'$ and the constants $\theta_0$ and $\theta_0$ are related to the normalization constant $\bar{c}_1 = -2i\eta \exp(-\theta_0)\exp(i\theta_0)$. This is a complex pulse whose amplitude is $4\eta$ sech$(\cdot)$, traveling at speed $2\eta/\omega_2$, with a phase traveling at $2\xi/\omega_1$. We emphasize that the imaginary part of the eigenvalue $\zeta_1$ determines the maximum amplitude of $E$ and, up to $\omega_2$, the speed of the envelope, while the real part of $\zeta_1$ (up to the factor of $\omega_1$) determines the phase velocity. In the literature, this solution is known by various names—soliton, kink, $2\pi$ pulse, and is similar to the $2\pi$ pulse of Sect.2.2.1 except that here the phase $\phi$ has not been specialized to zero. It is important to remark that the linear space and time dependence of this phase term corroborates early assumptions of [2.13,14,38] and experimental results of [2.40-42].

When more than one eigenvalue occurs, a multi-soliton wave results. One particularly interesting case occurs when two solitons have the same envelope velocity. This pulse has zero area and is given various names in the literature—the "0-$\pi$" pulse, the "doublet", the "breather". It can be viewed as a two-soliton bound state or as a localized waveform with an internal degree of freedom. Its general features can be best described in the simplified situation of no Doppler broadening. In this case, $g(\zeta) = \delta(\zeta)$, and we have seen that the dynamics reduces to the sine-Gordon equation (2.21). The pulse is real and the eigenvalues either lie on the imaginary axes and represent solitons, or in conjugate pairs $(\zeta, -\zeta^*)$ and represent "breathers". These breathers translate at a constant speed without decay and emerge from interactions suffering only a phase shift—just as solitons. In addition they pulsate, as can be seen from the analytical formula [2.18]

$$E(t,x) = 8\eta \operatorname{sech}\theta \frac{\sin\phi + (\eta/\xi)\tanh\theta \cos\phi}{1+(\eta/\xi)^2 \operatorname{sech}^2\theta c \cos^2\phi} ,$$

where

$$\phi \equiv \phi_0 + \omega_1 x - 2\xi t$$

$$\theta \equiv \theta_0 + \omega_2 x - 2\eta t \quad .$$

A typical graph is sketched in Fig.2.3. Such a breather has 4 degrees of freedom: $\mathrm{Re}\{\zeta_1\} = -\mathrm{Re}\{\zeta_2\} = \xi$ which defines a phase speed and is related to the pulsating effect; $\mathrm{Im}\{\zeta_1\} = \mathrm{Im}\{\zeta_2\} = \eta$ which defines the amplitude; the complex number $\bar{c}_1$ which locates the pulse. As can be inferred from formulas in Sect.2.3.3, the energy of the breather is slightly less than that of a two-soliton state; hence, it is often treated as a two-soliton bound state. The pulsating effect can be viewed as a manifestation of an internal degree of freedom. The viewpoint seems prevalent among those who study a quantized sine-Gordon equation as a model nonlinear field theory since the extra degree of freedom is a possible source of mass spectra [2.32,33, 67].

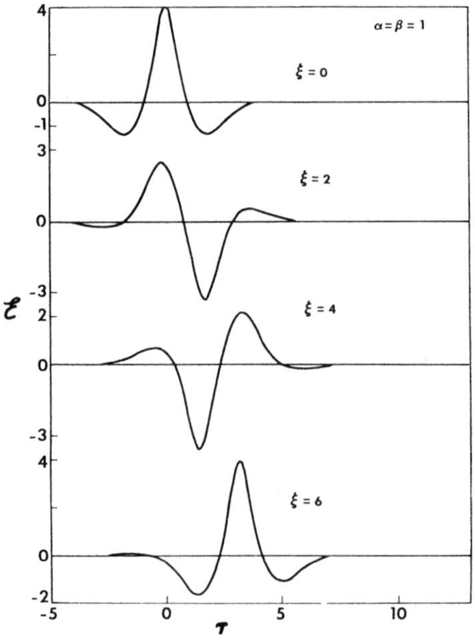


Fig.2.3. Graph of breather solution

Generally, the oscillators comprising our measuring device at $x = 0$ will not all be returned to the ground state after the pulse passes, but will be left ringing in some mixed state. In this general case, the structure of the solution can still be obtained from the Marchenko equations. Even though $R(\zeta,x) \neq 0$, the kernel $\hat{R}$ is still dominated by the discrete components since the discrete sum is exponential for large x while the contribution from the continuous spectrum is much smaller. (For most broadening functions g, this continuous contribution is exponentially small over most regions of space. At worst, it decays algebraically.) Thus, for large x a general pulse $\mathcal{E}$ shapes itself into a sum of solitons and breathers, ordered according to their speeds, together with a decaying contribution from the continuous spectrum whose only real significance is to ensure causality [2.43]. By far the most important features of the $\mathcal{E}$ field in an attenuator are the discrete components. (In an amplifying medium the continuous components become more significant. In this case, various asymptotic [2.25,27,28] calculations which use stationary phase methods may prove useful.)

### 2.3.3 Hamiltonian Description and Constants of the Motion

Systems which support solitons possess infinite families of conservation laws. These conserved quantities can be used to predict, from the initial data itself, the speeds of those solitons which eventually emerge from the initial data [2.68]. Such applications of conservation laws were quite useful for interpreting the early numerical results in the quantum optics of self-induced transparency. More recently, it has been shown that these infinite families of conservation laws are manifestations of the complete integrability of the nonlinear system [2.69-71]. In this final section we discuss connections between the physical information which the early studies extracted from the conservation laws and the mathematically precise formulation which uses the inverse scattering transform to display the complete integrability of the equations of motion.

We begin by reminding the reader of some of the properties of a completely integrable Hamiltonian system in finite dimensions. Let $S_{2n}$ denote a (2n-dimensional) phase space, and consider a Hamiltonian system on $S_{2n}$ generated by Hamiltonian $H : S_{2n} \to R$,

$$\frac{d}{dz} z = T\nabla_z H \quad , \tag{2.90}$$

where $z = (Q,P) \in S_{2n}$, the matrix $T \equiv \begin{pmatrix} 0 & I \\ \hline -I & 0 \end{pmatrix}$, and I denotes the $n \times n$ identity matrix. A constant of the motion E for this Hamiltonian system is a real function on phase space, $E : S_{2n} \to R$, which is invariant along solution curves of (2.90); i.e., if z(t) denotes a solution of (2.90), E[z(t)] is constant in time t. A Hamiltonian system in $S_{2n}$ is called *completely integrable* if it possesses n constants of the motion, $\{E_1, E_2, \ldots, E_n\}$, which are I) independent and II) pairwise in involution[2] [2.72-73].

Completely integrable systems have many striking properties [2.73,74]. The solution trajectories lie on the surfacs $\{E_i = \text{constant}, i = 1, \ldots, n\}$ which, if compact, are (topologically) n tori (n-dimensional donuts in phase space $S_{2n}$). Intimately connected with these tori is a coordinate system $(\theta, J)$ which is related to the $(Q,P)$ coordinates by a canonical transformation [2.73]. Indeed the surfaces $J$ = constant, that is $\{J_1 = C_1, J_2 = C_2, \ldots, J_n = C_n\}$ for constants $\{C_1, C_2, \ldots, C_n\}$, define these same tori. The *action variables* (J) select a torus while the *angle variables* ($\theta$) locate the exact position of the phase point on this torus.

Since the Hamiltonian flow (2.90) leaves these tori invariant, the action variables J must be constants of the motion for this Hamiltonian system. Analytically, this fact arises because the Hamiltonian H, when expressed in terms of the $(\theta, J)$ variables, depends only on J and is independent of $\theta$. Since the map from $(Q,P) \to (\theta, J)$ is canonical, the equations of motion take the form

$$\frac{d}{dt}\begin{pmatrix} \theta \\ -- \\ J \end{pmatrix} = \begin{pmatrix} 0 & I \\ -I & 0 \end{pmatrix}\begin{pmatrix} \nabla_\theta H \\ \nabla_J H \end{pmatrix} = \begin{pmatrix} \omega \\ -- \\ 0 \end{pmatrix} \quad .$$

Notice in particular that $J = 0$ and that $\omega = \nabla_J H$ is a function of J only; hence, $\theta_k$ grows linearly in t with slope $\omega_k$. Moreover, since the variables, are angular variables on the torus, the quantities $\{\omega_j, j = 1,2, \ldots n\}$ represent frequencies. Thus, the motion in phase space is explicitly shown to be quasi periodic in t, with basic frequencies $\omega = \nabla_J H$. Quasi periodic is reasonably regular; for example, initial states "nearly recur" as time progresses—recurrences of the type observed in the original FPU computer experiments [2.9]. Although limitations in space prohibit us from covering this connection, a discussion may be found in [2.75].

Since the action variables select a torus on which the trajectory lies and thereby fix its radii, these variables are related to the amplitudes of vibration. In addition, they define the vibration frequencies through the formula $\omega = \nabla_J H$. In this manner, they carry the most important information about the dynamics of

---

[2] A family $\{E_1, \ldots, E_n\}$ of real functions on phase space consists of *independent members* if the functions $\{E_i\}$ have linearily independent gradients (with respect to $\nabla_z$). Two functions E and F are said to be in involution if their Poisson brackets vanish,

$$\{E,F\}_J = (\nabla_z E, T\nabla_z F) = \sum_{k=1}^{n} \left[\frac{\partial E}{\partial Q_k}\frac{\partial F}{\partial P_k} - \frac{\partial E}{\partial P_k}\frac{\partial F}{\partial Q_k}\right] = 0 \quad .$$

the system. In contrast, the initial values for the angle variables, which precisely locate the phase point on the torus, carry far less important information. Actually, the action variables measure the area enclosed by the basic cycles of the torus through the formula

$$J_k = \frac{1}{2\pi} \oint_{k^{th}\text{cycle}} P \cdot dQ \quad .$$

Essentially due to this loop definition these action variables are not only constants of the motion for Hamiltonian system (2.90), but they are also remarkably stable to adiabatic (slow) perturbations of the dynamical system. In fact, they remain nearly constant under such perturbations, constant to a far greater accuracy than other constants of the motion such as the energy [2.72,76]. This stability property allows the action variables to form the basis for many perturbation calculations in mechanics.

Action angle variables which arise in oscillators with one degree of freedom and in linear systems of coupled harmonic oscillators are well known [2.76,77]. In [2.78,79] we have attempted to discuss the action-angle coordinates of systems which support solitons in a manner analogous to those more familiar systems.

With this summary of the properties of completely integrable Hamiltonian systems, we turn to the sine-Gordon equation which, as we have seen, arises as a limit of SIT in the absence of inhomogeneous broadening. Thus, this system provides a simple model to illustrate the use of higher conservation laws in quantum optics [2.18, 80]. Our goal here is to explain these early calculations in the light of complete integrability.

The sine-Gordon equation for the electric field $E$ takes the form[3]

$$\frac{d}{d\tau} E = \sin \int_{-\infty}^{z} E(z',\tau)dz) \quad .$$

This equation defines an infinite dimensional Hamiltonian system[4]

[3]We have changed the names of our coordinates from $x \to \tau$ and $t \to z$, essentially for convenience of presentation.

[4]Notice that the antisymmetric matrix T is replaced by the antisymmetric operator $\nabla = \partial/\partial z$ and the Poisson bracket structure generated by T is replaced by one generated by $\nabla$,

$$\{F,G\}_\nabla \equiv \left(\frac{\delta F}{\delta E} , \nabla \frac{\delta G}{\delta E}\right) = \frac{1}{2} \int_{-\infty}^{\infty} \left(\frac{\delta F}{\delta E} \frac{\delta}{\delta z} \frac{\delta G}{\delta E} - \frac{\delta G}{\delta E} \frac{\partial}{\partial z} \frac{\delta F}{\delta E}\right) dz \quad .$$

$$\frac{d}{d\tau} = \nabla \frac{\delta H}{\delta E} \quad , \tag{2.91}$$

with Hamiltonian H given by

$$H(E) = \int_{-\infty}^{\infty} \left\{ \cos \left[ \int_{-\infty}^{z} E(z')dz' \right] - 1 \right\} dz \quad .$$

(Here the field $E$ must satisfy the boundary condition $\int_{-\infty}^{\infty} E(z')dz' = 2n\pi$ for integer n.) Just as in the full SIT case, this sine-Gordon equation can be integrated by the inverse scattering method, using essentially the same linear eigenvalue problem and scattering data.

Hamiltonian system (2.91) possesses an infinite family of constants of the motion. In fact, calculation of the evolution of the scattering data in $\tau$ shows that the inverse of the transmission coefficient, $a(\zeta)$, is constant in $\tau$ for all values of $\zeta$. Actually, this family does not consist of independent members since $a(\zeta)$, $\mathrm{Im}\{\zeta\} > 0$, is determined by $|a(\zeta)|$ on the real axis together with the zeros of $a(\zeta)$ in the upper half $\zeta$ plane by the formula [2.24]

$$\ln a(\zeta) = \sum_{j=1}^{N} \ln \left( \frac{\zeta - \zeta_j}{\zeta - \bar{\zeta}_j} \right) + \frac{1}{2\pi i} \int_{-\infty}^{\infty} \frac{\ln |a(\xi')|^{-2}}{\zeta - \xi'} d\xi' \quad , \quad \mathrm{Im}\{\zeta\} > 0 \quad . \tag{2.92}$$

However, the family

$$\{\zeta_j (j = 1,2, \ldots, N) \quad \text{and} \quad \ln|a(\xi)|^{-2} \quad (\text{for all real } \xi)\} \tag{2.93}$$

does constitute an infinite, independent set of constants of the motion. Moreover, a calculation shows this family to be in involution with respect to Poisson bracket structure $\{,\}_\nabla$.

The existence of such an infinite family leads one to suspect that the system is completely integrable.[5] In infinite dimensions an infinite number of constants of the motion is necessary but not sufficient for complete integrability since in this case it is difficult to see that one really has enough constants; nevertheless this sine-Gordon system is completely integrable [2.24,81,82]. To prove this, one uses $\zeta_j$ and $\ln|a(\zeta)|^{-2}$ to define action variables, explicitly displays an angle variable for each action variable, and proves that the transformation from $E$ to these action-angle coordinates is a canonical map (one-to-one, invertible, and preserves the Poisson bracket structure). That is, one explicitly constructs a

[5]It is just this suspicion that led Gardner as well as Zakharov and Faddeev to develop the Hamiltonian formalism for KdV. [Private communications].

coordinate system equivalent to $E$ for which one-half the degrees of freedom are constants of the motion for Hamiltonian system (2.91).

The physical information carried by the constants of the motion in family (2.93) seems quite clear. The eigenvalues $\zeta_j$ yield the amplitudes and speeds of the distortionless pulses. The function $\ln|a(\xi)|^{-2}$ is associated in some nonlinear manner with the amplitude of the ringing component of the field. To see this, realize that conservation of probability (2.84) may be rewritten to yield $|a(\xi)|^{-2} = 1 - |R(\xi)|^2$; thus, $\ln|a(\xi)|^{-2}$ determines the magnitude of the reflection coefficient $|R|^2$, which, in turn, is related to the amplitude of the ringing components of $E$, a connection which becomes very direct as the pulse $E$ becomes weak. In fact, for weak pulses, $R(\xi)$ reduces to the Fourier transform of $E$ [2.25,26,65,78].

Although the physical information carried by the motion invariants of family (2.93) is quite direct, these constants are expressed in spectral language. For connection with the earlier work on "higher conservation laws", it would be desirable to express these constants explicitly in terms of $E$. Unfortunately, explicit expressions in terms of $E$ for the members of this family of action variables are not known; however, very explicit formulas are available for a related family of motion invariants.

Clearly the family of invariants (2.93) determines the constant of motion $\ln a(\zeta)$ by formula (2.92). In turn, expansions of $\ln a(\zeta)$ near $\zeta = 0$ and near $\zeta = \infty$ yield the asymptotic representations

$$\ln a(\zeta) \simeq \sum_{k=1} D_{2k-1}\zeta^{2k-1} \quad \text{near} \quad \zeta \simeq 0 \quad ,$$

$$\ln a(\zeta) \quad \sum_{k=1}^{\infty} C_{2k-1}(1/\zeta)^{2k-1} \quad \text{near} \quad \zeta \simeq \infty \ ,$$

where the expansion coefficients are given by

$$D_k \equiv \sum_{j=1}^{N} \frac{1}{k}\,[(\zeta_j^*)^{-k} - (\zeta_j)^{-k}] + \frac{1}{\pi i}\int_0^\infty d\xi\, \frac{\ln|a(\xi)|^2}{(\xi)^{k+1}}$$

$$C_k \equiv \sum_{j=1}^{N} \frac{1}{k}\,[(\zeta_j^*)^{k} - (\zeta_j)^{k}] + \frac{1}{\pi i}\int_0^\infty d\xi\, \ln(|a(\xi)|^{-2})\xi^{k-1} \ , \qquad k = 1,3\ \ldots \quad .(2.94)$$

The collections $\{C_k, k = 1,3, \ldots\}$ and $\{D_k, k = 1,3, \ldots\}$ constitute infinite families of constants of the motion which are determined by the action variables (2.93). For this system, the converse has not yet been proven. That is, it is not known whether the family $\{C_k\}$ and/or the family $\{D_k\}$ determine the action variables and therefore are equivalent to the action variables. It would be interesting to establish this equivalence because it is these families $\{C_k\}$ and $\{D_k\}$ (and not the action variables) that possess explicit representations in terms of the field $E$, that contain familiar constants such as the energy, and that were first used to extract information from "higher constants of motion". On the other hand, the exact amount of information carried by the families $\{C_k\}$ and $\{D_k\}$ is unknown, while it is clear that the action variables constitute one-half (and the most important half at that) of the information needed to explicitly integrate the equation. The formulas (2.94) for the constants $\{C_k\}$ and $\{D_k\}$ show that this equivalence question can be formulated as a moment problem [2.24,78]. Do the moments of the function $\ln|a(\xi)|$ determine $\ln a(\zeta)$? For a somewhat simpler system, the Toda lattice, we have shown [2.78] that a family analogous to $\{C_k\}$ determines the action variables. However, that proof relies upon a representation of $\ln a(\zeta)$ as a convergent power series expansion rather than a representation as an asymptotic series, and it does not seem to extend to this sine-Gordon system.

Next, we indicate how the family $\{D_k\}$ is given in terms of the $E$ field. This family $\{D_k\}$ is determined by $\ln a(\zeta)$ near $\zeta \simeq 0$. Using techniques similar to those employed in low energy nuclear physics in the calculation of "scattering lengths" [2.24,81,82], we compute the identity

$$\frac{d}{d\zeta}\ln a(\zeta) = -i\int_{-\infty}^{\infty}\left\{\left[\frac{\phi_1(z,\zeta)\psi_2(z,\zeta)+\phi_2(z,\zeta)\psi_1(z,\zeta)}{a(\zeta)}\right]-1\right\}dz \quad . \tag{2.95}$$

Here, $\phi$ and $\psi$ are solutions of eigenvalue problem (2.80) subject to the boundary conditions

$$\phi(z,\zeta)_{z\to-\infty}\simeq\begin{pmatrix}1\\0\end{pmatrix}e^{-i\zeta z} \quad \text{and} \quad \psi(z,\zeta)_{z\to+\infty}\simeq\begin{pmatrix}0\\-1\end{pmatrix}e^{i\zeta z} \quad .$$

At $\zeta = 0$, we can solve for $\phi$ and $\psi$ explicitly and find

$$\phi(z,\ ) = \begin{bmatrix}\cos\int_{-\infty}^{z}\frac{E(z',\tau)}{2}\,dz \\[2ex] -\sin\int_{-\infty}^{z}\frac{E(z,\tau)}{2}\,dz'\end{bmatrix} \quad , \quad \psi(z,\tau) = (-1)^n\begin{bmatrix}\sin\int_{-\infty}^{z}\frac{E(z',\tau)}{2}\,dz' \\[2ex] \cos\int_{-\infty}^{z}\frac{E(z',\tau)}{2}\,dz'\end{bmatrix} \quad ,$$

where $\int_{-\infty}^{\infty}E(z',\tau)dz' = 2n\pi$. Inserting these into (2.95) yields

$$\left.\frac{d}{d\zeta}\ln a(\zeta)\right|_{\zeta=0} = -i\int_{-\infty}^{\infty}\left[\cos\int_{-\infty}^{z}E(z',\tau)dz' - 1\right]dz \quad ,$$

and shows explicitly $D_1 = -iH$ where H is the Hamiltonian. Similar, but more complicated, manipulation would yield expressions for $D_k$ in terms of the field $E$.

Representations for the family $\{C_k\}$ in terms of $E$ follow from the large $\zeta$ behaviour of $\ln a(\zeta)$. The techniques used are similar to WKB methods of quantum mechanics. We follow [2.20] and define

$$F(z,\zeta) \equiv \ln[\phi_1(z,\zeta)e^{i\zeta z}] \quad .$$

Notice, $\lim_{z\to+\infty} F(z,\zeta) = \ln a(\zeta)$. Manipulation of eigenvalue problem (2.80) shows F satisfies

$$(2i\zeta)\frac{d}{dz}F = E^2 + \left(\frac{dF}{dz}\right)^2 + E\frac{d}{dz}\left(\frac{1}{E}\frac{d}{dz}F\right) \quad .$$

From large $\zeta$ this yields the asymptotic representation

$$\ln a(\zeta) = \sum_{k=1}^{\infty} C_{2k-1}(1/\zeta)^{2k-1} \quad ,$$

with $\{C_k\}$ given in terms of $E$. The first two terms in this expansion are given by

$$C_1 = \frac{1}{2i}\int_{-\infty}^{\infty} E^2(z,\tau)dz$$

$$C_3 = \frac{1}{(2i)^3}\int_{-\infty}^{\infty} (|E_z|^2 + |E|^4)dz \quad , \tag{2.96}$$

and so forth. These formulas enable us to evaluate the constants $\{C_k\}$ in terms of the initial pulse profile $E(z,\tau = 0)$, and they are just the densities of the higher conservation laws [2.18].

This observation amplifies the importance of the equivalence problem mentioned earlier. Can we use the constants $\{C_k\}$ as specified by the initial data, to determine the action variables and, thus, the physical information carried by them — the amplitudes and speeds of the solitons, the amplitudes and frequencies of the ringing components?

In applications [2.83,84], one assumes that the ringing contribution is negligible and, in addition, that one knows apriori — say from "area theorem" — the number of solitons which emerge from the initial data. Then one uses the constants $\{C_k\}$ to estimate the speeds of the solitons which emerge. For example, if two solitons are known to emerge, the pair

$$C_1 = \frac{1}{2i}\int_{-\infty}^{\infty} |E(z,\tau = 0)|^2 dz \simeq [(\zeta_1^*) - (\zeta_1)] + [(\zeta_2^*) - (\zeta_2)]$$

$$3C_3 = \frac{-3}{(2i)^3} \int_{-\infty}^{\infty} [| \ _z|^2 + | \ |^4]dz \simeq [(\zeta_1^*)^3 - (\zeta_1)^3] + [(\zeta_2^*)^3 - (\zeta_2)^3]$$

may be treated as two (nonlinear) algebraic equations in two unknowns ($\zeta_1$ and $\zeta_2$). Solving these yields the speeds of the emerging solitons in terms of the initial pulse $E$ [2.66].

This method was exploited by BEREZIN and KARPMAN [2.68] and shown by them to be quite accurate for the Korteweg-de Vries equation. When considering the accuracy of this method in general, one immediately realizes that the omission of the continuous "ringing" components may not be a good approximation. While it is true that these continuous components decay in $\tau$, frequently they do so only algebraically — and locally in $(z,\tau)$. Experience [2.18,68] shows that this omission is more accurate the smoother the initial data. Initial profiles which have a jump discontinuity in the field or in one of its low derivatives yield considerable ringing which cannot be neglected. (This omission of the ringing contribution is much better for SIT than sine-Gordon since in SIT the continuous components decay exponentially in $\tau$ for most z. It should be remarked that SIT systems do not process global constants of the motion in the $E$ field alone. However, local conservation laws do exist and permit similar calculations. Again experience shows that for the SIT system under reasonably smooth initial data the procedure agrees rather well with numerical data. Indeed, initially it provided a synthesis of extensive numerical data.) We feel viewing these calculations in the light of completely integrable Hamiltonian dynamics clarifies them considerably by making the types of approximations involved more apparent.

## Appendix A: *Formal Derivation of the Marchenko Equations*

In this appendix we present a derivation of the Marchenko equations which is more direct than the derivations which are generally available in the literature. Various parts of this appendix may be found in [2.5,8,22,23,85-88]. The derivation is directly related to the important work [2.89] and can be found in [2.90].

We begin by considering the following "scattering problem" as presented in spectral language

$$\left[i\begin{pmatrix}1 & 0\\ 0 & -1\end{pmatrix}\partial_t - \frac{i}{2}\begin{pmatrix}0 & E(t)\\ E^*(t) & 0\end{pmatrix}\right]\begin{pmatrix}\psi_1(t,\zeta)\\ \psi_2(t,\zeta)\end{pmatrix} = \begin{pmatrix}\psi_1(t,\zeta)\\ \psi_2(t,\zeta)\end{pmatrix} \qquad (2A.1)$$

$$\psi(t,\zeta) \simeq \begin{pmatrix} 0 \\ T(\zeta) \end{pmatrix} e^{i\zeta t} \quad \text{as} \quad t \to -\infty$$

$$\simeq \begin{pmatrix} R(\zeta)e^{-i\zeta t} \\ e^{i\zeta t} \end{pmatrix} \quad \text{as} \quad t \to -\infty \quad ,$$

and the same scattering problem as represented in causal language

$$\begin{bmatrix} \partial_t - \partial_y & \frac{-E(t)}{2} \\ \frac{E^*(t)}{2} & \partial_t + \partial_y \end{bmatrix} \begin{bmatrix} \hat{\psi}_1(t,y) \\ \hat{\psi}_2(t,y) \end{bmatrix} = 0 \quad t < y \tag{2A.2}$$

$$\hat{\psi}(t,y) = 0 \qquad\qquad t > y$$

$$\hat{\psi}(t,y) \simeq \begin{pmatrix} \hat{R}(t+y) \\ \delta(t-y) \end{pmatrix} \quad \text{as} \quad t \to -\infty$$

$$\simeq \begin{pmatrix} 0 \\ \hat{T}(t-y) \end{pmatrix} \quad \text{as} \quad t \to -\infty$$

Of course, these two representations are equivalent, being related by the transform

$$\hat{\psi}(t,y) \equiv \frac{1}{2\pi} \int_{\Gamma_a} e^{-i\zeta y} \psi(t,\zeta) d\zeta \quad ,$$

where $\Gamma_a$ is a contour of integration running from $-\infty$ to $\infty$ passing above all poles of $\psi(t,\zeta)$. This choice of contour ensures that the causal representation vanishes for $t > y$. (The poles of $\psi$ occur at the poles of $T(\zeta)$, that is at the zeros of $[\bar{a}(\zeta)]^* = [T(\zeta)]^{-1}$.) For simplicity, let $E$ vanish outside a finite interval in t. In this case, the asymptotic boundary conditions merely mean "outside the support of $E$", and it is quite useful to picture the causal representation of the scattering problem as in Fig.2.4.

It turns out that the causal representation provides the most convenient means to derive the Marchenko equations which are the basic equations of the inverse method. On the other hand, the spectral representation provides more information about the structure, behaviour and properties of the solution of these equations, and therefore of $E$.

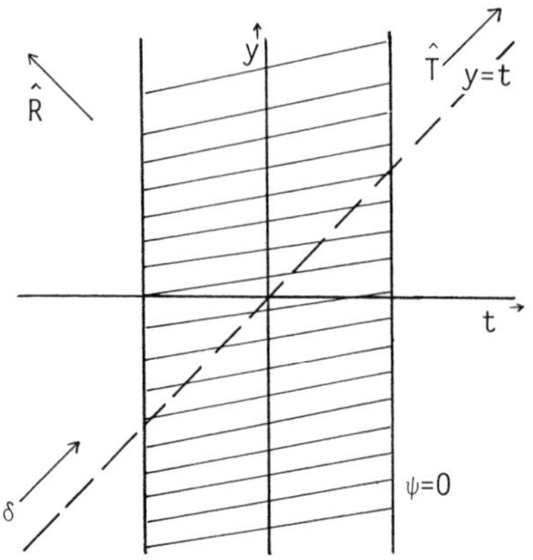


Fig.2.4. The Causal Scattering Problem. The shaded region denotes the support of $E$. The field ψ is impinging from the left as a delta function. R is the reflected component. T denotes the transmitted component. ψ vanishes in the region t > y

The derivation of the Marchenko integral equations proceeds in two steps. First, consider any solution $\phi^F$ of the "free problem" [(2A.2) with $E = 0$]. Define a transformation *operator* $\hat{K}$ which transforms any such free solution into a solution $\phi$ of the full problem [(2A.2) with general $E \neq 0$] which agrees with $\phi^F$ as $t \to -\infty$

$$\phi(t,y) = \phi^F(t,y) + \int_{-\infty}^{t} K(t,y')\phi^F(y',y)dy' \quad .$$

The demand that $\phi$ solves the full problem provided $\phi^F$ solves the free problem quickly shows that the matrix K must satisfy

$$\begin{pmatrix} 1 & 0 \\ 0 & -1 \end{pmatrix}\partial_t K + \partial_{y'} K \begin{pmatrix} 1 & 0 \\ 0 & -1 \end{pmatrix} - \frac{1}{2}\begin{pmatrix} 0 & E \\ E^* & 0 \end{pmatrix} K = 0 \qquad t > y' \tag{2A.3a}$$

$$K = 0 \qquad t < y'$$

$$\lim_{y' \to -\infty} K(t,y') = 0$$

$$\begin{pmatrix} 1 & 0 \\ 0 & -1 \end{pmatrix} K(t,t) - K(t,t)\begin{pmatrix} 1 & 0 \\ 0 & -1 \end{pmatrix} = \frac{1}{2}\begin{pmatrix} 0 & E(t) \\ E^*(t) & 0 \end{pmatrix} \quad . \tag{2A.3b}$$

Conversely, this characteristic problem (2A.3), which is pictured in Fig.2.5, uniquely defines K. Moreover, if the transformation kernel K is known, the boundary condition (2A.3b) yields the field $E$ in terms of $K_{12}$ on the diagonal.

$$E(t) = 4K_{12}(t,t) \quad . \tag{2A.4}$$

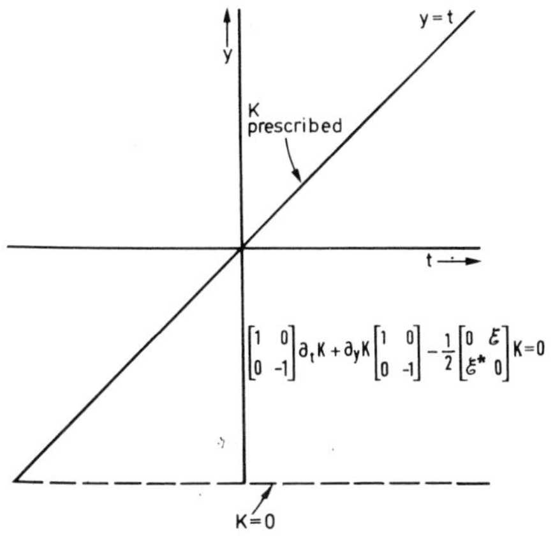


Fig.2.5. Characteristic problem for K

Thus, if we can find the kernel K from the spectral data, we have found the pulse $\mathcal{E}$.

The second step in the derivation of the Marchenko equations is to find an integral equation for K with a kernel which is given in terms of the scattering data $\sum$. This second step begins by treating the transformation kernel K as known and using it to map the causal representation of the scattering problem. Here the free wave is given by $\begin{pmatrix}\hat{R}(t+y)\\ \delta(t-y)\end{pmatrix}$ and the full wave (which agrees with the free wave as $t\to-\infty$) by $\hat{\psi}$,

$$\hat{\psi}(t,y) = \begin{pmatrix}\hat{R}(t+y)\\ \\ \delta(t-y)\end{pmatrix} + \int_{-\infty}^{t} K(t,y')\begin{pmatrix}\hat{R}(y'+y)\\ \\ \delta(y'-y)\end{pmatrix} dy' \quad .$$

Using the fact that $\hat{\psi}(t,y)$ vanishes for $t > y$, we reduce this equation to the identity

$$\begin{pmatrix}0\\ \\ 0\end{pmatrix} = \begin{pmatrix}\hat{R}(t+y)\\ \\ 0\end{pmatrix} + K(t,y)\begin{pmatrix}0\\ \\ 1\end{pmatrix} + \int_{-\infty}^{t} K(t,y')\begin{pmatrix}\hat{R}(y'+y)\\ \\ 0\end{pmatrix} dy' \quad , \quad t > y \quad . \tag{2A.5}$$

This identiy can be viewed as a linear integral equation (the Marchenko equation) whose solution determines the kernel K from the reflected wave $\hat{R}$.[6] This derivation

---

[6]It is important to realize that K in (2A.5) is of the form

$$K = \begin{matrix} K_{22}^{*} & K_{12} \\ \\ -K_{12}^{*} & K_{22} \end{matrix} \qquad \text{which follows from (2A.3)} \quad .$$

of the Marchenko equation seems the most direct available in that it minimizes detailed manipulations; nevertheless, to find the structure of K (and, therefore, of $E$), it seems necessary to employ the spectral representation of the reflected wave $\hat{R}$.

Since $\psi_1(t,\zeta) \simeq R(\zeta)\exp(-i\zeta t)$ as $t \to -\infty$, its transform yields the important representation

$$\hat{R}(t+y) = \frac{1}{2\pi} \int_{\Gamma_a} R(\zeta) e^{-i\zeta(t+y)} d\zeta \quad ,$$

where we remind the reader that the contour $\Gamma_a$ has been chosen to lie above all poles of $R(\zeta)$. Using the calculus of residues to deform the contour $\Gamma_a$ to the real axis yields the equivalent representation

$$R(t+y) = \frac{1}{2\pi} \int_{-\infty}^{\infty} R(\zeta) e^{-i\zeta(t+y)} d\zeta - i \sum_{j=1} e^{-\zeta_j(t+y)} \frac{d}{d\zeta} R(\zeta)\bigg|_{\zeta=\zeta_j} \quad . \qquad (2A.6)$$

In this manner, the scattering data $\sum$ determine R, and hence $E$ through the Marchenko equations.

Representation (2A.6) holds for all values of x; however, combining it with (2.85) which governs the evolution in x expresses $\hat{R}(t+y,x)$ in terms of $\sum(x=0)$

$$R(t+y,x) = \frac{1}{2\pi} \int_{-\infty}^{\infty} d\zeta \exp\left\{-i\left[\zeta(t+y) + \frac{x}{2} \int_{\Gamma_u} \frac{g(\zeta')}{(\zeta-\zeta')} d\zeta'\right]\right\} R(x=0,\zeta)$$

$$- i \sum_{j=1}^{N} \exp\left\{-i\left[\zeta_j(t+y) + \frac{x}{2} \int_{\Gamma_u} \frac{g(\zeta')}{(\zeta_j-\zeta')} d\zeta'\right]\right\} \bar{C}_j(x=0) \quad . \qquad (2A.7)$$

*Acknowledgment.* It is our pleasure to thank Professor H. Flaschka for reading a preliminary form of this manuscript and making many helpful comments.

## References

2.1 J.S. Russell: Proc. Roy. Soc. Edingburgh, 319 (1844)
2.2 D.J. Korteweg, G. de Vries: Philos. Mag. *39*,422 (1895)
2.3 R. Miura: SIAM Rev. (to be published)
2.4 B.B. Kadomtsev, V.I. Karpman: Sov. Phys. Usp. *14*, 40 (1971)
2.5 A.C. Scott, F. Chu, D.W. McLaughlin: Proc. IEEE *61*,1443 (1973
2.6 A.V. Bäcklund: Math. Ann. *9*, 297 (1876)
2.7 A.V. Bäcklund: Math. Ann. *19*, 387 (1882)

2.8 G.B. Whitham: *Linear and Nonlinear Waves* (John Wiley, New York 1974)
2.9 E. Fermi, J.R. Pasta, S. Ulam: In *Collected Works of Enrinco Fermi*, Vol.II (Univ. of Chicago Press, Chicago, Ill. 1965)
2.10 A.C. Newell: To appear in Proc. N.S.F. Conference on Solitons, held in Tucson, Arizona (1976)
2.11 M. Toda: Phys. Rep. *18*C, 2 (1975)
2.12 N.J. Zabusky, M.D. Kruskal: Phys. Rev. Lett. *15*, 240 (1965)
2.13 S.L. McCall, E.L. Hahn: Bull. Am. Phys. Soc. *10*, 1189 (1965)
2.14 S.L. McCall, E.L. Hahn: Phys. Rev. Lett. *18*, 908 (1967)
2.15 G.L. Lamb, Jr: Phys. Lett. *25*A, 181 (1967)
2.16 F.T. Arechi, R. Bonifacio: IEEE J. *QE-1*, 169 (1965)
2.17 A. Seeger, H. Donth, A. Kochendörfer: A. Phys. *134*, 173 (1953)
2.18 G.L. Lamb, Jr.: Rev. Mod. Phys. *43*, 99 (1971)
2.19 C.S. Gardner, J.M. Greene, M.D. Kruskal, R.M. Miura: Phys. Rev. Lett. *19*, 1095 (1967)
2.20 V.E. Zakharov, A.B. Shabat: Sov. Phys. JETP *34*, 62 (1972)
2.21 P.D. Lax: Comm. Pure Appl. Math. *21*, 467 (1968)
2.22 M.J. Ablowitz, D.J. Kaup, A. Newell, H. Segur: Stud. Appl. Math. *53*, 249 (1974)
2.23 V.E. Zakharov, A.B. Shabat: Funct. Anal. Appl. 226 (1975)
2.24 H. Flaschka, A.C. Newell: "Integrable Systems of Nonlinear Evolution Equations", in *Dynamical Systems, Theory and Application*, Lecture Notes in Physics, Vol.38, ed. by J. Moser (Springer, Berlin, Heidelberg, New York 1975)
2.25 M. Ablowitz, H. Segur: Preprint, Clarkson College of Technology, Potsdam, N.Y. (1975)
2.26 S.V. Manakov: Sov. Phys. JETP *38*, 693 (1974)
2.27 S.V. Manakov: Sov. Phys. JETP *40*, 269 (1975); Zh. Exp. i. Teoret. Fisiki *67*, 543 (1974)
2.28 M.J. Ablowitz, A.C. Newell: J. Math. Phys. *14*, 1277 (1973)
2.29 P. Lax: Comm. Pure Appl. Math. *28*, 141 (1975)
2.30 H.P. McKean, P. van Moerbeke: Inventiones Math. *30*, 217 (1975)
2.31 S.P. Novikov: Funkts. Analiz. *8:3*, 54 (1974)
2.32 L.D. Faddeev: Preprint, Institute for Advanced Study (1975)
2.33 R. Rajaraman: Phys. Rep. *21*, 227 (1975)
2.34 H. Segur: Rocky Mount. J. Math. *8*, 15 (1978)
2.35 D.J. Kaup: Stud. Appl. Math. *55*, 9 (1976)
2.36 V.E. Zakharov, S.V. Manakov: Sov. Phys. JETP *69*, 1654 (1975)
2.37 L. Allen, J.H. Eberly: *Optical Resonance and Two Level Atoms* (Wiley, New York 1975)
2.38 S.L. McCall, E.L. Hahn: Phys. Rev. *183*, 457 (1969)
2.39 M. Sargent III, M.O. Scully, W.E. Lamb, Jr.: *Laser Physics* (Addison-Wesley, Reading, Mass. 1974)
2.40 H.M. Gibbs, R.E. Slusher: Phys. Rev. Lett. *24*, 683 (1970)
2.41 H.M. Gibbs, R.E. Slusher: Phys. Rev. A*5*, 1634 (1972)
2.42 H.M. Gibbs, R.E. Slusher: Phys. Rev. A*6*, 2326 (1972)
2.43 M. Ablowitz, D. Kaup, A. Newell: J. Math. Phys. *15*, 1852 (1974)
2.44 L.P. Eisenhart: *A Treatise on the Differential Geometry of Curves and Surfaces* (Dover, New York 1960)
2.45 J. Clairin: Annales de Toulouse ze ser.*5*, 437 (1903)
2.46 G.L. Lamb, Jr.: J. Math. Phys. *15*, 2157 (1974)
2.47 T.W. Barnard: Phys. Rev. A*7*, 373 (1973)
2.48 L. Matulic, J.H. Eberly: Phys. Rev. A*6*, 822 (1972)
2.49 T.K. Gustafson, P.L. Kelley, R.Y. Chiao, R.G. Brewer: Appl. Phys. Lett. *12*, 165 (1968)
2.50 F. Tappert: Private communication
2.51 D.J. Benney, A.C. Newell: J. Math. Phys. *46*, 133 (1967)
2.52 G.J. Morales, Y.C. Lee: Preprint in U.C.L.A. (1975)
2.53 A. Hasegawa, F. Tappert: Appl. Phys. Lett. *23*, 142 (1973)
2.54 V.E. Zakharov, A.B. Shabat: Sov. Phys. JETP *37*, 823 (1973)
2.55 T. Kakutani, T. Kawahara: J. Phys. Soc. Jpn. *29*, 1068 (1970)
2.56 T. Kakutani, H. Ono, T. Taniuti, C.C. Wei: J. Phys. Soc. Jpn. *24*, 1159 (1968)

2.57 H. Kever, G.K. Morikawa: Phys. Fluids *12*, 2090 (1969)
2.58 C.S. Gardner, G.K. Morikawa: Comm. Pure Appl. Math. *18*, 35 (1965)
2.59 H. Washimi, T. Taniuti: Phys. Rev. Lett. *17*, 996 (1966)
2.60 E. Ott, R.L. Sudan: Phys. Fluids *12*, 2388 (1969)
2.61 A. Jeffrey, T.K. Kakutani: SIAM Rev. *14*, 582 (1972)
2.62 L. Spitzer, Jr.: *Physics of Fully Ionized Gases* (Interscience, New York 1956)
2.63 H. Ikezi: Phys. Fluids *16*, 1668 (1973)
2.64 D.W. McLaughlin, J. Corones: Phys. Rev. A*10*, 2051 (1974)
2.65 M. Ablowitz, D. Kaup, A. Newell, H. Segur: Phys. Rev. Lett. *31*, 125 (1973)
2.66 C.S. Gardner, J.M. Greene, M.D. Kruskal, R.M. Miura: Comm. Pure Appl. Math. *27*, 97 (1974)
2.67 R.F. Dashen, B. Hasslacher, A. Neveu: Phys. Rev. D*11*, 3424 (1975)
2.68 Y.A. Berezin, V.I. Karpman: Sov. Phys. JETP *24*, 1049 (1967)
2.69 C.S. Gardner: J. Math. Phys. *12*, 1548 (1971)
2.70 V.E. Zakharov, L.D. Faddeev: Funk. Analiz Prilozh. *5*, 18 (1970); translation in Funct. Anal. Appl. *5*, 280 (1971)
2.71 V.E. Zakharov, S.V. Manakov: Teor. Mat. Fiz. *19*, 332 (1974)
2.72 V.J. Arnold: Russ. Math. Surveys *18*, 85 (1963)
2.73 V.I. Arnold, A.A. Avez: *Ergodic Problems of Classical Mechanics* (Benjamin, New York 1968)
2.74 J. Ford: In *Fundamental Problems in Statistical Mechanics*, Vol.3, ed. by G. Cohen (North Holland, Amsterdam 1975)
2.75 V.E. Zakharov: Sov. Phys. JETP *38*, 108 (1974)
2.76 L.D. Landau, E.M. Liftshitz: *Mechanics* (Pergamon Press, Oxford 1960)
2.77 H. Goldstein: *Classical Mechanics* (Addison-Wesley, Reading Mass. 1957)
2.78 H. Flaschka, D.W. McLaughlin: Progr. Theor. Phys. *55* (1976)
2.79 H. Flaschka, D.W. McLaughlin: To appear in Proc. N.S.F. Conference on Bäcklund Transformation, held at Vanderbuilt Univ. (1974)
2.80 F.A. Hopf, M.O. Scully: Phys. Rev. *179*, 399 (1969)
2.81 L.A. Takhtadzhyan: Sov. Phys. JETP *39*, 228 (1974)
2.82 D.W. McLaughlin: J. Math. Phys. *16*, 96 (1975)
2.83 D.D. Schnack, G.L. Lamb, Jr.: "Higher Conservation Laws and Coherent Pulse Propagation", in 3rd Rochester Conf. on Coherence and Quantum Optics (1972)
2.84 R.T. Deck, G.L. Lamb, Jr.: Phys. Rev. *12*A, 1503 (1975)
2.85 I. Kay: Comm. Pure Appl. Math. *13*, 371 (1960)
2.86 M.M. Sondhi, B. Gopinoth: J. Acoust. Soc. Amer. *49*, 1867 (1970)
2.87 G.N. Balamis: J. Math. Phys. *13*, 1001 (1972)
2.88 L.D. Faddeev: J. Math. Phys. *4*, 72 (1963)
2.89 A.B. Shabat: Diff. Eq. (Soviet Journal) *8*, 164 (1972)
2.90 A.C. Scott, F. Chu, S. Reibel: Preprint, Math. Res. Center, University of Wisconsin, Madison, Wisconsin (1976)

# 3. The Double Sine-Gordon Equations: A Physically Applicable System of Equations

R. K. Bullough, P. J. Caudrey, and H. M. Gibbs

With 13 Figures

We treat two physical problems — the propagation of ultra-short optical pulses in a resonant 5-fold degenerate medium, and the creation — and propagation of spin waves in the anisotropic magnetic liquids $^3$HeA and $^3$HeB at temperatures below the transition to the A phase at 2.6 mK. The optical pulse problem is governed by a 'double' sine-Gordon equation, the $^3$HeA problem by the sine-Gordon and the $^3$HeB by a double sine-Gordon with changed sign. The double sine-Gordons are not integrable: we study them analytically and numerically and predict a "wobbling" $4\pi$ pulse in the optical case, a result supported by experiments. We solve certain initial value problems and show how they could be used for the creation of spin excitations in $^3$HeA and $^3$HeB. Finally we develop singular perturbation theory in terms of sine-Gordon scattering data in order to describe the $4\pi$ wobbler and a similar $0\pi$ pulse.

## 3.1 Physical Background

This chapter is designed to illustrate the actual behaviour of solitons or soliton-like objects in two physical contexts. Except in the last section, the mathematical techniques required of the reader are relatively slight. It is hoped that the chapter will make clear how both 'integrable' and 'nonintegrable' systems arise in the study of physical problems and by contrasting the behaviour of one integrable system, the sine-Gordon equation, with a closely related nonintegrable one, the double sine-Gordon equations, make clear how such systems differ. The chapter briefly considers particular initial value problems of physical interest for each system and it illustrates how inverse scattering techniques can be used both to solve initial value problems for integrable systems and provide perturbation theories for related nonintegrable ones. The double sine-Gordon equations are a novel system. They are introduced here to place the theory of solitons in a wider context.

This chapter is concerned with the two physical problems: the problem in nonlinear optics of the propagation of resonant short pulses of light in a medium with degenerate atomic transitions, and the problem of the propagation of spin waves[1] in the very low temperature magnetic phases of liquid $^3$He. The former problem concerns optical self-induced transparency (SIT) in a degenerate atomic medium [3.1,2]. SIT in a nondegenerate medium has already been described by Lamb and McLaughlin in the previous chapter. The problem of $^3$He spin waves is not treated elsewhere in this book and the theory in the form we described it is largely due to ourselves [3.3-6]. Comparable work by MAKI and KUMAR [3.7,8] parallelled and slightly preceded our work, however, and the analysis in [3.7] influenced our thinking. As in their more recent work [3.9], MAKI and KUMAR have been very much concerned with the possibilities for experiment. This is also one concern of the present chapter.

A successful experiment has actually been performed in the optical context [3.10] and this experiment is described briefly in this article. No equally relevant experiment has yet been performed in the spin wave case[2]; and although with MAKI and KUMAR [3.8] we suggest some possible ones in what follows, we are obliged to recognise that such experiments may be very difficult to perform. One reason is that the solitonlike spin wave excitations we shall describe are likely to be created in large numbers where they can be created at all; another is that the theory is a plane wave theory and may provide a poor description of the real situation in $^3$He where 'textures' [3.9,11] associated with the three-dimensional geometry of the system play an important role. In SIT, in contrast, transverse self-focussing is important but not crucial [3.12,13], plane wave theory is qualitatively sufficient, and single solitonlike excitations are relatively easily created and studied in the laboratory.

Even so, this chapter is predominantly concerned with a physically applicable system of nonlinear partial differential equations, rather than their experimental consequences. These equations are the double sine-Gordon equations

$$u_{xx} - u_{tt} = \pm\left(\sin u + \frac{1}{2}\lambda \sin \frac{1}{2} u\right) \tag{3.1}$$

with boundary conditions $u \to c(\mathrm{mod}\ 4\pi)$, $u_x$, $u_{xx}$, $u_{xxx}$, etc. $\to 0$ as $|x| \to \infty$. The number c is an appropriate zero of the right side: these zeros are $u = 0$, or $2\pi$ for the +ve sign and $\lambda = 1$; and $u = \delta \equiv 2\cos^{-1}(-1/4)$ or $4\pi - \delta$ for the -ve sign and $\lambda = 1$. For $\lambda = 0$ and +ve sign (3.1) is the usual sine-Gordon equation with boundary conditions $u \to 0(\mathrm{mod}\ 2\pi)$ as $|x| \to \infty$; for the same boundary conditions, $\lambda = 0$, and -ve sign, (3.1) is unstable—see, e.g., remarks in [3.14] and in the previous chapter. Similar unstable cases can be extracted from (3.1) when $\lambda = 1$

---

[1]We use "spin waves" to mean waves of magnetic spin excitations: these could be propagating or not, but in the latter case they are often thought of as "domain walls". Our usage always refers to a nonlinear regime and so generalises the idea of "spin waves", necessarily harmonic waves, arising in linear theory.

[2]We have not been able so far to consult Refs.28 and 29 of MAKI's paper [3.9], however.

[3.2]. The stable sine-Gordon equation governs the propagation of *strictly resonant* 'sharp line' (that is not 'inhomogeneously broadened') short optical pulses in nondegenerate absorbers (see, e.g., Chap.2 or Chap.6; also see the references therein, notably [3.1] or, e.g., [3.13-16]). The unstable case describes propagation in amplifiers [3.1]. As we shall see the stable sine-Gordon equation also governs the propagation of plane wave spin waves in the A phase of $^3$He below 2.6 mK. The first result here is due to MAKI and TSUNETO [3.7] (and see [3.19]).

Below 2.6 mK $^3$He has essentially two liquid phases, A and B [3.17]. Both are anisotropic magnetic liquids [3.3-5,7,8,18,19]. It turns out, as we shall see below, that plane wave spin waves in the B phase are governed by the double sine-Gordon equation (3.1) with $\lambda = 1$ and -ve sign [3.3-6,8,9]. It is also so that the propagation of sharp line resonant optical pulses in degenerate atomic media with a transition of Q(2) symmetry [3.1,2,14] are governed by the double sine-Gordon equation (3.1) with $\lambda = 1$ and +ve sign. In principle all four cases for $\lambda = 1$, namely +ve sign and $c = 0$ or $2\pi$, and -ve sign with $c = \delta$ or $4\pi - \delta$ describe resonant SIT in Q(2) degenerate absorbers [3.2] and there are four more corresponding cases for degenerate amplifiers [3.1].[3]

The double sine-Gordon equations form part of a sequence of multiple sine-Gordon equations which are in principle applicable [3.4]: for example the triple s-G

$$u_{xx} - u_{tt} = \sin u + \frac{1}{3}\sin\frac{1}{3}u + \frac{2}{3}\sin\frac{2}{3}u \tag{3.2}$$

describes the propagation of strictly resonant sharp line optical pulses through unexcited absorbing media with Q(3) symmetry [3.1,14] and the extrapolation to Q(J) is obvious. For these symmetries the selection rules for the optical transitions are $\Delta J = 0$, $\Delta M_J = 0$: J is an angular momentum quantum number. In the experiments [3.10] J is the hyperfine structure quantum number F = 2: the transition is 2F + 1 = 5-fold degenerate; the Wigner-Eckhart theorem shows that the distinct matrix elements are $\pm p, \pm\frac{1}{2}p, 0$ for $\Delta M_F = 0$ and, respectively, $M_F = \pm 2, \pm 1, 0$; the distinct magnitudes are $p, \frac{1}{2}p, 0$ and only p and $\frac{1}{2}p$ are significant. Equation (3.1) with +ve sign and $\lambda = 1$ follows after scaling as we show below.

Equations of double sine-Gordon type are relevant to other physical problems: for example, they arise in quasi one-dimensional charge density wave (CDW) condensate theories of the organic linear conductors like TTF-TCNQ (tetrathiafulvalene-tetracyanoquinodimethane) [3.21]. These systems consist of one-dimensional organic

---

[3]The alternative word 'attenuator' is used for the absorber in [3.2]. All absorbers (attenuators) are stable in the absence of spontaneous emission. We have discussed elsewhere [3.20] a theory of degenerate superfluorescence (and see later on in Sect.3.2) in which e.g., a medium in the unstable amplifying state associated with $c = 0$ decays to the stable states $u = \delta$ or $4\pi - \delta$ partly via the unstable $c = 2\pi$ state.

polymer chains. The sliding CDWs first introduced by PEIERLS [3.22] and considered by FRÖHLICH [3.23] translate freely; but in practice lattice commensurability breaks translational invariance [3.24] and the CDW moves in a periodic 'pinning' potential with simplest representation cos Mu in which u is the condensate phase. Interchain coupling through Coulomb interactions introduces a second potential which at low temperatures can be taken proportional to cos u. In a phase dominated context the potential $V(u) \propto \cos Mu + \lambda \cos u$, is assigned to a physically reasonable Lagrangian with density $\mathscr{L}[u] = \frac{1}{2}u_t^2 - \frac{1}{2}u_x^2 - V(u)$ [3.21]. Equations of motion are therefore

$$u_{xx} - u_{tt} = \sin u + \lambda \sin \frac{u}{M} \qquad (3.3)$$

after scaling. Since $M = 2\nu^{-1}$ where $\nu$ is the number of electrons per atom (or molecular group), M = 4 for $D(TCNQ)_2$ (D stands for a 'donor' group), M = 3 for $D_3(TCNQ)_2$; for TTF-TCNQ itself, $\nu = 0.58$ which might be approximated by $\nu = 1/2$ and M = 4. It is easy to see from phase plane studies of the Lorentz covariant equations (3.3) in the rest frame that these equations must behave in many ways very like the double s-G (3.1) with +ve sign.[4]

A form of the double s-G itself has arisen in a theory of "poling" in poly (vinylidene fluoride), $PVF_2$ [3.25]. LUTHER [3.26] has obtained the *quantised* double s-G as a mapping of the eight vertex model in a field; the eight vertex model is important in statistical mechanics and contains within it the Ising model (in this case in an electric field). It seems probable that there will be other applications of the double sine-Gordon-like equations in statistical mechanics. The propagation of fluxons in large area (say 1 mm × 1 mm) Josephson junctions is governed by the s-G (the case $\lambda = 0$) — see Chap.2 or [3.14] — and spin waves in the A phase of liquid $^3$He can be interpreted as a consequence of an internal Josephson effect [3.7,14]. The source term in Josephson's junction equations is typically *periodic*, however, [3.27], and (3.3) may have some relevance in this connection.

These remarks all suggest that the study of (3.1) is of real physical interest either directly, to the optical and spin wave problems, or indirectly, as typifying the behaviour of equations like (3.3). However, in mathematical terms, except when $\lambda = 0$, these equations stand outside the main focus of interest in this book; except when $\lambda = 0$ the systems are not integrable systems and do not have soliton solutions. The situation is suggested by two results, namely that nonlinear Klein-Gordon equations

$$u_{xt} = F(u) \qquad (3.4)$$

[4] For phase plane studies of the double s-G in the rest frame see [1.14].

have auto-Bäcklund transformation if, and only if, $F''(u) + \alpha^2 F(u) = 0$ for some $\alpha$ [3.28,29], and that they have an infinity of polynomial conserved densities (p.c. ds, see Chap.1) if and only if $F''(u) + \alpha^2 F(u) = 0$ [3.30]. That the equations are not integrable for either choice of sign is finally confirmed by numerical work [3.4-6,31,32]; there is now no doubt that a pair of solitary wave solutions (actually kink solutions) of (3.1) with -ve sign and $\lambda = 1$ emit radiation after their collision [3.32,33]; there is no doubt that solutions in the +ve sign case also emit radiation (Fig.3.1 shows one example).[5] We have argued in Chap.1 that this behaviour means that the equations cannot be solved by a form of the inverse scattering method (see the note in proof on p.138).

It might be concluded from this that the study of equations (3.1) has no place in this volume. However, the equation with +ve sign at least is close in some functional sense to the completely integrable sine-Gordon equation obtained for $\lambda = 0$. Singular perturbation theory about the s-G has already been successfully used to describe the behaviour of two of the simpler solutions of (3.1) [3.32,34-36] and we describe this theory briefly at the end of this chapter.

In what follows we treat the problem of degenerate SIT in Sect.3.2 and the $^3$He spin wave problem in Sect.3.3; we then briefly sketch the perturbation theory in Sect.3.4. To close Sect.3.1 it is perhaps worth showing in connection with Chap.5 that this direct method yields the solitary wave solutions of the double s-G but does not yield 2-soliton solutions.

To show this we define the differential operator (see Chap.5)

$$D_x^m D_t^n(a \cdot b) = \left\{\left(\frac{\partial}{\partial x} - \frac{\partial}{\partial x'}\right)^m \left(\frac{\partial}{\partial t} - \frac{\partial}{\partial t'}\right)^n \left[a(x,t)b(x',t')\right]\right\}_{\substack{x'=x\\ t'=t}} \quad .$$

The argument proceeds as follows: put

$$u = 4\tan^{-1}(f/g) \quad ,$$

then

$$u_x = 4D_x(f \cdot g)/(f^2 + g^2)$$

$$u_{xt} = 4\left[-(f^2 - g^2)D_xD_t(f \cdot g) + fgD_xD_t(f \cdot f - g \cdot g)\right] \times (f^2 + g^2)^{-2}$$

$$\sin\frac{1}{2}u = 2fg(f^2 + g^2)^{-1} \quad ; \quad \sin u = -4fg(f^2 - g^2)(f^2 + g^2)^{-2} \quad .$$

Take (3.1) with +ve sign: change $\frac{1}{2}(x - t) \to x$, $\frac{1}{2}(x + t) \to t$ so that

[5] In [3.28] where the integrability of the double s-G is discussed we tended to conclude otherwise. Further work has demonstrated that, despite the evidence of Fig.1 in this reference, significant radiation occurs in a number of situations.

$$u_{xt} = \sin u + \frac{1}{2}\lambda \sin\frac{1}{2}u \quad , \tag{3.5}$$

and write $4\lambda$ for $\lambda$ and choose $\lambda > 0$. The equation implies

$$-(f^2 - g^2)D_xD_t(f\cdot g) + fgD_xD_t(f\cdot f - g\cdot g) - \lambda fg(f^2 + g^2) + fg(f^2 - g^2) = 0 \quad .$$

This suggests we put

$$D_xD_t(f\cdot g) = \mu fg$$

$$D_xD_t(f\cdot f - g\cdot g) = (\mu - 1)(f^2 - g^2) + \lambda(f^2 + g^2) \quad .$$

If we let

$$f = \varepsilon f^{(1)} + \varepsilon^3 f^{(3)} + \ldots$$

$$g = 1 + \varepsilon^2 g^{(2)} + \varepsilon^4 g^{(4)} + \ldots$$

and equate powers of $\varepsilon$ we find $\mu = 1 + \lambda$ from $\varepsilon^0$ and

$$f^{(1)}_{xt} = (1 + \lambda)f^{(1)}$$

from $\varepsilon^1$. Then if $f^{(1)} = e^{\theta}$, $\theta = \omega t - kx + \delta$ with $\omega k = -1 - \lambda$ ($\omega$, $\lambda > 0$ mean $k < 0$ therefore) we find

$$g^{(2)} = [-\lambda/4(1 + \lambda)]$$

and all other $f^{(n)}$, $g^{(n)}$ can be chosen to vanish. Thus

$$u = = 4\tan^{-1}\left[-\sqrt{1 + \lambda^{-1}}/\sinh\theta'\right]$$

$$\theta' = \theta + \ln\left[\frac{1}{2}\,\varepsilon\sqrt{\lambda(1 + \lambda)^{-1}}\right] \tag{3.6}$$

is a solution. For $x \to -\infty$, $u \to 0$; for $x \to +\infty$, $u \to 4\pi$. It is therefore a $4\pi$-kink solution of permanent shape. However, for a 2-soliton solution we would make the natural choice

$$f^{(1)} = e^{\theta 1} + e^{\theta 2} \quad ; \quad \theta_i = \omega_i t - k_i x + \delta_i (i = 1,2)$$

with $\omega_i k_i = -1 - \lambda$. The series does not terminate in this case in general but does so when $\lambda = 0$ (the case of the s-G).

## 3.2 Theory of Degenerate SIT

We consider a model medium of '2-level atoms': the levels are each 5-fold, $F = 2$, degenerate. Selection rules are $\Delta F = 0$, $\Delta M_F = 0$ so there are 5 possible transitions at the same resonant frequency labelled by $M_F = \pm 2, \pm 1, 0$. There is therefore a 5-pseudo spin description for each atom with a spin up labelling occupation of an upper state and spin down labelling occupation of a lower state [3.14,31]. But we have noted in Sect.3.1 that only two significant magnitudes of transition matrix elements arise, p and $\frac{1}{2}$ p. Accordingly the 5-pseudo spin problem can be reduced to a 2-pseudo spin problem with these matrix elements [3.31]. It is easy to see [3.1,2,14] that there is a Bloch equation for each of these atomic pseudo spins (single spin Bloch equations in the theory of nondegenerate SIT are introduced in Chap.2. Just as the electric field envelope $\mathcal{E}$ couples the pseudo spins of different atoms in the nondegenerate case, now it also couples the two pseudo spins in each of the degenerate atoms, and this is the only coupling (only transverse field couplings are considered). The electric field envelope $\mathcal{E}$ is a field envelope (time scale $\lesssim 10^{-9}$ s) modulating a harmonic carrier wave on exact resonance with the frequency of the atomic optical transition. From long usage [3.13-16] we call this frequency $\omega_s$ but make no use of this symbol here: $\hbar\omega_s$ is thus the energy spacing of the two resonant but degenerate atomic levels if such a quantity should be required.

In general Bloch equations involve the coupled motion of an atomic inversion N, a component of the envelope of atomic dipole moment Q in phase with the electric field envelope $\mathcal{E}$, and an envelope P of dipole moment in phase quadrature.

On exact resonance in the sharp line case Q is a constant of the motion which must be zero for an initially unexcited medium (cf., e.g. [3.16]). The equations of motion for the degenerate case therefore involve $P^{(i)}$, $N^{(i)}$ $i = 1,2$ only. They prove to be [3.1,2,5,13-16] (and compare Chap.2)

$$P_t^{(i)} = -\kappa^{(i)}N^{(i)}\mathcal{E}$$

$$N_t^{(i)} = \kappa^{(i)}P^{(i)}\mathcal{E} \quad ; \quad \kappa^{(i)} = 1,\ 1/2 \quad ; \quad i = 1,2$$

$$c\mathcal{E}_x + \mathcal{E}_t = \alpha\left(P^{(1)} + \frac{1}{2}P^{(2)}\right) \tag{3.7}$$

in suitable units. In these units (see [3.13-16]) the coupling constant $\alpha$ has the dimension of frequency squared and $\sqrt{\alpha}$ plays an interesting role in a quantised s-G theory based on (3.7) [3.4,5,13]. By taking suitable linear combinations of atomic state vectors, one finds that the equations of motion for many atomic level schemes can be put into the form appearing in (3.7): typically the new features are then $i = 1,2,\ldots,N$; the $\kappa^{(i)}$ cease to be simple rational fractions; the dipoles $P^{(i)}$ assume weights $w^{(i)}$ in the Maxwell equation in $\mathcal{E}$ and the $P^{(i)}$. Some examples appear in [3.31]. A 3-spin problem with weights is considered before (3.17) below.

The c-number equations (3.7) are reduced to (3.1) (+ve sign and $\lambda = 1$) by setting $P^{(i)} = -\sin\kappa^{(i)}u$, $N^{(i)} = \cos\kappa^{(i)}u$, $\mathcal{E} = u_t$, from which

$$u_{tt} + u_{xt} = \alpha(\sin u + \frac{1}{2}\sin\frac{1}{2}u) \quad . \tag{3.8}$$

---

$\hbar = h/2\pi$ (normalized Planck's constant)

Then, after the change of variables $\xi = \sqrt{\alpha}(2x - t)$, $\tau = \sqrt{\alpha}t$ one finds

$$u_{\xi\xi} - u_{\tau\tau} = \sin u + \frac{1}{2}\sin\frac{1}{2}u \tag{3.9}$$

exactly.

This Lorentz covariant equation can be solved for single kinks by working in the rest frame. Trajectories in the phase space $u$, $u_\xi$ are

$$\frac{1}{2}u_\xi^2 + \cos u + \cos\frac{1}{2}u = C \quad . \tag{3.10}$$

C is a constant. For C = 2 the trajectory connects unstable singular points at $u_\xi = 0$, $u = 0$ and $u_\xi = 0$, $u = 4\pi$. This trajectory describes the solution corresponding to (3.6) in the changed coordinates [we transferred from Lorentz covariant to light cone coordinates at (3.5) in reaching (3.6)].

In a similar way one can solve (3.1) in all its cases: a comprehensive solution is

$$u = u_0 + 4\tan^{-1}[\cosh\delta_3 \exp(\theta - \delta_1) + \sinh\delta_3]$$
$$+ 4\tan^{-1}[\cosh\delta_3 \exp(\theta - \delta_2) - \sinh\delta_3]$$
$$\theta = \omega(t - xv^{-1}) + \theta_0 = \frac{\omega}{2\sqrt{\alpha}v}[(2v - 1)\tau - \xi] + \theta_0$$
$$= u_t = \omega\frac{du}{d\theta} \quad . \tag{3.11}$$

Four cases are of particular interest here: these are[6]

I) $u_0 = 0$, $\delta_1 = -\delta_2 = \ln(\sqrt{5} + 2) \equiv \Delta$, $\delta_3 = 0$

$$u = 4\tan^{-1}(e^{\theta-\Delta}) + 4\tan^{-1}(e^{\theta+\Delta})$$
$$\mathscr{E} = 2\omega\,\mathrm{sech}(\theta - \Delta) + 2\omega\,\mathrm{sech}(\theta + \Delta)$$
$$v = \omega^2\left(\omega^2 + \frac{5}{4}\alpha\right)^{-1} \tag{3.12}$$

which satisfies (3.1) for $\lambda = 1$ and +ve sign with boundary conditions (b.cs) $u \to 0 (\mathrm{mod}\ 4\pi)$, $|x| \to \infty$.

II) $u_0 = 2\pi$, $\delta_1 = i\pi - \delta_2 = \ln(\sqrt{3} + 2) = \Delta'$, $\delta_3 = 0$

$$u = 2\pi + 4\tan^{-1}(e^{\theta-\Delta'}) - 4\tan^{-1}(e^{\theta+\Delta'})$$
$$\mathscr{E} = 2\omega\,\mathrm{sech}(\theta - \Delta') - 2\omega\,\mathrm{sech}(\theta + \Delta')$$
$$v = \omega^2[\omega^2 + 3/4\alpha]^{-1} \tag{3.13}$$

which satisfies (3.1) for $\lambda = 1$ and +ve sign with b.cs $u \to 2\pi (\mathrm{mod}\ 4\pi)$, $|x| \to \infty$ (actually $u \to 2\pi$ for $x \to \pm\infty$).

---

[6]There is an error in [3.4,5] where $\Delta$ in (I) and $\Delta'$ in (II) are given respectively as $\ln(\frac{1}{2} + \frac{1}{2}\sqrt{5})$ and $\ln(\frac{1}{2} - \frac{1}{2}\sqrt{3})$.

III) $u_0 = \delta - 2\pi,\ \delta_1 = 0,\ \delta_2 = \infty,\ \delta_3 = \frac{1}{2}\ln 5/3$

$$u = 2\pi - \delta + 4\tan^{-1}\left(\frac{4}{\sqrt{15}}e^{\theta} + \frac{1}{15}\right)$$

$$\mathcal{E} = \frac{2\sqrt{15}\omega}{4\cosh\theta + 1} \quad , \quad V = \frac{\omega^2}{\omega^2 + \frac{15}{16}\alpha} \tag{3.14}$$

which satisfies (3.1) with -ve sign, $\lambda = 1$, and b.cs $u \to +\delta$, $x \to +\infty$, $u \to 4\pi - \delta$, $x \to -\infty$.

IV) $u_0 = 2\pi - \delta,\ \delta_1 = 0,\ \delta_2 = -\infty,\ \delta_3 = -\frac{1}{2}\ln 5/3$

$$u = \delta - 2\pi + 4\tan^{-1}\left(\frac{4}{\sqrt{15}}e^{\theta} - \frac{1}{\sqrt{15}}\right)$$

$$\mathcal{E} = \frac{2\sqrt{15}\omega}{4\cosh\theta - 1} \quad , \quad V = \frac{\omega^2}{\omega^2 + \frac{15}{16}\alpha} \tag{3.15}$$

which satisfies (3.1) with -ve sign, $\lambda = 1$, and b.cs $u \to -\delta$, $x \to +\infty$, $u \to +\delta$, $x \to -\infty$.

Other solutions corresponding to different b.cs can be found: in particular fo- $\lambda = 0$ one finds $u = 4\tan^{-1}(e^{\theta})$, or $u = -\pi + 4\tan^{-1}(e^{\theta})$ with $V = \omega^2(\omega^2 \pm \alpha)^{-1}$, respectively. These are kink solutions corresponding to the nondegenerate absorber ($N \to -1$ as $|x| \to \infty$) and amplifier ($N \to +1$ as $|x| \to \infty$), respectively. One sees that in the amplifier $V > 1$ (i.e., $V > c$), a point discussed in [3.14] for example. Solutions for the degenerate amplifiers of (3.1) can also be found: for example (3.12) is unchanged except that $V = \omega^2[\omega^2 - 5\alpha/4]^{-1} > 1$.

Notice that the solution (3.12) for u describes a kink with jump $4\pi$: it is the solution (3.6) again, but in the different coordinates and now expressed as the sum of two $2\pi$ kinks of the s-G (case $\Delta = 0$) separated by $2\Delta = 2\ln(\sqrt{5} + 2)$. Likewise (3.13) is the difference of two $2\pi$ kinks of the s-G separated by $2\Delta' = 2\ln(\sqrt{3} + 2)$. The solutions (3.14,15) are kinks with jumps $4\pi - 2\delta$ and $2\delta$, respectively, and cannot be expressed as combinations of $2\pi$-kinks of the s-G.

The electric fields in (3.12,13) are the sum and difference of hyperbolic secant pulses. The hyperbolic secant pulse in the nondegenerate adsorber has "area" $\Theta = 2\pi$: the "area theorem" for resonant optical pulses in nondegenerate SIT is described in Chap.2 and there is an interesting analysis by Newell in terms of the inverse scattering transform in Chap.6. There is a corresponding area theorem in the degenerate case even though there is no scattering transform: this area theorem is

$$\Theta_x = \pm\beta(\sin\Theta + \frac{1}{2}\sin\frac{1}{2}\Theta) \tag{3.16a}$$

where as in the nondegenerate case

$$\Theta(x) \equiv \int_{-\infty}^{\infty} \mathcal{E}(x,t)dt \quad . \tag{3.16b}$$

It is important to realise that x,t are the laboratory coordinates for SIT as these appear in (3.7): note that the change of variables to $\xi$, $\tau$ made after (3.8) is necessary to obtain the covariant form (3.9).

For the sharp line problem $\beta \propto \delta(0)$ where $\delta(x)$ is the $\delta$ function [3.1,14,31]. Hence the only acceptable areas are the zeros of $\sin\Theta + \frac{1}{2}\sin\frac{1}{2}\Theta$, namely 0, $2\pi$, $\delta$ and $4\pi - \delta$, (mod $4\pi$). A little thought shows that this area theorem corresponds to the absorber with b.cs $u \to 0 (\mathrm{mod}\ 4\pi)$, $|x| \to \infty$ and that the areas $\delta$ and $4\pi - \delta$ are reached in the amplifier described by $P^{(i)} = \sin\kappa^{(i)}u$, $N^{(i)} = \cos\kappa^{(i)}u$ [compare the +ve signs with the choice by (3.8)], and $u = 0$. This amplifier has the solution (3.12) with $v > 1$ but is unstable to change of area from $4\pi$ to $4\pi - \delta$. There is an area theorem for each set of b.cs and the fundamental areas are $\Theta = 0, 2\pi, 4\pi$; $\Theta = 0, 2\pi, 4\pi$; $\Theta = 0, 4\pi - 2\delta, 4\pi$; $\Theta = 0, 2\delta, 4\pi$. The solutions (3.12-15) are the distortionless solutions corresponding to each case.

In the optical problem the collective states of the optical medium $u = 2\pi$, $\delta$ and $4\pi - \delta$ are excited states and as noted in Sect.3.1 are rendered unstable by the spontaneous emission of radiation. Although in the far infrared such states could last for times $\sim 1$ s [3.2,37], no experiments have been done on them. The state $u = 0$, if this is an absolute ground state, does not suffer from this instability. We therefore focus attention on the solution (3.12).

Evidently the optical pulse (3.12) is double peaked and the energy $\mathcal{E}^2$ associated with it is double peaked. The analytical form suggests that still further approximate solutions of (3.9) can be found by changing $\Delta$. It is easy to see that the form is stable even to large changes of $\Delta$: consider the single $2\pi$ pulse $\mathcal{E} = 2\omega\,\mathrm{sech}\theta$ entering the 2-spin degenerate medium. Such a pulse rotates the 'heavy' spin (with matrix element p) by $2\pi$, but the 'light' spin (with matrix element $\frac{1}{2}$ p) by $\pi$. It thus leaves a trail of excitation $u = 2\pi$ (light spin up—heavy spin down) behind it, loses energy (and amplitude), and slows down. Now consider a second $2\pi$ sech pulse entering the medium $u = 2\pi$ excited by the first $2\pi$ pulse. This second pulse restores both atomic spins to spin down, gains energy, accelerates and (as numerical work shows) passes the slowed $2\pi$ sech just like any s-G soliton. Thereafter it loses energy to the medium $u = 0$ before it and slows down whilst the previously slowed $2\pi$ sech, seeing the state $u = 2\pi$ before it, gains energy and accelerates. On this large scale we have called the motion the "leap frogging" of optical pulses [3.2,10]. But any two kinks like (3.12) with $\Delta \neq \ln(2 + \sqrt{5})$ will go into oscillation with an apparent finite binding energy: the pair (3.12) itself is a bound pair of $2\pi$ sechs at equilibrium. Notice that the Hamiltonian density for (3.8) is

$$\mathcal{H} = \frac{1}{2}u_\tau^2 + \frac{1}{2}u_\xi^2 + V(u)$$

$$V(u) = \left(2 - \cos u - \cos\frac{1}{2}u\right) \quad . \tag{3.17}$$

$V(u)$ has deep minima at $u = 0$ and $4\pi$ [where $V(u) = 0$] and a shallower minimum at $u = 2\pi$ [where $V(u) = 2$]. It is perhaps helpful to view the perturbed oscillatory solution as a kink interpolating between the two deep minima but capable of oscillations associated with the shallower minimum [3.35]. This kink is necessarily stable since it is stabilised topologically by the jump of $4\pi$.[7]

The conclusion from this analysis is that, for both small and large changes of $\Delta$, the $4\pi$ kink (3.12) has an internal oscillation approximately interpretable as the oscillations of a bound pair of sine-Gordon $2\pi$ kinks. The numerical integration of (3.9) confirms this expectation totally [3.2,4,31]. We show in Fig.3.1 the behaviour of a single $4\pi$ kink with $\Delta$ a little different from $\ln(\sqrt{5}+2)$. We plot the kink (in u) because this way one can see the emission of radiation at this 'wobbler' wobbles. But in Fig.3.2 we plot the electric field $\mathscr{E}$ and show the collision of two $4\pi$-optical pulses each initially in equilibrium ($\Delta = \ln(\sqrt{5}+2)$). In Fig.3.2 the strict collision property of solitons as described in Chap.1 seems well exemplified; but study of the details shows that some radiation is emitted after the collision and our opinion now is that this is significant.[8]

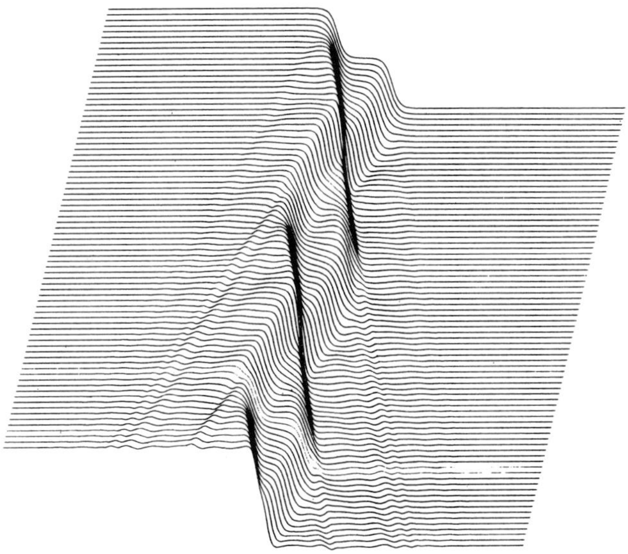

Fig.3.1. The single "wobbling" $4\pi$ kink of the double s-G with +ve sign (and $\lambda = 1$). Note the emission of radiation especially when the two $2\pi$ kinks meet

[7]For a layman's introduction to topological solitons see [3.38]. But beware that only the sine-Gordon solitons considered there are solitons in the way that word is used in this book.

[8]Unfortunately all numerical work even for the integrable systems like the s-G shows some emission of radiation. The problem is always to decide its significance (see [3.28]).

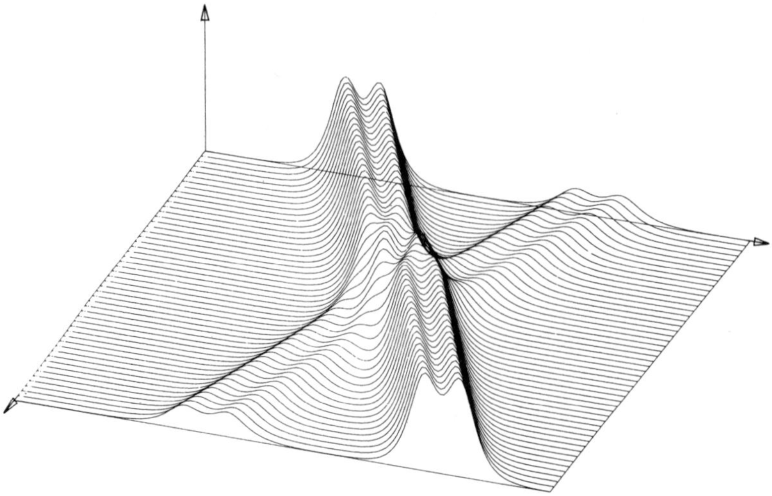

Fig.3.2. Collision of two $4\pi$ kinks of the double s-G each in the equilibrium (nonwobbling) state before collision. The electric field (the derivative of u) is plotted

It remains to observe such behaviours in the laboratory. Fortunately Na vapour which is excellent for SIT studies [3.39] has an $F = 2 \rightarrow F' = 2$ hyperfine levels transition amongst its $D_1$ $2S_{1/2} \rightarrow 2P_{1/2}$ transitions. There are additional transitions: the lower $2S_{1/2}(F = 1)$ level is 1772 MHz away from the F = 2 and this level spacing is well resolved by a 5 ns pulse. The upper $2P_{1/2}(F' = 2)$ has an $F' = 1$ state only 192 MHz away. This is just about resolved by a 5 ns pulse. At first sight the resolved problem is Q(2) and the unresolved problem involves a three level atomic system with the F = 2 a shared lower level. However, Doppler broadening of about 1700 MHz line width due to atomic motions means that each transition $F = 2 \rightarrow F' = 2$, $F = 2 \rightarrow F' = 1$ will go into exact resonance in two different atomic groups. Thus on the one hand both transitions are excited, but on the other one can assign distinct spins to each transition. The result is transitions with three distinct matrix elements p, $\frac{1}{2}$ p, $\sqrt{3/2}$ p and weights 3, 2 and 2. The resulting equation of motion is the triple s-G

$$u_{\xi\xi} - u_{\tau\tau} = 3 \sin u + \sin \frac{1}{2} u + \sqrt{3} \sin\sqrt{3/2}\, u \qquad (3.18)$$

instead of (3.9).

Equation (3.18) is of interest because the nonlinear term is not periodic. Its zeros arise at $1.1\pi$, $2.2\pi$, $3.0\pi$, $4.2\pi$, ..., $7.8\pi$, ... . Its phase plane trajectory from u = 0 is that shown in Fig.3.3. Notice the alteration of high and low minima at each second zero. There is an area theorem like (3.16a) with now $\pm\beta[3 \sin\Theta + \sin \frac{1}{2} \Theta + \sqrt{3} \sin\sqrt{3/2}\, \Theta]$ on the right side. Areas $4.2\pi$ and $7.8\pi$ at the deep minima in Fig.3.3 are not only stable for the area theorem but leave least energy as excitation in the atoms: numerical work [3.31] shows, for example, that

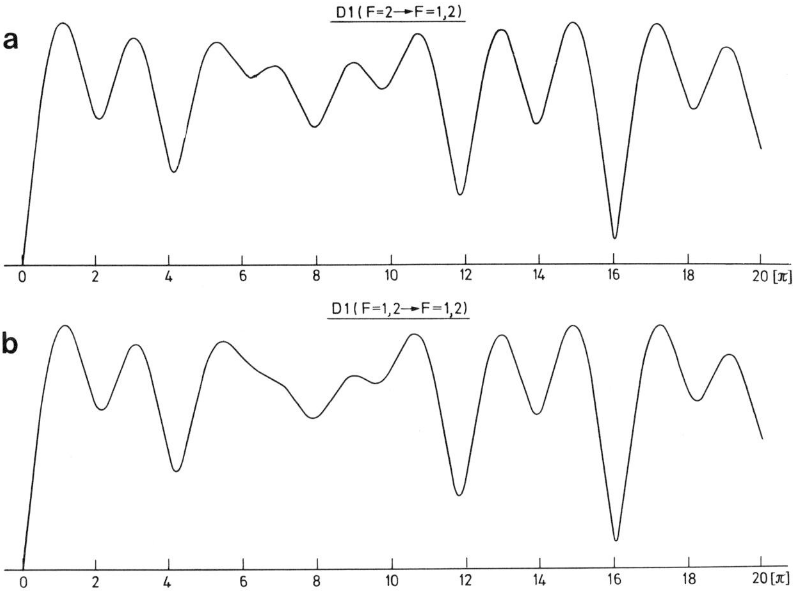


Fig.3.3. (a) Phase plane trajectory from u = 0 for (3.18) corresponding to Na $D_1(F = 2 \rightarrow F' = 1, 2)$; (b) the corresponding trajectory for Na $D_1(F = 1 \rightarrow F' = 1, 2)$

a $6.2\pi$ pulse sheds a pulse of area $2\pi$ at the boundary of the medium and that the remaining pulse evolves as a $4.2\pi$ double peaked wobbler. The experiment [3.10] consisted in observing this wobbler.

The procedure was to send an approximately $4\pi$, 5 ns plane wave pulse,[9] a few $Wcm^{-2}$ at peak intensity, into a 1 cm long cell of NA vapour and observe the output. The effective length of the cell was changed by changing the vapour density of Na within it. In this way different intensity profiles were presented to the detector. Figure 3.4a shows numerical values of the successive intensity profiles for the Q(2) system governed by (3.9): data are for the hyperfine Na $D_1$ transitions $F = 2 \rightarrow F' = 2$ each, however, with certain damping constants included. These constants have values $T_1 = 24$ ns, $T_2 = 48$ ns, close to the proper values for 3-level

[9]All pulses are of finite transverse aperture and both whole beam self focussing and filamentation (see [3.42] below) are possible in principle. The former was excluded by the choice of large enough diameter beams. Agreement with plane wave simulations suggests neither focussing nor filamentation occurs. For the experts the approximately 5 ns pules were produced by amplification of the output of a Spectra Physics 580 CW dye laser by an $N_2$-pumped single pass dye amplifier. Uniform plane wave conditions were actually well satisfied: the input beam was a few mm in diameter with small divergence, and the output at the cell exit was magnified four times and imaged on a 280 μm diameter Si avalanche photodiode. The output of the latter was observed with subnanosecond resolution by a sampling oscilloscope, boxcar and recorder.

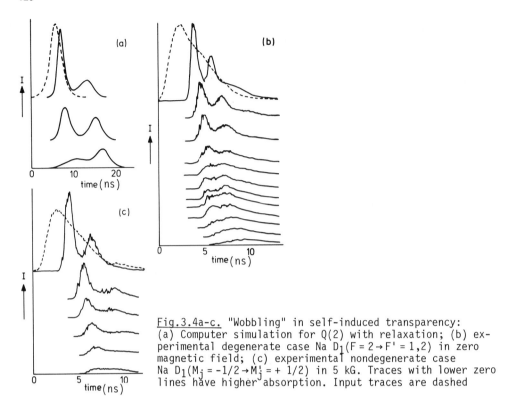

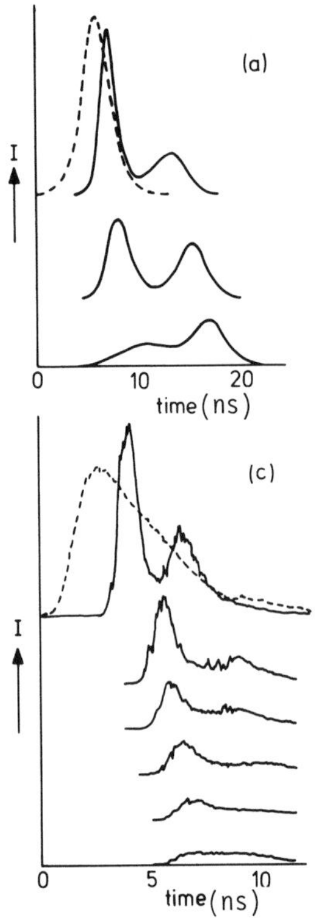


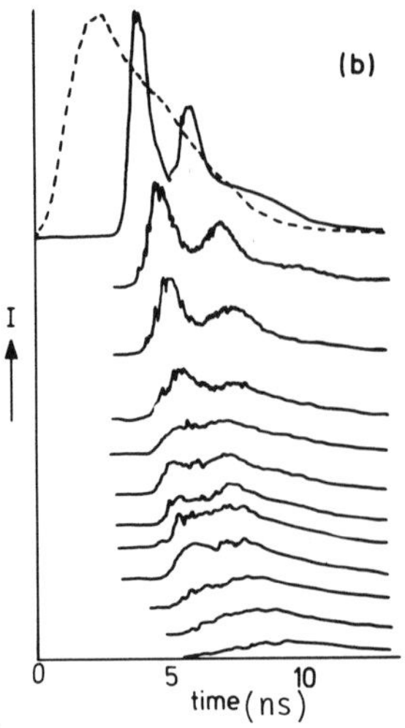


Fig.3.4a-c. "Wobbling" in self-induced transparency: (a) Computer simulation for Q(2) with relaxation; (b) experimental degenerate case Na $D_1(F=2\rightarrow F'=1,2)$ in zero magnetic field; (c) experimental nondegenerate case Na $D_1(M_j=-1/2\rightarrow M_j'=+\,1/2)$ in 5 kG. Traces with lower zero lines have higher absorption. Input traces are dashed

Na.[10] This computer simulation shows that, in the two-peaked wobbler, the peak leading in time wobbles down first whilst the trailing peak wobbles up — even when damping is included. Fig.3.4b shows the sequence of outputs actually observed. Damping mechanisms (homogeneous broadening or spontaneous emission) damp both leading and trailing peaks but not sufficiently to destroy the leading peak's relative wobble down which is clearly to be seen in Fig.3.4b. In contrast in Fig.3.4c the experiment is repeated with a magnetic field of 5 kG imposed. This field splits the hyperfine degeneracies and the system is governed by the s-G. The $4\pi$ pulse input now breaks up into $2\pi$ pulses with the larger amplitude (equals higher speed) pulse leading. The damping mechanisms damp the weaker trailing pulse preferentially (we believe this can be demonstrated by perturbation theory: certainly experiment

---

[10]Data are actually for the $2S_{1/2}$ $F=2\rightarrow 2P_{1/2}$ $F=1,2$ transitions which are damped by spontaneous emission. The radiative relaxation times vary slightly from transition to transition, but all of them are close to the high-magnetic field values $T_1 = 24$ ns, $T_2 = 48$ ns used in the simulations. Reference can be made to the Ph.D. thesis of G.V. Salamo (City University, New York, 1974) which shows the branching ratio from each F, $M_F$ substate. The point is that every excited substate has a 16 ns lifetime, but the branching to the lower state ($\Delta M_F = 0$) as opposed to all other states ($\Delta M_F = \pm 1$) is not the same. Thus strictly speaking $T_1$ and $T_2$ are different. We believe the variation has negligible effect.

confirms this conclusion[11]) and this sequence appears to wobble down on the *trailing* pulse. Of course the observation of actual pulse breakup would clinch the matter, but clear pulse breakup in nondegenerate SIT has only been obtained in other circumstances — again in Na vapour [3.40].

This gratifying result in nonlinear optics is one of several recently obtained: these include (a) the observation of well separated pulse breakup in nondegenerate Na vapour mentioned [3.40]; (b) observation of self-focussing and some optical filamentation in nondegenerate SIT with finite incident transverse profiles [3.12, 41,42]; (c) observation of the *interaction* of oppositely directed optical pulses in Na [3.13,42], and (d) observation of single pulse 'superfluorescent' emission [3.43] from Cs vapour [3.44]. The breakup (a) is a beautiful example of soliton breakup; the self-focussing (b) is predictable [3.13], but the nondegenerate (and degenerate) SIT equations are apparently unstable to self-defocussing [3.13,45]; the observation of filamentation is suggestive of transverse solitonlike breakup (see Sect.1.4 and the articles referred to there); the interaction (c) of oppositely directed optical pulses is consistent with the fact that the SIT equations (3.7) are *not* invariant under $x \to -x$ and apparently become a nonintegrable system when envelope pulses travel in opposite directions (for the relevant equations in this case see, e.g. [3.13]); the single pulse superfluorescence (d) is connected with the theory of the s-G under boundary and initial conditions very different from those conditions — namely $u \to 0 (\mathrm{mod}\ 2\pi)$, $u_x$, $u_{xx} \to 0$; $u(x,0) = u_0(x)$, $u_t(x,0) = u_1(x)$ — for which the s-G has soliton solutions. A relevant set is $u(0,t) = u_t(0,t) = 0$; $u(x,0) = \varepsilon_0$, $u_t(x,0) = 0$, $0 < x < L$; $10^{-2} \leqq \varepsilon_0 \leqq 10^{-10}$ and the unstable case $u_{xt} = -\sin u$ of the s-G applies [3.43].

## 3.3 Spin Waves in Liquid $^3$He

The experimental results described in Sect.3.2 illustrate how nonlinear optics has been able to influence soliton theory. Its influence on the evolution of the subject has been summarised in Sect.1.3. Nuclear magnetic resonance (NMR) is the experimental tool which has so far been used to elucidate the properties of the magnetic liquids $^3$He A and $^3$He B. The motion of nuclear spins is governed by Bloch equations and these describe nonlinear oscillators in a spatially homogeneous situation. One becomes concerned with solitonlike excitations when translational invariance is broken. It is this aspect we examine in what follows.

---

[11]For example if one has a $5\pi$ pulse showing good break up and one increases the absorption (and hence the losses) one sees that the second pulse dies first. The same has been seen with weak focussing to compensate losses so enhancing the pulse separation.

Elsewhere [3.3] we have introduced an 'adiabatic' Hamiltonian density $\mathscr{H}$ which generalises that proposed by LEGGETT [3.19] to describe the unusual NMR behaviours observed in $^3$HeA and $^3$HeB. This Hamiltonian density is

$$\mathscr{H}(\underline{x}) = \frac{1}{2}\gamma^2\chi^{-1}\sigma_w^2 - \gamma\sigma_w B_0 + \frac{1}{4}\gamma^{-2}\bar{c}^2\nabla\Theta \cdot \underline{\underline{\rho}} \cdot \nabla\theta + H_D(\theta) \quad ; \tag{3.19}$$

$\gamma = e/mpc$ is the gyromagnetic ratio, $\chi$ is the magnetic susceptibility, $\bar{c}$ is a velocity ($\leq v_F \sim 10^3$ cm s$^{-1}$) and $H_D(\theta)$ is the dipole interaction; $\underline{\underline{\rho}}$ = diag (2,1,2) for $^3$HeB and diag (2,2,1) for $^3$HeA because of the different preferred directions, y and z, in the two phases [3.3] and the fraction "1/4" is a dummy taken to ensure that $\bar{c}$ is the spin wave speed in each case [3.3,4]; $\sigma_w$ is a spin one operator in spin space labelled by (u, v, w) and for technical reasons [3.3] (u, v, w) corresponds to (z, x, y); $B_0$ is a constant uniform magnetic field applied along the y direction which is also the w direction; $\mathscr{H}(x)$ needs scaling by $\rho_s\hbar$ where $\rho_s$ is the superfluid number density. The "order parameters" which define the states of superfluid $^3$He are complicated and it is sufficient for the nonexpert reader to apprehend that (3.19) is a usuable Hamiltonian density describing the motion of spin waves in these substances.

For the equations of motion we define the frequency $\Omega_\chi \equiv 4\gamma^2\chi^{-1}$, the longitudinal NMR frequency $\Omega_\ell$, and the coupling constant $\gamma_0 = \Omega_\chi\Omega_\ell^{-1}$ (for technical reasons [3.3,4] $\Omega_\ell = \Omega_{\ell A}$ for $^3$He A, but $\Omega_\ell = 4\Omega_{\ell B}/\sqrt{15}$ for $^3$He B). We introduce the natural commutation relations

$$[\sigma_w(\underline{x},t),\theta(\underline{x},t)] = i\delta(\underline{x} - \underline{x}') \quad . \tag{3.20}$$

Heisenberg's equations are

$$\begin{aligned}\theta_t &= \gamma^2\chi^{-1}\sigma_w - \gamma B_0\\ \sigma_{w,t} &= -\,\delta H_D/\delta\theta + \frac{1}{2}\chi\gamma^{-2}\bar{c}^2\nabla \cdot \underline{\underline{\rho}} \cdot \nabla\theta\end{aligned} \tag{3.21a}$$

and

$$\theta_{tt} = \frac{1}{2}\bar{c}^{-2}\nabla \cdot \underline{\underline{\rho}} \cdot \nabla\theta - \gamma^2\chi^{-1}\gamma H_D/\delta\theta \tag{3.21b}$$

with

$$[\theta,\theta_t] = i\gamma^2\chi^{-1}\delta(\underline{x} - x') \quad . \tag{3.21c}$$

Results for the A phase are [3.3,7,19]

$$H_D(\theta) = -\frac{3}{5}g_{DA}(T)\cos^2\theta + \text{const} \tag{3.22a}$$

and for the B phase [3.3,8,19]

$$H_D(\theta) = \frac{4}{5}g_{DB}(T)(\cos\theta + \cos 2\theta) + \text{constant} \quad . \tag{3.22b}$$

In both cases $g_D$ is a number depending on temperature T which will be absorbed in $\Omega_\ell$. The second form (3.22b) is expressed in a way which immediately illustrates the two-spin description of the double s-G — but note the change of sign compared with (3.17).

We shall consider plane waves propagating up $B_0$ the direction of y and rotating in the x-z plane: this is because spins tend to lie parallel to the direction of the so-called orbital vector $\hat{\underline{\ell}}$ in the A phase and $\hat{\underline{\ell}}$ sets perpendicularly to $B_0$; in the B phase there is no orbital vector and $B_0$ defines the natural axis for rotations parametrised by the angle $\theta$. It is convenient to relabel the y axis, the direction of $B_0$, as x.

For the A phase we then find

$$\theta_{tt} - \bar{c}_A^{2}\theta_{xx} = -\Omega_{\ell A}^2 \sin\theta\cos\theta \qquad (3.23)$$

$\left[\Omega_{\ell A}^2 = \frac{6}{5}\, g_{DA}(t)\gamma^2\chi^{-1}\right]$. For the B phase

$$\theta_{tt} - \bar{c}_B^{2}\theta_{xx} = \frac{4}{15}\,\Omega_{\ell B}^2(\sin\theta + 2\sin 2\theta) \qquad (3.24)$$

$\left[\Omega_{\ell B}^2 = 3g_{DB}(T)\gamma^2\chi^{-1}\right]$. We set $u = -2\theta$ (A phase) and $u = 2\theta$ (B phase). We use t for $\Omega_\ell t$. Then we easily reach the operator equations

$$u_{xx} - u_{tt} = \sin u \qquad (3.25)$$

(A phase) and

$$u_{xx} - u_{tt} = -\left(\sin u + \frac{1}{2}\sin\frac{1}{2}u\right) \qquad (3.26)$$

(B phase) with the commutation relation

$$[u,u_t] = i\gamma_0\delta(x - x') \qquad (3.27)$$

in both cases.

The 'gap matrix' which defines $\Theta$ [3.4] is a c number so $\theta$ is a c number. Nevertheless, the spin wave elementary excitations below 2.6 mK could arguably be quantised. In this case (3.25,26) must be solved as operator equations. The spectrum of (3.25) is obtained by Luther in Chap.12: it is that quoted in (1.104) in Chap.1. The quantised double s-G, with either of the signs in (3.1), has not been solved even for its spectrum although its S-matrix is known to involve particle production [3.46].[12] (Footnote 12 see p.125.)

The c-number equation (3.26) derives from the Hamiltonian density

$$\mathscr{H}(x) = \gamma_0^{-1}\left[\frac{1}{2}u_x^2 + \frac{1}{2}u_t^2 + 2\left(\cos\frac{1}{2}u + \frac{1}{4}\right)^2\right] . \qquad (3.28)$$

The dipole interaction minima occur at $u = \delta = 2\cos^{-1}(-1/4)$ and at $4\pi - \delta$. There are two kink solutions of permanent shape namely (3.14,15). Bearing in mind that (3.26) is Lorentz covariant in variables (x,t) we can re-express these solutions in the alternative forms

$$u = 2\pi + 4\tan^{-1}\left(\sqrt{\frac{3}{4}}\tan\frac{1}{2}\Theta\right)$$

$$u = 4\tan^{-1}\left(\sqrt{\frac{5}{3}}\tan\frac{1}{2}\Theta\right)$$

$$\Theta = \chi(x - Vt), \quad \chi = \sqrt{\frac{15}{16}}\,(1 - V^2)^{-1/2} \quad . \tag{3.29}$$

The former is the $4\pi - 2\delta$ kink which interpolates between the dipole energy minima $\delta$ and $4\pi - \delta$: the latter is the $2\delta$ interpolating between $-\delta$ and $+\delta$.

These two kinks can only bump in collision: they cannot pass through each other as solitons since $2\delta \neq 4\pi - 2\delta$. Figures which show the derivatives of these kinks, corresponding to $\mathscr{E}$ in (3.14,15) as they bump in collision are plotted in [3.3,4, 14,31] for example. Similar results for the triple s-G appear in [3.14,37].

These two kinks have rest energies and effective rest masses, namely 5.1097 $\gamma_0^{-1}$ units for the $4\pi - 2\delta$ and 11.3929 $\gamma_0^{-1}$ units for the $2\delta$. When $2\delta$-kink-antikink pairs collide they can pass through each other maintaining the b.cs $u \to -\delta$, $|x| \to \infty$ providing the pair emerges as a $-(4\pi - 2\delta)$ antikink-kink pair. Numerical work [3.4,5,8,33,47] shows that this is exactly what they do for all incident velocities V, taken in the centre of mass frame. In each case there is additional radiation emitted after the collision.

When $4\pi - 2\delta$ kink-antikink pairs collide there are two complications: whilst the b.cs $u \to \delta$, $|x| \to \infty$ are consistent with both $4\pi - 2\delta$ kink-antikink and $-2\delta$ antikink-kink pairs there is a threshold $V = 0.8938$ below which the $4\pi - 2\delta$ pair has insufficient energy to create the rest mass of the $2\delta$ pair: in this case the $4\pi - 2\delta$ pair bumps, and because of radiative losses the actual threshold is raised to $V = 0.925$ [3.4,33]. There is also a lower threshold close to $V = 0.36$ [3.33,48]. Here the approaching $4\pi - 2\delta$ pair loses a little energy by radiation and settles into a long lived bound breatherlike state. Figure 3.5 shows a $4\pi - 2\delta$ kink and antikink each with $V = 0$ falling into such a breatherlike state under their own mutual attraction. Figure 3.6 shows the behaviour for $V = 0.36$.

---

[12] If the spin waves are quantised there might have been an interesting consequence. We computed $\gamma_0$ to be $\sim 10^5$ [3.4,5]. In the A-phase sine-Gordon case (and perhaps therefore in the B-phase double sine-Gordon case) this very large value of $\gamma_0$ destroys any breather spectrum. This result follows from (1.105) since the number of quantised breather state $N < 1$ for large enough $\gamma_0$ [the observed renormalised $\gamma_0$ replaces the $\gamma_0$ in (1.105)]. Despite the very low temperatures at which these elementary spin excitations exist our conclusion is, nevertheless, that the spin waves are not quantised. This is consistent with the c-number form of the order parameter [3.4,5], the long correlation length $\bar{c}\Omega_\ell^{-1}$, the observation of ringing by WHEATLEY [3.17] discussed below, and the suggestion also mentioned below that the ringing may itself depend significantly on the breathers [3.48]. We note in any case that (3.19) is to be scaled by $\rho_s$, the superfluid number density per unit length (that is $\rho_s\Omega_\ell\bar{c}^{-1}$). In particular, if this number multiplies the Hamiltonian density (3.28) following, it scales $\gamma_0$ *down* by this factor, an intuitively correct result. It is not clear (to us) what number should be chosen for $\rho_s\Omega_\ell\bar{c}^{-1}$ in the one-dimensional case, however. Nor is it clear that this particular scaling provides an internally consistent treatment of the one-dimensional theory.
For derivations of the eigenspectrum of the quantised sine-Gordon equation see especially Chap.12 of this volume, and the references therein; [Ref.1.42, pp.114-117]; BULLOUGH et al. [3.13]; or the earlier work by FADDEEV [1.41] and DHN [1.40]. Also see the review by FADDEEV and KOREPIN [1.218]; Faddeev's recent work referenced in Chap.12; and BULLOUGH [1.221].

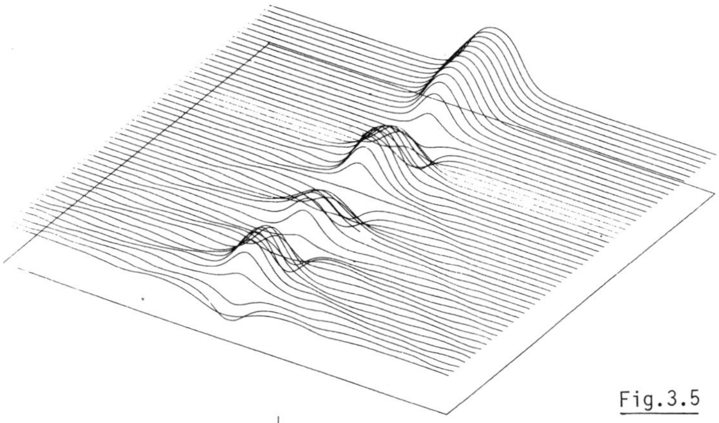

Fig.3.5

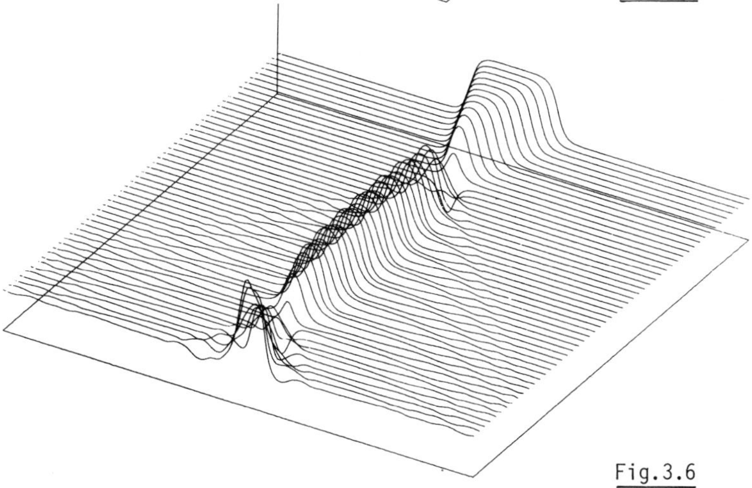

Fig.3.6

Fig.3.5. The long lived breatherlike state of the double s-G (3.1) (taken with -ve sign and $\lambda = 1$) obtained by letting a $4\pi - 2\delta$ kink and antikink fall into each other ($V = 0$). Notice the small amount of radiation emitted at velocities $\pm 1(\pm \bar{c})$. Time t runs towards the observer; the variable x runs from left to right. Boundary conditions simulate $u \to \delta = 2\cos^{-1}(-1/4)$ as $|x| \to \infty$

Fig.3.6. Behaviour of the $4\pi - 2\delta$ kink-antikink pair of the double s-G at the threshold ($V = 0.36$) for breather formation. Notice the irregular behaviour and the enhanced emission of radiation compared with Fig.3.5. (Time t runs forward)

It seems possible that these breathers when created could be driven by a spatially uniform harmonic longitudinal field. If $B_0$ in (3.19) is replaced by $B_0 + B_1 \cos t$ one finds that (3.26) becomes

$$u_{xx} - u_{tt} = -\left(\sin u + \frac{1}{2}\sin\frac{1}{2}u\right) - \gamma\omega\Omega_\ell^{-2}B_1 \sin(\omega\Omega_\ell^{-1}t) \quad . \tag{3.30}$$

In the corresponding A phase sine-Gordon case, in the presence of damping and with the equations read as c-number equations, KAUP and NEWELL [3.49] have shown through singular perturbation theory that the breathers of the s-G phase lock at a frequency $\omega$ in a window *below* the NMR frequency $\Omega_{\ell A}$. The situation is not quite the same in the B-phase case because the potential $V(u) = 2(\cos \frac{1}{2} u + \frac{1}{4})^2$ is not symmetrical about the minima at $\delta$ or $4\pi - \delta$; and the breather spectrum is also unknown. But still the frequency $\omega$ may phase lock below $\Omega_{\ell B}$. We have concluded elsewhere [3.48] that for the boundary conditions there assumed (and which at best can only approximate the physical situation so the conclusion must be qualified accordingly) the breather solutions must play a role in the NMR response. Certainly satellite NMR frequencies from 'composite solitons' have been reported [3.9] but these are essentially static features. We know of no evidence yet for complications in the NMR response which could be ascribed to the presence of driven breathers.

Spin waves could be created in both $^3$He A and $^3$He B by an initial magnetic shock. We suppose $B_0$ in (3.19) is again homogeneous for $t > 0$, but that at $t = 0$ an additional inhomogeneous field $\Delta B_0(x)$, superposed for $t < 0$ is then removed (this method is also suggested in [3.8]). We show elsewhere [3.48] how the problem may be roughly equivalent in the A phase to solving (3.25) (in c-number form) with $u$, $u_x$, $u_{xx}$, etc $\to 0$, $|x| \to \infty$; and

$$u(x,0) = 0 \; ; \quad u_t(x,0) = 2\tau \; , \; |x| < \ell \; ; \quad u_t(x,0) = 0 \; , \; |x| > \ell \; ; \qquad (3.31)$$

($\ell$ is a length in dimensionless units and $\tau$ is a frequency in similar units). KAUP [3.50] has partially solved this particular initial value problem and we refer the reader to his article for the details of a simple but neat application of the spectral transform method. MAKI and KUMAR [3.8] have given a comparable analysis applied directly to the $^3$He spin wave problem. The essential point for present purposes is that for $\tau < 1$ stationary breathers but not kinks are created: each stationary breather is of the form given in (1.31). Kaup's analysis shows that the number of such breathers is the integral part of $(\ell\tau/\pi) + 1/2$. In real units $\ell\tau = (L\lambda_A^{-1})\times(\gamma\Delta B_0\Omega_{\ell A})^{-1}$ where the correlation length $\lambda_A = 2\pi\Omega_{\ell A}^{-1}\bar{c}_A$. Since the length of the sample $L \leqq 1$ cm and $\bar{c}_A \lesssim 3\times 10^3$ cm s$^{-1}$, whilst for $\Delta B_0 \sim \Delta B_{0c}$, a critical value (see [3.47] and below) $\gamma\Delta B_0 \sim \Omega_{\ell A} \sim 10^6$ rad s$^{-1}$, $\ell\tau$ can be large $\gtrsim 300$ (for $\bar{c}_A \sim 10^2$ cm s$^{-1}$, $\ell\tau \sim 3000$). There are two eigenvalues of the associated scattering problem involved with each breather so the number of these $\geqq 600$ (for $\bar{c}_A \sim 10^2$ cm s$^{-1}$ the number of eigenvalues $\geqq 6000$). We shall deliberately interest ourselves here in the case of a narrow region of magnetic excitation, however, so $L \ll 1$ cm and $\ell\tau \lesssim 10$. This choice will surely be a difficult one to achieve in actual experiments but certainly the theoretical predictions are now most interesting. Figure 3.7 shows the case of $\tau < 1$ and 5 stationary breathers. The figure does not cover the period of initial creation of these breathers (for this interesting period see some of the figures in [3.47] and Fig.3.8: time in Fig.3.7 starts at $t = 60\sqrt{2}$, and the figure shows that even over such a long period of time the stationary breathers are persistent solutions of standing wave type which have no tendency to move outwards.

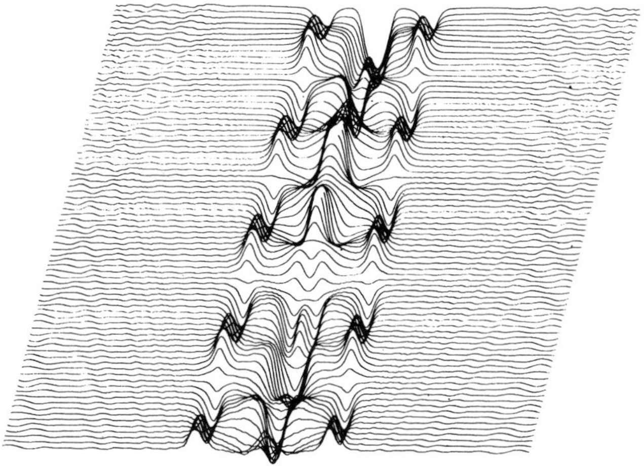

Fig.3.7. Five standing wave breathers of the sine-Gordon equation created from initial data $u(x,0) = 0$; $u_t(x,0) \leqq 2$, $|x| < 15$, $u_t(x,0) = 0$, $|x| > 15$. Time starts at $t = 60\sqrt{2}$ and runs away from the observer. Note that there is a tendency for these breathers to radiate

We show below by phase plane analysis in the *time* frame that the value $\tau = 1$ where $\gamma\Delta B_0 = \gamma\Delta B_{0c} \equiv \Omega_{\ell A}$ has a 'critical' character: we can expect it to remain critical in the spatially inhomogeneous problem also and it does. For $\tau>1$ kink-antikink pairs begin to travel out in opposite directions. In terms of the scattering data when $\tau>1$ and kinks and antikinks move out, pairs of eigenvalues $\zeta$, $-\zeta^*$ move successively onto the imaginary axis and split up there. The maximum number of kink-antikink pairs which can be produced is therefore $\sim \ell\tau/\pi = \gamma\Delta B_0 L\bar{c}_A^{-1}\pi^{-1}$. An important point is that whilst standing wave breathers remain in $|x| < \ell$, kinks radiate to the left $(x \to -\infty)$ and antikinks radiate to the right $(x \to +\infty)$. Kinks and antikinks do not radiate in the same direction.

Corresponding results for the B phase are more complicated. First of all no analytical solution to the initial value problem for the c-number equation (3.26) is avaiable. Second there are critical fields at two values $\gamma\Delta B_{0c_1} = \sqrt{3/5}\Omega_{\ell B}$ and $\gamma\Delta B_{0c_2} = \sqrt{5/3}\Omega_{\ell B}$ corresponding to critical values of $\tau$ at $2\tau = 3/2$ and $2\tau = 5/2$. These critical fields are well known in the spatially homogeneous case [3.18] and preliminary experiments designed to observe them are reported by WHEATLEY [3.17]. In the inhomogeneous case we have reported numerical evidence for these critical fields but also a sequence of behaviours rather different from the s-G A phase case [3.48]. There is not the space to present all these numerical results in detail here. But one result of particular interest is that above the first of the thresholds, where $\Delta B_0 = \Delta B_{0c_1}$, we find that both $4\pi - 2\delta$ kinks *and antikinks* can radiate outwards in the same direction.

The conclusions we reached from the results in [3.48] are that, unlike the case of the sine-Gordon equation, the initial value problem (3.31) for the double sine-Gordon equation with -ve sign, $\lambda = 1$, $u \to \delta$, $|x| \to \infty$, $u(x,0) = \delta$, is sensitive to $\ell$ as well as $\tau$. There is a threshold at $u_t = 2\tau = 3/2$: below this threshold only breathers are formed but these may radiate out; on or above this threshold radiating kinks and radiating kink-antikink pairs are possible (a result critically different from the sine-Gordon case as we noted) and with these kink-antikinks pairs there may be radiating breatherlike excitations. The reader is referred to our article [3.48] and to [3.32] where the numerical results on which these conclusions are based are displayed in various figures. Here we show only the result of crossing the second critical threshold at $u_t = 5/2$.

The boundary condition on Fig.3.8 is $u = \delta$, $|x| \to \infty$. From the right of Fig.3.8 one sees that $\delta \to 4\pi - \delta \to 4\pi + \delta \to 8\pi - \delta \to 4\pi + \delta \to 8\pi - \delta$. The sequence of kinks is therefore: (right to left) $-(4\pi - 2\delta)$antikink, $-2\delta$ antikink, $-(4\pi-2\delta)$ antikink, $4\pi - 2\delta$ kink, $-(4\pi - 2\delta)$ antikink. The sequence confirms the conclusions one can easily reach, namely that I) with boundary conditions $u \to \delta$, $|x| \to \infty$, a $-(4\pi - 2\delta)$ antikink (or a $2\delta$ kink) is first emitted; II) a $-(4\pi - 2\delta)$ antikink is followable by either a $-2\delta$ antikink or a $4\pi - 2\delta$ kink. There is no evidence in this example of the formation of breatherlike excitations, but it seems possible that by changing $\ell$ the final $-(4\pi - 2\delta)$ antikink pair could bind and appear as an outgoing breather. We have no evidence yet of outgoing $2\delta$ kink-antikink pairs. We noted that if such pairs collide they convert to $-(4\pi - 2\delta)$ antikink pairs and always provide enough rest mass energy as kinetic energy to overcome any tendency to breather formation. Thus an outgoing $2\delta$ kink-antikink pair is possibly unstable to the formation of an outgoing $-(4\pi - 2\delta)$ antikink-kink pair.

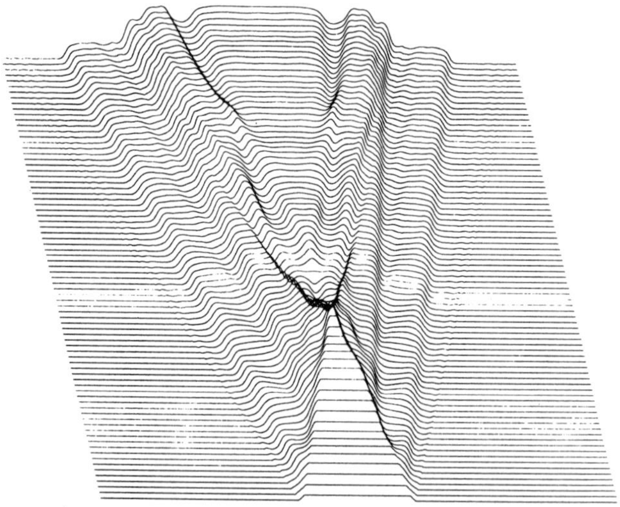

Fig.3.8. A result of crossing the second critical threshold of the double sine-Gordon equation at $u_t = 5/2$; $u_t = 2.6$, $|x| < 15$. Boundary conditions simulate $u \to \delta$, $|x| \to \infty$; and reading from right to left the sequence of kink-antikink emission is: - $-(4\pi-2\delta)$ antikink, $-2\delta$ antikink, $-(4\pi-2\delta)$ antikink, $4\pi-2\delta$ kink, $-(4\pi-2\delta)$ antikink. Note the $-2\delta$ antikink and $2\delta$ kink. No breathers of any sort seem to be formed in this example

The "critical fields" we have mentioned for both the A-phase sine-Gordon and the B-phase double sine-Gordon problems have the property that in the homogeneous spatially independent problem the induced magnetisation does not "ring". One can see this easily from the phase plane in the *time* frame: for example, the c-number equation (3.26) has a first integral

$$-\frac{1}{2}u_t^2 - \cos u - \cos\frac{1}{2}u = C \qquad (3.32)$$

and this is actually a trajectory in the phase plane of (3.10) (when $u_t \to u_\xi$). Trajectories which connect unstable singular points arise for -C = 2 and C = 0. At $u = \delta$ when $C = -2$ $u_t = 5/2$: at $u = \delta$ when $C = 0$ $u_t = 3/2$. On these trajectories $u_t$ moves to zero without repeated oscillations (without ringing) as u moves to $4\pi$ and $2\pi$, respectively. Since $u_t$ determines the magnetisation [multiply the first equation of (3.21a) by $\chi\gamma^{-1}$] we call these trajectories ringing free: the magnetic fields which can shock the system in the state $u = \delta$ into following these trajectories are $\Delta B_{0c_2} = \gamma^{-1}\sqrt{5/3}\Omega_{\ell B}$ and $\Delta B_{0c_1} = \gamma^{-1}\sqrt{3/5}\Omega_{\ell B}$ and these are the values quoted above. For the sine-Gordon A-phase problem a similar analysis shows that $u_t = 2\tau = 2$ defines the single ringing free trajectory and the system in the state $u = 0$ is shocked into following this trajectory for $\Delta B_{0c} = \gamma^{-1}\Omega_{\ell A}$.

The NMR behaviour is a time-dependent spatially independent phenomenon. Thus one thing it can be used for is to determine critical fields for no ringing, and this has been done for the A phase by WHEATLEY [3.17] in detail and for the B phase less completely. One conclusion of the present analysis important to the physics therefore is that the critical fields for no ringing NMR in the homogeneous case remain critical for breakup into solitons, both in the A phase and the B phase, in the spatially inhomogeneous case. However, if breakup into solitons occurs (or into quasi-solitons of course in the B-phase case) such breakup would be accompanied by an apparent ringing signal at least at specific points $x > \ell$ outside the original region of inhomogeneous magnetic field. It is not easy to see that an averaged signal will necessarily be ringing free, and perhaps this has some bearing on an imperfect agreement between theory and experiment which was found by WHEATLEY in studies of the A phase [3.17]. At least the lesson of our analysis is that boundary conditions are important.

Despite the possibility of forming outgoing breathers above and below threshold in the B phase, it ought to be possible, in principle at least, to detect both kinks *and* antikinks moving in the same direction on or above the threshold at $u_t = 3/2$. Apparently below the second threshold at $u_t = 5/2$ $2\delta$ kinks are not created. The sequence of kink emission for $3/2 < u_t(x,0) < 5/2$ *must* therefore be kink-antikink-kink, etc., (or antikink-kink, etc.) and for large enough $\ell$ we expect many such kink-antikink pairs. In addition the results in [3.48] show that radiating kink-antikink pairs (antikink-kink) pairs may form outgoing breatherlike excitations.

All this variety of behaviour for the nonintegrable double sine-Gordon equation contrasts with the relative simplicity of the integrable sine-Gordon equation of the A phase case. Here above the single threshold at $u_t = 2$ only a simple sequence consisting purely of kinks (or purely of antikinks) can be emitted—as numerical and analytical analysis shows [3.48]. We have therefore suggested [3.48] that this fact might be used: a magnetic detector would in principle detect kinks and antikinks in the B phase case but only kinks in the A phase case and in this way provide evidence of the correctness of the symmetries assigned to the "order parameters" in the two different phases. Our results [3.48] for the B phase show, however, that small changes (~ one or two correlation lengths $\lambda_B = c_B \Omega_{\ell B}^{-1}$ in the initial distribution of $u_t$) change, for example, breather emission at threshold to single kink emission. In order to exploit the details of the behaviour in $^3$He B one needs small fields ~20 Gauss removed over well defined small regions ~100 microns and the detailed magnetic response detected. Such experiments may be beyond present capability. It may be more practicable to excite breathers by magnetic shocks and, e.g., phase lock to them with longitudinal rF fields. In the case of a large enough value of the coupling constant $\gamma_0$ there could be many thermally excited kinks and breathers even at temperatures $\lesssim$2.6 mK—as the spectrum (1.104) indicates.[13] However, we now believe that the effective value of $\gamma_0$ is very much smaller than the value we originally found.[13]

This completes our contrast of the integrable sine-Gordon and nonintegrable double sine-Gordon equations and our survey of the opportunities for experiments on real physical systems which could exploit the differences between them. Analytically it is tempting to interpret these various results for the double sine-Gordon equations in terms of the scattering data of an appropriate scattering problem in a direct comparison with the sine-Gordon case. So far this has been done only for two particular solutions of (3.1): these are both for the case $\lambda = 1$ and +ve sign, and no comparable work has yet been done for the case of -ve sign. We sketch the analysis for the +ve sign solutions in the next section. Rather more mathematical technique is demanded of the reader than in the rest of this chapter and he or she may care to read Chap.6 beforehand in order to get a good grip on the inverse scattering method and how it can be used for singular perturbation theory.

[13]The spectrum (1.104) for the sine-Gordon equation shows that the energies for creating kinks and breathers are proportional to $\gamma_0^{-1}$. The computed value of $\gamma_0$ was ~$10^5$ [3.4,5]. However, the effective $\gamma_0$ is presumably very substantially reduced by scaling against the superfluid number density $\rho_c$ (see footnote 12).

## 3.4 Perturbation Theory for the Double Sine-Gordon Equation

In Sect.3.2 we showed how the $4\pi$-kink solution (3.12) of the double s-G with $\lambda = 1$ and +ve sign could be put into stable internal 'wobbles'. A wholly similar analysis is possible for the solution (3.13) corresponding to excited state boundary conditions $u = 2\pi$, $|x| \to \infty$. The solution for u is a bound kink-antikink pair: the electric field has zero area so optically it is a $0\pi$-pulse. The expectation is that it is a breather of the double s-G. However, it is plainly unstable: for $\Delta' > \ln(\sqrt{3}+2)$ the first $2\pi$ sech accelerates whilst the second $2\pi$ sech slows down. The potential V(u) of (3.17) is unchanged: but the state $u = 2\pi$ can lower its energy by interpolating $u = 2\pi \to u = 0 \to u = 2\pi$ and breaks up into a kink and antikink: this again illustrates the instability. The second argument, however, intuitively suggests that oscillation of the kink-antikink pair is possible as in a true breather: numerical work [3.31,32] shows that this is precisely what does happen if $\Delta' < \ln(\sqrt{3}+2)$. Figure 3.9 shows the breakup regime. Figure 3.10 shows the oscillatory breatherlike regime induced by compression of the spacing $2\Delta'$. In this section we sketch the confirmation of this behaviour which is provided by singular perturbation theory. The analysis is primarily due to our two colleagues A.L. Mason and P.W. Kitchenside.

Fig.3.9. Numerical integration of the double sine-Gordon equation for +ve sign (and $\lambda = 1$) in the breakup regime. Boundary conditions are $u \to 2\pi$, $|x| \to \infty$

The s-G is completely integrable and for small $\lambda$ it is natural to regard the term $\frac{1}{2}\lambda\sin\frac{1}{2}u$ in (3.1) in the case of +ve sign as a perturbation of the usual s-G. A convenient starting point is a two-kink solution of the s-G which may be written in terms of scattering data in the form

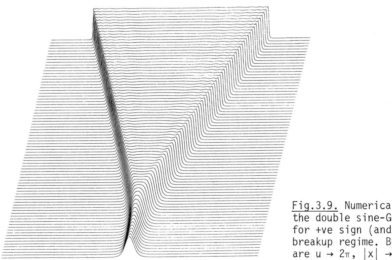

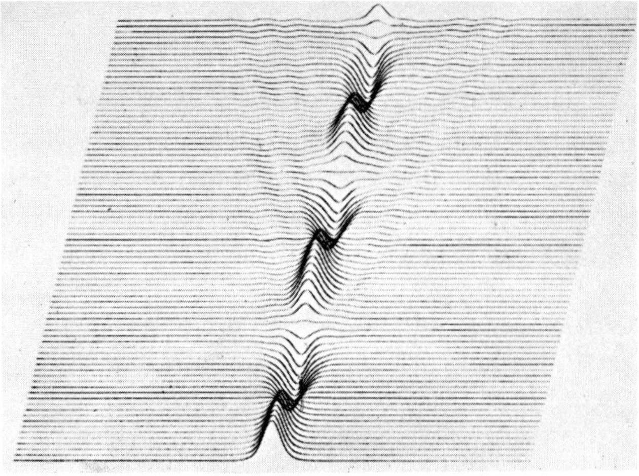

Fig.3.10. As Fig.3.9, but in the oscillatory regime

$$u = 4\tan^{-1}\left[\frac{1-\left(\frac{\zeta_1-\zeta_2}{\zeta_1+\zeta_2}\right)^2 \frac{\gamma_1}{2\zeta_1}\frac{\gamma_2}{2\zeta_2}\exp[i2(k_1+k_2)x]}{\frac{\gamma_1}{2\zeta_1}\exp(i2k_1x)+\frac{\gamma_2}{2\zeta_2}\exp(i2k_2x)}\right]$$

with

$$k_i = \frac{1}{2}\left(\zeta_i - \frac{1}{4\zeta_i}\right) , \quad i = 1,2 \quad . \tag{3.33}$$

Here $\gamma_1$ and $\gamma_2$ are the residues of the reflection coefficient b/a (when this quantity is analytically continued into the upper half plane), evaluated at the complex eigenvalues $\zeta_1$ and $\zeta_2$, respectively.

For the s-G, the time evolution of the scattering data is rather simple, and in particular the spectrum of the scattering problem is a constant of the motion. However, for a nonintegrable system like (3.1) for nonzero $\lambda$ the situation is more complicated. Newell in Chap.6 (and [3.51]) shows that the appropriate equations for the scattering data of (3.33) are[14]

$$\zeta_{i,t} = -\frac{\lambda}{8\gamma_i\dot{a}_i^2}\int_{-\infty}^{\infty} \sin\frac{u}{2}\,(\phi_1^2+\phi_2^2)_{\zeta_i}\,dx \qquad (i = 1,2)$$

$$\gamma_{i,t} = i2\gamma_i\omega_i - \frac{\gamma_i\ddot{a}_i}{\dot{a}_i}\,\zeta_{i,t} - \frac{\lambda}{8\dot{a}_i^2}\int_{-\infty}^{\infty} \sin\frac{u}{2}\cdot\frac{\partial}{\partial\zeta}\,(\phi_1^2+\phi_2^2)_{\zeta_i}\,dx$$

---

[14] This is not quite true. Newell considers evolution equations of the form $u_t=K[u]$ where K[u] is a functional of u, $u_x$, $u_{xx}$, etc. The s-G takes the form in the dependent variable $u_x$ in light cone coordinates: $u_{xt} = \sin u$. We are working in Lorentz covariant coordinates and the problem must be transcribed for these. The appropriately transcribed scattering problem is given by KAUP [3.49].

$$\left(\frac{b}{a}\right)_t = i2\omega \left(\frac{b}{a}\right) - \frac{\lambda}{8a^2} \int_{-\infty}^{\infty} \sin \frac{u}{2} \cdot |\phi_1^2 + \phi_2^2|_{\zeta_i} \, dx$$

$$\omega = \frac{1}{2}\left(\zeta + \frac{1}{4\zeta}\right) , \quad \gamma_i = \frac{b_i}{\dot{a}_i} , \tag{3.34}$$

where $\dot{a}_i$ denotes differentiation of a with respect to $\zeta$ evaluated at $\zeta = \zeta_i$. The expression $(\phi_1^2 + \phi_2^2)$ denotes the *squared eigenfunction* of the s-G scattering problem evaluated at $\zeta$.

Equations (3.34) show the simple form the time evolution of the scattering data takes for an integrable system like the s-G (case $\lambda = 0$). But in the general form given they are also exact for *all* values of $\lambda$ as the analysis given in Chap.6 shows. They are merely difficult to handle. They are, however, simplified considerably if we discard the contributions of the continuous part of the spectrum. This seems intuitively plausible if we start with (3.33) which involves only two discrete eigenvalues and $\lambda$ is taken sufficiently small. In fact, as we shall show, for many characteristics of the motion $\lambda$ can be *large* (as in, e.g., the physically important case for which $\lambda = 1$).

Since we wish to discuss the breather associated with (3.13), we use the breather parametrisation of the s-G $\zeta_1 = -\zeta_2^*$, $\gamma_1 = -\gamma_2^*$. The specifications

$$\zeta_1 = \frac{1}{2} e^{i\phi} , \quad \gamma_1 = \tan\phi \; e^{i\psi}$$

in terms of new functions $\phi$ and $\psi$ ensure that (3.33) is symmetrical about the origin in the rest frame. By putting $\phi = \cos^{-1}(\lambda/4)$ and $\theta \equiv \psi - \phi = 0$ in (3.33) we find

$$u = 4\tan^{-1}\left(\sqrt{\frac{\lambda}{4-\lambda}} \cosh \frac{1}{2} \sqrt{4 - \lambda}\, x\right) , \tag{3.35}$$

this is a form of the solution (3.13) in its rest frame not necessarily restricted to $\lambda = 1$. The subsequent time evolution can now be determined from (3.34) under the assumption that the effects of the continuum can be neglected. This has been done by MASON [3.35]. The details are heavy and the reader wishing to carry out similar computations himself is referred to [3.32,52]. Providing the continuum can be neglected, that is to say providing the problem can be described by two discrete eigenvalues $\zeta_i$, the results are exact and the equations take the form

$$C_t = \frac{\lambda}{2} P \frac{J(a)+a}{1+a} \; ; \quad a \equiv \cos^2\theta \tan^2\phi \quad , \quad J \equiv \frac{a}{\sqrt{1+a}} \tan^{-1} \frac{a}{\sqrt{1+a}}$$

$$P_t = P^2 - \frac{\lambda}{4} + C \; ; \quad P \equiv \tan\theta\cos\phi \quad , \quad C \equiv \cos^2\phi \quad . \tag{3.36}$$

These equations show that for initial conditions $\theta = 0$, $0 < \phi < \cos^{-1}\lambda/4$, the resulting motion is oscillatory: for $\phi > \cos^{-1}\lambda/4$, $\phi$ eventually reaches $\pi/2$ [3.32, 35,36]. At this point the solution becomes a rapidly separating kink and antikink. The breather form is now inappropriate and the analysis has to be repeated using kink parameters. A suitable parametrisation is now

$$\zeta_1 = \frac{i}{2} e^{-q} \quad , \frac{\gamma_1}{2\zeta_1} = \coth q \; e^{p} \tag{3.37}$$

the equations of motion being

$$P_t = P + C - \frac{\lambda}{4} \quad ; \quad P = \coth p \sinh q \quad , \quad C = -\sinh^2 q$$

$$C_t = \frac{\lambda}{2} P \frac{J(a)+a}{1+a} \quad ; \quad a = \coth^2 q \sinh^2 p \quad . \tag{3.38}$$

The fact that the equations of motion (3.38) take the same form as (3.36) with the different parametrisation enables a smooth match to be made through the breakup regime.

Equations (3.36,38) were solved numerically for the initial conditions

$$\theta = 0 \;, \quad \cos^2\phi = \frac{1}{2} \lambda (1 - \varepsilon)^{-1} \tag{3.39}$$

with $\varepsilon = 0.1$ or $-0.1$ and $\lambda = 1.0$. When $\varepsilon = -0.1$, the motion is oscillatory as shown in Fig.3.11. For $\varepsilon = 0.1$ breakup occurs (Fig.3.12) and this requires the change of parameterisation when C becomes negative. A comparison of Fig.3.12 with Fig.3.9 shows good agreement for breakup in this no-continuum approximation. Indeed the figures themselves accurately superpose for the range of evolution shown in Fig. 3.12. However, truncation errors cause the parametrisation to break down due to the asymptotic properties of the hyperbolic functions in (3.38) and this explains why Fig.3.12 is carried no further.

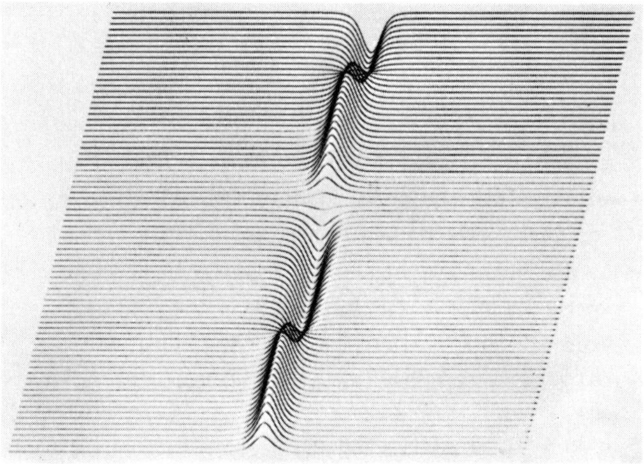

Fig.3.11. The oscillatory regime described by (3.36)

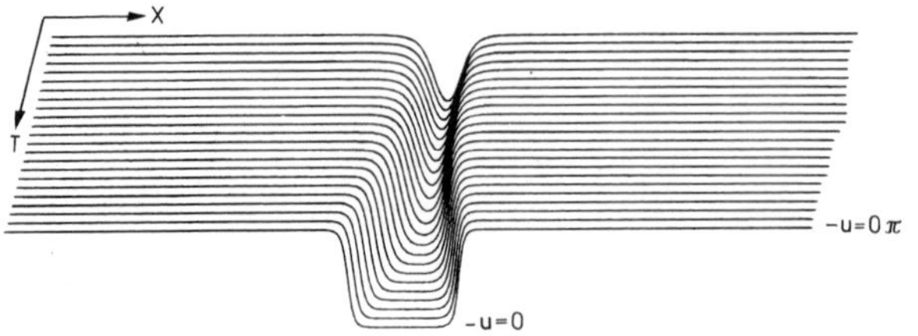


Fig.3.12. Breakup described by (3.36) and (3.38)

For the oscillatory regime agreement is almost perfect until u reaches its maximum amplitude. Subsequently a slight periodic emission of ripples of radiation can be seen in Fig.3.10 and this represents energy being pumped into the continuum. The net result is to shorten the period of the breatherlike oscillation in Fig.3.10 compared with that found by perturbation theory as displayed in Fig.3.11. We have not yet found a procedure with the relative simplicity of equations (3.36,38) which allows us to account for the growth of the continuum, and this situation is indicative of the state of perturbation theoretical studies at the present time. It is noteworthy that in all other respects, agreement between theory and direct numerical integration is almost perfect even for the large value $\lambda = 1$.

We can gain other analytical insights. For example, numerical solutions for the breakup case show that the quantity $Y \equiv C + P^2$ tends rapidly to zero. This makes it easy to find an asymptotic solution describing the behaviour of widely separated kink-antikink pairs. Equation (3.38) for $P_t$ yields the approximation

$$P = \int_0^\infty Y\,dt - \frac{\lambda}{4}t \equiv Z - \frac{\lambda}{4}t \tag{3.40}$$

and we can obtain the approximate equation

$$(1 + P^2)\frac{d}{dt}\ln Y - \frac{\lambda}{4}P\ln Y$$

$$= 2P\left[1 + P^2 + \frac{\lambda}{4}\left(\ln 2\sqrt{1 + P^2} - 1\right)\right] \tag{3.41}$$

for Y neglecting terms of second order in Y. Using (3.40) for P we can solve (3.41) explicitly and eventually find that

$$\left(\ln 2\sqrt{1 + P^2} - \frac{1}{2}\ln Y\right)/\sqrt{1 + P^2}$$

$$= \left[\left(\frac{4}{\lambda}\right)^2 + \left(\frac{4Z}{\lambda} - t\right)^2\right]^{1/2} - \gamma \tag{3.42}$$

where $\gamma$ is a constant of integration. Following the work of A.L. Mason to whom this analysis is due we have chosen to write the solution in this rather unlikely form because the left side gives the position of the centre of the kink resulting from the solution of

$$4\tan^{-1}\left(\frac{\cosh x\sqrt{1-c}}{\sqrt{\frac{1-c}{c+p^2}}}\right) = \pi \quad . \tag{3.43}$$

We can compare the motion of the kink with that of a relativistic particle that picks up constant energy per unit distance travelled. The motion of this particle is governed by the equation

$$m_0(1 - v^2)^{-1/2} = \varepsilon X + \gamma$$

which is easily solved to give

$$X = \left[\frac{m_0^2}{\varepsilon^2} + (t - k)^2\right]^{1/2} - \gamma \tag{3.44}$$

a result which is totally consistent with (3.42).

The motion described by (3.42) is of course also the motion of the single $2\pi$ kink entering the medium in the state $u = 0$ and slowing down as it leaves a trail of excitation behind it as described in Sect.3.2, that is to say it is the motion of the first of the very well separated pair of $2\pi$ kinks in the $4\pi$ wobbler.

We can treat the $4\pi$-wobbler solution of (3.1) in a similar fashion. It is convenient to replace u by $u-2\pi$ which changes the sign of $\lambda$ in (3.1). For a centre of mass frame two kink collision u is now an odd function of x.

For the centre of mass frame we have the zero momentum condition

$$\zeta_1\zeta_2 = -\frac{1}{4} \quad \text{where} \quad \zeta_i = i\eta_i(\eta_i > 0) \quad , \tag{3.45}$$

so $k = k_1 = k_2$ [whilst $k/\zeta = (\zeta - 1/4\zeta)/2$]. The condition that u is odd is

$$1 = \left(\frac{\zeta_1-\zeta_2}{\zeta_1+\zeta_2}\right)^2 \frac{\gamma_1}{2\zeta_1}\,\frac{\gamma_2}{2\zeta_2} \tag{3.46}$$

in which for a 2-kink solution $\gamma_1$ and $\gamma_2$ have the same sign and are pure imaginary. We may therefore write

$$\zeta_1 = \frac{i}{2}\,e^{q} \quad , \quad \zeta_2 = \frac{i}{2}\,e^{-q} \tag{3.47}$$

and also

$$\left(\frac{\zeta_1-\zeta_2}{\zeta_1+\zeta_2}\right)\frac{\gamma_1}{2\zeta_1} = e^{p} \quad , \quad \frac{\zeta_1-\zeta_2}{\zeta_1+\zeta_2}\,\frac{\gamma_2}{2\zeta_2} = e^{-p} \tag{3.48}$$

using the experience gained from the $0\pi$ problem. With these choices

$$u = 4\tan^{-1}\left[\frac{\sinh(\cosh qx)}{\cosh p \coth q}\right] . \tag{3.49}$$

The equations of motion (3.34) with continuum neglected yield

$$P_t = P^2 - C + \frac{\lambda}{4}$$

$$C_t = -\frac{\lambda}{2}\,PI \tag{3.50}$$

in which $P \equiv \tanh p \sinh q$, $C = \sinh^2 q$

$$I \equiv \frac{J(a)-a}{a-1} \quad , \quad J(a) = \left(\frac{a}{a-1}\right)^{1/2} \coth^{-1}\left(\frac{a}{a-1}\right)^{1/2}$$

$$a \equiv \coth^2 q \cosh^2 p \tag{3.51}$$

and $\lambda > 0$. Note the changes of sign compared with (3.36) and (3.38).

From the parametrisation (3.47), $\infty > P > -\infty$, $C > 0$ and we can certainly choose $P = 0$, $C = \lambda/4$ which is $\tanh p = 0$, $\cosh q = (1 + \lambda/4)$. This choise yields the time independent solution

$$u = 4\tan^{-1}\left[\left(1+\frac{4}{\lambda}\right)^{1/2} \sinh\left(1+\frac{\lambda}{4}\right)^{1/2} x\right] \tag{3.52}$$

and this can be recognised as the stationary $4\pi$-kink solution (3.6) or (3.12) of the double sine-Gordon equation once it is recalled that we are working in the region $-2\pi < u < 2\pi$.

If we choose initial conditions close to this solution, that is $\lambda/4 \approx C \gg P \sim 0$, and $\lambda$ is still small enough, then $I \approx -1$ and

$$C_t = \frac{\lambda}{2} P \quad , \quad P_t = \frac{\lambda}{4} - C \tag{3.53}$$

since P can be neglected. This pair of equations has solution

$$P = A \cos\left(\sqrt{\frac{\lambda}{2}}\, t + \alpha\right)$$

$$C = \frac{\lambda}{4} + \sqrt{\frac{\lambda}{2}}\, A \sin\left(\sqrt{\frac{\lambda}{2}}\, t + \alpha\right) \tag{3.54}$$

which is simple harmonic motion of angular frequency $\sqrt{\lambda/2}$. Working in light cone coordinates (i.e., with the double s-G in the form (3.5) NEWELL [3.34], in an early application of his perturbation theory, also found a $4\pi$ wobbler with internal oscillations with period $\sqrt{\lambda/2}$ for small $\lambda$.

This calculation again presupposes that the contributions of the continuum can be neglected. Figure 3.13 shows the result of integrating the equations (3.50) numerically. Agreement with the direct numerical integration of the double s-G shown in Fig.3.1 is almost perfect: the differences are the obvious emission of radiation in Fig.3.1 and some slight steepening of the kink profiles in the neighbourhood of points where the two $2\pi$ kinks change places. The period of oscillation ('wobble') is correctly reproduced in Fig.3.13 so that the result here is different from that of the $0\pi$.

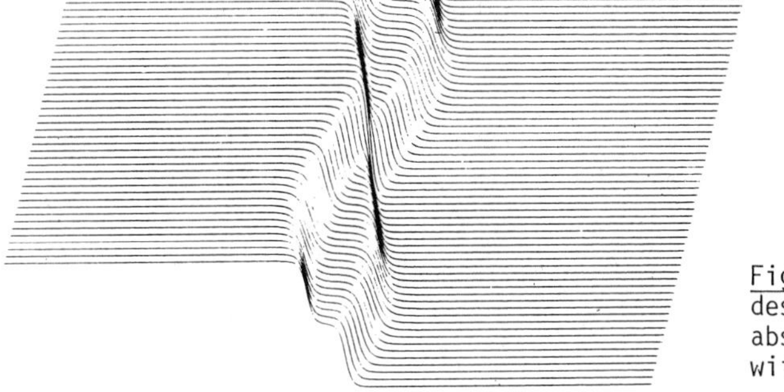

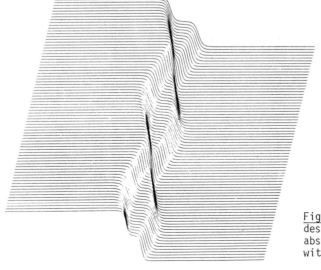

Fig.3.13. The wobbling $4\pi$ kink described by (3.50). Note the absence of radiation compared with Fig.3.1

These results show, on the one hand, that the manipulations involved in applications of singular perturbation theory are nontrivial, that nevertheless, on the other hand, the theory yields excellent agreement with the results of direct numerical integration, that the analytical side of the perturbation theory provides insights which numerical integration does not, and that the contributions of the continuum are in certain circumstances important (namely those obtaining in the case of the $0\pi$ problem). Further work on these continuum contributions is needed in the problems treated here.

It should be plain that these results are first results only and that, for example, $4\pi$-wobbler collisions are prime objects for further work. Nor is it clear even how to treat the kink-antikink collisions of the double s-G with -ve sign described in Sect.3.3. As far as we are aware perturbation theory has not yet been applied to cases more complicated than those discussed here. McLAUGHLIN and SCOTT [3.53] treated the damped driven sine-Gordon equation and NEWELL [3.54] has discussed the direct relation between the perturbation theory and those based on (3.34). As we have already mentioned, the remarks in Sect.3.4 can only sketch the argument and the reader must go to the work of KITCHENSIDE [3.32], MASON [3.52], or NEWELL [3.34] for details. Our aim in this article has been to show what can now be achieved through singular perturbation theory and to give some indication of the character of the work. We can expect much more work of this kind in the future.

*Note in Proof*

The Theorem 10 of [3.30] that the nonlinear Klein-Gordon equation $u_{xt} = F(u)$ of (3.1) has an infinity of polynomial conserved densities if and only if $F''(u)+\alpha^2 F(u) = 0$ for some $\alpha \neq 0$ may be incomplete. MIKHAILOV [3.55] has recently found a $3\times 3$ inverse method for integrating

$$u_{tt} - u_{xx} + 2\exp(4u) - 2\exp(-2u) = 0 \quad , \tag{3.55}$$

in connection with work on a generalised Toda lattice. The system (3.55) appears to be a completely integrable one with an infinity of nontrivial first integrals.

Independently FORDY and GIBBONS [3.56] have found an $N\times N$ inverse scattering scheme to solve a sequence of nonlinear Klein-Gordon equations $u_{i,xt} = F_i(u)$ in $N-1$ coupled field $u = (u_1,\text{--},u_{N-1})$ and auto-Bäcklund transformations for them. For $N = 3$ they found that the system

$$\theta_{xt} = e^{2\theta} - e^{-\theta}\cosh 3\phi \quad , \quad \phi_{xt} = e^{-\theta}\sin^3\phi \tag{3.56}$$

is a completely integrable Hamiltonian system soluble by their scattering transform: this system (3.56) has an auto-BT and appears to have an infinity of nontrivial *polynomial* conserved densities. Notice now that a solution of (3.56) is $\phi \neq 0$ with $\theta$ satisfying

$$\theta_{xt} = \exp(2\theta) - \exp(-\theta) \quad , \tag{3.57}$$

essentially (3.55) and certainly an example of (3.1). The equation (3.57) has no aBT (apparently) since the aBT of (3.56) maps between (3.57) (with $\phi \equiv 0$) to (3.56) (with $\phi \neq 0$). But it does appear to have an infinity of nontrivial p.c.ds even though $\phi \equiv 0$. It seems then that the theorem of [3.28,29] that $u_{xt} = F(u)$ for a single field u has an aBT if and only if $F''(u) + \alpha^2 F(u) = 0$ is not contraverted. But Theorem 10 of [3.30] on p.c.ds is contraverted. It is not clear at this stage whether the peculiar relationship of (3.57) to (3.56) excludes (3.57) from the considerations of Theorem 9 of [3.30] and so excludes it from Theorem 10. Notice, however, that Theorem 9 was based on the isolation of two cases I) $F(u) = A\exp(\alpha u) + B\exp(-\alpha u)$ and II) $F(u) = A\exp(\alpha u) + B\exp(-\frac{1}{2}\alpha u)$ [the case of (3.57)]. Thus it seems possible that the proof in [3.30] that there is an upper bound ($< 24$) on the rank and hence the number of p.c.ds in the case II) is incorrect. We are naturally scrutinizing the very tedious proof of the existence of this bound with some interest; but at the time of writing we have not shown that Theorem 9 is false.

## References


3.1 G.L. Lamb: Rev. Mod. Phys. *43*, 99 (1971)

3.2 S. Duckworth, R.K. Bullough, P.J. Caudrey, J.D. Gibbon: Phys. Lett. *57*A, 19 (1976)

3.3 R.K. Bullough, P.J. Caudrey, P.W. Kitchenside: "Bumping spin waves in the B-Phase of liquid $^3$He", J. Phys. C (1980) To be published

3.4 R.K. Bullough, P.J. Caudrey: "The Multiple Sine-Gordon Equations in Non-Linear Optics and in Liquid $^3$He", in *Nonlinear Evolution Equations Solvable by the Spectral Transform*, ed. by F. Calogero (Pitman, London 1978) pp.180-224

3.5 R.K. Bullough, P.J. Caudrey: In *Coherence and Quantum Optics IV*, ed. by L. Mandel, E. Wolf (Plenum Press, New York 1978) pp.767-780
R.K. Bullough: "Solitons in Physics", *Proc. NATO Advanced Study Institute on Nonlinear Equations in Physics and Mathematics*, Istanbul, August, 1977, ed. by A.O. Barut (D. Reidel, Dordrecht, Holland 1978) pp.99-141

3.6 P.W. Kitchenside, R.K. Bullough, P.J. Caudrey: In *Solitons and Condensed Matter Physics*, Springer Series in Solid State Sciences, Vol.8, ed. by A.R. Bishop (Springer, Berlin, Heidelberg, New York 1978) pp.291-296

3.7 K. Maki, T. Tsuneto: Phys. Rev. B*11*, 2539 (1979)

3.8 K. Maki: Phys. Rev. B*11*, 4264 (1975)
K. Maki, P. Kumar: Phys. Rev. B*14*, 118, 3920 (1976)

3.9 K. Maki, P. Kumar: Phys. Rev. Lett. *38*, 558 (1977)
K. Maki: In *Solitons and Condensed Matter Physics*, Springer Series in Solid State Sciences, Vol.8, ed. by A.R. Bishop (Springer, Berlin, Heidelberg, New York 1978) pp.278-290

3.10 R.K. Bullough, P.J. Caudrey, J.D. Gibbon, S. Duckworth, H.M. Gibbs, B. Bölger, L. Baede: Opt. Commun. *18*, 200 (1976)

3.11 G.E. Volovik, V.P. Mineev: Pisma ZETF *24*, 605 (1976); ZEFT *72*, 2256 (1977); Phys. Rev. B (1979) (to be published)

3.12 H.M. Gibbs, B. Bölger, L. Baede: Opt. Commun. *18*, 199 (1976)
F.P. Mattar, M.C. Newstein, P.E. Serafin, H.M. Gibbs, B. Bölger, G. Forster, P.E. Toschek: In *Coherence and Quantum Optics IV*, ed. by L. Mandel, E. Wolf (Plenum Press, New York 1978) pp.143-164

3.13 R.K. Bullough, P.M. Jack, P.W. Kitchenside, R. Saunders: "Solitons in laser physics", Phys. Scr. *20*, 364 (1979)
3.14 R.K. Bullough: "Solitons", in *Interaction of Radiation and Condensed Matter*, Vol.1, IAEA-SMR-20/51 (International Atomic Energy Agency, Vienna 1977) pp.381-469
3.15 R.K. Bullough, P.J. Caudrey, J.C. Eilbeck, J.D. Gibbon: Opto-electronics *6*, 121 (1974)
3.16 J.C. Eilbeck, P.J. Caudrey, J.D. Gibbon, R.K. Bullough: J. Phys. A: Math. Gen. *6*, 1337 (1973)
3.17 J.C. Wheatley: Rev. Mod. Phys. *47*, 415 (1975)
3.18 A.J. Leggett: Ann. Phys. *85*, 11 (1974)
3.19 A.J. Leggett: Rev. Mod. Phys. *47*, 331 (1975)
3.20 J.A. Hermann, R.K. Bullough: "Superfluorescence from degenerate media", Paper T.9 X Intern. Quantum Electronics Conf. Atlanta, Georgia, May 29 - June 1, 1978 (J. Opt. Soc. Am., May 1978 issue) and to be published (1979)
3.21 M.R. Rice: In *Solitons and Condensed Matter Physics*, Springer Series in Solid State Sciences, Vol.8, ed. by A.R. Bishop, T. Schneider (Springer, Berlin, Heidelberg, New York 1978) pp.246-253
3.22 See R.E. Peierls: *Quantum Theory of Solids* (Oxford University Press, London 1955) p.108
3.23 H. Fröhlich: Proc. Roy. Soc. A*223*, 296 (1954)
3.24 P.A. Lee, T.M. Rice, P.W. Anderson: Solid State Commun. *14*, 703 (1974)
3.25 A.J. Hopfinger, A.J. Lewanski, T.J. Sluckin, P.L. Taylor: In *Solitons and Condensed Matter Physics*, Springer Series in Solid State Sciences, Vol.8, ed. by A.R. Bishop, T. Schneider (Springer, Berlin, Heidelberg, New York 1978) pp.330-333
3.26 A. Luther: In *Solitons and Condensed Matter Physics*, Springer Series in Solid State Sciences, Vol.8, ed. by A.R. Bishop, T. Schneider (Springer, Berlin, Heidelberg, New York 1978) pp.78-84
3.27 B.D. Josephson: "Weakly Coupled Super-Conductors", in *Superconductivity*, ed. by R.D. Parks (Marcel Dekker, New York 1969) pp.423-448
3.28 R.K. Dodd, R.K. Bullough: Proc. Roy Soc. (London) A*351*, 499 (1976)
3.29 D.W. McLaughlin, A.C. Scott: J. Math. Phys. *14*, 1817 (1973)
3.30 R.K. Dodd, R.K. Bullough: Proc. Roy. Soc. (London) A*352*, 481 (1977)
3.31 S. Duckworth: Ph.D. Thesis, University of Manchester (1976)
3.32 P.W. Kitchenside: Ph.D. Thesis, University of Manchester (1979)
3.33 J. Shiefman, P. Kumar: Phys. Scr. *20*, 435 (1979)
3.34 A.C. Newell: J. Math. Phys. *18*, 922 (1977)
3.35 A.L. Mason: In *Nonlinear Equations in Physics and Mathematics*, ed. by A.O. Barut (D. Reidel, Dordrecht, Holland 1978) pp.205-218
3.36 P.W. Kitchenside, A.L. Mason, R.K. Bullough, P.J. Caudrey: In *Solitons and Condensed Matter Physics*, Springer Series in Solid State Sciences, Vol.8, ed. by A.R. Bishop, T. Schneider (Springer, Berlin, Heidelberg, New York 1978) pp.48-51
3.37 R.K. Dodd, R.K. Bullough, S. Duckkworth: J. Phys. A: Math. Gen. *8*, L 64 (1975)
3.38 C. Rebbi: Sci. Amer. *240*(2), 76-91 (1978)
3.39 G.J. Salamo, H.M. Gibbs, G.G. Churchill: Phys. Rev. Lett. *33*, 273 (1974)
3.40 B. Bölger, L. Baede, H.M. Gibbs: Opt. Commun. *18*, 67 (1976)
3.41 W. Kreiger, G. Gaida, P.E. Toschek: Z. Phys. B*25*, 297 (1976)
H.M. Gibbs, B. Bölger, F.P. Mattar, M.C. Newstein, G. Forster, P.E. Toschek: Phys. Rev. Lett. *37*, 174 (1976)
3.42 H.M. Gibbs, B. Bölger: In *Coherence and Quantum Optics IV*, ed. by L. Mandel, E. Wolf (Plenum Press, New York 1978) pp.759-765
3.43 R. Saunders, R.K. Bullough: In *Cooperative Effects in Matter and Radiation*, ed. by C.M. Bowden, D.W. Howgate, H.R. Robl (Plenum Press, New York 1977) pp.209-256
R. Bonifacio, M. Gronchi, L.A. Lugiato, A.M. Ricca: ibid., pp.193-208; also references therein
R.K. Bullough, R. Saunders, C. Feuillade: In *Coherence and Quantum Optics IV*, ed. by L. Mandel, E. Wolf (Plenum Press, New York 1978) pp.263-279 and references therein

3.44 Q.H.F. Vrehen, H.M.J. Hikspoors, H.M. Gibbs: In *Coherence and Quantum Optics IV*, ed. by L. Mandel, E. Wolf (Plenum Press, New York 1978) pp.543-553 and references therein
3.45 M.J. Ablowitz, Y. Kodama: "Transverse instability of one-dimensional transparent optical pulses in resonant media", Phys. Lett. *70*A, 83 (1979)
3.46 B. Schroer: Private communication
3.47 R.K. Bullough, P.J. Caudrey: Rocky Mount. J. Math. *8*, 53 (1978)
3.48 P.W. Kitchenside, P.J. Caudrey, R.K. Bullough: "Soliton like spin waves in $^3$He B", Phys. Scr. *20*, 673 (1979)
3.49 D.J. Kaup, A.C. Newell: Phys. Rev. B*18*, 5162 (1978)
3.50 D.J. Kaup: Stud. Appl: Math. *LIV*, 165 (1975)
3.51 D.J. Kaup, A.C. Newell: Proc. Roy. Soc. (London) A*361*, 413 (1978)
3.52 P.W. Kitchenside, A.L. Mason, P.J. Caudrey, R.K. Bullough: To be published (1980)
3.53 D.W. McLaughlin, A.C. Scott: "Soliton perturbation theory" in *Nonlinear Evolution Equations Solvable by the Spectral Transform*, ed. by F. Calogero (Pitman, London 1978); Phys. Rev. A*18*, 1632 (1978)
3.54 A.C. Newell: In *Solitons and Condensed Matter Physics*, Springer Series in Solid State Sciences, Vol.8, ed. by A.R. Bishop, T. Schneider (Springer, Berlin, Heidelberg, New York 1978) pp.52-64
A valuable additional reference is:
V.I. Karpman: "Soliton perturbation in the presence of perturbation", Phys. Scr. *20*, 462 (1979)
3.55 A.B. Mikhailov: JETP Lett. *30*(7), 443 (1979)
3.56 A.P. Fordy, J. Gibbons: "A class of integrable nonlinear Klein-Gordon equations in many dependent variables", Dublin Inst. Adv. Study preprint DIAS-STP-80-03

# 4. On a Nonlinear Lattice (The Toda Lattice)

M. Toda

In order to elucidate certain characteristic feature of nonlinear waves, a one-dimensional lattice model is described. The model affords exact solutions such as multi-soliton state, periodic waves, and solution to the initial value problem by the inverse scattering method. The Bäcklund transformation and the relation to the Korteweg-de Vries equation are also discussed.

## 4.1 Nonlinear Lattices

There was a generally accepted belief that nonlinear couplings between normal modes of harmonic oscillators would lead to complete energy sharing between these modes. FERMI et al. [4.1] used a computer to verify this belief for one-dimensional lattices with nonlinear interactions between adjacent particles. To their surprise, these nonlinear lattices yielded very little energy sharing at all and, on the contrary, exhibited recurrence of the initial state.

FORD [4.2] investigated nonlinear oscillator systems by perturbation theory and numerical computation, and concluded that nonlinear systems would not be ergodic, even though they might share energy if certain resonance conditions were satisfied. Moreover, FORD and WATERS [4.3] demonstrated the existence of normal modes of nonlinear systems, where a normal mode was defined as motion for which each oscillator moved with essentially constant amplitude and at a given frequency. This was a surprising and revealing result which became the starting point of my research into nonlinear lattices.

If a system is nonergodic, it may admit analytic solutions to the equations of motion. Therefore it was hoped that we might find a certain nonlinear lattice, which would be susceptible to analysis. Such a system was expected to have some bearing on physical systems and to admit solutions to the equations of motion which were not incredibly complex. Thus I was led to search for a tractable nonlinear lattice, and after a while, I found a model system which proved to be quite interesting to investigate.

It is a great honor and pleasure to describe here the line of thought which led me to invent this system and present some of the recent achievements and future prospects.

### 4.1.1 Equations of Motion

The equations of motion of the lattice are of the form

$$m\ddot{Y}_n = -\phi'(Y_n - Y_{n-1}) + \phi'(Y_{n+1} - Y_n) \tag{4.1}$$

where $Y_n$ stands for the displacement of the $n^{th}$ particle, and $\phi(r)$ denotes the interaction potential between adjacent particles.

There was no special strategy, except for the hope that I might, by trial and error, find both an interaction potential and solutions at the same time.

### 4.2.1 The Dual Transformation

It was very fortunate that I had been working on the theory of harmonic lattices and had previously found a transformation which related a system with a variety of masses with another with a variety of force constants [4.4]. This transformation could linearize the interaction part of the equations of motion of a nonlinear lattice by introducing a nonlinear momentum.

In this transformation we use the mutual displacements

$$r_n = y_n - Y_{n-1} \tag{4.2}$$

as the generalized coordinates, and introduce the generalized momenta $s_n$ canonically conjugate to $r_n$. The canonical equations of motion for $s_n$,

$$\dot{s}_n = -\frac{d\phi(r_n)}{dr_n} \quad , \tag{4.3}$$

are assumed to yield the inverse

$$r_n = -\frac{1}{m}\chi(\dot{s}_n) \tag{4.4}$$

as a one-valued function of $\dot{s}_n = ds_n/dt$. Then (4.1) is transformed to the equation

$$\frac{d}{dt}\chi(\dot{s}_n) = s_{n-1} - 2s_n + s_{n+1} \quad . \tag{4.5}$$

Since (4.1) and (4.5) are equivalent expressions of a system, this transformation was called the dual transformation. It is one of the consequences of this transformation that the momentum is

$$m\dot{Y}_n = s_n - s_{n+1} \quad . \tag{4.6}$$

Therefore, we have

$$Y_n = (S_n - S_{n+1})/m \tag{4.7}$$

with

$$S_n = \int^t s_n(t)dt \quad . \tag{4.8}$$

The equations of motion take the following simple form:

$$\chi(\ddot{S}_n) = S_{n-1} - 2S_n + S_{n+1} \quad . \tag{4.9}$$

## 4.2 The Exponential Interaction

The nonlinear normal mode or the periodic behaviour of nonlinear systems was so striking that I tried to find a periodic wave and a nonlinear lattice which might afford such a solution. The dual transformation provided a key to this problem. Being aided by the fact that (4.5) could be considered as a recursion formula for $s_n$, at last I found the following answer, which turned out to be quite fruitful in many respects [4.5].

The potential function I adopted consists of the exponential repulsive part A exp(-br) and the attractive part ar where a,b and A are constants such that $ab > 0$, and $A > 0$. If the parameter A is so chosen that the minimum of the potential is just at $r = 0$, the interaction potential takes the form

$$\phi(r) = (a/b)\exp(-br) + ar + \text{const.} \quad . \tag{4.10}$$

Then, (4.1) gives the equations of motion

$$m\ddot{Y}_n = a\{\exp[-b(Y_n - Y_{n-1})] - \exp[-b(Y_{n+1} - Y_n)]\} \tag{4.11}$$

and (1.3) and (1.5) yield

$$\exp(-br_n) - 1 = (1/a)\dot{s}_n \tag{4.12}$$

$$\frac{\ddot{s}_n}{a+\dot{s}_n} = \frac{b}{m}(s_{n-1} - 2s_n + s_{n+1}) \quad . \tag{4.13}$$

### 4.2.1 Cnoidal Waves

The periodic wave solution of (4.13) is [4.5]

$$s_n = \mp \frac{2mK\nu}{b} Z\left\{2\left(\frac{n}{\lambda} \mp \nu t\right)K\right\} \quad , \tag{4.14}$$

where Z(u) is the Jacobian Z function defined by

$$Z(u) = \int_0^u dn^2 u du - \frac{E}{K} u \qquad (4.15)$$

with the dispersion relation

$$2K\nu = \sqrt{\frac{ab}{m}} \left( \frac{1}{sn^2(2K/\lambda)} - 1 + \frac{E}{K} \right)^{-\frac{1}{2}} \qquad (4.16)$$

between the frequency $\nu$ and the wavelength $\lambda$. In the above formulas, sn and dn represent the Jacobian elliptic functions; K and E are the complete elliptic integral of the first and the second kind. These have the same modulus k, which is responsible for the amplitude of the wave. Equation (4.12) gives the waveform

$$\exp(-br_n) - 1 = \frac{(2K\nu)^2}{ab/m} \left[ dn^2\left\{ 2\left(\frac{n}{\lambda} \mp \lambda t\right)K \right\} - \frac{E}{K} \right] \quad . \qquad (4.17)$$

Since the function dn can be written in terms of cn, the periodic wave can be called the cnoidal wave for the lattice following the terminology of KORTEWEG and DE VRIES [4.6], who had a similar periodic wave for shallow water.

If we keep the modulus k constant, (4.16) gives a spectral band which extends from $\nu = 0$ for $1/\lambda = 0$ to the maximum frequency $\nu_m = \sqrt{ab/mEK}/2$ for the wave number $1/\lambda_m = \frac{1}{2}$.

The cnoidal wave (4.14) can be considered as a wave in a periodic lattice. If this lattice consists of N particles, we have independent waves with the wave numbers

$$1/\lambda = 0 \ , \quad 1/N \ , \quad 2/N, \ \ldots, \ (N-1)/N \quad . \qquad (4.18)$$

Waves with the wave numbers $1/\lambda = N/N, (N+1)/N, \ldots$ are equivalent to those with $1/\lambda = 0, 1/N, \ldots,$ because of the periodicity 2K of the functions $sn^2$ and $dn^2$ in (4.16,17).

### 4.2.2 Solitons

KORTEWEG and DE VRIES found solitary wave solutions for their equation of motion for shallow water waves. Solitary waves for our lattice are obtained by taking the limit $k \to 1$, keeping

$$\alpha = 2K/\lambda \qquad (4.19)$$

finite [4.7]. Equations (4.17) and (4.16) reduce, in this limit, to

$$\exp(-br_n) - 1 = \frac{m}{ab} \beta^2 \mathrm{sech}^2(\alpha n \mp \beta t) \qquad (4.20)$$

with

$$\beta = \sqrt{\frac{ab}{m}} \sinh\alpha \quad . \qquad (4.21)$$

Equations (4.20) and (4.21) represent a pulse-like wave of compression for $b > 0$, with width $h/\alpha$, height $\beta^2$ and speed $c = h\beta/\alpha$, where h is the lattice spacing. The higher the pulse, the smaller the width and the larger its speed. It has been shown numerically that this pulse is quite stable [4.8]. That the solitary waves of the KdV equation are stable when they interact has been shown numerically by ZABUSKY and KRUSKAL [4.9] and they coined the name "soliton" for the stable solitary wave. In this sense (4.20) represents a lattice soliton.

The soliton can also be expressed as

$$s_n = \mp \frac{\beta m}{b} \tanh(\alpha n \mp \beta t) \quad , \tag{4.22}$$

or as

$$S_n = \frac{m}{b} \log\{1 + A \exp[2(\alpha n \mp \beta t)]\} \quad . \tag{4.23}$$

### 4.2.3 The Harmonic Limit

One of the merits of our lattice is that the system and the exact solutions admit both the harmonic limit ($b \to 0$) and the hard sphere limit ($b \to \infty$). This is valid for other solutions which will follow in the subsequent sections. A harmonic lattice is obtained in the limit $b \to 0$ keeping the force constant $\kappa = ab$ finite. In the limit $b \to 0$, $k \to 0$, the cnoidal wave (4.17) reduces to the sinusoidal wave

$$r_n = - \frac{(\pi\nu)^2}{ab/m} \frac{k^2/b}{2} \cos 2\pi \left(\frac{n}{\lambda} \mp \nu t\right) \tag{4.24}$$

with

$$\nu = \frac{1}{\pi} \sqrt{\frac{ab}{m}} \sin(\pi/\lambda) \quad . \tag{4.25}$$

This limit exists when $k^2/b$ is kept finite. In the limit $b \to 0$ we expect that $\exp(-br_n) \to 1$ which can be achieved in (4.20) only if $\alpha$ goes to zero because of (4.21). Therefore we have no solitary wave solution in the harmonic limit.

### 4.2.4 Two-Soliton Solutions

The solution which contains two solitons can be obtained by assuming

$$S_n = \frac{m}{b} \log \psi_n \tag{4.26}$$

with [4.10]

$$\psi_n = A \cosh(\alpha n - \beta t) + B \cosh(\mu n - \gamma t) \quad , \tag{4.27}$$

where A, B, $\beta$ and $\gamma$ are functions of $\alpha$ and $\mu$ to be determined by the equations of motion. Alternatively we may use the formula [4.11]

$$\psi_n = 1 + A_1 \exp[2(\alpha_1 n - \beta_1 t)] + A_2 \exp[2(\alpha_2 n - \beta_2 t)]$$

$$+ A_3 \exp[2(\alpha_1 + \alpha_2)n - 2(\beta_1 + \beta_2)t] \quad , \tag{4.28}$$

where $A_1$, $A_2$, $A_3$, $\beta_1$ and $\beta_2$ are to be determined as functions of $\alpha_1$ and $\alpha_2$ by inserting (4.26) and (4.28) into the equations of motion (4.13). Thus we get the following two cases.

*Case I)*

$$\beta_1 = \sqrt{\frac{ab}{m}} \sinh\alpha_1 \quad ,$$

$$\beta_2 = \sqrt{\frac{ab}{m}} \sinh\alpha_2 \quad ,$$

$$A_3 = \left(\frac{\sinh\frac{1}{2}(\alpha_1-\alpha_2)}{\sinh\frac{1}{2}(\alpha_1+\alpha_2)}\right)^2 A_1 A_2 \quad . \tag{4.29}$$

In this case two solitons propagate in the same direction; and one overtakes and passes over the other.

*Case II)*

$$\beta_1 = \sqrt{\frac{ab}{m}} \sinh\alpha_1 \quad ,$$

$$\beta_2 = -\sqrt{\frac{ab}{m}} \sinh\alpha_2 \quad ,$$

$$A_3 = \left(\frac{\cosh\ (\alpha_1-\alpha_2)}{\cosh\ (\alpha_1+\alpha_2)}\right)^2 A_1 A_2 \quad . \tag{4.30}$$

In this case two solitons propagate in the opposite directions: Head-on collision and passing over take place.

In either case, the asymptotic forms for two solitons are

$$\exp(-br_n) - 1 = \frac{m}{ab}\, \beta_i^2 \operatorname{sech}^2\left(\alpha_i n - \beta_i t + \delta_i^{\mp}\right) \tag{4.31}$$

for i = 1,2, with

$$\delta_1^- = \log A_1 \quad ,$$

$$\delta_2^- = \log \frac{A_3}{A_1} \tag{4.32}$$

for $t \to -\infty$, and

$$\delta_1^+ = \log \frac{A_3}{A_2}$$

$$\delta_2^+ = \log A_2 \tag{4.33}$$

for t → +. . Generally we have the relation

$$\delta_1^- + \delta_2^- = \delta_1^+ + \delta_2^+ \tag{4.34}$$

which has the following meaning.

A soliton is a compression pulse with excess mass (h = lattice spacing)

$$\begin{aligned} M &= - \sum_n Y_n - Y_{n-1}) \frac{m}{h} \\ &= \frac{m}{h} (Y_{-\infty} - Y_\infty) \quad . \end{aligned} \tag{4.35}$$

However, (42.6) and (4.7) yield

$$\begin{aligned} mY_n = S_n - S_{n+1} &\to 0 \qquad (n \to -\infty) \\ &\to -2 \frac{m}{b} \alpha \quad (n \to +\infty) \quad . \end{aligned} \tag{4.36}$$

Therefore we have

$$M = \frac{2m}{bh} \alpha \tag{4.37}$$

which holds for $\alpha = \alpha_1$, and $\alpha_2$. The momentum of the soliton is

$$\begin{aligned} P &= \sum_n m\dot{Y}_n \\ &= m(s_{-\infty} - s_\infty) \\ &= \frac{2m}{b} \beta \\ &= Mc \end{aligned} \tag{4.38}$$

where $c = h\beta/\alpha$ is the speed.

Since the positions of the two solitons for $t \to \mp \infty$ are given by

$$n_i^\mp = (\beta_i t - \delta_i^\mp)/\alpha_i \tag{4.39}$$

their common center of mass is at

$$\begin{aligned} n^\mp &= (M_1 n_1^\mp + M_2 n_2^\mp)/(M_1 + M_2) \\ &= \frac{1}{\alpha_1+\alpha_2} \left[ (\beta_1 + \beta_2)t - \delta_1^\mp - \delta_2^\mp \right] \quad . \end{aligned} \tag{4.40}$$

Therefore, the relation (4.34) implies that the center of mass of the two solitons moves on a straight line (and with a constant speed) in the x-t plane.

It is shown that a cnoidal wave can be considered as a succession of solitons [4.12].

The general N-soliton solution was obtained by HIROTA [4.13] and by FLASCHKA [4.15]. We shall come back to this solution in the next section.

## 4.3 Matrix Formalism

For brevity, we shall write the equations of motion (4.11) in a dimensionless canonical form as

$$\dot{Q}_n = P_n \quad ,$$
$$\dot{P}_n = \exp[-(Q_n - Q_{n-1})] - \exp[-(Q_{n+1} - Q_n)] \quad . \tag{4.41}$$

The equations of motion can be written as [4.14]

$$\dot{L} = BL - LB \tag{4.42}$$

where L and B are matrices defined by

$$(L\phi)_n = b_n\phi(n) + a_n\phi(n - 1) + a_{n+1}\phi(n + 1) \quad ,$$
$$(B\phi)_n = a_n\phi(n - 1) - a_{n+1}\phi(n + 1) \quad . \tag{4.43}$$

Equation (4.42) gives (4.43) by choosing $a_n$ and $b_n$ as

$$a_n = \frac{1}{2}\exp[-(Q_n - Q_{n-1})/2]$$
$$b_n = \frac{1}{2}P_n \quad . \tag{4.44}$$

It can be shown that the eigenvalues $\lambda$ of the equation

$$L\phi = \lambda\phi \tag{4.45}$$

are independent of time, if $\phi$ evolves according to the equation

$$\dot{\phi} = B\phi \quad . \tag{4.46}$$

### 4.3.1 The Inverse Method

For an infinite lattice [4.15] where the motion is in a finite region, we can speak of the scattering of an incident wave $z^{-n} = \exp(-ikn)$ ($k$ = wave number) and its scattered wave $R(k)z^n$, so that

$$\phi \simeq z^{-n} + R(z,t)z^n \quad (n \to +\infty) \quad . \tag{4.47}$$

In addition, (4.45) will have discrete eigenvalues

$$\lambda_j = \frac{z_j+z_j^{-1}}{2} \quad (|z_j| < 1) \tag{4.48}$$

which are independent of time. The associated eigenfunctions have the asymptotic form

$$\phi_j(n) \simeq c_j(t)z_j^n \qquad (n \to +\infty) \tag{4.49}$$

where $c_j(t)$ is the normalization constant defined in such a way that

$$\sum_{n=-\infty}^{\infty} \phi_j(n)^2 = 1 \quad . \tag{4.50}$$

If the initial wave is given, we can calculate $R(z,0)$, $z_j$ and $c_j(0)$. We call these the initial scattering data. It can be shown that the scattering data at time t are given as

$$R(z,t) = R(z,0)\exp[t(z^{-1} - z)] \quad , \tag{4.51}$$

$$\lambda_j(t) = \lambda_j(0) \quad , \tag{4.52}$$

$$c_j(t)^2 = c_j(0)^2 \exp[t(z_j^{-1} - z_j)] \quad . \tag{4.53}$$

We construct the kernel

$$F(m) = \frac{1}{2\pi i} \oint R(z,t)z^{m-1}dz + \sum_j c_j(t)^2 z_j^m \quad , \tag{4.54}$$

where the path of integration encircles the origin of the complex z plane, and the discrete integral equation

$$\kappa(n,m) + F(n+m) + \sum_{n'=n+1}^{\infty} \kappa(n,n')F(n'+m) = 0 \quad . \tag{4.55}$$

If this equation is solved for $\kappa(n,m)$, we have the solution of the equations of motion in the form [4.15]

$$\exp[-(Q_n - Q_{n-1})] = [K(n,n)/K(n-1,n-1)]^2 \quad , \tag{4.56}$$

$$[K(n,n)]^{-2} = 1 + F(2n) + \sum_{n'=n+1}^{\infty} \kappa(n,n')F(n'+n) \quad . \tag{4.57}$$

We see also that [4.16]

$$\begin{aligned} P_n &= s_n - s_{n+1} \quad , \\ s_n &= \kappa(n-1,n) \quad . \end{aligned} \tag{4.58}$$

Such a method of solving the initial value problem of nonlinear systems is generally called the inverse (scattering) method.

### 4.3.2 Multi-Soliton Solutions

When we assume no reflection

$$R(z,0) = 0 \tag{4.59}$$

we have the N-soliton solution given by [4.13,15]

$$S_n = \log \det B_n \tag{4.60}$$

where $B_n$ is an $N \times N$ matrix whose elements are

$$\left(B_n\right)_{jk} = \delta_{jk} + c_j c_k \frac{(z_j z_k)^{n+1}}{1 - z_j z_k} \exp[-(\beta_j + \beta_k)t] \qquad (j,k = 1,2, \ldots, N) \tag{4.61}$$

with

$$\begin{aligned} z_j &= \pm \exp(-\alpha_j) \quad , \\ \beta_j &= \pm \sqrt{\frac{ab}{m}} \sinh\alpha_j \end{aligned} \tag{4.62}$$

where $\alpha_j$ and $c_j = c_j(0)$ are arbitrary constants. The above solution consists of N solitons with the asymptotic forms

$$\exp\left[-(Q_n - Q_{n-1})\right] - 1 = \beta_j^2 \operatorname{sech}^2\left(\alpha_j n - \beta_j t + \delta_j^{\mp}\right) \tag{4.63}$$

where the phase shifts $\delta_j^{\mp}$ for $t \to \mp \infty$ satisfy the relation

$$\sum_j \delta_j^- = \sum_j \delta_j^+ \tag{4.64}$$

which represents the conservation of the total momentum [c.f. (4.34)].

## 4.4 The Continuum Limit

We shall briefly present some of the results of the continuum limit. For sufficiently smooth waves, our equations of motion (4.11) reduce to the differential equation [4.7]

$$Y_{tt} = c_0^2\left[(1 - bhY_x)Y_{xx} + \frac{h^2}{12} Y_{xxxx}\right] \tag{4.65}$$

with

$$c_0 = h\sqrt{ab/m} \quad , \quad x = nh \quad , \tag{4.66}$$

and after further transformation it reduces to the KdV equation

$$u_\tau - 6uu_\xi + u_{\xi\xi\xi} = 0 \tag{4.67}$$

with

$$u = 2br_n \ ,$$

$$\xi = (x - c_0 t)/h \quad , \quad \tau = c_0 t/24h \quad . \tag{4.68}$$

If we write

$$u = w_\xi \tag{4.69}$$

the KdV equation is written as

$$w_\tau - 3w_\xi^2 + w_{\xi\xi\xi} = 0 \quad . \tag{4.70}$$

## 4.5 Bäcklund Transformations

When fixed end or periodic boundary conditions are imposed, the Hamiltonian of our system can be written as

$$H(Q,P) = \frac{1}{2}\sum_n P_n^2 + \sum_n \frac{\exp(Q_{n-1})}{\exp(Q_n)} + \text{const.} \tag{4.71}$$

As is readily seen, the transformation $(Q,P) \to (Q',P')$, where

$$P_n = \frac{\exp(Q_n)}{\exp(Q_n')} + \frac{\exp(Q_{n-1}')}{\exp(Q_n)} - \alpha \quad ,$$
$$P_n' = \frac{\exp(Q_n)}{\exp(Q_n')} + \frac{\exp(Q_n')}{\exp(Q_{n+1})} - \alpha \quad . \tag{4.72}$$

($\alpha$ = const.) keeps the Hamiltonian unchanged except for a constant,

$$H'(Q',P') = H(Q,P) + \text{const.} \quad , \tag{4.73}$$

the additional constant being determined by the boundary conditions. This transformation is canonical, and is derived from the generating function [4.17]

$$W(Q,Q') = \sum_n \left\{\exp[-(Q_n' - Q_n)] - \exp[-(Q_{n+1} - Q_n')] - \alpha(Q_n' - Q_n)\right\} \tag{4.74}$$

by the usual procedure of the canonical transformation, that is, by

$$P_n = \frac{\partial W}{\partial Q_n} \quad , \quad P_n' = -\frac{\partial W}{\partial Q_n'} \quad . \tag{4.75}$$

Suppose that (Q,P) is a known solution, then (Q',P') is also a solution for the same system. Such transformations are known for other nonlinear wave equations, and are called Bäcklund transformations.

Since $P_n = \dot{Q}_n$ and $P'_n = \dot{Q}'_n$, the Bäcklund transformation for our lattice can be written as [4.17,18]

$$\dot{Q}_n = \exp[-(Q'_n - Q_n)] + \exp[-(Q_n - Q'_{n-1})] - \alpha \quad ,$$

$$\dot{Q}'_n = \exp[-(Q'_n - Q_n)] + \exp[-(Q_{n+1} - Q'_n)] - \alpha \ . \tag{4.76}$$

As a simple example, we start with the trivial solution

$$Q_n = \dot{Q}_n = 0 \tag{4.77}$$

to obtain a solution of (4.76) given by

$$\exp Q'_n = \frac{\cosh(\kappa n - \beta t)}{\cosh[\kappa(n+1) - \beta t]} \tag{4.78}$$

with

$$\alpha = 2\cosh\kappa \quad , \qquad \beta = -\sinh\kappa \quad . \tag{4.79}$$

Equation (4.78) can be written as

$$\exp[-(Q'_n - Q'_{n-1})] = 1 + \sinh^2\kappa \cdot \mathrm{sech}^2(\kappa n - \beta t) \quad . \tag{4.80}$$

Thus it is seen that when (Q,P) is transformed to (Q',P'), a new soliton is added by this transformation. It can be shown that this transformation is equivalent to the Bäcklund transformation proposed by CHEN and LIU [4.19].

It can be shown [4.20] that the above transformation reduces, in the continuum limit, to the Bäcklund transformation for the KdV equations [4.21]

$$\begin{aligned} w_\xi - w'_\xi &= -2\eta^2 - (w - w')^2/2 \quad , \\ w_\tau - w'_\tau &= [-2w_{\xi\xi} - 2w_\xi(w - w') - 4\eta^2(w - w')]_\xi \end{aligned} \tag{4.81}$$

when we write

$$w = 2Q_{n+\frac{1}{2}} = Q_n + Q_{n+1} \quad ,$$

$$w' = 2Q'_n \quad . \tag{4.82}$$

## 4.6 Concluding Remarks

Recently, the general solution to the initial value problem of the periodic Toda lattice has been obtained [4.22,23]. As further problems, we have to deal with a lattice including impurities, which is related to heat conduction [4.24] and also to the stability of trajectories in phase space. We have also the problem related to the boundary conditions and to external forces. Besides these problems in classical mechanics, quantization of the waves in the nonlinear lattice is expected to elucidate the relation with lattice phonons.

## References


4.1 E. Fermi, J. Pasta, S. Ulam: *Collected Papers of ENRICO FERMI*, Vol.II (University of Chicago Press 1965) p.978
4.2 J. Ford: J. Math. Phys. *2*, 387 (1961)
4.3 J. Ford, J. Waters: J. Math. Phys. *4*, 1293 (1963)
4.4 M. Toda: J. Phys. Soc. Jpn. *20*, 2095 (1965); Progr. Theor. Phys. Suppl. *36*, 113 (1966)
4.5 M. Toda: J. Phys. Soc. Jpn. *22*, 431 (1967)
4.6 D.J. Korteweg, G. De Vries: Philos. Mag. *18*, 35 (1895)
4.7 M. Toda: J. Phys. Soc. Jpn. *23*, 501 (1967)
4.8 N. Saitô: Progr. Theor. Phys. Suppl. *45*, 201 (1970)
4.9 N.J. Zabusky, M.D. Kruskal: Phys. Rev. Lett. *15*, 240 (1965)
4.10 M. Toda: Intern. Conf. Statistical Mech. Kyoto (1968); J. Phys. Soc. Jpn. Suppl. *26*, 235 (1969)
4.11 M. Toda, M. Wadati: J. Phys. Soc. Jpn. *34*, 18 (1973)
4.12 M. Toda: Progr. Theor. Phys. Suppl. *45*, 174 (1970)
4.13 R. Hirota: J. Phys. Soc. Jpn. *35*, 286 (1973)
4.14 H. Flaschka: Phys. Rev. B*9*, 1924 (1974)
4.15 H. Flaschka: Progr. Theor. Phys. *51*, 703 (1974)
4.16 M. Toda: Phys. Rep. *18*C, 1 (1975)
4.17 M. Toda, M. Wadati: J. Phys. Soc. Jpn. *39*, 1204 (1975)
4.18 M. Wadati, M. Toda: J. Phys. Soc. Jpn. *39*, 1196 (1975)
4.19 H. Chen, C. Liu: J. Math. Phys. *16*, 1428 (1975)
4.20 M. Toda: Progr. Theor. Phys. Suppl. *59*, 1 (1976); in *Intern. Symp. on Mathematical Problems in Theoretical Physics*, ed. by H. Araki, Lecture Notes in Physics, Vol.39 (Springer, Berlin, Heidelberg, New York 1975) p.387
4.21 H.D. Wahlquist, F.B. Estabrook: Phys. Rev. Lett. *31*, 1386 (1973)
4.22 M. Kac, P. Van Moerbeke: J. Math. Phys. *72*, 2879 (1974)
4.23 E. Date, S. Tanaka: Progr. Theor. Phys. *55*, 457 (1976)
4.24 M. Toda: Phys. Scr. *20*, 424 (1979)

# 5. Direct Methods in Soliton Theory

R. Hirota

The main purpose of this chapter is to present a direct and systematic way of finding exact solutions and Bäcklund transformations of a certain class of nonlinear evolution equations. The nonlinear evolution equations are transformed, by changing the dependent variable(s), into bilinear differential equations of the following special form

$$F\left(\frac{\partial}{\partial t}-\frac{\partial}{\partial t'},\frac{\partial}{\partial x}-\frac{\partial}{\partial x'}\right)f(t,x)f(t',x')\Big|_{t=t',x=x'}=0\quad,$$

which we solve exactly using a kind of perturbational approach.

Examples shown in this article are the Korteweg-de Vries (KdV, equation, higher order KdV equations, model equations for shallow water waves, the Boussinesq equation, the Toda equation, discrete-time Toda equation, etc.

## 5.1 Preliminaries

In order to illustrate the present method we consider the KdV equation [5.1,2]

$$u_t + 6uu_x + u_{xxx} = 0\quad, \tag{5.1}$$

with the boundary condition $u = 0$ at $|x| = \infty$. Here subscripts indicate partial differentiation.

We solve (5.1) using the usual perturbation method. Let $u = w_x$, then integrating (5.1) with respect to x, we have

$$w_t + 3w_x^2 + w_{xxx} = 0\quad, \tag{5.2}$$

where an integration constant is chosen to be zero. We expand w as a power series in a small parameter $\varepsilon$

$$w = \varepsilon w_1 + \varepsilon^2 w_2 + \dots\quad. \tag{5.3}$$

Substituting (5.3) into (5.2) and collecting terms with the same power of $\varepsilon$, we have

$$\left(\frac{\partial}{\partial t} + \frac{\partial^3}{\partial x^3}\right)w_1 = 0 \quad , \tag{5.4}$$

$$\left(\frac{\partial}{\partial t} + \frac{\partial^3}{\partial x^3}\right)w_2 = -3(w_1)_x^2 \quad , \tag{5.5}$$

$$\left(\frac{\partial}{\partial t} + \frac{\partial^3}{\partial x^3}\right)w_3 = -6(w_1)_x(w_2)_x \quad , \tag{5.6}$$

and so on. Solving these equations successively we have a formal solution in terms of perturbation series. The difficulty encountered in the process is that the perturbation series do not converge rapidly or diverge.

Recently a powerful summation technique called the Padê Approximant [5.3] was developed. The [N/M] Padê Approximant to a function $f(\varepsilon)$ is the ratio of two polynomials, the numerator of degree N and the denominator of degree M. The Padê Approximant has been applied to many branches of physics as a systematic method of extracting more information from power series expansions [5.4].

We consider a Padê Approximant to $w(\varepsilon)$. Although we could construct a Padê Approximant to $w(\varepsilon)$ directly from the power series solution (5.3), it is of more interest to transform the original equation into a special form convenient for the Padê Approximant. For this purpose we replace w by G/F and find the equations to be satisfied by F and G. Then power series solutions of F and G would provide a Padê Approximation to w.

Substituting $w = G/F$ into (5.2) we have

$$\begin{aligned}&(G_tF - GF_t)/F^2 + 3(G_xF - GF_x)^2/F^4 \\ &\quad + (G_{xxx}F - 3G_{xx}F_x - 3G_xF_{xx} - GF_{xxx})/F^2 \\ &\quad + 6(FG_xF_x^2 + FGF_xF_{xx} - GF_x^3)/F^4 = 0 \quad ,\end{aligned} \tag{5.7}$$

which apparently has a more complicated form than the original equation for w. There are now two dependent variables F and G in one equation. But, noting that (5.7) is rewritten as

$$\begin{aligned}&[G_tF - GF_t + 3\lambda(G_xF - GF_x) + G_{xxx}F - 3G_{xx}F_x + 3G_xF_{xx} - GF_{xxx}]/F^2 \\ &\quad + 3(G_xF - GF_x)[G_xF - GF_x - 2(FF_{xx} - F_x^2) - \lambda F^2]/F^4 = 0 \quad ,\end{aligned} \tag{5.8}$$

we decouple (5.7), by introducing an arbitrary function $\lambda$. We obtain the two equations

$$G_tF - GF_t + 3\lambda(G_xF - GF_x) + G_{xxx}F - 3G_{xx}F_x + 3G_xF_{xx} - GF_{xxx} = 0 \quad , \tag{5.9}$$

$$2(FF_{xx} - F_x^2) + \lambda F^2 - (G_xF - GF_x) = 0 \quad , \tag{5.10}$$

which are then expressed as

$$\left[\frac{\partial}{\partial t} - \frac{\partial}{\partial t'} + 3\lambda\left(\frac{\partial}{\partial x} - \frac{\partial}{\partial x'}\right) + \left(\frac{\partial}{\partial x} - \frac{\partial}{\partial x'}\right)^3\right]G(x,t)F(x',t')\Big|_{x=x',t=t'} = 0 \quad , \tag{5.11}$$

and

$$\left[\left(\frac{\partial}{\partial x} - \frac{\partial}{\partial x'}\right)^2 + \lambda\right]F(x,t)F(x',t)\Big|_{x=x'} - \left(\frac{\partial}{\partial x} - \frac{\partial}{\partial x'}\right)G(x,t)F(x',t)\Big|_{x=x'} = 0 \quad . \tag{5.12}$$

It is convenient to introduce the operators $D_t$, $D_x$ and various products of them by

$$D_t^n D_x^m f \cdot g = \left(\frac{\partial}{\partial t} - \frac{\partial}{\partial t'}\right)^n \left(\frac{\partial}{\partial x} - \frac{\partial}{\partial x'}\right)^m f(x,t)g(x',t')\Big|_{x=x',t=t'} \quad . \tag{5.13}$$

With this notation, (5.11,12) are expressed simply as

$$(D_t + 3\lambda D_x + D_x^3)G \cdot F = 0 \quad , \tag{5.14}$$

$$(D_x^2 + \lambda)F \cdot F - D_x G \cdot F = 0 \quad . \tag{5.15}$$

We note that (5.14,15) are invariant under the transformation of F and G which keeps $w(= G/F)$ invariant. Let $F = hF'$ and $G = hG'$, then (5.14,15) are transformed to [see (VI.3) below]

$$(D_t + 3\lambda' D_x + D_x^3)G' \cdot F' = 0 \quad , \tag{5.16}$$

$$(D_x^2 + \lambda')F' \cdot F' - D_x G' \cdot F' = 0 \quad , \tag{5.17}$$

where $\lambda' = \lambda + (D_x^2 h \cdot h)/h^2$. (5.18)

We rewrite (5.15) as

$$\lambda = (G/F)_x - 2(\log F)_{xx} \quad , \tag{5.19}$$

which shows that the asymptotic value of $\lambda$ is determined by the boundary condition on $u[= (G/F)_x]$ and the asymptotic form of F at $|x| = \infty$. In the following we assume $\lambda$ to be zero.

For $\lambda = 0$, we find

$$G = 2F_x \tag{5.20}$$

is a solution to (5.19). Hence we have

$$u = 2(\log F)_{xx} \quad . \tag{5.21}$$

Substituting (5.20) into (5.14) we have

$$(D_t + D_x^3)F_x \cdot F = 0 \quad , \tag{5.22}$$

which is equivalent [see (III.1) below] to

$$D_x(D_t + D_x^3)F \cdot F = 0 \quad . \tag{5.23}$$

## 5.2 Properties of the D Operator

In this section we list some properties of the operators $D_t$, $D_x$ introduced in the previous section. We have

$$D_t^n D_x^m a \cdot b = \left(\frac{\partial}{\partial t} - \frac{\partial}{\partial t'}\right)^n \left(\frac{\partial}{\partial x} - \frac{\partial}{\partial x'}\right)^m a(t,x)b(t',x')\Big|_{t=t',x=x'} \quad .$$

It is convenient to introduce an operator $D_z$ and a differentiation $\partial/\partial z$ by

$$D_z = \delta D_t + \varepsilon D_x \quad , \quad \frac{\partial}{\partial z} = \delta \frac{\partial}{\partial t} + \varepsilon \frac{\partial}{\partial x} \quad ,$$

where $\delta$ and $\varepsilon$ are constants. The following properties are easily seen from the definition

(I) $\qquad D_z^m a \cdot 1 = \left(\frac{\partial}{\partial z}\right)^m a \quad .$

(II) $\qquad D_z^m a \cdot b = (-1)^m D_z^m b \cdot a \quad ,$

(II.1) $\qquad D_z^m a \cdot a = 0 \quad$ for odd m.

(III) $\qquad D_z^m a \cdot b = D_z^{m-1}(a_z \cdot b - a \cdot b_z) \quad ,$

(III.1) $\qquad D_z^m a \cdot a = 2D_z^{m-1} a_z \cdot a \quad$ for even m,

(III.2) $\qquad D_x D_t a \cdot a = 2D_x a_t \cdot a \quad ,$

$$= 2D_t a_x \cdot a \quad .$$

(IV) $\qquad D_x^m \exp(p_1 x) \cdot \exp(p_2 x) = (p_1 - p_2)^m \exp[(p_1 + p_2)x] \quad .$

Let $F(D_t,D_x)$ be a polynomial of $D_t$ and $D_x$, we have

(IV.1) $\qquad F(D_t,D_x)\exp(\Omega_1 t + p_1 x) \cdot \exp(\Omega_2 t + p_2 x)$

$$= F(\Omega_1 - \Omega_2, p_1 - p_2)/F(\Omega_1 + \Omega_2, p_1 + p_2)$$

$$\times F(D_t,D_x)\exp[(\Omega_1 + \Omega_2)t + (p_1 + p_2)x] \cdot 1 \quad .$$

(V) $\qquad \exp(\varepsilon D_x)a(x) \cdot b(x) = a(x + \varepsilon)b(x - \varepsilon) \quad .$

(VI) $\exp(\varepsilon D_z)ab \cdot cd = [\exp(\varepsilon D_z)a \cdot c][\exp(\varepsilon D_z)b \cdot d]$

$= [\exp(\varepsilon D_z)a \cdot d][\exp(\varepsilon D_z)b \cdot c]$ ,

(VI.1) $D_z ab \cdot c = \left(\frac{\partial a}{\partial z}\right)bc + a(D_z b \cdot c)$ ,

(VI.2) $D_z^2 ab \cdot c = \left(\frac{\partial^2 a}{\partial z^2}\right)bc + 2\left(\frac{\partial a}{\partial z}\right)D_z b \cdot c + a(D_z^2 b \cdot c)$ ,

(VI.3) $D_z^3 ac \cdot bc = (D_z^3 a \cdot b)c^2 + 3(D_z a \cdot b)D_z^2 c \cdot c$ ,

(VI.4) $D_x^m \exp(px)a \cdot \exp(px)b = \exp(2px)D_x^m a \cdot b$ .

(VII) $\exp(\delta D_t)[\exp(\varepsilon D_x)a \cdot b] \cdot [\exp(\varepsilon D_x)c \cdot d]$

$= \exp(\varepsilon D_x)[\exp(\delta D_t)a \cdot c] \cdot [\exp(\delta D_t)b \cdot d]$

$= [\exp(\delta D_t + \varepsilon D_x)a \cdot d][\exp(-\delta D_t + \varepsilon D_x)c \cdot b]$ .

The following are useful for transforming nonlinear differential equations into the bilinear forms.

(VIII) $\exp(\varepsilon\partial/\partial z)[a/b] = [\exp(\varepsilon D_z)a \cdot b]/[\cosh(\varepsilon D_z)b \cdot b]$ ,

(VIII.1) $\frac{\partial}{\partial z}\left(\frac{a}{b}\right) = \frac{D_z a \cdot b}{b^2}$ ,

(VIII.2) $\frac{\partial^2}{\partial z^2}\left(\frac{a}{b}\right) = \frac{D_z^2 a \cdot b}{b^2} - \left(\frac{a}{b}\right)\frac{D_z^2 a \cdot b}{b^2}$ ,

(VIII.3) $\frac{\partial^3}{\partial z^3}\left(\frac{a}{b}\right) = \frac{D_z^3 a \cdot b}{b^2} - 3\left[\frac{D_z a \cdot b}{b^2}\ \frac{D_z^2 b \cdot b}{b^2}\right]$ .

(IX) $2\cosh(\varepsilon\partial/\partial z)\log f = \log[\cosh(\varepsilon D_z)f \cdot f]$ ,

(IX.1) $\frac{\partial^2}{\partial z^2}\log f = \frac{D_z^2 f \cdot f}{2f^2}$ ,

(IX.2) $\frac{\partial^4}{\partial z^4}\log f = \frac{D_z^4 f \cdot f}{2f^2} - 6\left[\frac{D_z^2 f \cdot f}{2f^2}\right]^2$ .

The following are useful for transforming the bilinear differential equations back into the original forms of nonlinear differential equations.

(X) $\exp(\varepsilon D_x)a \cdot b = \{\exp[2\cosh(\varepsilon\partial/\partial x)\log b]\}[\exp(\varepsilon\partial/\partial x)(a/b)]$ .

Let $\psi = a/b$, $u = 2(\log b)_{xx}$, we have

(X.1) $(D_x a \cdot b)/b^2 = \psi_x$ ,

(X.2) $(D_x^2 a \cdot b)/b^2 = \psi_{xx} + u\psi$ ,

(X.3) $(D_x^3 a \cdot b)/b^2 = \psi_{xxx} + 3u\psi_x$ ,

(X.4) $(D_x^4 a \cdot b)/b^2 = \psi_{xxxx} + 6u\psi_{xx} + (u_{xx} + 3u^2)\psi$ .

(XI) $\exp(\varepsilon D_x) a \cdot b = \exp[\sinh(\varepsilon\partial/\partial x)\log(a/b) + \cosh(\varepsilon\partial/\partial x)\log(ab)]$ .

let $\phi = \log(a/b)$ and $\rho = \log(ab)$, we have

(XI.1) $(D_x a \cdot b)/ab = \phi_x$ ,

(XI.2) $(D_x^2 a \cdot b)/ab = \rho_{xx} + (\phi_x)^2$ ,

(XI.3) $(D_x^3 a \cdot b)/ab = \phi_{xxx} + 3\phi_x\rho_{xx} + (\phi_x)^3$ ,

(XI.4) $(D_x^4 a \cdot b)/ab = \rho_{xxxx} + 4\phi_x\phi_{xxx} + 3(\rho_{xx})^2 + 6(\phi_x)^2\rho_{xx} + (\phi_x)^4$ .

All of these properties are easily verified so we prove only (X). We have

$$2\cosh(\varepsilon\partial/\partial x)\log b = \log b(x+\varepsilon) + \log b(x-\varepsilon) \quad ,$$

$$\exp(\varepsilon\partial/\partial x)(a/b) = a(x+\varepsilon)/b(x+\varepsilon) \quad ,$$

and from (V)

$$\exp(\varepsilon D_x) a \cdot b = a(x+\varepsilon)b(x-\varepsilon) \quad ;$$

hence

$$\exp(\varepsilon D_x) a \cdot b = \exp[2\cosh(\varepsilon\partial/\partial x)\log b][\exp(\varepsilon\partial/\partial x)(a/b)] \quad ,$$

which proves (X). Equations (X.1-4) are obtained by expanding (x) as a power series in $\varepsilon$ and equating terms with the same power of $\varepsilon$. Other properties of the D operators are described in another paper by the author [5.5].

## 5.3 Solutions of the Bilinear Differential Equations

We solve (5.23) by expanding F as power series in a parameter [5.6]

$$F = 1 + \varepsilon f_1 + \varepsilon^2 f_2 + \dots \quad . \tag{5.24}$$

Substituting (5.24) into (5.23) and collecting terms with the same power of $\varepsilon$, we have

$$2\frac{\partial}{\partial x}\left(\frac{\partial}{\partial t} + \frac{\partial^3}{\partial x^3}\right)f_1 = 0 \quad , \tag{5.25}$$

$$2\frac{\partial}{\partial x}\left(\frac{\partial}{\partial t} + \frac{\partial^3}{\partial x^3}\right)f_2 = -D_x(D_t + D_x^3)f_1 \cdot f_1 \quad , \tag{5.26}$$

$$2\frac{\partial}{\partial x}\left(\frac{\partial}{\partial t}+\frac{\partial^3}{\partial x^3}\right)f_3 = -D_x(D_t + D_x^3)(f_2 \cdot f_1 + f_1 \cdot f_2) \quad , \tag{5.27}$$

and so on.

We have two types of solutions: *I*) a polynomial solution and *II*) an exponential solution.

For case *I*) we find

$$f_1 = a_0 + a_1x + a_2x^2 + a_3x^3 + a_4x^4 + bt - 24a_4tx \tag{5.28}$$

is a solution of (5.25). Substituting (5.28) into (5.26) we find that the r.h.s. of(5.26) vanishes, so that $f_2$ can be chosen to be zero if

$$a_4 = 0 \quad , \quad 3a_1a_3 = a_2^2 \quad , \quad \text{and} \quad b = 12a_3 \quad .$$

Hence, we have an exact solution of (5.23)

$$F = 1 + \varepsilon[a_0 + a_1x + (3a_1a_3)^{\frac{1}{2}}x^2 + a_3(x^3 + 12t)] \quad . \tag{5.29}$$

If we impose the further boundary condition on u

$$u = 0 \quad \text{at} \quad x = 0 \quad , \tag{5.30}$$

we find $a_1 = 0$ and F is expressed, by choosing $\varepsilon$ to be unity, as

$$F = a_3[x^3 + 12(t + \text{const.})] \tag{5.31}$$

which together with $u = 2(\log F)_{xx}$, constitutes the solution bounded by a wall [5.7]

$$u = -6x(x^3 - 24t)/(x^3 + 12t)^2 \quad . \tag{5.32}$$

For case *II*) we have from (5.25)

$$f_1 = \sum_{i=1}^{N} a_i \exp(\Omega_i t + p_i x) \quad , \tag{5.33}$$

where $\Omega_i + p_i^3 = 0$, $p_i$ and $a_i$ are constants.

Substituting (5.33) into (5.26) we find that the r.h.s. of (5.26) excludes terms like $\exp 2(\Omega_i t + p_i x)$ because of the property (IV) of the D operator, and using (IV.1) we obtain

$$f_2 = \sum_{i>j}^{N} \exp(A_{ij} + \eta_i + \eta_j) \tag{5.34}$$

where $\exp(\eta_i) = a_i \exp(\Omega_i t + p_i x)$, and

$$\exp(A_{ij}) = -\frac{(p_i-p_j)[\Omega_i-\Omega_j+(p_i-p_j)^3]}{(p_i+p_j)[\Omega_i+\Omega_j+(p_i+p_j)^3]} \quad , \tag{5.35}$$

$$= (p_i - p_j)^2/(p_i + p_j)^2 \quad .$$

Substituting (5.34) into (5.27) we find that the r.h.s. of (5.27) excludes terms like $\exp(2\eta_i + \eta_k)$ because of the property (VI.4) and the relation (5.25), and we obtain

$$f_3 = \sum_{i>j>k}^{N} \exp(A_{ijk} + \eta_i + \eta_j + \eta_k) \tag{5.36}$$

where

$$\exp(A_{ijk}) = \exp(A_{ij} + A_{ik} + A_{jk}) \quad . \tag{5.37}$$

In the present case the power series terminates at $f_N$, and we have an exact solution which can be expressed as

$$F = \sum_{\mu=0,1} \exp\left(\sum_{i>j}^{(N)} A_{ij}\mu_i\mu_j + \sum_{i=1} \mu_i\eta_i\right) \quad , \tag{5.38}$$

where $\sum_{\mu=0,1}$ is the summation over all possible combinations of $\mu_1 = 0,1$, $\mu_2 = 0,1, \ldots, \mu_N = 0,1$, and $\sum_{i>j}^{(N)}$ indicates the summation over all possible pairs chosen from N elements. The parameter $\varepsilon$ is absorbed in the constant term $a_i$. Equation (5.38) together with $u = 2(\log F)_{xx}$ gives the N-soliton solutions of the KdV equation [5.8].

## 5.4 N-Soliton Solution of KdV-Type Equations

In a previous section we transformed the KdV equation into the bilinear differential equation

$$D_x(D_t + D_x^3)f \cdot f = 0 \quad . \tag{5.39}$$

In the present section, we consider a generalized form of (5.39)

$$F(D_t,D_x)f \cdot f = 0 \quad , \tag{5.40}$$

where F is a polynomial or exponential function of $D_t, D_x$, and satisfies the condition

$$F(D_t,D_x) = F(-D_t, -D_x) \quad , \tag{5.41}$$

$$F(0,0) = 0 \quad . \tag{5.42}$$

Repeating the procedure of the previous section, we find that the following form of f gives N-soliton solutions

$$f = \sum_{\mu=0,1} \exp\left(\sum_{i>j}^{(N)} A_{ij}\mu_i\mu_j + \sum_{i=1}^{N} \mu_i\eta_i\right) \tag{5.43}$$

where

$$\eta_i = \Omega_i t + p_i x + \eta_i^0 \quad , \tag{5.44}$$

$$F(\Omega_i, p_i) = 0 \quad , \tag{5.45}$$

$$\exp(A_{ij}) = -F(\Omega_i - \Omega_j, p_i - p_j)/F(\Omega_i + \Omega_j, p_i + p_j) \quad , \tag{5.46}$$

provided that the following identity holds

$$\sum_{\sigma=\pm 1} F\left(\sum_{i=1} \sigma_i\Omega_i \quad , \quad \sum_{i=1} \sigma_i p_i\right)$$

$$\times \prod_{i>j}^{(n)} F(\sigma_i\Omega_i - \sigma_j\Omega_j, \sigma_i p_i - \sigma_j p_j)\sigma_i\sigma_j = 0 \quad , \tag{5.47}$$

for $n = 1,2, \ldots, N$.

Note that the identity (5.47) holds for $N = 2$ without any further condition on F, which implies that there is at least a two-soliton solution for each of the nonlinear evolution equations which can be transformed into the generalized bilinear differential equation (5.40).

Equation (5.40) can be transformed back into the ordinary form of nonlinear evolution equation by using the properties (V), (IX), (X), etc. We list in the following the nonlinear evolution equations, their dependent variable transformations, and bilinear forms in those cases where F is known to satisfy the identity for arbitrary N.

*I) Boussinesq equation* [5.9]

$$u_{tt} - u_{xx} - 3(u^2)_{xx} - u_{xxxx} = 0 \quad , \tag{5.48}$$

$$u = 2(\log f)_{xx} \quad , \tag{5.49}$$

$$(D_t^2 - D_x^2 - D_x^4)f \cdot f = 0 \quad . \tag{5.50}$$

*II) Kadomtsev-Petviashvili equation* [5.10]

$$u_{tx} + u_{yy} + 6(uu_x)_x + u_{xxxx} = 0 \quad , \tag{5.51}$$

$$u = 2(\log f)_{xx} \quad , \tag{5.52}$$

$$(D_t D_x + D_y^2 + D_x^4)f \cdot f = 0 \quad . \tag{5.53}$$

*III) Model equations for shallow water waves* [5.11,12]

$$\text{(a)} \quad u_t - u_{xxt} - 3uu_t + 3u_x \int_x^\infty u_t dx' + u_x = 0 \quad , \tag{5.54}$$

$$u = 2(\log f)_{xx} \quad , \tag{5.55}$$

$$D_x(D_t - D_tD_x^2 + D_x)f \cdot f = 0 \quad . \tag{5.56}$$

$$\text{(b)} \quad u_t - u_{xxt} - 4uu_t + 2u_x \int_x^\infty u_t dx' + u_x = 0 \quad , \tag{5.57}$$

$$u = 2(\log f)_{xx} \quad , \tag{5.58}$$

$$[D_x(D_t - D_tD_x^2 + D_x) + \frac{1}{3} D_t(D_\tau + D_x^3)]f \cdot f = 0 \tag{5.59}$$

with $D_x(D_\tau + D_x^3)f \cdot f = 0 \quad ,$ (5.60)

where $\tau$ is an auxiliary variable.

*IV) Higher order KdV equations* [5.13,14]

$$\text{(a)} \quad u_t + 45u^2u_x + 15(u_xu_{xx} + u_{xxx}u) + u_{xxxxx} = 0 \quad , \tag{5.61}$$

$$u = 2(\log f)_{xx} \quad , \tag{5.62}$$

$$D_x(D_t + D_x^5)f \cdot f = 0 \quad . \tag{5.63}$$

$$\text{(b)} \quad u_t + 30u^2u_x + 10(2u_xu_{xx} + u_{xxx}u) + u_{xxxxx} = 0 \quad , \tag{5.64}$$

$$u = 2(\log f)_{xx} \quad , \tag{5.65}$$

$$[D_x(D_t + D_x^5) - \frac{5}{3} D_\tau(D_\tau + D_x^3)]f \cdot f = 0 \tag{5.66}$$

$$\text{with } D_x(D_\tau + D_x^3)f \cdot f = 0 \quad , \tag{5.67}$$

where $\tau$ is an auxiliary variable.

*V) Toda equation* [5.15]

$$\frac{\partial^2}{\partial t^2} \log[1 + V_n(t)] = V_{n+1}(t) + V_{n-1}(t) - 2V_n(t) \quad , \tag{5.68}$$

$$V_n(t) = \frac{\partial^2}{\partial t^2} \log f_n(t) \tag{5.69}$$

$$= f_{n+1}(t)f_{n-1}(t)/f_n(t)^2 - 1 \quad , \tag{5.70}$$

$$[D_t^2 - 4 \sinh^2(D_n/2)]f \cdot f = 0 \quad . \tag{5.71}$$

*VI) Discrete analogue of the KdV equation* [5.16]

$$\Delta_t \frac{W_n(t)}{1+W_n(t)} = W_{n-\frac{1}{2}}(t) - W_{n+\frac{1}{2}}(t) \quad , \tag{5.72}$$

$$W_n(f) = f_{n+\frac{1}{2}}(t)f_{n-\frac{1}{2}}(t)/f_n\left(t + \frac{\delta}{2}\right)f_n\left(t - \frac{\delta}{2}\right) - 1 \quad , \tag{5.73}$$

where $\Delta_t$ is a central difference operator defined by

$$\Delta_t F(t) = \delta^{-1}\left[F\left(t + \frac{\delta}{2}\right) - F\left(t - \frac{\delta}{2}\right)\right] \quad , \tag{5.74}$$

$$\sinh\tfrac{1}{4}(D_n + \delta D_t)[2\delta^{-1}\sinh(\delta D_t/2) + 2\sinh(D_n/2)]f \cdot f = 0 \quad . \tag{5.75}$$

*VII) Discrete-time Toda equation* [5.17]

$$\Delta_t^2 \log[1 + V_n(t)] = \hat{V}_{n+1}(t) + \hat{V}_{n-1}(t) - 2\hat{V}_n(t) \quad , \tag{5.76}$$

$$\text{where } \hat{V}_n(t) = \delta^{-2} \log[1 + \delta^2 V_n(t)] \quad , \tag{5.77}$$

$$V_n(t) = f_{n+1}(t)f_{n-1}(t)/f_n(t)^2 - 1 \quad , \tag{5.78}$$

$$[4\delta^{-2} \sinh^2(\delta D_t/2) - 4\sinh^2(D_n/2)]f \cdot f = 0 \quad , \tag{5.79}$$

where $\Delta_t^2$ is defined by

$$\Delta_t^2 F(t) = \delta^{-2}[F(t + \delta) + F(t - \delta) - 2F(t)] \quad . \tag{5.80}$$

These are, we believe, only a few of the examples of nonlinear evolution equations which exhibit N-soliton solutions. The following question remains for future study. "What is the condition on F for which there are N-soliton solutions?" or "Under what conditions does F satisfy the identity (5.47)?"

## 5.5 Bäcklund Transformations in Bilinear Form

We consider the Bäcklund transformation which relates pairs of solutions of nonlinear evolution equations. The Bäcklund transformation has been used to construct N-soliton solutions [5.18,19] and to obtain the higher order conservation laws [5.20]. Its relation to the inverse scattering transform has been discussed in [5.21-24].

The Bäcklund transformation of the bilinear differential equation was introduced in [5.5]. In the previous section, we considered a bilinear differential equation of the form

$$F(D_t, D_x)f \cdot f = 0 \quad , \tag{5.81}$$

for which we shall show a method of constructing the Bäcklund transformation. The method consists of the following procedure. First, we consider an equation (or one similar to it)

$$[F(D_t,D_x)f'\cdot f']ff - f'f'[F(D_t,D_x)f\cdot f] = 0 \quad . \tag{5.82}$$

Clearly if f is a solution of (5.81), f' is another solution and vice versa. Then, by developing (5.82) we derive equations which relate f' and f. The equations so obtained are the Bäcklund transformation for the bilinear equation $F(D_t,D_x)f\cdot f = 0$.

The Bäcklund transformation is linear with respect to each dependent variable, and can be converted to either

I) new nonlinear evolution equations which exhibit N-soliton solutions, or
II) an inverse scattering form for the nonlinear evolution equation.

We shall illustrate the Bäcklund transformation in bilinear form with the KdV equation.

Before doing so, let us introduce a mathematical formula, which is a key relation for finding a Bäcklund transformation in bilinear form. It is written as

$$\exp(D_1)[\exp(D_2)a\cdot b]\cdot[\exp(D_3)c\cdot d]$$

$$\equiv \exp\tfrac{1}{2}(D_2 - D_3)\quad \{\exp[\tfrac{1}{2}(D_2 + D_3) + D_1]a\cdot d\}\cdot\{\exp[\tfrac{1}{2}(D_2 + D_3) - D_1]c\cdot b\} \quad , \tag{5.83}$$

where $D_i = \varepsilon_i D_x + \delta_i D_t$, and $\varepsilon_i$ and $\delta_i$ are constants, for i = 1,2,3. Note that the positions of b and d are exchanged with respect to a and c in (5.83). For this reason, we call it an "exchange formula". The exchange formula is easily proved by using the property (V).

Expanding the exchange formula in power series in $\varepsilon_1, \varepsilon_2, \ldots, \delta_3$, equating like powers and combining the resultant identities, we find a variety of operator identities. We list some of them relevant to the present discussion.

$$(D_x^2 a\cdot b)cd - ab(D_x^2 c\cdot d) = D_x[(D_x a\cdot d)\cdot cb + ad\cdot(D_x c\cdot b)] \quad . \tag{5.84}$$

$$(D_x D_t f'\cdot f')ff - f'f'(D_x^2 f\cdot f) = 2D_x(D_t f'\cdot f)\cdot ff' \quad . \tag{5.85}$$

$$(D_x^2 f'\cdot f')ff - f'f'(D_x^2 f\cdot f) = 2D_x(D_x f'\cdot f)\cdot ff' \quad . \tag{5.86}$$

$$(D_x^4 f'\cdot f')ff - f'f'(D_x^4 f\cdot f) = 2D_x(D_x^3 f'\cdot f)\cdot ff' + 6D_x(D_x^2 f'\cdot f)\cdot D_x(f\cdot f') \quad . \tag{5.87}$$

These relations will be used in finding the Bäcklund transformation for the KdV equation.

Now we consider the KdV equation in the following bilinear form

$$D_x(D_t + c_0 D_x + D_x^3)f\cdot f = 0 \quad , \tag{5.88}$$

where $c_0$ is a constant and is introduced for the purpose of later discussions.

Let f be a solution of (5.88) and f' another one. If we can find equations which relate f to f' by converting the following equation

$$[D_x(D_t + c_0D_x + D_x^3)f' \cdot f']ff - f'f'[D_x(D_t + c_0D_x + D_x^3)f \cdot f] = 0 \quad , \tag{5.89}$$

it is then a Bäcklund transformation.

By using the identities (5.85-87) and the relation

$$D_x(D_xf' \cdot f) \cdot (D_xf \cdot f') = 0 \quad , \tag{5.90}$$

we convert (5.89) to

$$\begin{aligned} &2D_x\{[D_t + (c_0 + 3\lambda)D_x + D_x^3]f' \cdot f\} \cdot (ff') \\ &\quad + 6D_x[(D_x^2 - \mu D_x - \lambda)f' \cdot f] \cdot (D_xf \cdot f') = 0 \quad , \end{aligned} \tag{5.91}$$

where $\lambda$ and $\mu$ are arbitrary constants. Hence, if f is a solution of (5.88), then f' would be another one of the same equation provided that f' satisfies the following equations

$$[D_t + (c_0 + 3\lambda)D_x + D_x^3]f' \cdot f = 0 \quad , \tag{5.92}$$

$$(D_x^2 - \mu D_x - \lambda)f' \cdot f = 0 \quad . \tag{5.93}$$

Equations (5.92,93) are the Bäcklund transformation for (5.88).

Now we have the Bäcklund transformation in bilinear form. In the previous section we discussed a method of finding N-soliton solutions of the equations written in the bilinear form. So it is quite reasonable to expect that the present Bäcklund transformation is a bilinear form of a certain nonlinear evolution equation which has an N-soliton solution.

In fact, by suitable dependent variable transformations, we can generate, from (5.92,93), two types of nonlinear evolution equations that exhibit N-soliton solutions. What is more, (5.92,93) are transformed into the well-known inverse scattering form for the KdV equation by choosing proper dependent variables.

I) *Modified KdV equation describing a weakly nonlinear lattice*

First we consider the case where f' and f are complex conjugate to each other, $f' = \hat{f} + i\hat{g}$, $f = \hat{f} - i\hat{g}$, then (5.92,93) reduce to

$$(D_t + D_x^3)\hat{g} \cdot \hat{f} = 0 \quad , \tag{5.94}$$

$$D_x^2(\hat{f} \cdot \hat{f} + \hat{g} \cdot \hat{g}) + 2\alpha\beta^{-1/2}D_x\hat{g} \cdot \hat{f} = 0 \quad , \tag{5.95}$$

where we put $\mu = i\alpha\beta^{-1/2}$ and $\lambda = c_0 = 0$.

Equations (5.94,95) are transformed back to a modified KdV equation describing a weakly nonlinear lattice [5.25]

$$u_t + 6\alpha u u_x + 6\beta u^2 u_x + u_{xxx} = 0 \quad , \quad (\beta > 0) \tag{5.96}$$

through the dependent variable transformation

$$u = i\beta^{-1/2}\phi_x \quad , \tag{5.97}$$

$$\phi = \log(\hat{f} + i\hat{g})/(\hat{f} - i\hat{g}) \quad . \tag{5.98}$$

II) *Modified KdV equation that exhibits a solution of shock wave type*

Let $\lambda = 0$, $c_0 = -6$, $\mu = -2$ and $\tau = -t$, then (5.92,93) reduce to

$$(D_\tau + 6D_x - D_x^3)f' \cdot f = 0 \quad , \tag{5.99}$$

$$(D_x^2 + 2D_x)f' \cdot f = 0 \quad , \tag{5.100}$$

which are transformed back to a modified KdV equation with a solution of shock wave type [5.26]

$$v_\tau + 6v^2 v_x - v_{xxx} = 0 \quad , \tag{5.101}$$

through the dependent variable transformation

$$v = 1 + \phi_x \quad , \tag{5.102}$$

$$\phi = \log(f'/f) \quad . \tag{5.103}$$

Reversing the procedure generating these equations, we find that they can be transformed to the bilinear form, (5.92,93), by suitable variable transformations. Following the procedure given in the previous section, we expand f' and f as a power series in a parameter $\varepsilon$

$$f' = 1 + \varepsilon f_1' + \varepsilon^2 f_2' + \dots \quad , \tag{5.104}$$

$$f = 1 + \varepsilon f_1 + \varepsilon^2 f_2 + \dots \quad , \tag{5.105}$$

and determine the coefficients by the usual perturbation method. N-soliton solutions of these equations are given by WADATI [5.25] and by PERELMAN et al. [5.26], respectively.

III) *Inverse scattering transform for the KdV equation*

Let $\psi = f'/f$, $u = 2(\log f)_{xx}$, then (5.92,93) are transformed to

$$\psi_t + 3\lambda\psi_x + \psi_{xxx} + 3u\psi_x = 0 \quad , \tag{5.106}$$

$$\psi_{xx} + u\psi = \lambda\psi \quad , \tag{5.107}$$

where $c_0$ and $\mu$ are chosen to be zero. Equations (5.106,107) constitute the well-known inverse scattering transform for the KdV equation found by GARDNER et al. [5.27].

IV) *MIURA's nonlinear transformation*

MIURA's famous transformation [5.28] which relates the solution of the KdV equation to the solution of the modified KdV equation was a key to finding the inverse scattering transform for the KdV equation [5.27].

For $\lambda = \mu = c_0 = 0$, (5.93) becomes

$$D_x^2 f' \cdot f = 0 \quad . \tag{5.108}$$

Using (XI.2), we find

$$\begin{aligned}(D_x^2 f' \cdot f)/(f'f) &= [\log(f'f)]_{xx} + \{[\log(f'/f)]_x\}^2 \\ &= 2(\log f)_{xx} + [\log(f'/f)]_{xx} + \{[\log(f'/f)]_x\}^2 \quad .\end{aligned} \tag{5.109}$$

Then, (5.108) is transformed to

$$u = (2v)^2 - 2iv_x \tag{5.110}$$

by the relation

$$u = 2(\log f)_{xx} \quad , \tag{5.111}$$

$$v = (1/2i)[\log(f'/f)]_x \quad , \tag{5.112}$$

where v satisfies the modified KdV equation

$$v_t + 24v^2 v_x + v_{xxx} = 0 \quad . \tag{5.113}$$

Equation (5.110) is the MIURA transformation.

V) *Bäcklund transformation in ordinary form*

The Bäcklund transformation in the bilinear form may be rewritten in the ordinary form, that is, with a potential defined by

$$w = (1/2) \int_{-\infty}^{x} u \, dx = (\log f)_x \tag{5.114}$$

and its derivatives. In order to rewrite (5.92,93) for $\mu = c_0 = 0$, we introduce the new variables

$$\phi = \log(f'/f) \tag{5.115}$$

$$\rho = \log(f'f) \quad . \tag{5.116}$$

Then using (XI.1-3), we have from (5.92,93), respectively,

$$\rho_{xx} + (\phi_x)^2 = \lambda \ , \tag{5.117}$$

$$\phi_t + 3\lambda\phi_x + \phi_{xxx} + 3\phi_x\rho_{xx} + (\phi_x)^3 = 0 \ . \tag{5.118}$$

Noticing the relations,

$$\phi_x = w' - w \tag{5.119}$$

$$\rho_x = w' + w \tag{5.120}$$

we can reduce (5.117,118), respectively,

$$(w' + w)_x + (w' - w)^2 = \lambda \tag{5.121}$$

$$(w' - w)_t + 3\lambda(w' - w)_x + (w' - w)_{xxx}$$
$$+ 3[(w' - w)(w' + w)_x]_x + [(w' - w)^3]_x = 0 \ . \tag{5.122}$$

Equations (5.121,122) are equivalent to the Bäcklund transformation first found by WAHLQUIST and ESTABROOK [5.19].

It is to be noted that (5.122) is written in a conserved form. From (5.122) and (5.121) we may obtain an infinite sequence of conserved quantities [5.28] using a systematic method developed by SATSUMA [5.20].

The present scheme used in obtaining the Bäcklund transformation for the KdV equation is applicable to the equations I) - VII) presented in Sect.5.4. We list in the following the bilinear differential equations and their Bäcklund transformations, which are obtained by developing the equation

$$[F(D_t,D_x)f'\cdot f']ff - f'f'[F(D_t,D_x)f\cdot f] = 0 \tag{5.123}$$

with the help of the exchange formula (5.83).

I) *Boussinesq equation* [5.29,30]

$$(D_t^2 - D_x^2 - D_x^4)f\cdot f = 0 \ . \tag{5.124}$$

$$\begin{cases} (D_t + aD_x^2)f'\cdot f = 0 \ , & (5.125) \\ (aD_tD_x + D_x + D_x^3)f'\cdot f = 0 \ , & (5.126) \end{cases}$$

where $a^2 = -3$.

II) *Kadomtsev-Petviashvili equation* [5.31,32]

$$(D_tD_x + D_y^2 + D_x^4)f\cdot f = 0 \ . \tag{5.127}$$

$$\begin{cases}(D_y + aD_x^2)f' \cdot f = 0 \quad , & (5.128)\\ (-aD_yD_x + D_t + D_x^3)f' \cdot f = 0 \quad , & (5.129)\end{cases}$$

where $a^2 = 3$.

III) *Model equations for shallow water waves*

(a) $$D_x(D_t - D_tD_x^2 + D_x)f \cdot f = 0 \quad . \qquad (5.130)$$

$$\begin{cases}(D_x^3 - D_x)f' \cdot f = \lambda f'f \quad . & (5.131)\\ (3D_xD_t - 1)f' \cdot f = \mu D_x f' \cdot f \quad . & (5.132)\end{cases}$$

(b) $$[D_x(D_t - D_tD_x^2 + D_x) + (1/3)D_t(D_\tau + D_x^3)]f \cdot f = 0 \qquad (5.133)$$

with $D_x(D_\tau + D_x^3)f \cdot f = 0$ . (5.134)

$$\begin{cases}(D_\tau + 3\lambda D_x + D_x^3)f' \cdot f = 0 \quad , & (5.135)\\ D_x^2 f' \cdot f = \lambda f'f + \mu D_x f' \cdot f \quad , & (5.136)\\ [(1 - 3\lambda)D_t - D_x^2D_t + D_x]f' \cdot f = 0 \quad . & (5.137)\end{cases}$$

IV) *Higher order KdV equations* [5.33]

(a) $$D_x(D_t + D_x^5)f \cdot f = 0 \quad . \qquad (5.138)$$

$$\begin{cases}D_x^3 f' \cdot f = \lambda f'f \quad , & (5.139)\\ [D_t - (15/2)\lambda D_x^2 - (3/2)D_x^5]f' \cdot f = 0 \quad . & (5.140)\end{cases}$$

(b) $$[D_x(D_t + D_x^5) - (5/6)D_x^3(D_\tau + D_x^3)]f \cdot f = 0 \qquad (5.141)$$

with $D_x(D_\tau + D_x^3)f \cdot f = 0$ (5.142)

$$\begin{cases}(D_\tau + 3\lambda D_x + D_x^3)f' \cdot f = 0 \quad , & (5.143)\\ D_x^2 f' \cdot f = \lambda f'f \quad , & (5.144)\\ (D_t + 15\lambda^2 D_x + D_x^5)f' \cdot f = 0 \quad . & (5.145)\end{cases}$$

V) *Toda equation* [5.34,35]

$$[D_t^2 - 4\sinh^2(D_n/2)]f \cdot f = 0 \quad . \qquad (5.146)$$

(a)
$$\begin{cases}[D_t \exp(-D_n/2) - 2\lambda\sinh(D_n/2)]f'\cdot f = 0 \ , & (5.147)\\ [D_t + \lambda^{-1}(\exp(-D_n) - 1)]f'\cdot f = 0 \ . & (5.148)\end{cases}$$

(b)
$$\begin{cases}D_t f'\cdot f + 2\alpha\sinh(D_n/2)g'\cdot g = 0 \ , & (5.149)\\ D_t g'\cdot g + 2\alpha^{-1}\sinh(D_n/2)f'\cdot f = 0 \ , & (5.150)\\ [\beta_1 \sinh(D_n/2) + \cosh(D_n/2)]g'\cdot g = f'f \ , & (5.151)\\ [\beta_2 \sinh(D_n/2) + \cosh(D_n/2)]f'\cdot f = g'g \ , & (5.152)\end{cases}$$

where $\alpha^{-1}(\beta_1^2 - 1) = \alpha(\beta_2^2 - 1)$ . (5.153)

VI) *Discrete analogue of the KdV equation* [5.16]

$$\sinh\tfrac{1}{4}(D_n + \delta D_t)[2\delta^{-1}\sinh(\delta D_t/2) + 2\sinh(D_n/2)]f\cdot f = 0 \ . \tag{5.154}$$

$$\begin{cases}[2\delta^{-1}\sinh(\delta D_t/2) + 2\lambda\sinh(D_n/2)]f'\cdot f = 0 \ , & (5.155)\\ \cosh(D_n/2)f'\cdot f = \lambda\cosh(\delta D_t/2)f'\cdot f \ . & (5.156)\end{cases}$$

VII) *Discrete-time Toda equation* [5.36]

$$\{[2\delta^{-1}\sinh(\delta D_t/2)]^2 - [2\sinh(D_n/2)]^2\}f\cdot f = 0 \ . \tag{5.157}$$

$$\begin{cases}\delta^{-1}\sinh(\delta D_t/2)f'\cdot f + \alpha\sinh(D_n/2)g'\cdot g = 0 \ , & (5.158)\\ \delta^{-1}\sinh(\delta D_t/2)g'\cdot g + \alpha^{-1}\sinh(D_n/2)f'\cdot f = 0 \ , & (5.159)\\ [\cosh(D_n/2) + \beta_1\sinh(D_n/2)]g'\cdot g = [\cosh(\delta D_t/2) + \gamma_1\sinh(\delta D_t/2)]f'\cdot f \ , & (5.160)\\ [\cosh(D_n/2) + \beta_2\sinh(D_n/2)]f'\cdot f = [\cosh(\delta D_t/2) + \gamma_2\sinh(\delta D_t/2)]g'\cdot g \ , & (5.161)\end{cases}$$

where $\delta$, $\alpha$ and $\beta_i$, $\gamma_i$ for i = 1,2, are arbitrary constants satisfying the relation

$$\alpha^{-1}[(\beta_1 + \gamma_1\alpha\delta)^2 - (1 - \delta^2)] = \alpha[(\beta_2 + \gamma_2\alpha^{-1}\delta)^2 - (1 - \delta^2)] \ . \tag{5.162}$$

We note that the Bäcklund transformation for the discrete analogue of the KdV equation is obtained by developing the following equation

$$\begin{aligned}&\left\{\sinh\tfrac{1}{4}(D_n + \delta D_t)\left[\delta^{-1}\sinh\left(\tfrac{\delta}{2}D_t\right) + \sinh\left(\tfrac{1}{2}D_n\right)\right]f'\cdot f'\right\}\left[\cosh\tfrac{1}{4}(D_n - \delta D_t)f\cdot f\right]\\ &- \left[\cosh\tfrac{1}{4}(D_n - \delta D_t)f'\cdot f'\right]\left\{\sinh\tfrac{1}{4}(D_n + \delta D_t)\left[\delta^{-1}\sinh\left(\tfrac{\delta}{2}D_t\right) + \sinh\left(\tfrac{1}{2}D_n\right)\right]f\cdot f\right\} = 0 \ .\end{aligned} \tag{5.163}$$

The examples I) - VII) above indicate that for each bilinear equation of the form $F(D_t,D_x)f\cdot f = 0$ that exhibits an N-soliton solution, we can also find a Bäcklund transformation in a bilinear form. In this connection, it is of interest to find other examples of F that satisfy the identity (5.47) and look for their Bäcklund transformations.

In this article, we have discussed nonlinear evolution equations which are transformed into bilinear equations of the form

$$F(D_t,D_x)f\cdot f = 0 \tag{5.164}$$

and their Bäcklund transformations. Other classes of nonlinear evolution equations which exhibit N-soliton solutions are partially discussed in [5.6] and [5.34].

## References


5.1 D.J. Korteweg, G. de Vries: Philos. Mag. *39*, 422 (1895)
5.2 N.J. Zabusky, M.D. Kruskal: Phys. Rev. Lett. *15*, 240 (1965)
5.3 G.A. Baker, Jr., J.L. Gammel (eds.): *The Padé Approximant in Theoretical Physics* (Academic Press, New York 1970)
5.4 P.R. Graves-Morris (ed.): *Padé Approximants and their Applications* (Academic Press, New York 1973)
5.5 R. Hirota: Prog. Theor. Phys. *52*, 1498 (1974)
5.6 R. Hirota: "Direct Methods of Finding Exact Solutions of Nonlinear Evolution Equations", in *Bäcklund Transformations*, ed. by R.M. Miura, Lecture Notes in Mathematics (Springer, Berlin, Heidelberg, New York 1976) Vol.515
5.7 H.E. Moses: J. Math. Phys. *17*, 73 (1976)
5.8 R. Hirota: Phys. Rev. Lett. *27*, 1192 (1971)
5.9 R. Hirota: J. Math. Phys. *14*, 810 (1973)
5.10 J. Satsuma: J. Phys. Soc. Jpn. *40*, 286 (1976)
5.11 R. Hirota, J. Satsuma: J. Phys. Soc. Jpn. *40*, 611 (1976)
5.12 M.J. Ablowitz, D.J. Kaup, A.C. Newell, H. Segur: Stud. Appl. Math. *53*, 249 (1974)
5.13 K. Sawada, T. Kotera: Prog. Theor. Phys. *51*, 1355 (1974)
5.14 P.J. Caudrey, R.K. Dodd, J.D. Gibbon: Proc. Roy. Soc. Lond. A *351*, 407 (1976)
5.15 R. Hirota: J. Phys. Soc. Jpn. *35*, 286 (1973)
5.16 R. Hirota: J. Phys. Soc. Jpn. *43*, 1424 (1977)
5.17 R. Hirota: J. Phys. Soc. Jpn. *43*, No.6 (1977)
5.18 G.L. Lamb, Jr.: Rev. Mod. Phys. *43*, 99 (1971)
5.19 H.D. Wahlquist, F.B. Estabrook: Phys. Rev. Lett. *31*, 1386 (1973)
5.20 J. Satsuma: Prog. Theor. Phys. *52*, 1396 (1974)
5.21 H.H. Chen: Phys. Rev. Lett. *33*, 925 (1974)
5.22 M. Wadati, H. Sanuki, K. Konno: Prog. Theor. Phys. *53*, 419 (1975)
5.23 G.L. Lamb, Jr.: J. Math. Phys. *15*, 2157 (1974)
5.24 H.H. Chen, C.C. Liu: J. Math. Phys. *16*, 1428 (1975)
5.25 M. Wadati: J. Phys. Soc. Jpn. *38*, 673 (1975)
5.26 T.L. Perelman, A.Kh. Fridman, M.M. El'Yashevich: Phys. Lett. *47*A, 321 (1974)
5.27 C.S. Gardner, J.M. Greene, M.D. Kruskal, R.M. Miura: Phys. Rev. Lett. *19*, 1095 (1967)
5.28 R.M. Miura: J. Math. Phys. *9*, 1202 (1968)
5.29 R. Hirota, J. Satsuma: Prog. Theor. Phys. *57*, 797 (1977)

5.30 H.H. Chen: "Relations between Bäcklund Transformations and Inverse Scattering Problems" in *Bäcklund Transformations*, ed. by R.M. Miura, Lecture Notes in Mathematics, Vol.515 (Springer, Berlin, Heidelberg, New York 1976)
5.31 V.S. Druyma: Sov. Phys.-JETP Lett. *19*, 387 (1974)
5.32 V.E. Zakharov, A.B. Shabat: Func. Anal. Appls. *8*, 43 (1974)
5.33 J. Satsuma, D. Kaup: J. Phys. Soc. Jpn. *43*, 692 (1977)
5.34 R. Hirota, J. Satsuma: Prog. Theor. Phys. Suppl. No.*59*, 64 (1976)
5.35 M. Wadati, M. Toda: J. Phys. Soc. Jpn. *39*, 1196 (1975)
5.36 R. Hirota: In preparation

# 6. The Inverse Scattering Transform

A. C. Newell

With 1 Figure

A detailed description of the inverse scattering transform associated with the generalized Zakharov-Shabat and Schrödinger eigenvalue problems is given. The close analogy with the ideas of the Fourier transform is emphasized and the general expansions for the unknown functions in terms of the squared eigenfunctions and their derivatives are developed for both eigenvalue problems. The results for the Schrödinger equation are new.[1] The partial differential equations which are solvable by the inverse scattering transforms associated with these eigenvalue problems are identified, almost by inspection, and classified according to the nature of the dispersion relation. Of particular interest are those classes which are integrable but which do not possess conserved quantities, and also those equations which are integrable but for which the spectrum is not invariant. Several examples, the coherent pulse propagation problem, the nonlinear Schrödinger equation and the sine-Gordon equation, are used to illustrate some of the important points. Finally, a singular perturbation theory for examining the effects of perturbations over long times is given. Last-minute revisions and additions have been made to Sects.6.12 and 13 to reflect some recent developments.

## 6.1 General Discussion

One of the significant advances in mathematical physics over the past decade has been the discovery by GARDNER, GREENE, KRUSKAL, and MIURA (GGKM) [6.1,2], and ZABUSKY and KRUSKAL [6.3] of 1) a new nonlinear transform and 2) the soliton. The transform (the Inverse Scattering Transform or IST) works in precisely the same way as the Fourier transform does in linear problems; namely, it transforms the dependent variable which satisfies a given partial differential equation to a set of new dependent variables whose evolution in time is described by an infinite sequence of ordinary differential equations. For special classes of partial differential equations, these equations are separable and hence trivially integrable.

Compared with the Fourier transform, there are two major differences. The first is that the basis is no longer fixed [like $\exp(\pm ikx)$] but moves in a way which

---

[1]This article was written in April 1976.

depends on the unknown variable. The second difference is that the spectrum (and here we are considering partial differential equations over infinite spatial intervals) no longer simply consists of the continuum of real wave numbers k but includes in addition a finite number of isolated complex wave numbers. It is the complex wave numbers which give rise to the entities known as solitons. They are truly nonlinear quantities and have no linear analogue.

Indeed, the general solution of any member of one of the aforementioned classes of partial differential equations can be qualitatively described in terms of the various spectral components. The soliton (a term coined by ZABUSKY and KRUSKAL) is a solitary wave, that is, a localized, stable, permanent waveform (it may contain internal oscillations), with the crucial additional property that its identity (amplitude, velocity, shape, internal frequency) is preserved even after collision with one of the other solution components. This invariance is derived from the invariance of the eigenvalue with which it is associated. The only interaction memory is a shift in the position of the soliton relative to where it would have been if it had traveled without collision. In a two-soliton collision, the phase shift is a simple function of the two eigenvalues. (On the other hand, the solitary wave, for example, the hyperbolic tangent solution of the equation $\phi_{tt} - \phi_{xx} + \phi - \lambda\phi^3 = 0$, is subject to distortion on collision.) The component of the solution associated with the continuous spectrum is in general not localized, does not have a permanent waveform and, as it disperses, decays algebraically in time in close analogy with the long-time behavior of linear dispersive waves. There are, however, some features of this solution component which are distinctly nonlinear, particularly the structure of the self-similar regions [6.4].

One of the major reasons for the widespread and interdisciplinary interest in IST is that the special classes of integrable equations include such a wide variety of useful and universal equations which are central to many areas of mathematical physics and whose solutions are important to our general understanding of nonlinear wave phenomena. To illustrate the point, we will consider a list of these equations and at the same time discuss the historical development of IST over the last ten years.

It is perhaps fitting that the IST method itself was first developed through the studies of the Korteweg-de Vries equation,

$$q_t + 6qq_x + q_{xxx} = 0 \tag{6.1}$$

which arises so very naturally as the leading approximation in all conservative wave systems which are *weakly* dispersive and *weakly* nonlinear. It was first suggested by Korteweg and deVries as being relevant to the description of long surface gravity waves whose slope is small and approximately equal to the cube of the depth to wavelength ratio. The wave is large enough so that it initially attempts to break but the water is deep enough so that dispersive effects are eventually important. Imagine then an initial local disturbance. The first response of the system is to divide the arbitrary initial disturbance into leftward and rightward traveling waves according to the D'Alembert solution of the linear wave equation. Both the left and the right traveling profiles (which are now separated) eventually distort due to the combined effects of nonlinearity and dispersion. It is this evolution and this balance which is described by (6.1). There are many natural contexts in which similar dynamics can arise; long internal gravity waves [6.5], ion-acoustic waves in a cold plasma [6.6], waves of vortex tubes [6.7], and longitudinal vibration of a discrete mass striking [6.8]. The equation came to the attention of ZABUSKY and KRUSKAL [6.3] through their studies on lattices and on the FERMI-PASTA-ULAM [6.9] results on heat conduction in solids.

It is also fitting that the second equation to which IST was applied was also of universal character. In a beautiful 1972 paper, ZAKHAROV and SHABAT [6.10] showed how the nonlinear Schrödinger equation

$$q_t - iq_{xx} \pm 2iq^2q^* = 0 \qquad (6.2)$$

fits in the formalism. These authors used heavily the ideas of LAX [6.11] who has reformulated the principal results of GGKM in an operator-theoretic notation, and in addition found several other of the integrable equations in the Korteweg-deVries family. This universal equation (discovered in various contexts by several authors in the late fifties and early sixties [6.12-16] describes the slow temporal and spatial evolution of the envelope $q[\varepsilon(X - c_gT),\varepsilon^2T]$ (X, T real space and time $c_g$ the group velocity) of an *almost monochromatic* wave train (centered on wave number k) in a *weakly* nonlinear, *strongly* dispersive system. For a context-free and very general derivation, see [6.16] or [6.17]. It simply comes up everywhere, from the modulation of high intensity electromagnetic signals in media where the refractive index is amplitude dependent to the breakup of deep water gravity wave envelopes. (It should be pointed out that in more than one dimension, waves with the linear dispersion relation $\omega = c|\underline{k}|$ are strongly dispersive; the (vector) group velocity depends on the direction of $\underline{k}$ and the dispersion tensor $\partial^2\omega/\partial k_r\partial k_s$ is nonzero.) One of the most dramatic applications of (6.2) is in the context of deep water gravity waves. Indeed, one can show that if the minus sign in (6.2) obtains, then the monochromatic wave-train solution q = q(t) is unstable to x-dependent perturbations (an instability first discovered by BENJAMIN and FEIR [6.18]) and the wave-train breaks up into separated and localized pulses. In some sense, this provides an explanation of a fact known to everybody with a surfboard—namely, that every tenth (seventh, eleventh) wave is the largest. Imagine the following experiment. A paddle is oscillated in a periodic manner (say the Fourier spectrum contains a number of distinct frequencies) at the water surface and excites several wave packets (of width $\varepsilon$) each being centered on a distinct frequency. Since the system is strongly dispersive, the packets move with different group velocities and separate in a time scale $\varepsilon^{-1}$. For times on the order of $\varepsilon^{-2}$, each packet feels the combined effects of dispersion (which tends to split up the packet) and nonlinearity (just like a simple weakly nonlinear oscillator, the nonlinearity is manifested as a third order self-modal $\omega + \omega - \omega = \omega$ interaction of strength $\varepsilon^2$), and the envelope $q(x,t)$ is modulated according to (6.2). What happens is that the initial envelope $q(x,0)$ is decomposed into a series of *solitons* given by

$$q(x,t) = 2\eta \operatorname{sech} 2\eta(\theta_0 - \eta x - 4\xi\eta t)\exp\left[-2i\phi_0 - 4i(\xi^2 - \eta^2)t - 2i\xi x\right] , \qquad (6.3)$$

in which the parameter $\zeta = \xi + i\eta$ is one of the complex eigenvalues discussed before, and *radiation* which disperses and decays. We have assumed that the excitation is of finite duration. If the excitation is continuous, the solitons regroup into an almost monochromatic wave, which again splits up into solitons and this process continues to recur [6.19]. In a sense (6.2) is even more canonical than (6.1). Equation (6.2) is irreducible whereas if dispersion dominates nonlinearity, an almost monochromatic solution of (6.1) will evolve according to (6.2).

Once ZAKHAROV and SHABAT had demonstrated that the method of GGKM could work for equations other than the Korteweg-deVries equation and therefore was not simply a fluke, there was renewed investigation of many other equations of the class which shared the important property common to (6.1,6,2), namely, an infinite number of conserved quantities. In rapid succession WADATI [6.20] solved the modified Korteweg-deVries equation,

$$q_t + 6q^2q_x + q_{xxx} = 0 \qquad (6.4)$$

and ABLOWITZ, KAUP, NEWELL and SEGUR (AKNS) [6.21-23], LAMB [6.24] and a little later FADDEEV and TAKHTAJAN [6.25] solved the sine-Gordon equation,

$$u_{xt} = + \sin u \qquad (6.5a)$$

$$u_{TT} - u_{XX} + \sin u = 0 \quad . \qquad (6.5b)$$

The method used by AKNS was novel in the sense that it was the spectral analogue of the Lax formulation and had the distinct advantage that all the necessary computations were simple and algebraic. In fact, the method generated, in a fairly natural way, classes of integrable equations and showed how each member of these classes could be identified with the dispersion relation of its associated linear version [6.23].

The fact that (6.4) has an infinite number of conservation laws and conserved quantities had been established by MIURA [6.26] who discovered a remarkable transformation between solutions of (6.1) and (6.4). Indeed, this transformation not only holds between (6.1) and (6.4) but also between solutions of corresponding members of the Korteweg-deVries and modified Korteweg-deVries families [6.23], the correspondence being established by the common dispersion relation [6.27]. The fact that the sine-Gordon equation (6.5) has an infinite number of conservation laws and conserved quantities was established by KRUSKAL [6.28] who also found that the sine, sinh and Klein-Gordon equations (which turn out to be the only integrable ones) are the only ones of the class

$$u_{tt} - u_{xx} + V(u) = 0 \tag{6.6}$$

which possess this property. In a sense, then, the sine-Gordon equation is the simplest nonlinear model of type (6.6), (having the property of Lorentz invariance) which is integrable and possesses soliton solutions (the sinh-Gordon equation does not have any soliton solution), properties which suggest its relevance as a field theory model. It also occurs in many other contexts and in Sect.6.8, we will discuss how it arises as the singular limit of the coherent pulse propagation problem [6.29]. The equations which describe this phenomenon are the Maxwell-Bloch equations which are also integrable [6.30] but which do not have an infinite number of conserved quantities except in the singular limit. A system which is related to the sine-Gordon equation and which is also exactly integrable [6.31] is the classical massive Thirring model,

$$\begin{aligned} p_{2x} &= imp_1 + 2igp_1p_1^*p_2 \\ p_{1t} &= imp_2 - 2igp_2p_2^*p_1 \quad . \end{aligned} \tag{6.7}$$

Having found the natural action and angle variables, one can discuss certain general features of the quantized model [6.31,25].

At approximately the same time, a parallel study on differential-difference equations was being developed by FLASCHKA [6.32] who solved the equations for the Toda lattice

$$Q_{ntt} = \exp\left[-(Q_n - Q_{n+1})\right] - \exp\left[-(Q_{n-1} - Q_n)\right] \tag{6.8}$$

(where $Q_n$ is the position of the $n^{th}$ particle in the lattice) one of several models vital to our eventual understanding of the heat conduction process. These ideas have been extended by MOSER [6.33], CALOGERO [6.34] and ABLOWITZ and LADIK [6.35, 36] who have also developed the theory for partial difference equations, a development which could be very significant in the theory of numerical computation.

In late 1973, ZAKHAROV and MANAKOV [6.37] introduced a matrix operator of higher order and found a way to express the 3-wave interaction problem

$$\frac{\partial A_j}{\partial t} + \underset{-}{c}_j \cdot \nabla A_j = \theta_j A_k^* A_\ell^* \tag{6.9}$$

($j$, $k$, $\ell$ cycled over 1, 2, 3) in the Lax formalism and once again an exact solution was found for a set of canonical equations. These equations are central to all *weakly* nonlinear systems which support a *continuum* of dispersive waves since the quadratic nonlinearities can cause a resonance between three resonant wave trains with amplitudes, $A_j$, $j = 1, 2, 3$, whose wave vectors, $\underset{-}{k}_j$ and frequencies $\omega_j$ satisfy the conservation of energy and momentum (the resonant conditions) conditions $\underset{-}{k}_1 + \underset{-}{k}_2 + \underset{-}{k}_3 = 0$, $\omega_1 + \omega_2 + \omega_3 = 0$. Interactions of this kind are important in Rossby and baroclinic waves [6.38-41], in internal gravity waves [6.42], plasma waves [6.43] and many other areas of continuum physics. Both KAUP [6.44] and

ZAKHAROV and MANAKOV [6.45] independently developed the inverse formulae for solving these equations in the one-dimensional case. A closely related system is a model which simulates the interactions between long waves [amplitude $A(x,t)$] and short waves [envelope amplitude $B(x,t)$]. The equations are an extension of the nonlinear Schrödinger equation to account for the effects of a long wave and take the form

$$\frac{\partial A}{\partial t} = 2S \frac{\partial}{\partial x} BB^*, \quad \frac{\partial B}{\partial t} - i \frac{\partial^2 B}{\partial x^2} = - \frac{\partial A}{\partial x} B + iA^2 B - 2iSB^2 B^* \quad . \tag{6.10}$$

This exactly solvable model was developed [6.46] in order to investigate further the very novel idea of BENNEY [6.47] who suggested that long waves can be driven by the recurring instability of (say, wind driven) short waves which together with the long wave form a resonant triad.

This partial list of extremely important equations illustrates the tremendous relevance and importance of the inverse scattering transform. It also very strongly suggests that the soliton is ubiquitous in nature (in the oceans, atmosphere, plasmas, lattices, superconductors, superfluids and in fundamental particle structures), at least in those situations which are to a good approximation one-dimensional. There has been some effort [6.48-50] to extend these ideas to higher dimensions, and in particular the paper of ZAKHAROV and SHABAT [6.48] introduces some pioneering ideas and gives an alternative and wider (in the sense that it includes an extra spatial dimension) class of integrable systems. Unfortunately, although these authors can write many equations in the Lax formalism, that is, they can express the equation as the commutation of two operators, their ansatz that the eigenvectors admit a triangular representation is not always true (even in one dimension) and depends heavily on certain properties which all the eigenvectors do not ordinarily possess. At this time, a full inverse scattering theory in more than one spatial dimension has not been developed. There are some properties of local solutions in the higher dimensional analogue of the well-known equations which suggest that an extra spatial dimension can introduce unexpected features. The most dramatic of these is the result of ZAKHAROV and his co-workers [6.51-53] in which they show that solutions of the two-dimensional nonlinear Schrödinger equation [add $-iq_{yy}$ to (6.2)] become singular in finite time. This is somehow contrary to intuition as one would expect that the additional geometric dispersion in a problem of two or more spatial dimension would overcome the nonlinear focusing effect. Clearly things are not all that simple. We also mention the remarkable observation of MILES [6.54,55] (see also [6.56,57] for an extension of these ideas) concerning soliton resonances for the KADOMTSEV-PETVIASHVILI [6.58] equation (often called the weakly two-dimensional Korteweg-de-Vries equation as it takes account of slow variations in the perpendicular spatial direction on the time scale on which dispersive and nonlinear effects become important)

$$\frac{3}{4} u_{yy} + \left(u_t + \frac{3}{2} uu_x + \frac{1}{4} u_{xxx}\right)_x = 0 \quad . \tag{6.11}$$

MILES noted that the phase shift $\delta$ of the two soliton solution tends to $\pm\infty$ when the parameters describing the solitons satisfy certain linear-like resonance conditions. Again this fact suggests that irregular behavior can occur.

The purpose of this review, however, is to focus on the close relation between the inverse scattering transform and Fourier analysis, and we shall be working only in one spatial dimension. We follow the philosophy outlined in [6.59]. While there is a broad array of scattering problems we could use for our discussion (some of them are mentioned in Sect.6.13) and which in their own right have unique and interesting features, throughout this article we will be working with the generalized Zakharov-Shabat eigenvalue problem,

$$v_{1x} + i\zeta v_1 = q(x,t)v_2 \quad ,$$
$$v_{2x} - i\zeta v_2 = r(x,t)v_1 \quad , \tag{6.12}$$

and the Schrödinger equation

$$v_{xx} + [\zeta^2 + q(x,t)]v = 0 \quad . \tag{6.13}$$

There are two key ideas. The first is to notice that the infinitesimal rate of change of the scattering data can be written as an inner product between the infinitesimal change in the potential and the squares of the eigenfunctions of the appropriate eigenvalue problem and their derivatives. In fact, these expansions form the basis for the perturbation theory we shall describe in Sect.6.12. The second is to notice that the squared eigenfunctions form a basis and thus the preceding relations can be inverted to give the rate of change of the potential as an expansion in terms of the squared eigenfunctions with the coefficients being given by the changes in the scattering data. The potential function itself can also be written in such a way. These expressions are precisely the analogues of the Fourier expansions,

$$u(x,t) = \int_{-\infty}^{\infty} A(k,t)\, e^{ikx}\, dk \quad ,$$
$$u_t(x,t) = \int_{-\infty}^{\infty} A_t(k,t)\, e^{ikx}\, dk \quad . \tag{6.14}$$

The next question we address is the one of how to find the widest class of evolution equations which are solvable by the transform identified with a particular eigenvalue problem. From the formulation described in the previous paragraph, the question is equivalent to finding those operators $\Omega$ which, when applied to the expansion for the potential, act on each component of this expansion separately. For such operators $\Omega$, if $\underline{u}$ is the potential, the equation

$$\underline{u}_t = \Omega \cdot \underline{u} \tag{6.15}$$

is integrable, because by simply identifying the coefficients in the expansions of the left and right hand sides of (6.15), we obtain a system of ordinary differential equations for the rate of change of the scattering data at some particular wave number k in terms of the scattering data and the transform $\hat{\Omega}(k)$ of $\Omega$ evaluated at the same wave number k. $\hat{\Omega}(k)$ is called the dispersion relation. For example, in the expansions (6.14), the class of integrable equations is given by $\Omega = (\partial/\partial x)^n$ and in these cases $\hat{\Omega}(k) = (ik)^n$.

It turns out that there are three different types of evolution equations which may be classified according to the nature of the dispersion relation (or relations). The first occurs when there is only one dispersion relation $\hat{\Omega}(k)$ which is either an entire function or the ratio of entire functions whose poles never coincide with the spectrum of the eigenvalue problem. In this case, the partial differential equation possesses an infinite number of conserved quantities, each of which acting as a Hamiltonian generates a flow in which all the others are constants of the motion. The second case occurs when there are essentially two dispersion relations (one for q and the other for $q^*$, its complex conjugate) and in these cases, the flow is irreversible, the irreversibility being associated with the difference in the dispersion relations. The particular example we discuss in this connection is the problem of coherent pulse propagation through a resonant medium. As an important, but singular limit, we discuss what happens when the two dispersion relations approach each other and have a pole on the real wave number axis. The third case we examine is the case when the pole of the dispersion relation $\hat{\Omega}(k)$ lies on the discrete spectrum and show that in these circumstances, the associated eigenvalue can move. The corresponding soliton still preserves its identity but it is a changing identity.

Our approach in this chapter then, is to start with a scattering problem and discover and characterize in a convenient manner all the equations which are integrable via the inverse scattering transform associated with this particular eigenvalue problem. Some exciting progress has also been made on the inverse problem; namely given an evolution equation, how one can find if it is integrable and if so which eigenvalue problem is associated with it. There are formal procedures which all stem from or are variants of the ideas of WAHLQUIST and ESTABROOK (see [6.60-62, 57]). Each of these reduces to finding a nontrivial solution for a Lie algebra (usually not closed) and it turns out that if the equation is integrable, then there is a one-dimensional infinity of solutions of this algebra characterized by a parameter which turns out to play the role of the eigenvalue in the scattering problem. Although, as we have indicated, progress has been made, there is as yet no convenient characterization of those evolution equations which are integrable.

Much of the material in this chapter has already appeared or is about to appear in the literature [6.59,63], and so for the most part this chapter is an in-depth survey, from one perspective, of what has already been done. The material in Sect.6.11, in which we show how to carry through the squared eigenfunction expansions for the Schrödinger equation, is new and has not appeared elsewhere.

## 6.2 The Generalized Zakharov-Shabat Eigenvalue Problem

Consider the eigenvalue problem

$$\begin{aligned} v_{1x} + i\zeta v_1 &= q(x,t)v_2 \ , \\ & \qquad\qquad -\infty < x < \infty \ . \\ v_{2x} - i\zeta v_2 &= r(x,t)v_1 \ , \end{aligned} \tag{6.16}$$

In this section we plan to outline the properties of the scattering data of (6.16) given that both $q(x,t)$ and $r(x,t)$ are absolutely integrable over $(-\infty,\infty)$. We define the solutions $\bar{\phi}(x,t;\zeta)$, $\phi(x,t;\zeta)$, $\bar{\psi}(x,t;\zeta)$ and $\psi(x,t;\zeta)$ by their asymptotic properties [which are shown in (6.18)] at $-\infty(\phi,\bar{\phi})$ and $+\infty(\psi,\bar{\psi})$. Since only two of the four can be linearly independent, there exist, for real $\zeta$, the following relations between the quantities:

$$\begin{aligned} \phi &= a\bar{\psi} + b\psi \quad , \quad \psi = \bar{b}\phi - a\bar{\phi} \quad , \\ \bar{\phi} &= -\bar{a}\psi + \bar{b}\bar{\psi} \quad , \quad \bar{\psi} = \bar{a}\phi + b\bar{\phi} \quad . \end{aligned} \tag{6.17}$$

From the property of the Wronskian of $\phi$ and $\bar{\phi}$ and their behavior at $-\infty$, $a\bar{a} + b\bar{b} = 1$.

| | $x = -\infty$ | $x = +\infty$ |
|---|---|---|
| $\phi$ | $\begin{bmatrix}1\\0\end{bmatrix}e^{-i\zeta x}$ | $\begin{bmatrix}a(\zeta,t)e^{-i\zeta x}\\ b(\zeta,t)e^{i\zeta x}\end{bmatrix}$ |
| $\bar{\phi}$ | $\begin{bmatrix}0\\-1\end{bmatrix}e^{i\zeta x}$ | $\begin{bmatrix}\bar{b}(\zeta,t)e^{-i\zeta x}\\ -\bar{a}(\zeta,t)e^{i\zeta x}\end{bmatrix}$ |
| $\psi$ | $\begin{bmatrix}\bar{b}(\zeta,t)e^{-i\zeta x}\\ a(\zeta,t)e^{i\zeta x}\end{bmatrix}$ | $\begin{bmatrix}0\\1\end{bmatrix}e^{i\zeta x}$ |
| $\bar{\psi}$ | $\begin{bmatrix}\bar{a}(\zeta,t)e^{-i\zeta x}\\ -b(\zeta,t)e^{i\zeta x}\end{bmatrix}$ | $\begin{bmatrix}1\\0\end{bmatrix}e^{-i\zeta x}$ |

(6.18)

The properties of these scattering functions are spelled out in more detail in [6.23,59,63], but here we shall write down the important ones.

1) If $q(x,t)$ and $r(x,t)$ are absolutely integrable, then $\phi e^{i\zeta x}$, $\psi e^{-i\zeta x}$ and $a(\zeta,t) = \phi_1\psi_2 - \phi_2\psi_1$ ($\bar{\phi}e^{-i\zeta x}$, $\bar{\psi}e^{i\zeta x}$ and $\bar{a}(\zeta,t) = \bar{\phi}_1\bar{\psi}_2 - \bar{\phi}_2\bar{\psi}_1$) are analytic in the upper (lower) half $\zeta$ plane and $b(\zeta,t)$, $\bar{b}(\zeta,t)$ exist for real $\zeta$. If, further, $\int_{-\infty}^{\infty}|x|^n|q(x,t)\,dx$, $\int_{-\infty}^{\infty}|x|^n|r(x,t)\,dx$ exist, then $b$ and $\bar{b}$ are $n$ times differentiable. Furthermore, if $q(x,t)$ and $r(x,t)$ are on compact support, the regions of analyticity of all these quantities can be extended to the whole $\zeta$ plane, and in particular $b(\zeta,t)$ [$b(\zeta,t)$] are defined at the discrete eigenvalues of (6.16).

2) Since $a(\zeta,t) = \phi_1\psi_2 - \phi_2\psi_1$ $(\bar{a} = \bar{\phi}_1\bar{\psi}_2 - \bar{\phi}_2\bar{\psi}_1)$ is analytic in the upper (lower) half $\zeta$ plane, its zeros mark those values of $\zeta_k(\bar{\zeta}_k)$ for which $\phi(\bar{\phi})$ is proportional to $\psi(\bar{\psi})$ for all x. Since both $\phi$ and $\psi$ have decaying behavior at $-\infty$ and $+\infty$, respectively, when $\mathrm{Im}\{\zeta\} > 0$, these values $\zeta_k$ are associated with bounded eigenfunctions. We write $\phi(\zeta_k,t) = b_k(t)\psi(\zeta_k,t)$ and $\bar{\phi}(\bar{\zeta}_k,t) = \bar{b}_k(t)\bar{\psi}(\bar{\zeta}_k,t)$ and if q and r are on compact support $b(\zeta_k,t) = b_k(t), \bar{b}(\bar{\zeta}_k,t) = \bar{b}_k(t)$. The analyticity of $a(\zeta,t)[a(\bar{\zeta},t)]$ and the fact that both tend to unity as $\zeta\to\infty$ in their respective half planes insures that the numbers $N$, $\bar{N}$ of discrete eigenvalues $\{\zeta_k\}_{k=1}^{N}$, $\{\bar{\zeta}_k\}_{k=1}^{\bar{N}}$ are finite. We *assume* that no eigenvalue lies on the real axis. The reader may find it worthwhile to verify some of the above statements by taking the example,

$$-r(x,0) = q(x,0) = \begin{cases} 0 & x < 0 \quad x > L \\ Q & 0 < x < L \end{cases} .$$

3) We call the set $S\left\{(\zeta_k,b_k)_{k=1}^{N},\ (\bar{\zeta}_k,\bar{b}_k)_{k=1}^{\bar{N}},\ a(\zeta,t),\ b(\zeta,t),\ \bar{a}(\zeta,t)\bar{b}(\zeta,t)\right\}$ the scattering data. It turns out that only a subset need be prescribed in order to recover r and q. These inversion formulae are given in [6.23,63] and require the prescription of either

$$S_+ = S_+\left\{(\zeta_k,\gamma_k)_{k=1}^{N},\ (\bar{\zeta}_k,\bar{\gamma}_k)_{k=1}^{\bar{N}},\ \frac{b}{a}(\xi,t),\ \frac{\bar{b}}{\bar{a}}(\xi,t),\ \xi \text{ real}\right\} \tag{6.19a}$$

where $\gamma_k = b_k/a'_k$, $\bar{\gamma}_k = \bar{b}_k/\bar{a}'_k$, primes indicating differentiation with respect to $\zeta$ or

$$S_- = S_-\left\{(\zeta_k,\beta_k)_{k=1}^{N},\ (\bar{\zeta}_k,\bar{\beta}_k)_{k=1}^{\bar{N}},\ \frac{\bar{b}}{a}(\xi,t) = \frac{b}{\bar{a}}(\xi,t),\ \xi \text{ real}\right\} \tag{6.19b}$$

with $\beta_k = 1/b_k a'_k$, $\bar{\beta}_k = 1/\bar{b}_k\bar{a}'_k$. It also turns out (see [6.63] for discussion) that the arbitrary prescription of $S_-$ and $S_+$ does not always lead to absolutely integrable and unique potentials r and q and that some restrictions are needed [such as are found on the initial scattering data if $r(x,0) = \pm q^*(x,0)$].

4) In [6.63], and also in this work, we show that the transformation from the potentials q,r to the scattering data S is a canonical one in which the two-form $\int_{-\infty}^{\infty} \delta q\wedge\delta r\, dx$ is preserved and for which the conjugate coordinates in scattering space are

$$A = A\left\{(2i\zeta_k,\ln b_k)_{k=1}^{N},\ (2i\bar{\zeta}_k,\ln \bar{b}_k)_{k=1}^{\bar{N}},\ \left[\frac{1}{\pi}\ln a\bar{a},\ \ln b(\zeta)\right]\right\} . \tag{6.20}$$

These results were given in [6.63] but are introduced and proved in the present work in a much more powerful way. For a certain class of Hamiltonians, the conjugate coordinates are action-angle variables and the flow is trivially integrable.

5) Whenever r is linearly related to q or $q^*$, simplifications occur. First, consider $r = \alpha q^*$ where $\alpha$ is real. Then $\bar{\psi}(\zeta,x) = S\psi^*(\zeta^*,x)$, $\bar{\phi}(\zeta,x) = -S\phi^*(\zeta^*,x)/\alpha$, $\bar{a}(\zeta) = a^*(\zeta^*)$, $\bar{b}(\zeta) = -b^*(\zeta^*)/\alpha$, $N = \bar{N}$, $\bar{\zeta}_k = \zeta^*_k$ and $\bar{b}_k = -b^*_k/\alpha$. Here $S = \begin{pmatrix} 0 & 1 \\ \alpha & 0 \end{pmatrix}$.

Also if $r = \alpha q$, $\alpha$ any real or complex constant, $\bar{\psi}(\zeta,x) = S\psi(-\zeta,x)$, $\bar{\phi}(\zeta,x) = S\phi(-\zeta,x)/\alpha$, $\bar{a}(\zeta) = a(-\zeta)$, $\bar{b}(\zeta) = -b(-\zeta)/\alpha$, $N = \bar{N}$, $\bar{\zeta}_k = -\zeta_k$ and $\bar{b}_k = -b_k/\alpha$. If $r = \alpha q$, $\alpha$, q, r real, then $a^*(\zeta^*) = a(-\zeta)$, $b^*(\zeta^*) = b(-\zeta)$; also if $\zeta_k$ is an eigenvalue ($\mathrm{Re}\{\zeta_k\} \neq 0$) so is $-\zeta_k^*$. In the context of the nonlinear optics problem, the single eigenvalue $\zeta_k = i\eta_k$ gives rise to the $2\pi$ pulse or kink whereas the $(\zeta_k,-\zeta_k^*)$ pair gives rise to a breather or $0\pi$ pulse.

6) Two very useful relations are

$$\frac{\partial}{\partial\zeta} \ln a = -i \int_{-\infty}^{\infty} \left(\frac{\phi_1\psi_2+\phi_2\psi_1}{a} - 1\right)dx, \quad \mathrm{Im}\{\zeta\} > 0 \tag{6.21}$$

and

$$\int_{-\infty}^{\infty} [B(\zeta_1)C(\zeta_2) - B(\zeta_2)C(\zeta_1)]dx = \frac{1}{2i(\zeta_1-\zeta_2)} [2A(\zeta_1)A(\zeta_2) - B(\zeta_1)C(\zeta_2) - B(\zeta_2)C(\zeta_1)]_{-\infty}^{\infty} . \tag{6.22}$$

In (6.22), A,B,C satisfy the equations

$$A_x = qC + rB, \quad B_x + 2i\zeta B = 2qA, \quad C_x - 2i\zeta C = 2rA \quad . \tag{6.23}$$

Relation (6.21), which is derived from the formula

$$i(u_1w_2 + u_2w_1) = \frac{\partial}{\partial x}\left(-\frac{\partial u_1}{\partial\zeta} w_2 + \frac{\partial u_2}{\partial\zeta} w_1\right) \tag{6.24}$$

[when $(u_1,u_2)$ and $(w_1,w_2)$ are solutions of (6.16)], provides the means of writing the Hamiltonian of the integrable systems as a functional of r and q. In (6.32a) we show that taking the variation of (6.21) we obtain, $\delta \ln a = \int_{-\infty}^{\infty} (\delta q\phi_2\psi_2/a - \delta r\phi_1\psi_1/a)dx$. Relation (6.22) is the basis for proving the orthogonality relation between the squared eigenfunctions and their adjoints.

7) Two important asymptotic relations which reveal the central importance of the functions $\ln a(\zeta)$ [and $\ln \bar{a}(\zeta)$] are,

$$\ln a(\zeta) \approx -\sum_{n=1}^{\infty} \zeta^{-n} C_n , \quad \mathrm{Im}\{\zeta\} > 0$$

$$\ln \bar{a}(\zeta) \sim \sum_{n=1} \zeta^{-n} C_n , \quad \mathrm{Im}\{\zeta\} < 0 \quad . \tag{6.25}$$

Indeed we have shown in [6.63] how the function $\ln a(\zeta)$ can be considered both as a functional of r and q and of the scattering data and how it generates the class of Hamiltonians for which the corresponding flows are integrable. The functional form of the $C_n$'s may be calculated from recurrence relations. The first three are:

$$C_1 = \frac{1}{2i} \int_{-\infty}^{\infty} qr\, dx , \quad C_2 = \frac{1}{(2i)^2} \int_{-\infty}^{\infty} qr_x\, dx , \quad C_3 = \frac{1}{(2i)^3} \int_{-\infty}^{\infty} (qr_{xx} - q^2r^2)dx \quad . \tag{6.26}$$

## 6.3 Evolution of the Scattering Data

We now calculate the change in the scattering data due to infinitesimal changes in the potentials r and q. Write (6.16) as

$$V_x = PV, \; P = \begin{bmatrix} -i\zeta & q \\ r & i\zeta \end{bmatrix} . \tag{6.27}$$

The fundamental solution matrix,

$$\Phi = \begin{bmatrix} \phi_1 & \bar{\phi}_1 \\ \phi_2 & \bar{\phi}_2 \end{bmatrix} \tag{6.28}$$

is used to compute (by variation of parameters) the solution to the variation of (6.27)

$$(\delta V)_x = P\delta V + \delta PV, \; \delta P = \begin{bmatrix} -i\delta\zeta & \delta q \\ \delta r & i\delta\zeta \end{bmatrix} . \tag{6.29}$$

In particular, when $V = \Phi$, we obtain

$$\delta\Phi(x) = \Phi \left\{ \int_{-L}^{x} \begin{bmatrix} -\delta q\phi_2\bar{\phi}_2 + \delta r\phi_1\bar{\phi}_1 & -\delta q\bar{\phi}_2^2 + \delta r\bar{\phi}_1^2 \\ \delta q\phi_2^2 - \delta r\phi_1^2 & \delta q\phi_2\bar{\phi}_2 - \delta r\phi_1\bar{\phi}_1 \end{bmatrix} \right.$$
$$+ \delta\zeta \int_{-L}^{x} \begin{bmatrix} i(\phi_1\bar{\phi}_2 + \bar{\phi}_1\phi_2) & 2i\bar{\phi}_1\bar{\phi}_2 \\ -2i\phi_1\phi_2 & -i(\phi_1\bar{\phi}_2 + \bar{\phi}_1\phi_2) \end{bmatrix}$$
$$\left. + \delta\zeta \begin{bmatrix} iL & 0 \\ 0 & -iL \end{bmatrix} \right\} \tag{6.30}$$

where we have used (6.18) in finding the variation with respect to $\zeta$ of $\phi$ and $\bar{\phi}$ at $x = -L$. Now, let $x = L$ and let $L \to +\infty$ and find the variation of the scattering coefficients $\delta a$, $\delta\bar{a}$, $\delta b$, $\delta\bar{b}$, using the fact that as $x \to +\infty$

$$\Phi(x) \to \begin{bmatrix} a\,e^{-i\zeta x} & \bar{b}\,e^{-i\zeta x} \\ b\,e^{i\zeta x} & -\bar{a}\,e^{i\zeta x} \end{bmatrix} .$$

It is convenient to define the bilinear form,

$$I(u,v) = \int_{-\infty}^{\infty} (-\,\delta q u_2 v_2 + \delta r u_1 v_1)dx = \int_{-\infty}^{\infty} \begin{bmatrix} \delta r \\ -\delta q \end{bmatrix} \cdot \begin{bmatrix} u_1 v_1 \\ u_2 v_2 \end{bmatrix} dx . \tag{6.31}$$

We find

$$\delta a = -I(\phi,\psi) + \delta\zeta \lim_{L\to\infty} i \int_{-L}^{L} (a - \phi_1\psi_2 - \phi_2\psi_1)dx , \tag{6.32a}$$

$$\delta b = I(\phi,\bar{\psi}) + \delta\zeta \lim_{L\to\infty} i \int_{-L}^{L} (\phi_1\bar{\psi}_2 + \phi_2\bar{\psi}_1)dx \quad , \tag{6.32b}$$

$$\delta\bar{b} = -I(\bar{\phi},\psi) + \delta\zeta \lim_{L\to\infty} i \int_{-L}^{L} -(\bar{\phi}_1\psi_2 + \bar{\phi}_2\psi_1)dx \quad , \tag{6.32c}$$

$$\delta\bar{a} = -I(\bar{\phi},\bar{\psi}) + \delta\zeta \lim_{L\to\infty} i \int_{-L}^{L} -(\bar{a} + \bar{\phi}_1\bar{\psi}_2 + \bar{\phi}_2\bar{\psi}_1)dx \quad . \tag{6.32d}$$

By using the identity (6.24), we can identify the second terms on the right-hand sides of (6.32) with the derivatives $a_\zeta$, $b_\zeta$, $\bar{b}_\zeta$ and $\bar{a}_\zeta$, respectively. We note in particular that (6.32) may be written as the total change in $a(\zeta)$ being equal to the sum of the changes in r, q and $\zeta$.

Using (6.32), we next obtain the change in the scattering data $S_-$:

$$S_- = S_- \left\{(\zeta_k,\beta_k)_{k=1}^{N},\ (\bar{\zeta}_k,\bar{\beta}_k)_{k=1}^{\bar{N}},\ \frac{\bar{b}}{a}\ ,\ \frac{b}{\bar{a}}\ ,\ \zeta \text{ real}\right\} \tag{6.33}$$

where

$$\beta(\zeta) = (\zeta - \zeta_k)\,\frac{\bar{b}}{a}\ ,\ \bar{\beta}(\zeta) = (\zeta - \bar{\zeta}_k)\,\frac{b}{\bar{a}} \quad . \tag{6.34}$$

First if $\zeta$ is real and $\delta\zeta = 0$,

$$\delta\left(\frac{\bar{b}}{a}\right) = \frac{1}{a^2}\, I(\psi,\psi) = \frac{1}{a^2}\int_{-\infty}^{\infty} \begin{bmatrix} \delta r \\ -\delta q \end{bmatrix} \cdot \begin{bmatrix} \psi_1^2 \\ \psi_2^2 \end{bmatrix} dx \quad , \tag{6.35}$$

$$\delta\left(\frac{b}{\bar{a}}\right) = \frac{1}{\bar{a}^2}\, I(\bar{\psi},\bar{\psi}) = \frac{1}{\bar{a}^2}\int_{-\infty}^{\infty} \begin{bmatrix} \delta r \\ -\delta q \end{bmatrix} \cdot \begin{bmatrix} \bar{\psi}_1^2 \\ \bar{\psi}_2^2 \end{bmatrix} dx \quad . \tag{6.36}$$

In order to calculate $\delta\beta_k$, $\delta\zeta_k$ in a convenient manner, assume q and r to be on compact support and consider the analytic extension of (6.35) to $\zeta = \zeta_k$, $\mathrm{Im}\{\zeta_k\} > 0$. (We remind the reader that we are also assuming that the zeros of a and $\bar{a}$ are simple). Then expanding (6.35) as a Taylor series near $\zeta = \zeta_k$, we obtain to first order in $(\zeta - \zeta_k)$

$$\beta\delta\zeta_k + \delta\beta(\zeta - \zeta_k) + \ldots = \frac{1}{a_k'^2}\,[I_k(\psi,\psi) + (\zeta - \zeta_k)I_k'(\psi,\psi) + \ldots]$$
$$\left[1 - \frac{a_k''}{a_k'}(\zeta - \zeta_k) + \ldots\right] \quad .$$

Now $\beta(\zeta,t) = \beta(\zeta_k(t),t) + (\zeta - \zeta_k)\beta_k'(\zeta_k(t),t) + \ldots$ and thus $\delta\beta = \delta\beta_k - \delta\zeta_k\beta_k' + (\zeta - \zeta_k)(\delta\beta_k' - \delta\zeta_k\beta_k'') + \ldots$ . (Here the prime refers to the $\zeta$ derivative.) Equating powers of $(\zeta - \zeta_k)$ we obtain for k = 1 ... N,

$$\delta\zeta_k = \frac{b_k}{a_k'}\, I_k(\psi,\psi) = \frac{b_k}{a_k'}\int_{-\infty}^{\infty} \begin{bmatrix} \delta r \\ -\delta q \end{bmatrix} \cdot \begin{bmatrix} \psi_1^2 \\ \psi_2^2 \end{bmatrix}_{\zeta_k} dx \quad , \tag{6.37}$$

$$\delta\left(\frac{1}{b_j a_k'}\right) = \delta\beta_k = \frac{1}{a_k'^2}\left[I_k'(\psi,\psi) - \frac{a_k''}{a_k'} I_k(\psi,\psi)\right]$$

$$= \frac{1}{a_k'^2}\int_{-\infty}^{\infty} \begin{bmatrix}\delta r\\ -\delta q\end{bmatrix}\cdot\left[\frac{\partial}{\partial\zeta}\begin{bmatrix}\psi_1^2\\ \psi_2^2\end{bmatrix}_{\zeta_k} - \frac{a_k''}{a_k'}\begin{bmatrix}\psi_1^2\\ \psi_2^2\end{bmatrix}_{\zeta_k}\right]dx \quad . \tag{6.38}$$

Note in particular that (6.37) implies that

$$\delta a[\zeta_k(t),t] = -b_k I_k(\psi,\psi) + \delta\zeta_k a_k' = 0$$

(use $\phi_k = b_k\psi_k$) which shows that the eigenvalue $\zeta_k$ remains a zero of $a(\zeta)$. In a similar fashion,

$$\delta\bar{\zeta}_k = \frac{\bar{b}_k}{\bar{a}_k'} I_k(\psi,\psi) = \frac{\bar{b}_k}{\bar{a}_k'}\int_{-\infty}^{\infty}\begin{bmatrix}\delta r\\ -\delta q\end{bmatrix}\cdot\begin{bmatrix}\bar{\psi}_1^2\\ \bar{\psi}_2^2\end{bmatrix}_{\bar{\zeta}_k} dx \tag{6.39}$$

and

$$\delta\left(\frac{1}{\bar{b}_k\bar{a}_k'}\right) = \frac{1}{\bar{a}_k^2}\left[I_k'(\bar{\psi},\bar{\psi}) - \frac{\bar{a}_k''}{\bar{a}_k'} I_k(\psi,\psi)\right]$$

$$= \frac{1}{\bar{a}_k'^2}\int_{-\infty}^{\infty}\begin{bmatrix}\delta r\\ -\delta q\end{bmatrix}\cdot\frac{\partial}{\partial\zeta}\begin{bmatrix}\bar{\psi}_1^2\\ \bar{\psi}_2^2\end{bmatrix}_{\bar{\zeta}_k} - \frac{\bar{a}_k''}{\bar{a}_k'}\begin{bmatrix}\bar{\psi}_1^2\\ \bar{\psi}_2^2\end{bmatrix}_{\bar{\zeta}_k} dx \quad . \tag{6.40}$$

Most important, we note that the changes in the scattering data are simply given by the inner products between the vector $(\delta r, -\delta q)^T$ and the set of squared eigenfunctions

$$E_- = \left\{\Psi_k = \begin{bmatrix}\psi_1^2\\ \psi_2^2\end{bmatrix}_{\zeta_k}, \quad \tau_k = \frac{\partial}{\partial\zeta}\begin{bmatrix}\psi_1^2\\ \psi_2^2\end{bmatrix}_{\zeta_k}, \quad k = 1 \dots N; \quad \bar{\Psi}_k = \begin{bmatrix}\bar{\psi}_1^2\\ \bar{\psi}_2^2\end{bmatrix}_{\bar{\zeta}_k}, \right.$$

$$\left. \bar{\tau}_k = \frac{\partial}{\partial\zeta}\begin{bmatrix}\bar{\psi}_1^2\\ \bar{\psi}_2^2\end{bmatrix}_{\bar{\zeta}_k}, \quad k = 1 \dots \bar{N}; \quad \Psi = \begin{bmatrix}\psi_1^2\\ \bar{\psi}_2^2\end{bmatrix}, \quad \bar{\Psi} = \begin{bmatrix}\bar{\psi}_1^2\\ \bar{\psi}_1^2\end{bmatrix}, \quad \zeta \text{ real}\right\} \quad . \tag{6.41}$$

In a similar manner, we can also determine the change of the scattering data

$$S_+ = \left\{(\zeta_k,\gamma_k)_{k=1}^{N}, \quad (\bar{\zeta}_k,\bar{\gamma}_k)_{k=1}^{\bar{N}}, \frac{b}{a}, \frac{\bar{b}}{\bar{a}}, \zeta \text{ real}\right\}$$

with $\gamma_k = b_k/a_k'$, $\bar{\gamma}_k = \bar{b}_k/\bar{a}_k'$, associated with the *dual* inverse problem. We find for $\zeta$ real,

$$\delta\left(\frac{b}{a}\right) = \frac{1}{a^2} I(\phi,\phi) = -\frac{1}{a^2}\int_{-\infty}^{\infty}\begin{bmatrix}\delta q\\ \delta r\end{bmatrix}\cdot\begin{bmatrix}\phi_2^2\\ -\phi_1^2\end{bmatrix}dx \quad , \tag{6.42}$$

$$\delta\left(\frac{\bar{b}}{\bar{a}}\right) = \frac{1}{\bar{a}^2} I(\bar{\phi},\bar{\phi}) = -\frac{1}{\bar{a}^2}\int_{-\infty}^{\infty}\begin{bmatrix}\delta q\\ \delta r\end{bmatrix}\cdot\begin{bmatrix}\bar{\phi}_2^2\\ -\bar{\phi}_1^2\end{bmatrix}dx \quad . \tag{6.43}$$

Then, for $k = 1 \ldots N$,

$$\delta\zeta_k = \frac{1}{\gamma_k a_k'^2} I_k(\phi,\phi) = -\frac{1}{\gamma_k a_k'^2}\int_{-\infty}^{\infty}\begin{bmatrix}\delta q\\ \delta r\end{bmatrix}\cdot\begin{bmatrix}\phi_2^2\\ -\phi_1^2\end{bmatrix}_{\zeta_k} \quad , \tag{6.44}$$

$$\delta\gamma_k = \frac{1}{a_k'^2}\left[I_k'(\phi,\phi) - \frac{a_k''}{a_k'} I_k(\phi,\phi)\right] \tag{6.45}$$

$$= -\frac{1}{a_k'^2}\int_{-\infty}^{\infty}\begin{bmatrix}\delta q\\ \delta r\end{bmatrix}\cdot\left\{\frac{\partial}{\partial\zeta}\begin{bmatrix}\phi_2^2\\ -\phi_1^2\end{bmatrix}_{\zeta_k} - \frac{a_k''}{a_k'}\begin{bmatrix}\phi_2^2\\ -\phi_1^2\end{bmatrix}_{\zeta_k}\right\}dx \quad .$$

Finally, for $k = 1 \ldots \bar{N}$,

$$\delta\bar{\zeta}_k = \frac{1}{\bar{\gamma}_k \bar{a}_k'^2} I_k(\bar{\phi},\bar{\phi}) = -\frac{1}{\bar{\gamma}_k \bar{a}_k'^2}\int_{-\infty}^{\infty}\begin{bmatrix}\delta q\\ \delta r\end{bmatrix}\cdot\begin{bmatrix}\bar{\phi}_2^2\\ -\bar{\phi}_1^2\end{bmatrix}_{\bar{\zeta}_k} dx \quad , \tag{6.46}$$

and

$$\delta\bar{\gamma}_k = \frac{1}{\bar{a}_k'^2}\left[I_k'(\bar{\phi},\bar{\phi}) - \frac{\bar{a}''_k}{\bar{a}_k'} I_k(\bar{\phi},\bar{\phi})\right]$$

$$= -\frac{1}{\bar{a}_k'^2}\int_{-\infty}^{\infty}\begin{pmatrix}\delta q\\ \delta r\end{pmatrix}\cdot\left\{\frac{\partial}{\partial\zeta}\begin{bmatrix}\bar{\phi}_2^2\\ -\bar{\phi}_1^2\end{bmatrix}_{\bar{\zeta}_k} - \frac{\bar{a}_k''}{\bar{a}_k'}\begin{bmatrix}\bar{\phi}_2^2\\ -\bar{\phi}_1^2\end{bmatrix}\right\}dx \quad . \tag{6.47}$$

Again, we note that the change of the scattering data $S_+$ can be written as the inner products between the vector $(\delta q,\delta r)^T$ and the *dual* squared eigenfunctions

$$E_+ = \left\{ \Psi_k^A = \begin{bmatrix}\phi_2^2\\ -\phi_1^2\end{bmatrix}_{\zeta_k} \; , \; \chi_k^A = \frac{\partial}{\partial\zeta}\begin{bmatrix}\phi_2^2\\ -\phi_1^2\end{bmatrix}_{\zeta_k} \; , \; k = 1 \ldots N \; ; \; \bar{\Psi}_k^A = \begin{bmatrix}\phi_2^2\\ -\phi_1^2\end{bmatrix}_{\bar{\zeta}_k} \; , \right.$$

$$\left. \bar{\chi}_k^A = \frac{\partial}{\partial\zeta}\begin{bmatrix}\bar{\phi}_2^2\\ -\bar{\phi}_1^2\end{bmatrix}_{\bar{\zeta}_k} \; , \; k = 1 \ldots \bar{N} \; ; \; \Psi^A = \begin{bmatrix}\phi_2^2\\ -\phi_1^2\end{bmatrix} , \; \bar{\Psi}^A = \begin{bmatrix}\phi_2^2\\ -\bar{\phi}_1^2\end{bmatrix} , \; \zeta \text{ real} \right\} . \tag{6.48}$$

The reason we use the notation $\Psi^A$ for the dual squared eigenfunctions is that the $\Psi^A$ are also the adjoint eigenfunctions.

## 6.4 The Squared Eigenfunctions and Fourier Expansions

The squared eigenfunctions $E_-$ satisfy the equations

$$(L - \zeta)(\Psi,\bar{\Psi}) = 0 \ , \ \zeta \text{ real or } \zeta = (\zeta_k,\bar{\zeta}_k) \tag{6.49}$$

and

$$(L - \zeta)(\tau_k,\bar{\tau}_k) = (\Psi_k,\bar{\Psi}_k), \ \zeta = (\zeta_k,\bar{\zeta}_k) \tag{6.50}$$

$$L = \frac{1}{2i}\begin{pmatrix} -\frac{\partial}{\partial x} - 2q\int_x^\infty dy\ r\ \cdot & -2q\int_x^\infty dy\ q\ \cdot \\ 2r\int_x^\infty dy\ r\ \cdot & \frac{\partial}{\partial x} + 2r\int_x^\infty dy\ q \end{pmatrix} \cdot \tag{6.51}$$

These expressions are found directly from (6.16) by multiplying two equations by $\psi_1,\psi_2$, respectively, and using the fact that $\psi_1\psi_2 = -\int_x^\infty (q\psi_2^2 + r\psi_1^2)dx$. It turns out that the *dual* squared eigenfunctions $E_+$ satisfy the *adjoint* equations

$$(L^A - \zeta)(\Psi^A,\bar{\Psi}^A) = 0 \quad \zeta \text{ real}, \ \zeta = \zeta_k,\bar{\zeta}_k \tag{6.52}$$

$$(L^A - \zeta)(\chi_k^A,\bar{\chi}_k^A) = (\Psi_k^A,\bar{\Psi}_k^A), \ \zeta = (\zeta_k,\bar{\zeta}_k) \ , \tag{6.53}$$

with

$$L^A = \frac{1}{2i}\begin{pmatrix} \frac{\partial}{\partial x} - 2r\int_{-\infty}^x dy\ q\ \cdot & 2r\int_{-\infty}^x dy\ r\ \cdot \\ -2q\int_{-\infty}^x dy\ q & -\frac{\partial}{\partial x} + 2q\int_{-\infty}^x dy\ r \end{pmatrix} \cdot \tag{6.54}$$

Using these equations, we may evaluate the various *inner products*

$$<\underline{u},\underline{v}> = \int_{-\infty}^\infty \underline{u}\cdot\underline{v}\ dx$$

between the eigenfunctions $E_-$ and the eigenfunctions $E_+$. The results are given in Appendix A. Once these expressions are obtained, we can invert the relations (6.35-40) and (6.42-47) and find expressions for the vectors $(\delta r, -\delta q)^T$ and $(\delta q,\delta r)^T$ in terms of the bases $E_+$ and $E_-$, respectively.

A proof that each set $E_+$ and $E_-$ forms a basis in the intersection of $L^{(1)}$ and all continuously differentiable functions was given by KAUP [6.64]. The proof that each set represents a basis in $L^{(1)} \cap L^{(2)}$ or in $L^{(1)}$ alone has not yet been given.

It turns out, from (6.35-40) and (6.42-47) and the inner product relations (6A.1-9) that the coefficients in these expressions are simply the changes of the scattering data $S_-$ and $S_+$, respectively. From the relations that

$$\int_{-\infty}^\infty (q\psi_2^2 + r\psi_1^2)dx = (\psi_1\psi_2)_{-\infty}^\infty$$

and

$$\int_{-\infty}^{\infty}\left[q\phi_2^2 + (-r)(-\phi_1^2)\right]dx = (\phi_1\phi_2)_{-\infty}^{\infty}$$

we may also express the vectors $(r,q)^T$ and $(q,-r)^T$ in the bases $E_+$ and $E_-$, respectively. We find

$$\begin{bmatrix}\delta r\\ -\delta q\end{bmatrix} = -\frac{1}{\pi}\int_{-\infty}^{\infty}\left[\delta\left(\frac{\bar{b}}{a}\right)\Psi^A - \delta\left(\frac{b}{\bar{a}}\right)\bar{\Psi}^A\right]d\zeta$$
$$+ 2i\sum_1^N\left(\delta\beta_k\Psi_k^A + \beta_k\delta\zeta_k\chi_k^A\right) ,$$
$$+ 2i\sum_1^N\left(\delta\bar{\beta}_k\bar{\Psi}_k^A + \bar{\beta}_k\delta\bar{\zeta}_k\bar{\chi}_k^{A}\right) \tag{6.55}$$

$$\begin{bmatrix}r\\ q\end{bmatrix} = \frac{1}{\pi}\int_{-\infty}^{\infty}\left(\frac{\bar{b}}{a}\Psi^A + \frac{b}{\bar{a}}\bar{\Psi}^A\right)d\zeta$$
$$- 2i\sum_1\beta_k\Psi_k^A + 2i\sum_1\bar{\beta}_k\bar{\Psi}_k^A , \tag{6.56}$$

$$\begin{bmatrix}\delta q\\ \delta r\end{bmatrix} = \frac{1}{\pi}\int_{-\infty}^{\infty}\left[\delta\left(\frac{b}{a}\right)\Psi - \delta\left(\frac{\bar{b}}{\bar{a}}\right)\bar{\Psi}\right]d\zeta$$
$$-2i\sum_1^N(\delta\gamma_k\Psi_k + \gamma_k\delta\zeta_k\tau_k) ,$$
$$-2i\sum_1^{\bar{N}}(\delta\bar{\gamma}_k\bar{\Psi}_k + \bar{\gamma}_k\delta\bar{\zeta}_k\bar{\tau}_k) , \tag{6.57}$$

and

$$\begin{bmatrix}q\\ -r\end{bmatrix} = -\frac{1}{\pi}\int_{-\infty}^{\infty}\left(\frac{b}{a}\Psi + \frac{\bar{b}}{\bar{a}}\bar{\Psi}\right)d\zeta + 2i\sum_1\gamma_k\Psi_k - 2i\sum_1\bar{\gamma}_k\bar{\Psi}_k . \tag{6.58}$$

Now, the fun starts, for once having obtained these expressions we may ask the question; what operator $\Omega$ acting on $\begin{bmatrix}r\\ q\end{bmatrix}$ acts on each $\zeta$ component of the expansion separately? For such an operator, the partial differential equation

$$\delta\begin{bmatrix}r\\ -q\end{bmatrix} + \Omega\begin{bmatrix}r\\ q\end{bmatrix} = 0 \tag{6.59}$$

leads, by simply identifying the coefficients in the expansions for the two quantities $\delta\begin{bmatrix}r\\ -q\end{bmatrix}$ and $\Omega\begin{bmatrix}r\\ q\end{bmatrix}$, to a separable system of ordinary differential equations for the time rate of change of the scattering data. Hence, the Fourier expansions (6.55-58) are "natural" for that set of operator equations (6.59) where the operator $\Omega$ is diagonal in the spectral representation of $L^A$; i.e., operates on each component separately.

We will identify these classes of equations in Sect.6.5 but first, we wish to make some further points relating to Sects.6.2-4.

*Remark 1.* We are now in a position to prove directly that the inverse scattering transformation is canonical in which the symplectic form $\int_{-\infty}^{\infty}\delta q\wedge\delta r\, dx$ is preserved.

By simply taking the scalar product of (6.55,57) with different variations $\delta_1$ and $\delta_2$ and using the inner product expressions given in Appendix A, we have

$$\int_{-\infty}^{\infty} \delta q \wedge \delta r \, dx = \int_{-\infty}^{\infty} \delta \frac{\ln a\bar{a}}{\pi} \wedge \delta \ln b(\zeta) d\zeta + \sum_{1}^{N} \delta 2i\zeta_k \wedge \delta \ln b_k + \sum_{1}^{\bar{N}} \delta 2i\bar{\zeta}_k \wedge \delta \ln \bar{b}_k \quad . \tag{6.60}$$

In this expression we have used the fact that for real $\zeta$, $\bar{b} = (1 - a\bar{a})/b$. When we develop the Hamiltonian formulation in Sect.6.6, the result (6.60) will identify the action ($\pi^{-1} \ln a\bar{a}$, $2i\zeta_k$, $2i\bar{\zeta}_k$) and angle [$\ln b(\zeta)$, $\ln b_k$, $\ln \bar{b}_k$] variables.

*Remark 2.* It is illustrative to take the linear limit of (6.55,56). Indeed, if we calculate the eigenfunctions to first order in r and q, we find $a \simeq \bar{a} \simeq 1$,

$$\psi^A \simeq - \begin{pmatrix} 0 \\ 1 \end{pmatrix} e^{-2i\zeta x} \quad , \quad \bar{\psi}^A \simeq \begin{pmatrix} 1 \\ 0 \end{pmatrix} e^{2i\zeta x}$$

and thus (6.55,56) simply reduce to the linear Fourier transform relations,

$$\delta r = \frac{1}{\pi} \int_{-\infty}^{\infty} \delta b \, e^{2i\zeta x} d\zeta \quad , \quad r = \frac{1}{\pi} \int_{-\infty}^{\infty} b \, e^{2i\zeta x} d\zeta \quad ,$$

$$\delta q = - \frac{1}{\pi} \int_{-\infty}^{\infty} \delta \bar{b} \, e^{-2i\zeta x} d\zeta \quad , \quad q = - \frac{1}{\pi} \int_{-\infty}^{\infty} \bar{b} \, e^{-2i\zeta x} d\zeta \quad . \tag{6.61}$$

*Remark 3.* Our third remark is a word of warning in that the approach to date is formal and it may well be that an arbitrary specification of the scattering data $S_+$ or $S_-$ can lead to singular behavior in and nonunique solutions for r(x,t) and q(x,t). Indeed, if $b = \bar{b} = 0$, real $\zeta$, $N = \bar{N} = 1$, and $S_- = [(\zeta_1,\beta_1),(\bar{\zeta}_1,\bar{\beta}_1)]$, we obtain a r and a q which have singular behavior of the form $[x - x_0(t)]^{-1}$. In fact, this singular behavior may well develop from initial data r(x,0), q(x,0) which is both nonsingular and absolutely integrable. What is happening is that the bases $E_-$ and $E_+$ are themselves becoming singular at this point in time. One of the prerequisites of the Lax approach is that it works with self-adjoint operators $L$ whose evolution in time is unitary which in turn leads to a skew adjoint $B$ describing the evolution of the eigenfunctions. Thus the eigenfunctions merely rotate and their magnitude is always under control. It is an open question as to what are the necessary conditions on r and q in order that the quantities evolve in a nonsingular way for all time. We know [6.23] that it is sufficient for $r = \alpha q^*$ ($\alpha$ real), $r = \alpha q$ ($\alpha$ complex, q real or pure imaginary) and $\int_{-\infty}^{\infty} |q| dx < \infty$ to guarantee that r and q evolve in a nonsingular way. These conditions (with a little more; see [6.23]) also guarantee that the solution of the inversion equations is unique; there are no nontrivial homogeneous solutions. A possible answer to the necessity condition is given by FLASCHKA and NEWELL [6.63].

*Remark 4.* It is illustrative and helpful for purposes of generalization to note that (6.55,56) may be written

$$\begin{bmatrix}\delta r\\ -\delta q\end{bmatrix} = -\frac{1}{\pi}\oint \frac{I(\psi,\psi)}{a^2}\,\Psi^A d\zeta + \frac{1}{\pi}\oint \frac{I(\bar\psi,\bar\psi)}{\bar a^2}\,\bar\Psi^A d\zeta \quad , \tag{6.62}$$

and

$$\begin{bmatrix} r\\ q\end{bmatrix} = \frac{1}{\pi}\oint \frac{\bar b}{a}\,\Psi^A d\zeta + \frac{1}{\pi}\oint \frac{b}{\bar a}\,\bar\Psi^A d\zeta \tag{6.63}$$

when $\bar b/a$ and $b/\bar a$ admit analytic continuation to the upper and lower planes, respectively. Here C ($\bar C$) is a contour from $-\infty$ to $\infty$ goin over (under) all zeros of $a(\zeta)$ [$\bar a(\zeta)$]. It is worth noting that these expressions when re-expanded become the correct expressions for the vectors $(\delta r, -\delta q)^T$ and $(r,q)^T$ even when $a(\zeta)$ and $\bar a(\zeta)$ have multiple zeros; for example, if $a(\zeta)$ has a double zero, we must also include the quantities $(\partial^2\Psi/\partial\zeta^2)_k$, $(\partial^3\Psi/\partial\zeta^3)_k$ in the basis expansions.

## 6.5 Evolution Equations of Class I

We now ask the question: what operator $\Omega$ acting on the vector $(r,q)^T$ in (6.56) will give $(\delta r, -\delta q)^T$? Clearly, if $\Omega$ acts separately on each $\zeta$ component in (6.56), the equation

$$\begin{bmatrix}\delta r\\ -\delta q\end{bmatrix} + \Omega\begin{bmatrix} r\\ q\end{bmatrix} = 0 \tag{6.64}$$

will lead to separable system of ordinary differential equations for the scattering data. It is obvious that any polynomial in $L^A$ is a candidate for $\Omega$. Indeed, all operators B which commute with $L^A$ have this property since $L^A B\psi^A = BL^A\psi^A = \zeta B\psi^A$ and hence $B\psi^A$ is a linear combination of the solutions of $L^A\psi^A = \zeta\psi^A$. For convenience of presentation, we divide the integrable equations into a number of classes. The first class is associated with a meromorphic function $\Omega(\zeta)$ of $\zeta$ which is nonsingular on the spectrum of L, $L^A$. Then the equation

$$\begin{bmatrix}\delta r\\ -\delta q\end{bmatrix} + 2\Omega(L^A)\begin{bmatrix} r\\ q\end{bmatrix} = 0 \tag{6.65}$$

is equivalent to

$$\begin{aligned} &\delta(\bar b/a) = 2\Omega(\zeta)\bar b/a, \quad \delta(b/\bar a) = -2\Omega(\zeta)b/\bar a\\ &\delta\zeta_k = 0 \quad , \quad \delta\beta_k = 2\Omega(\zeta_k)\beta_k \quad , \quad k = 1, \ldots, N\\ &\delta\bar\zeta_k = 0 \quad , \quad \delta\bar\beta_k = -2\Omega(\zeta_k)\bar\beta_k, \quad k = 1, \ldots, N \end{aligned} \tag{6.66}$$

by simply equating the coefficients of the basis functions $E_-$ in the expansions for $(\delta r, -\delta q)^T$ and $(r,q)^T$. The system (6.66) is trivially integrable.

*Remark 1.* Examples. If $\Omega(\zeta) = \sum_{r=0}^{3} a_r\zeta^r$, and $\delta = \partial/\partial t$ then the equations (6.65) are

$$\begin{bmatrix} r_t \\ -q_t \end{bmatrix} + 2a_0\begin{bmatrix} r \\ q \end{bmatrix} + \frac{2a_1}{2i}\begin{bmatrix} r_x \\ -q_x \end{bmatrix} + \frac{2a_2}{(2i)^2}\begin{bmatrix} r_{xx} - 2qr^2 \\ q_{xx} - 2q^2r \end{bmatrix} + \frac{2a_3}{(2i)^3}\begin{bmatrix} r_{xxx} - 6qr\, r_x \\ -q_{xxx} + 6qr\, q_x \end{bmatrix} = 0 \quad , \tag{6.67}$$

which incorporates the linear unidirectional wave equation, the nonlinear Schrödinger equation and the modified Korteweg-deVries equation. Note that for stability reasons, $\Omega(\zeta)$ must be pure imaginary for $\zeta$ real; otherwise either $\bar{b}/a$ or $b/\bar{a}$ would grow exponentially. This fact is also reflected in (6.67) for if $a_2$ is real and positive, r satisfies the forward heat equation and q the backward heat equation.

*Remark 2.* Dispersion relation. Note that the function $\Omega(\zeta)$ is directly related to the dispersion relation of the associated linear problems for r and q. In fact $\Omega(\zeta) = i/2\omega_r(2\zeta) = -i/2\omega_q(-2\zeta)$.

*Remark 3.* Conserved quantities. For all the systems described by (6.65), the functions $a(\zeta)$ and $\bar{a}(\zeta)$ are time invariant. In particular, the coefficients $\{C_n\}_{n=1}^{\infty}$ in their asymptotic expansions (6.25) are also conserved quantities.

*Remark 4.* Higher dimensions and pseudo directional derivatives. It was pointed out to me by Francesco Calogero at the Arizona 1976 meeting that the $\delta$ could also contain derivatives with respect to other "spatial" dimensions. For example, if $\underset{\sim}{y} = (y_1, \ldots, y_n)$ and

$$\delta = F(L^A)\frac{\partial}{\partial t} + \underline{G}(L^A)\cdot\nabla_{\underset{\sim}{y}} \quad ,$$

then the evolution equation is

$$F(L^A)\begin{bmatrix} r_t \\ -q_t \end{bmatrix} + \underline{G}(L^A)\cdot\nabla_{\underset{\sim}{y}}\begin{bmatrix} r \\ -q \end{bmatrix} + 2\Omega(L^A)\begin{bmatrix} r \\ q \end{bmatrix} = 0 \quad , \tag{6.68}$$

and the evolution of the scattering data is given by

$$F(\zeta)\left(\frac{\bar{b}}{a}\right)_t + \underline{G}(\zeta)\cdot\nabla_{\underset{\sim}{y}}\left(\frac{\bar{b}}{a}\right) = 2\Omega(\zeta)\,\frac{\bar{b}}{a} \quad , \tag{6.69a}$$

$$F(\zeta)\left(\frac{b}{\bar{a}}\right)_t + \underline{G}(\zeta)\cdot\nabla_{\underset{\sim}{y}}\left(\frac{b}{\bar{a}}\right) = -2\Omega(\zeta)\,\frac{b}{\bar{a}} \quad , \tag{6.69b}$$

$$F(\zeta_k)\zeta_{kt} + \underline{G}(\zeta_k)\cdot\nabla_{\underset{\sim}{y}}\zeta_k = 0 \;,\; \frac{\partial}{\partial t}[F(\zeta_k)\beta_k] + \nabla\cdot[\underline{G}(\zeta_k)\beta_k] = -2\Omega(\zeta_k)\beta_k \qquad k = 1, \ldots, N \tag{6.69c}$$

$$F(\bar{\zeta}_k)\bar{\zeta}_{kt} + \underline{G}(\bar{\zeta}_k)\cdot\nabla_{\underset{\sim}{y}}\bar{\zeta}_k = 0 \;,\; \frac{\partial}{\partial t}[F(\bar{\zeta}_k)\bar{\beta}_k] + \nabla\cdot[\underline{G}(\bar{\zeta}_k)\bar{\beta}_k] = -2\Omega(\bar{\zeta}_k)\bar{\beta}_k, \qquad k = 1, \ldots, \bar{N} \;. \tag{6.69d}$$

Notice the very interesting properties which the equations (6.69) possess. They are hyperbolic and can develop multivalued behavior in finite times. For example, if $F = 1$ and $G = \zeta^2$, an initial profile of $\zeta(y,0)$ (say, of triangular form) will eventually admit multiple valued solutions which can be interpreted in a meaningful way as soliton splitting. One might go further and ask if we include second and third variations of $\begin{bmatrix} r \\ -q \end{bmatrix}$, would it be possible to find equations for the eigenvalues of the form

$$\zeta_t + \zeta^2\zeta_y + \zeta_{yyy} = 0 \qquad (6.70)$$

which would correspond to local two-dimensional soliton solutions in (x,y) space?

These ideas are particularly attractive, not simply because of the physical consequences, but because the decomposition of (6.68) to (6.69) and possibly (6.70), by one transform followed by another transform to decompose (6.70) is a very natural extension of the ideas of separation of variables in linear problems.

*Remark 5.* We can include the effects of linear density gradients by noting that

$$\left(\frac{b}{a}\right)_\zeta = -\frac{1}{a^2}\int_{-\infty}^{\infty} 2ix\begin{bmatrix} q \\ -r \end{bmatrix}\cdot\begin{bmatrix} \phi_2^2 \\ -\phi_1^2 \end{bmatrix}dx \quad , \quad \left(\frac{\bar{b}}{\bar{a}}\right)_\zeta = -\frac{1}{\bar{a}^2}\int_{-\infty}^{\infty} 2ix\begin{bmatrix} q \\ -r \end{bmatrix}\cdot\begin{bmatrix} \bar{\phi}_2^2 \\ -\bar{\phi}_1^2 \end{bmatrix}dx \quad , \qquad (6.71)$$

which allows us to write the expansion for

$$2ix\begin{bmatrix} q \\ -r \end{bmatrix} = \frac{1}{\pi}\int_C \left(\frac{b}{a}\right)_\zeta \Psi d\zeta - \frac{1}{\pi}\int_{\bar{C}} \left(\frac{\bar{b}}{\bar{a}}\right)_\zeta \bar{\Psi} d\zeta \quad . \qquad (6.72)$$

For example, the equation

$$\begin{bmatrix} q_t \\ +r_t \end{bmatrix} + 4iL^2\begin{bmatrix} q \\ -r \end{bmatrix} + 2i\alpha x\begin{bmatrix} q \\ -r \end{bmatrix} = 0$$

which when $r = -q^*$ is the nonlinear Schrödinger equation for wave envelopes propagating in a density gradient leads to following time evolutions for b/a:

$$(b/a)_t + \alpha(b/a)_\xi = 4i\xi^2 b/a \quad . \qquad (6.73)$$

The effect of the density gradient is to make the basic reference frame the locus $x + \alpha t^2/2 =$ constant. When the envelope propagates into a region of increasing density $\alpha > 0$, it is eventually repelled.

## 6.6 Hamiltonian Structure of the Equations of Class I

Our goal in this section is to introduce the Hamiltonian structure of (6.65) and in particular to point out that IST is simply a canonical transformation in which the form of Hamilton's equations are preserved. Our first step will be to show that the second term of (6.65) may be written as the gradient with respect to q and r of some functional, the Hamiltonian. We will find that the Hamiltonians are all generated by the single spectral function $\ln a(\zeta)$. Because this function may also be expressed in terms of the scattering data, we immediately have an expression for the transformed Hamiltonian in terms of the action-angle variables. Closely related are the trace formulae which express the conserved functionals in terms of the action variables.

Consider the function

$$\Phi^A = \frac{1}{a(\zeta)}\begin{bmatrix}\phi_2\psi_2\\ -\phi_1\psi_1\end{bmatrix} \quad . \tag{6.74}$$

It may easily be shown that

$$(L^A - \zeta)\Phi^A = \frac{1}{2i}\begin{bmatrix}r\\ q\end{bmatrix} \quad . \tag{6.75}$$

Solve (6.75) for large $|\zeta|$, $\mathrm{Im}\{\zeta\} > 0$ and obtain the asymptotic expansion,

$$\Phi^A \sim -\frac{1}{2i}\sum_{m=0}^{\infty}\frac{1}{\zeta^{m+1}}(L^A)^m\begin{bmatrix}r\\ q\end{bmatrix} \quad . \tag{6.76}$$

But from (6.32a) (with $\delta\zeta = 0$),

$$\delta\ln a = -\int_{-\infty}^{\infty}\left(-\delta q\,\frac{\phi_2\psi_2}{a} + \delta r\,\frac{\phi_1\psi_1}{a}\right)dx \quad , \tag{6.77}$$

and hence

$$\mathrm{grad}_{q,r}\ln a = \Phi^A \quad , \quad \mathrm{grad}_{q,r} = (\delta/\delta q,\ \delta/\delta r)^T \quad . \tag{6.78}$$

But from (6.25)

$$\ln a(\zeta) \sim -\sum_0^{\infty}\frac{1}{\zeta^{m+1}}C_{m+1} \tag{6.79}$$

where the sequence $\{C_m\}_{m=1}^{\infty}$ are the conserved quantities. Hence, from (6.76,78,79),

$$(L^A)^m\begin{bmatrix}r\\ q\end{bmatrix} = 2i\,\mathrm{grad}\,C_{m+1} \quad . \tag{6.80}$$

Therefore, if $\Omega(\zeta) = \sum_0^{\infty} a_m\zeta^m$, then

$$\begin{bmatrix}r_t\\ -q_t\end{bmatrix} = -4i\sum_0^{\infty} a_m\,\mathrm{grad}\,C_{m+1}$$

and the *Hamiltonian* is

$$H_\Omega = -4i\sum_0^{\infty} a_m C_{m+1} \quad . \tag{6.81}$$

Now let us introduce the notion of a *Poisson bracket*. Let $I[q,r]$ be a functional of $q,r$ and their derivatives $\left[\text{e.g., } \int_{-\infty}^{\infty}(q_x r_x + q^2 r^2)dx\right]$. Then

$$\frac{dI}{dt} = \int_{-\infty}^{\infty} \left(\frac{\delta I}{\delta q} q_t + \frac{\delta I}{\delta r} r_t\right)dx = \int_{-\infty}^{\infty} \left(\frac{\delta H}{\delta q}\frac{\delta I}{\delta r} - \frac{\delta H}{\delta r}\frac{\delta I}{\delta q}\right)dx \tag{6.82}$$

and we define this last expression to be the Poisson bracket $\langle H,I\rangle$. Note if I is a motion invariant, $\langle H,I\rangle = 0$ and the two quantities H and I are said to be in involution. Each member of the family $\{C_m\}_{m=1}^{\infty}$, acting as a Hamiltonian, generates a flow for which all other potential Hamiltonians are conserved. Hence $\langle C_m, C_n\rangle = 0$, and the $\{C_m\}_{m=1}^{\infty}$ are all in involution.

Furthermore, the function $\ln a(\zeta)$ itself, estimated at a particular $\hat{\zeta}$ ($\hat{\zeta}$ does not belong to the spectrum), can act as a Hamiltonian and generate the flow,

$$\begin{bmatrix} r_t \\ -q_t \end{bmatrix} = (\Phi^A)_{\hat{\zeta}} \tag{6.83}$$

which together with (6.75) gives a closed system. The system can also be written

$$(L^A - \hat{\zeta})\begin{bmatrix} r_t \\ -q_t \end{bmatrix} = \frac{1}{2i}\begin{bmatrix} r \\ q \end{bmatrix} \tag{6.84}$$

and corresponds to the dispersion relation

$$\Omega(\zeta) = \frac{i}{4}\frac{1}{\zeta-\hat{\zeta}} \quad . \tag{6.85}$$

It can also be shown [6.63] that

$$\langle \ln a(\zeta_1) \ , \ \ln a(\zeta_2)\rangle = 0 \quad .$$

Thus, the spectral function $\ln a(\zeta)$ plays a central role in analysis. For further discussion of its operator-theoretic role we refer the reader to [6.63]. At this point, we are going to use its expression in terms of the scattering data in order to develop the *Trace Formulae*. From 6.23 we may write [here we will set $r = -q^*$, $\bar{a}(\zeta) = a^*(\zeta^*)$, $\bar{b}(\zeta) = b^*(\zeta^*)$, $\bar{b}_k = b_k^*$, $\bar{\zeta}_k = \zeta_k^*$, $\bar{N} = N$],

$$\ln a(\zeta) = \sum_{k=1}^{N} \ln \frac{\zeta-\zeta_k}{\zeta-\zeta_k^*} + \frac{1}{2\pi i}\int_{-\infty}^{\infty} \frac{\ln aa^*}{\zeta-\zeta} d\zeta$$

$$\sim -\sum_{m=1}^{\infty} \frac{1}{\zeta^m}\left(\sum_{k=1}^{N} \frac{\zeta_k^m - \zeta_k^{*m}}{m} + \frac{1}{2\pi i}\int_{-\infty}^{\infty} \xi^{m-1} \ln aa^* \, d\xi\right) \quad . \tag{6.86}$$

Hence, from (6.79),

$$C_m = \sum_{k=1}^{N} \frac{\zeta_k^m - \zeta_k^{*m}}{m} + \frac{1}{2\pi i}\int_{-\infty}^{\infty} \xi^{m-1} \ln aa^* \, d\xi \quad , \tag{6.87}$$

and we have the Hamiltonian expressed in terms of the action variables ($\pi^{-1} \ln aa^*$, $2i\zeta_k$, $2i\zeta_k^*$) to which correspond the angle variables ($\ln b(\xi)$, $\ln b_k$, $\ln b_k^*$). Indeed we note, if $\Omega(\zeta) = a_m\zeta^m$, then

$$H_m = -4ia_m C_{m+1} = -4ia_m \sum_1^N \frac{\zeta_k^{m+1}-\zeta_k^{*m+1}}{m+1} - \frac{2a_m}{\pi}\int_{-\infty}^{\infty} \xi^m \ln aa^* \, d\xi \quad . \tag{6.88}$$

Hence, since Hamilton's equations are preserved,

$$\frac{d \ln b_k}{dt} = \frac{\delta H_m}{\delta(2i\zeta_k)} = -2a_m\zeta_k^m = -2\Omega(\zeta_k)$$

$$\frac{d \ln b_k^*}{dt} = \frac{\delta H_m}{\delta(2i\zeta_k^*)} = 2a_m\zeta_k^{*m} = 2\Omega(\zeta_k^*)$$

$$\frac{d \ln(\xi)}{dt} = \frac{\delta H_m}{\delta(\pi^{-1} \ln aa^*)} = -2a_m\xi^m = -2\Omega(\xi)$$

and

$$\frac{d}{dt} 2i\zeta_k = \frac{d}{dt} 2i\zeta_k^* = \frac{d}{dt}\frac{1}{\pi} \ln aa^* = 0 \quad . \tag{6.89}$$

As an example, take $\Omega = -2i\zeta^2$ and $r = -q^*$. Then (6.81) reduces to the nonlinear Schrödinger equation $q_t - i(q_{xx} + 2q^2q^*) = 0$ with

$$H = -i\int_{-\infty}^{\infty} (qq_{xx}^* + q^2q^{*2})dx$$

$$= -8\sum_1^N \frac{\zeta_k^3-\zeta_k^{*3}}{3} - \frac{4}{\pi i}\int_{-\infty}^{\infty} \xi^2 \ln aa^* \, d\xi \quad .$$

*Remark 1.* Linear Limit. Suppose q and r are small; then, as we have shown [6.63] there are no bound states and

$$C_m = \frac{1}{(2i)^m}\int_{-\infty}^{\infty} q \frac{\partial^{m-1}}{\partial x^{m-1}} r \, dx \quad . \tag{6.90}$$

In this case (6.87) (with r not necessarily $-q^*$) becomes

$$\frac{1}{(2i)^m}\int_{-\infty}^{\infty} q \frac{\partial^{m-1}}{\partial x^{m-1}} r \, dx = \frac{1}{2\pi i}\int_{-\infty}^{\infty} \xi^{m-1} b\bar{b} \, d\xi \quad . \tag{6.91}$$

*Remark 2.* The flow for $H = \alpha\zeta_k$. If the Hamiltonian of the system is proportional to the eigenvalue $\zeta_k$, then since $\delta a(\zeta_k,t) = 0$, we have from (6.32a) that

$$\delta\zeta_k = \frac{1}{a_k'}\int_{-\infty}^{\infty} (\delta q\phi_2\psi_2 - \delta r\phi_1\psi_1)_{\zeta_k} \, dx \quad . \tag{6.92}$$

Thus

$$\begin{bmatrix} r_t \\ -q_t \end{bmatrix} = \frac{\alpha}{a_k'}\begin{bmatrix} \phi_2\psi_2 \\ -\phi_1\psi_1 \end{bmatrix}_{\zeta_k} = \alpha\beta_k\begin{bmatrix} \phi_2^2 \\ -\phi_1^2 \end{bmatrix}_{\zeta_k} = \alpha\beta_k\Psi_k^A \quad . \tag{6.93}$$

Hence equating coefficients in (6.55), (let us assume again for simplicity that $r = -q^*$, $\zeta_k = i\eta_k$, $\alpha = 4i\bar{\alpha}$)

$$\left(\frac{b^*}{a}\right)_t = \zeta_{jt} = \beta_{jt} = 0 , \quad j \neq k$$

$$\zeta_{kt} = 0 , \ \beta_{kt} = 2\bar{\alpha}\beta_k . \tag{6.94}$$

The flow is a simple one. Imagine the initial data to be decomposed into its soliton and radiation components. Then the soliton corresponding to $\eta_k$ simply translates at constant speed and everything else stays the same.

## 6.7 Systems with Two Dispersion Relations

In Sects.6.5,6 we discussed evolution equations in which the two dependent variables were carried by the same dispersion relation and we saw that this led to systems which had an infinite sequence of conservation laws.

In this section, we are going to discuss the effect of q and r being carried by two *different* dispersion relations.

To begin, let us arbitrarily prescribe the motion of the scattering data $S_-$ to be:

$$\delta\left(\frac{\bar{b}}{a}\right) = 2M\,\frac{\bar{b}}{a} , \quad \delta\left(\frac{b}{\bar{a}}\right) = -2\bar{M}\,\frac{b}{\bar{a}} , \tag{6.95}$$

$$\delta\beta_k = 2m_k\beta_k , \quad \delta\zeta_k = 2Z_k , \quad k = 1, \ldots, N$$

$$\delta\bar{\beta}_k = -2\bar{m}_k\bar{\beta}_k , \quad \delta\bar{\zeta}_k = -2\bar{Z}_k , \quad k = 1, \ldots, \bar{N} . \tag{6.96}$$

Then, from (6.55)

$$\begin{bmatrix}\delta r\\ -\delta q\end{bmatrix} = -\frac{2}{\pi}\int_{-\infty}^{\infty}\left(M\,\frac{\bar{b}}{a}\,\Psi^A + \bar{M}\,\frac{b}{\bar{a}}\,\bar{\Psi}^A\right)d\zeta + 4i\sum_{k=1}^{N}\left(m_k\beta_k\Psi_k^A + \beta_k Z_k\chi_k^A\right)$$
$$- 4i\sum_{1}^{\bar{N}}\left(\bar{m}_k\bar{\beta}_k\bar{\Psi}_k^A + \bar{\beta}_k\bar{Z}_k\bar{\chi}_k^A\right) . \tag{6.97}$$

Now we ask, is it possible to act on the vector $\begin{bmatrix}r\\ q\end{bmatrix}$ with an operator $\Omega$ such that

$$\Omega\Psi^A = -2M\Psi^A \quad \text{and} \quad \Omega\bar{\Psi}^A = -2\bar{M}\bar{\Psi}^A , \quad M \neq \bar{M}$$

and still find equations of practical or theoretical interest? It is convenient to introduce the functions $\tilde{\Psi}$, $\tilde{\Psi}^A$ which are defined as follows,

$$\tilde{\Psi}(x,\zeta) = \begin{bmatrix}\psi_1\bar{\psi}_1\\ \\ \psi_1\bar{\psi}_1\end{bmatrix} , \quad \tilde{\Psi}^A(x,\zeta) = \begin{bmatrix}\phi_2\bar{\phi}_2\\ \\ -\phi_1\bar{\phi}_1\end{bmatrix} . \tag{6.98}$$

We may show (for convenience in notation we assume q and r to be on compact support and hence the analyticity of $\bar{b}/a$ and $b/\bar{a}$),

$$\tilde{\Psi}^A(x,\eta) = -\frac{1}{2\pi i}\oint \frac{\bar{b}}{a}\frac{1}{\zeta-\eta}\Psi^A d\zeta - \frac{1}{2\pi i}\oint \frac{b}{\bar{a}}\frac{1}{\zeta-\eta}\bar{\Psi}^A d\zeta \tag{6.99}$$

where the contours C, $\bar{C}$ are defined in Sect.6.4. $\tilde{\Psi}^A(x,\eta)$ is defined for real $\eta$ and at the eigenvalues $\zeta_k$, $\bar{\zeta}_k$ but nowhere else. In order for $\tilde{\Psi}^A$ to tend to zero as $x\to\pm\infty$ for real $\eta$, we must require $b(\eta) = \bar{b}(\eta) = 0$. Now let us consider the weighted average of $\tilde{\Psi}^A(x,\eta)$ over real $\eta$

$$\int_{-\infty}^{\infty} g(\eta)\tilde{\Psi}^A(x,\eta)d\eta \quad , \tag{6.100}$$

where $g(\eta)$ is analytic in a strip including the real axis and admits Wiener-Hopf decomposition

$$g(\eta) = \frac{2}{\pi}[\bar{M}(\eta) - M(\eta)] \tag{6.101}$$

where $M(\bar{M})$ is analytic for $\mathrm{Im}\{\eta\} > 0$ ($\mathrm{Im}\{\eta\} < 0$). For example, given $g(\eta)$ analytic in the neighborhood of $\mathrm{Im}\{\eta\} = 0$, we may write

$$M(\eta) = -\frac{i}{4}\int_{C_u}\frac{g(\alpha)d\alpha}{\eta-\alpha} \quad , \quad \bar{M}(\eta) = -\frac{i}{4}\int_{C_A}\frac{g(\alpha)d\alpha}{\eta-\alpha} \tag{6.102}$$

where $C_u(C_A)$ are contours along the real axis going under (over) the pole at $\alpha = \eta$. As a particular and useful choice of $g(\eta)$ consider the Lorentzian linewidth,

$$g(\eta) = \frac{\beta}{\pi(\eta^2+\beta^2)} = \frac{2}{\pi}\left[\left(-\frac{i}{4}\right)\frac{1}{\eta-i\beta} - \left(-\frac{i}{4}\right)\frac{1}{\eta+i\beta}\right] \quad . \tag{6.103}$$

Using (6.99),

$$\begin{aligned}\int_{-\infty}^{\infty} g(\eta)\tilde{\Psi}^A(x,\eta)d\eta &= -\frac{2}{\pi}\oint\frac{\bar{b}}{a}\Psi^A\frac{1}{2\pi i}\int_{-\infty}^{\infty}\frac{\bar{M}-M}{\zeta-\eta}d\eta d\zeta \\ &\quad -\frac{2}{\pi}\oint\frac{-b}{\bar{a}}\bar{\Psi}^A\frac{1}{2\pi i}\int_{-\infty}^{\infty}\frac{\bar{M}-M}{\zeta-\eta}d\eta d\zeta \\ &= -\frac{2}{\pi}\oint M(\zeta)\frac{\bar{b}}{a}\Psi^A d\zeta - \frac{2}{\pi}\oint \bar{M}(\zeta)\frac{b}{\bar{a}}\bar{\Psi}^A d\zeta \quad ,\end{aligned} \tag{6.104}$$

where we closed the $\eta$ integrations in the upper (lower) half plane in those terms containing $M(\bar{M})$. Recall in the first term $\mathrm{Im}\{\zeta\} \geq 0$, in the second term $\mathrm{Im}\{\zeta\} \leq 0$.

Now, we see that (6.104) is simply (6.97) if we choose

$$m_k = M(\zeta_k) \quad , \quad \bar{m}_k = \bar{M}(\bar{\zeta}_k) \quad , \quad z_k = \bar{z}_k = 0 \quad . \tag{6.105}$$

Hence the system consisting of the equation

$$\begin{bmatrix} r_t \\ -q_t \end{bmatrix} = \int_{-\infty}^{\infty} g(\eta)\tilde{\Psi}^A(x,\eta)d\eta \quad , \tag{6.106}$$

and the equation which relates $\tilde{\Psi}^A$ back to r and q,

$$(L^A - \eta)\tilde{\Psi}^A = \frac{i}{2}\begin{bmatrix} r \\ q \end{bmatrix} \quad , \tag{6.107}$$

transforms under IST to the trivially integrable system (6.96). Note, since $\bar{b}$ is associated with r, b with q, $M(\zeta)$ carries the evolution of q and $\bar{M}(\zeta)$ of r.

## 6.8 Coherent Pulse Propagation

In order to illustrate the power of the method of IST in concrete terms, we will now consider a specific and important application. For in-depth discussions of the phenomenon briefly reviewed here we suggest [6.29,30]. Let $r = \alpha q^*$, $\alpha$ real, $q = E/2$, $2\phi_1\bar{\phi}_1 = \lambda$, $N = \phi_1\bar{\phi}_2 + \bar{\phi}_1\phi_2$. Then $2\phi_2\bar{\phi}_2 = \alpha\lambda^*$ and (6.106) becomes

$$E_t(x,t) = \int_{-\infty}^{\infty} g(\eta)\lambda(x,t,\eta)d\eta \quad , \tag{6.108}$$

which is the averaged Maxwell equation describing the change in the envelope of the electric field E(x,t) as it travels through a resonant, inhomogeneously broadened (due to Doppler shifting), nondegenerate medium of two-level atoms with energy level difference close [to within the width of $g(\eta)$] to the frequency of the carrier wave underlying the envelope E. Here t measures the distance X along the medium and x is T-X. The quantities $\lambda(x,t;\eta)$ and $N(x,t;\eta)$ satisfy the equations

$$\lambda_x + 2i\eta\lambda = EN \quad , \quad N_x = \alpha/2(E\lambda^* + E^*\lambda) \tag{6.109}$$

which for the case $\alpha = -1$ are the well-known Bloch equations [6.30] describing the influence of the electric field envelope E(x,t) on the polarization $\lambda(x,t,\eta)$ and population density $N(x,t,\eta)$. In fact, $N = (n_E - n_G)/(n_E + n_G)$ where $n_E(n_G)$ is the number of atoms in the excited (ground) state. As $x \to -\infty$, the boundary conditions are prescribed to be $\lambda = 0$ and $N = -1$ which simply means that the medium is in the ground state before the pulse arrives.

The dynamics of the system are shown in Fig.6.1. The initial pulse is decomposed into solitons which correspond to the discrete eigenvalues $\zeta = i\eta$ ($2\pi$ pulses) and the eigenvalue pairs $\zeta$, $-\zeta^*$ ($0\pi$ pulses) and radiation [measured by $b^*(\zeta,0)/a(\zeta,0)$]. These quantities can be found simply by solving (6.16) for $-r^* = q = E/2$, where $E(x,0)$ [$=E(0,T)$] is given. The solitons propagate through the medium in a lossless coherent fashion whereas the radiation is all eventually absorbed (Beer's law). However, the system does admit some very interesting properties, some of which are worth describing in more detail. Remarks 1, 2 and 3 pertain to the case $\alpha = -1$ whereas Remark 4 pertains to the case $\alpha = +1$.

*Remark 1.* There are no conserved quantities other than the spectrum. Indeed, a check of the time behavior of $\ln a(\zeta,t)$ shows that

$$\frac{\partial}{\partial t} \ln a(\zeta,t) = \frac{i}{4} \int_{-\infty}^{\infty} g(\eta) \frac{2bb^*}{\zeta-\eta} d\eta \ , \quad \mathrm{Im}\{\zeta\} > 0 \quad . \tag{6.110}$$

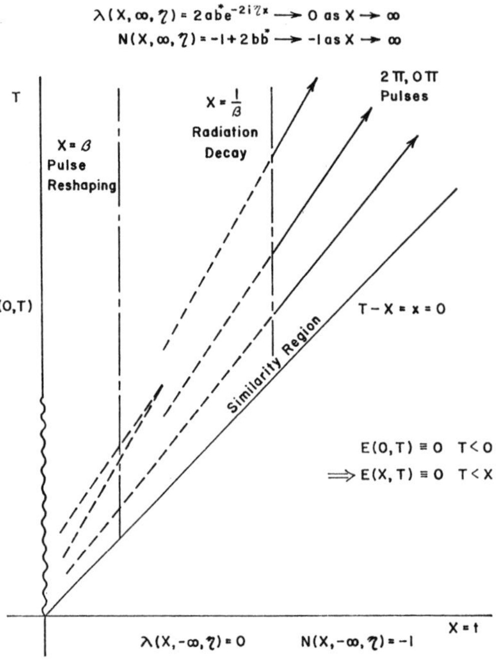


Fig.6.1. X - T diagram showing the propagation resulting from a pulse E(0,T) arriving at X = 0 at time T = 0

In particular, using (6.25) we see that

$$\frac{\partial}{\partial t} C_1 = - \frac{1}{8i} \frac{\partial}{\partial t} \int_{-\infty}^{\infty} EE^* dx = - \frac{i}{4} \int_{-\infty}^{\infty} g(\eta)(N+1)_{x=+\infty} d\eta$$

or

$$\frac{\partial}{\partial t} \int_{-\infty}^{\infty} EE^* dx = -4 \int_{-\infty}^{\infty} g(\eta) bb^* d\eta \ , \quad \text{since} \quad N|_{+\infty} = -1 + 2bb^* \ , \tag{6.111}$$

which in general is not zero. This result can also be derived directly from (6.108, 109). The reason that the energy is lost is that some of it (the radiation portion) is absorbed by the medium. The absorbtion is due to phase mixing and depends on the fact that $g(\eta) = 2/\pi\ (\bar{M} - M)$ is a smooth nonzero function which points out that the difference in the dispersion relations has a very real effect.

*Remark 2.* The generalized McCall-Hahn area theorem. If we take $g(\eta)$ to be given by the Lorentzian line width $\beta/[\pi(\eta^2 + \beta^2)]$ , we have [from integrating (6.95)] that

$$\frac{b^*(\zeta,t)}{a(\zeta,t)} = \frac{b^*(\zeta,0)}{a(\zeta,0)} \exp\left[\frac{-(\beta+i\zeta)t}{2(\zeta^2+\beta^2)}\right] \ . \tag{6.112}$$

For $\zeta \gg \beta$ the reflection coefficient decays and energy is lost to the medium in a distance proportional to $1/\beta$ (Beer's law). On the other hand for $\zeta = 0$,

$$\frac{b^*(0,t)}{a(0,t)} = \frac{b^*(0,0)}{a(0,0)} e^{-t/2\beta} \tag{6.113}$$

and we find that $b^*(0,t)/a(0,t)$ tends to zero in a distance proportional to $\beta$, the inhomogeneous broadening width. For real $E$, whence $b^*(0,t) = b(0,t)$, we may show that $b^*/a$ is simply $\tan A(t)/2$ where $A(t)$ is the (negative) area of the electric pulse $-\int_{-\infty}^{\infty} E(x,t)dx$. [This is easy to see, for at $\zeta = 0$ if $E/2 = q = -r = -u_x/2$, $\phi_1 = \cos u/2$, $\phi_2 = \sin u/2$; hence $b^*(0,t)/a(0,t) = b(0,t)/a(0,t) = \tan u/2(+\infty)$ where $u(x) = -\int_{-\infty}^{\infty} E\, dx$.] Equation (6.113) shows that the area $\int_{-\infty}^{\infty} E\, dx$ tends to a multiple of $2\pi$ in a distance $\beta$. If $\beta < 0$ [or Re $M(0) > 0$], corresponding to the system being initially in an excited state, the area tends to an odd multiple of $\pi$. A generalization of this theorem to higher nonlinear moments can be given.

From these two remarks, we can piece together the structure of the general solution. From Fig.2.1, we see that the pulse arrives at $X = 0$ at $T = 0$, and assuming $\beta < 1$, the first thing that happens is that the pulse is reshaped so that its area is a multiple of $2\pi$. This does not mean that the radiation component is lost, but simply that the total area of the soliton and radiation field is $2\pi n$. The next thing to occur is that the pulse separates out into a number of $2\pi$ and $0\pi$ pulses (if $E$ is real), and a radiation component which is absorbed in the Beer's length $1/\beta$. After a distance $X = t \sim 1/\beta$, the radiation is all absorbed but the $2\pi$ and $0\pi$ pulses continue to propagate in a coherent and lossless fashion.

*Remark 3.* The sine-Gordon limit. If we allow $\beta$ to approach zero, the two dispersion relations $M$ and $\bar{M}$ approach $-i/4\zeta$ and $g(\eta) \to \delta(\eta)$, the Dirac delta function. Formally, $\lambda(\eta = 0) = 2\phi_1\bar{\phi}_1\big|_{\eta=0} = \sin u$ and since $E = -u_x$, the Maxwell equation (6.108) becomes simply

$$u_{xt} = -\sin u \quad , \tag{6.114}$$

the Goursat version of the sine-Gordon equation with data supplied at $t = 0$ and $x = -\infty$. But as $\beta \to 0$, we see from (6.113), that $b(0) \to 0$ for any $t > 0$, and the pulse is reshaped instantaneously. What this says is that only a subclass of the class of initial data acceptable for the coherent pulse propagation problem is acceptable for the Goursat version of the sine-Gordon equation. A necessary condition is that $b(0,0) = 0$ [or $u(\pm\infty,0) = 2n\pi$]. A sufficient condition [6.65] on the initial data in order to guarantee the integrability of (6.114) by IST is that $b(\zeta,0) = O(\zeta^{\frac{1}{2}})$ as $\zeta \to 0$. We see then that there can be radiation in the initial data but it must be associated with wave numbers other than $\eta = 0$. Furthermore, in the sine-Gordon limit, the radiation never decays (there is no absorption) but rather disperses and travels together with the solitons.

Finally, we remark that if $\bar{M} = M = -1/4(\zeta - \hat{\zeta})$, $\hat{\zeta}$ real, then it is necessary that $b(\hat{\zeta}) = \bar{b}(\hat{\zeta}) = 0$ in order that the Goursat problem be well posed.

*Remark 4.* An operator theoretic viewpoint. FLASCHKA and NEWELL [6.63] have shown how the Lax equation $L_t = [B,L]$ transforms under IST to the equation

$$\hat{S}_t = T_+\hat{S} - \hat{S}T_- \tag{6.115}$$

for the scattering matrix $\hat{S}$. Since the Lax formalism requires L to be self-adjoint and B to be skew-adjoint we take $r = +q^*$ or $\alpha = +1$ whence (6.16) is self-adjoint. We find [using (6.95) to compute the time behavior of $b^*/a$, etc.],

$$\hat{S} = \begin{bmatrix} \frac{1}{a}, & -\frac{b^*}{a} \\ \frac{b}{a}, & \frac{1}{a} \end{bmatrix}, \quad T_+ = \begin{bmatrix} -\frac{i}{4}\left\langle\frac{1+bb^*}{\zeta\eta}\right\rangle + \frac{\pi}{4}gb^* & \\ -\frac{\pi}{4}gb & \frac{i}{4}\left\langle\frac{1+bb^*}{\zeta\eta}\right\rangle \end{bmatrix}, \quad T_- = \begin{bmatrix} -\frac{i}{4}\left\langle\frac{1+3bb^*}{\zeta\eta}\right\rangle & -\frac{\pi}{4}b^*g \\ \frac{\pi}{4}gb & \frac{i}{4}\left\langle\frac{1-bb^*}{\zeta\eta}\right\rangle \end{bmatrix}$$

where $\langle f\rangle = P\int_{-\infty}^{\infty} g(\eta)f(\eta)d\eta$. Now, if $T_+ = T_- = T$, then (6.115) is simply $\hat{S}_t = [T,\hat{S}]$ from which we can deduce there is a unitary U such that $U_t = TU$ and $\hat{S}(t)U = U\hat{S}(0)$. Hence we can easily identify many quantities (such as det $\hat{S}$) which are conserved by the motion. In general, however, $T_+ \neq T_-$ and only the spectrum is conserved and the principal reason for this is that $g(\zeta) \not\equiv 0$. Now if $g(\zeta) \equiv 0$ (a.e.) e.g., $g(\zeta) = \delta(\zeta)$, then, since we must have $b(0) = 0$ because $M = \bar{M} = -i/4\zeta$ has a pole at $\zeta = 0$, we find,

$$T = T_+ = T_- = \begin{bmatrix} -\frac{i}{4\zeta} & 0 \\ 0 & \frac{i}{4\zeta} \end{bmatrix} .$$

In this limit, the system (6.108,109) relaxes to the sinh-Gordon equation

$$u_{xt} = -\sinh u \ , \ \left[E(x,t) = -u_x\right] .$$

## 6.9 Moving Eigenvalues

The feature common to the classes of evolution equations discussed heretofore is the invariance of the discrete spectrum. However, there is nothing in the general theory which prohibits their motion. Indeed, it is clear that practically every operator equation of the form

$$\begin{bmatrix} r_t \\ -q_t \end{bmatrix} + \Omega\begin{bmatrix} r \\ q \end{bmatrix} = 0 \tag{6.116}$$

will transform, by IST, to an equivalent mechanical system in scattering space in which all the normal modes are strongly coupled and in which the eigenvalues move. Because of the strong coupling between the normal modes, the canonical transformation to the scattering data is not directly useful. What we want to do here (and this material is also presented in [6.59]) is to develop a set of operators $\Omega$ for which the eigenvalues move and for which the equations in scattering space are still separable.

It is desirable then that the operator $\Omega$ act on each $\zeta$ component in the expression $\begin{bmatrix} r \\ q \end{bmatrix}$ separately and in addition act on $\Psi_k^A$ for some k in such a way that

$$\Omega\Psi_k^A = \alpha\Psi_k^A + \chi_k^A \quad , \tag{6.117}$$

for then there will be a term in the expansion for $\Omega\begin{bmatrix} r \\ q \end{bmatrix}$ to balance the coefficient $\zeta_{kt}$ of $\chi_k^A$ in the expansion for $\begin{bmatrix} r_t \\ -q_t \end{bmatrix}$. Since we know already that

$$(L^A - \zeta_k)\chi_k^A = \Psi_k^A \quad , \tag{6.118}$$

one possible choice is the operator $(L^A - \zeta_k)^{-1}$. But from the definition (6.98) of $\tilde{\Psi}^A(x,\eta)$, we know that

$$(L^A - \zeta_k)\tilde{\Psi}^A(x,\zeta_k) = \frac{i}{2}\begin{bmatrix} r \\ q \end{bmatrix} \quad . \tag{6.119}$$

Recall that $\tilde{\Psi}^A(x,\eta)$ is defined for real $\zeta$ or at an eigenvalue. Thus,

$$(L^A - \zeta_k)^{-1}\begin{bmatrix} r \\ q \end{bmatrix} = -2i\tilde{\Psi}^A(x,\zeta_k) \quad . \tag{6.120}$$

Indeed, one directly sees from (6.99) that if $\eta = \zeta_k$, then in writing the contour integral along C as an integral along real $\eta$ plus the contributions from the poles, the residue at $\zeta = \zeta_k$ will contain both $\Psi^A(x,\zeta_k)$ and $\frac{\partial}{\partial\zeta}\Psi^A(x,\zeta_k) = \chi_k^A$ since $a(\zeta)(\zeta - \zeta_k)$ has a double zero at $\zeta = \zeta_k$.

From these observations, we find that the equation

$$\begin{bmatrix} r_t \\ -q_t \end{bmatrix} = 4i\Omega_j\tilde{\Psi}^A(x,\zeta_j) + 4i\bar{\Omega}_j\tilde{\Psi}^A(x,\bar{\zeta}_j) \tag{6.121}$$

corresponds to a flow in scattering space

$$\left(\frac{\bar{b}}{a}\right)_t = 2M(\zeta)\,\frac{\bar{b}}{a} \quad , \quad \left(\frac{b}{\bar{a}}\right)_t = -2M(\zeta)\,\frac{b}{\bar{a}} \quad ,$$

$$\beta_{k,t} = 2M(\zeta_k)\beta_k \quad , \quad \bar{\beta}_{k,t} = -2M(\bar{\zeta}_k)\bar{\beta}_k \quad , \quad \zeta_{k,t} = \bar{\zeta}_{k,t} = 0 \qquad k \neq j$$

$$\zeta_{j,t} = 2\Omega_j \quad , \quad \beta_{j,t} = \frac{2\bar{\Omega}_j\beta_j}{\zeta_j - \bar{\zeta}_j} + 2\Omega_j\beta_j' \quad ,$$

$$\bar{\zeta}_{j,t} = -2\bar{\Omega}_j \quad , \quad \bar{\beta}_{k,t} = \frac{2\Omega_j\bar{\beta}_j}{\zeta_j - \bar{\zeta}_j} - 2\bar{\Omega}_j\beta_j' \quad , \tag{6.122}$$

where

$$M(\zeta) = \frac{\Omega_j}{\zeta - \zeta_j} + \frac{\bar{\Omega}_j}{\zeta - \bar{\zeta}_j} \quad .$$

First, let us observe that the motion of $\beta_j$ (which tells the position of the soliton) may be written

$$\frac{\partial}{\partial t}\,\beta[\zeta_j(t),t] = \frac{2\bar{\Omega}_j}{\zeta_j - \bar{\zeta}_j}\,\beta[\zeta_j(t),t]$$

where

$$\frac{\partial}{\partial t}\beta[\zeta_j(t),t] = \frac{d}{dt}\beta[\zeta_j(t),t] - \zeta_{j,t}\left(\frac{\partial\beta}{\partial\zeta}\right)_{\zeta_j} \quad ; \tag{6.123}$$

this means that the motion of $\beta_j(\bar{\beta}_j)$ only depends on $\bar{\Omega}_j(\Omega_j)$ and on $\zeta_j(t)[\bar{\zeta}_j(t)]$. Furthermore the whole flow (6.122) is open to a much more simple interpretation.

Consider the time dependency of $a(\zeta,t)$. For simplicity, we will take $r = -q^*$, $\bar{\Omega}_j = -\Omega_j^*$, $\bar{\zeta}_k = \zeta_k^*$ (all k). Then, from (6.32), we can show that

$$a(\zeta,t) = a(\zeta,0)\,\frac{\zeta-\zeta_j(t)}{\zeta-\zeta_j(0)}\,\frac{\zeta-\zeta_j^*(0)}{\zeta-\zeta_j^*(t)} \quad , \tag{6.124}$$

which simply says that the $[\zeta - \zeta_j(0)/(\zeta - \zeta_j^*(0)]$ term in the expression (6.86) for $a(\zeta,0)$

$$a(\zeta,0\ ) = \prod \frac{\zeta-\zeta_k(0)}{\zeta-\zeta_k^*(0)}\,\exp\left(\frac{1}{2\pi i}\int_{-\infty}^{\infty}\frac{\ln aa^*}{\zeta-\zeta}\right)d\xi \tag{6.125}$$

is replaced by $[\zeta - \zeta_j(t)/(\zeta - \zeta_j^*(t)]$. We can also show that $\frac{\partial}{\partial t}b(\zeta,t) = 0$ and furthermore that $b_j[\zeta_j(t),t] = b_j$, a constant. Thus the only elements of the scattering data which move are $\zeta_j$ and $\zeta_j^*$ and their motion is prescribed by coefficients in the evolution equation (6.121).

From (6.32), we see using $\phi_k = b_k\psi_k$ and the fact that $b_k b_k^* = 1$, that (6.121) may be written

$$\begin{bmatrix} r_t \\ -q_t \end{bmatrix} = \mathrm{grad}_{q,r}\, H \quad , \tag{6.126}$$

with

$$H = 4i\Omega_j \ln b_j + 4i\Omega_j^* \ln b_j^* \tag{6.127}$$

where we understand $b_j$ and $b_j^*$ to be functionals of r and q whose variations with respect to these quantities are given by (6.32).

The result of all this is that all the scattering functions $2i\zeta_k$, $2i\zeta_k^*$, $k \neq j$, $\ln b_k$, $\ln b_k^*$, $\pi^{-1}\ln aa^*$, $\ln b(\zeta)$ which play the role of action-angle variables are now constants of the motion. The action-angle pairs $(2i\zeta_j, \ln b_j)$ and $(2i\zeta_j^*, \ln b_j^*)$, relating to the eigenvalue $\zeta_j$ switch roles, with $\ln b_j$ $(\ln b_j^*)$ playing the role of the *action* variable and $2i\zeta_j$ $(2i\zeta_j^*)$ the *angle* variable; whose motion is given by,

$$\frac{\partial}{\partial t}2i\zeta_j = \frac{\delta H}{\delta \ln b_j} = 4i\Omega_j \quad , \quad \frac{\partial}{\partial t}2i\zeta_j^* = \frac{\delta H}{\delta \ln b_j^*} = 4i\Omega_j^* \quad . \tag{6.128}$$

Again we emphasize that the motion of the data $E_+$ or $E_-$ are produced only be dependence on $\zeta_j(t)$ or $\zeta_j^*(t)$.

For example, if $r = -q^*$ and q is real, then if

$$M(\zeta) = \frac{1}{2}\left[\frac{i\hat{\eta}_t(t)}{\zeta-i\hat{\eta}(t)} + \frac{i\hat{\eta}_t(t)}{\zeta+i\hat{\eta}(t)}\right] , \tag{6.129}$$

the corresponding evolution equation is

$$r_{xxt} + (2rR_t)_x - 4\hat{\eta}^2(t)r_t = 4\hat{\eta}_t(t)r_x$$

$$R = \int_{-\infty}^{x} r^2 dy \quad . \tag{6.130}$$

If one of the eigenvalues associated with r(x,0) corresponds with $\hat{\eta}(0)$, then the subsequent behavior of the associated soliton is

$$r(x,t) = 2\hat{\eta}(t)\mathrm{sech}\left[2\hat{\eta}(t)x + \ln\left(\frac{\hat{\eta}(t)}{\hat{\eta}(0)}\right)\right] \quad . \tag{6.131}$$

Questions associated with nonuniqueness of the solution are discussed in [6.59].

We may combine the general features of Sects.6.5,7-9 and write the most general evolution equations which may be solved by IST [using (6.16) as the eigenvalue problem] in the form

$$\begin{bmatrix} \delta r \\ -\delta q \end{bmatrix} + 2\Omega(L^A)\begin{bmatrix} r \\ q \end{bmatrix}$$

$$= \int_{-\infty}^{\infty} g(\eta)\tilde{\Psi}^A(x,\eta)d\eta + \sum_1^N g_j\tilde{\Psi}^A(x,\zeta_j) + \sum_1^{\bar{N}} \bar{g}_j\bar{\tilde{\Psi}}^A(x,\bar{\zeta}_j) \quad . \tag{6.132}$$

## 6.10 The Sine-Gordon Equation

When we set r = -q ab initio, the variations in r and q are no longer independent and a different Hamiltonian formulation is required. We will assume that r and q are both real, in which case the following relations hold

$$\bar{\phi}(\zeta) = S\phi^*(\zeta^*) = S\phi(-\zeta), \; \bar{\psi}(\zeta) = S\psi^*(\zeta^*) = S\psi(-\zeta) \quad ,$$

$$S = \begin{bmatrix} 0 & 1 \\ -1 & 0 \end{bmatrix} \; , \; \bar{b}(\zeta) = b^*(\zeta^*) = b(-\zeta), \; \bar{a}(\zeta) = a^*(\zeta^*) = a(-\zeta)$$

$$\bar{\zeta}_k = -\zeta_k \; , \; \bar{a}'_k = \frac{\partial}{\partial\zeta}\bar{a}(\zeta)\Big|_{\bar{\zeta}_k} = -a'_k = -\frac{\partial}{\partial\zeta}a(\zeta)\Big|_{\zeta_k} \; , \; \bar{\beta}_k = -\beta_k, \bar{\gamma}_k = -\gamma_k \quad . \tag{6.133}$$

Then expressions (6.57,58) give

$$\delta q = -\frac{1}{\pi}\int_{-\infty}^{\infty} \delta\left(\frac{b}{a}\right)\Psi^-(\zeta)d\zeta + 2i\sum\left(\delta\gamma_k\Psi^-_k + \gamma_k\delta\zeta_k\tau^-_k\right) \quad , \tag{6.134}$$

$$q = -\frac{1}{\pi}\int_{-\infty}^{\infty} \frac{b}{a}\Psi^+(\zeta)d\zeta + 2i\sum\gamma_k\Psi^+_k \quad , \tag{6.135}$$

where

$$\Psi^-(\zeta) = \psi_2^2 - \psi_1^2 \; , \; \Psi^-_k = \left(\psi_2^2 - \psi_1^2\right)_k \; , \; \tau^-_k = \frac{\partial}{\partial\zeta}\left(\psi_2^2 - \psi_1^2\right)_{\zeta_k}$$

$$\Psi^+(\zeta) = \psi_2^2 + \psi_1^2 \; , \; \psi_k^+ = \left(\psi_2^2 + \psi_1^2\right)_k \quad . \tag{6.136}$$

The function $\Psi^-(\zeta)$ satisfies

$$L_F\psi^-(\zeta) = \left(-\frac{1}{4}\frac{\partial^2}{\partial x^2} - q^2 + q_x\int_x^\infty dy\ q\right)\Psi^-(\zeta) = \zeta^2\Psi^-(\zeta) \qquad (6.137)$$

and

$$\Psi_x^+(\zeta) = 2i\zeta\Psi^-(\zeta) \quad . \qquad (6.138)$$

The dual expansions (6.55,56) reduce to

$$\delta q = \frac{1}{\pi}\int_{-\infty}^{\infty}\delta\frac{\bar{b}}{a}\Phi^-(\zeta)d\zeta - 2i\sum(\delta\beta_k\Phi_k^- + \beta_k\delta\zeta_k\chi_k^-) \quad , \qquad (6.139)$$

and

$$q = -\frac{1}{\pi}\int_{-\infty}^{\infty}\frac{\bar{b}}{a}\Phi^+(\zeta)d\zeta + 2i\sum\beta_k\Phi_k^+ \qquad (6.140)$$

where

$$\Phi^-(\zeta) = \phi_2^2 - \phi_1^2 \quad , \quad \Phi^+(\zeta) = \phi_2^2 + \phi_1^2 \quad , \quad \Phi_k^- = \left(\phi_2^2 - \phi_1^2\right)_{\zeta_k}$$

$$\Phi_k^+ = \left(\phi_2^2 + \phi_1^2\right)_{\zeta_k} \quad , \quad \chi_k^- = \frac{\partial}{\partial\zeta}\left(\phi_2^2 - \phi_1^2\right)\Big|_{\zeta_k} \quad . \qquad (6.141)$$

The functions $\Phi^-(\zeta)$ and $\Phi^+(\zeta)$ satisfy

$$L_F^A\Phi^+(\zeta) = \left(-\frac{1}{4}\frac{\partial^2}{\partial x^2} - q^2 + q\int_{-\infty}^{x}dy\ q_y\right)\Phi^+(\zeta) = \zeta^2\Phi^+(\zeta)$$

and

$$\Phi_x^+(\zeta) = 2i\zeta\Phi^-(\zeta) \quad . \qquad (6.142)$$

It is easy to see from (6.139,140,142) that, for an entire function $F(\zeta^2)$, the evolution equation

$$q_t = i\frac{\partial}{\partial x}F(L_F^A)q \qquad (6.143)$$

separates in scattering space to

$$\left(\frac{\bar{b}}{a}\right)_t = 2\zeta F(\zeta^2)\frac{\bar{b}}{a} \quad , \quad \beta_{kt} = 2\zeta_k F(\zeta_k^2)\beta_k\ \zeta, \quad \zeta_{kt} = 0, \quad k = 1, \ldots, N \quad . \qquad (6.144)$$

For example, if $F(\zeta^2) = -4i\zeta^2$, then (6.143) is the modified Korteweg-deVries equation

$$q_t + 6q^2q_x + q_{xxx} = 0 \quad .$$

On the other hand, we may also extend this equation to the case of singular $F(\zeta^2)$. If $F(\zeta^2) = i/4\zeta^2$, then we will find that

$$q_t = \frac{1}{2}(\phi_2\bar{\phi}_2 - \phi_1\bar{\phi}_1)_{\zeta=0} = -\frac{1}{2}\sin u \qquad (6.145)$$

where $u(x,t) = -2\int_{-\infty}^{x}q(x,t)dx$. Here we have used a number of formulae given in [6.49].

The orthogonality relations given in Appendix B may be used to show that the following simplectic form

$$\int_{-\infty}^{\infty} \left\{ \frac{1}{2}\, \delta q \wedge \int_{-\infty}^{x} \delta q \, dy \right\} dx \tag{6.146}$$

is preserved and equal to

$$\int_0^{\infty} \delta\left(\frac{\ln aa^*}{-2\xi\pi}\right) \wedge \delta \mathrm{Arg}\, b(\xi) d\xi + \sum \delta(-\ln\zeta_k) \wedge \delta \ln b_k \quad . \tag{6.147}$$

We can now read off the action-angle variables to be

$\dfrac{\ln aa^*}{-2\xi\pi}$ for $\xi$ real and positive, $-\ln\zeta_k$

and

$$\mathrm{Arg}\, b(\xi) \quad \text{and} \quad \ln b_k \quad , \tag{6.148}$$

respectively.

We have also shown elsewhere [6.63] that

$$iF(L_F^A)q = \frac{\delta H}{\delta q}$$

where if the *dispersion relation*

$$\Omega(\zeta) = \zeta F(\zeta^2) = \sum_0^{\infty} a_{2m+1}\zeta^{2m+1} \quad , \tag{6.149}$$

the *Hamiltonian* is

$$H_\Omega = \sum_0^{\infty} a_{2m+1} C_{2m+1} \tag{6.150}$$

with the $\{C_r\}_{r=1}^{\infty}$ given by (6.25).

For cases where the dispersion relation is singular, the Hamiltonian is simply $\ln a(\zeta)$ and its derivatives estimated at some $\hat{\zeta}$. For example, if $F(\zeta^2) = i/4\zeta^2$ we have seen that the sine-Gordon equation results. We may show that the Hamiltonian is simply

$$H = \frac{i}{4}\left(\frac{\partial}{\partial\zeta} \ln a\right)_{\zeta=0} \tag{6.151}$$

which from (6.21) is

$$H = -\frac{1}{4}\int_{-\infty}^{\infty} (\phi_1\bar{\phi}_2 + \bar{\phi}_1\phi_2 + 1)_{\zeta=0}\, dx$$

$$= -\frac{1}{4}\int_{-\infty}^{\infty} \left[1 - \cos\left(-2\int_{-\infty}^{x} q\, dy\right)\right] dx \quad . \tag{6.152}$$

Using (6.86), we may express the Hamiltonian in scattering space as

$$H = \sum_{\mathrm{Re}\{\zeta_k\}\neq 0} \left(-\frac{i}{2\zeta_k} + \frac{i}{2\zeta_k^*}\right) - \sum_{\eta_k \text{ real}} \frac{1}{2\eta_k} + \frac{1}{4\pi i}\int_0^{\infty} \frac{\ln aa^*}{\xi^2}\, d\xi \quad , \tag{6.153}$$

where we have recognized two types of discrete eigenvalues $\zeta_k$, $\mathrm{Re}\{\zeta_k\} \neq 0$ and $i\eta_k$. The reason we must do this is that the former come in pairs $(\zeta_k, -\zeta_k^*)$ and lead to the more complicated solitons known as breathers. It is now easy to verify that

$$\frac{d\ \ln b_k}{dt} = \frac{\delta H}{\delta(-\ln\zeta_k)} = \frac{-i}{2\zeta_k} \ ,$$

$$\frac{d\ \ln b(\zeta)}{dt} = \frac{\delta H}{\delta\left(-\dfrac{\ln aa^*}{2\pi\xi}\right)} = -\frac{1}{2\xi} \ ,$$

$$\frac{d(-\ln\zeta_k)}{dt} = \frac{d}{dt}\left(\frac{\ln aa^*}{-2\pi\xi}\right) = 0 \ , \tag{6.154}$$

which after using (6.133) are precisely in agreement with (6.144).

## 6.11 Schrödinger Equation

The evolution equations which can be associated with the Schrödinger equation can be found in a manner similar to the methods already used. Consider the decomposition of

$$v_{2xx} + \left[\zeta^2 + q(x,t)\right]v_2 = 0 \ , \qquad \int_{-\infty}^{\infty} (1 + |x| + |x|^2)|q(x,t)|dx < \infty \tag{6.155}$$

into the system

$$v_{1x} + i\zeta v_1 = q(x,t)v_2$$

$$v_{2x} - i\zeta v_2 = -v_1 \ . \tag{6.156}$$

Define the solutions $\phi(x,t;\zeta)$, $\bar{\phi}(x,t;\zeta)$, $\psi(x,t;\zeta)$ and $\bar{\psi}(x,t;\zeta)$ to have the following asymptotic behaviors

| | $x = -\infty$ | $x = +\infty$ |
|---|---|---|
| $\phi$ | $\begin{bmatrix} 2i\zeta \\ 1 \end{bmatrix} e^{-i\zeta x}$ | $2i\zeta a\ e^{-i\zeta x}$<br>$a\ e^{-i\zeta x} + b\ e^{i\zeta x}$ |
| $\bar{\phi}$ | $\begin{bmatrix} 0 \\ 1 \end{bmatrix} e^{i\zeta x}$ | $2i\zeta\bar{b}\ e^{-i\zeta x}$<br>$\bar{a}\ e^{i\zeta x} + \bar{b}\ e^{-i\zeta x}$ |
| $\psi$ | $-2i\zeta\bar{b}\ e^{-i\zeta x}$<br>$a\ e^{i\zeta x} - \bar{b}\ e^{-i\zeta x}$ | $\begin{bmatrix} 0 \\ 1 \end{bmatrix} e^{i\zeta x}$ |
| $\bar{\psi}$ | $2i\zeta\bar{a}\ e^{-i\zeta x}$<br>$\bar{a}\ e^{-i\zeta x} - b\ e^{i\zeta x}$ | $\begin{bmatrix} 2i\zeta \\ 1 \end{bmatrix} e^{-i\zeta x}$ |

The scattering functions $a(\zeta,t)$, $\bar{a}(\zeta,t)$, $b(\zeta,t)$, $\bar{b}(\zeta,t)$ are the coefficients in the relations between the four solutions:

$$\phi = a\bar{\psi} + b\psi \ , \quad \psi = a\bar{\phi} - \bar{b}\phi$$

$$\bar{\phi} = \bar{a}\psi + \bar{b}\bar{\psi} \quad , \quad \bar{\psi} = \bar{a}\phi - b\bar{\phi} \tag{6.157}$$

and $a\bar{a} - b\bar{b} = 1$. It may also be shown that $a,\phi,\psi$ are analytic in the upper half $\zeta$ plane; $\bar{a}$, $\bar{\phi}$, $\bar{\psi}$ in the lower. The zeros of $a(\zeta,t)$ [$\bar{a}(\zeta,t)$] are the discrete eigenvalues of (6.155) in the upper (lower) half $\zeta$ plane. Since $\phi_2(-\zeta)$ is a solution of (6.155) if $\phi_2(\zeta)$ is, then relations such as $\bar{\phi}_2(\zeta) = \phi_2(-\zeta)$ hold. From these we may deduce $\bar{a}(\zeta) = a(-\zeta)$, $\bar{b}(\zeta) = b(-\zeta)$, $\bar{\zeta}_k = -\zeta_k$, $k = 1, \ldots, N$. Also, if $q(x,t)$ is real, $a^*(\zeta^*) = a(-\zeta)$, $b(-\zeta) = b^*(\zeta^*)$ and all $\zeta_k = i\eta_k$, $\eta_k$ real. We will assume that the zeros of a $(\zeta,t)$ are simple and do not lie on the real $\zeta$ axis.

The linear integral equations which allow the reconstruction of the potential $q(x,t)$ from the scattering data make use of the following combinations

$$S_+; \ \frac{b}{a}(\zeta,t) \ , \ \zeta \text{ real} \ ; \quad (\zeta_k,\gamma_k) \ , \quad k = 1, \ldots, N$$

$$S_-; \ \frac{b}{a}(\zeta,t) \ , \ \zeta \text{ real} \ ; \quad (\zeta_k,\beta_k) \ , \quad k = 1, \ldots, N \ . \tag{6.158}$$

Here $\gamma_k = b_k/a_k'$, $\beta_k = \bar{b}_k/a_k' = -1/b_k a_k'$ and $b_k$ is defined by the $\phi(\zeta_k,t) = b_k\psi(\zeta_k,t)$ and is the analytic extension of $b(\zeta,t)$ to $\zeta_k$ if there is one. The rate of change of the scattering data $S_+$ and $S_-$ is computed

$$\delta S_+: \ \delta\left(\frac{b}{a}\right) = \frac{1}{2i\zeta a^2}\int_{-\infty}^{\infty} -\delta q\phi_2^2 \, dx \quad ,$$

$$\gamma_k\delta\zeta_k = \frac{1}{2i\zeta_k a_k'^2}\int_{-\infty}^{\infty} -\delta q\phi_{2k}^2 \, dx \quad ,$$

$$\delta\gamma_k = -\frac{1}{2i\zeta_k a_k'^2}\int_{-\infty}^{\infty} \delta q\left[\left(\frac{\partial\phi_2^2}{\partial\zeta}\right)_k - \left(\frac{a_k''}{a_k'} + \frac{1}{\zeta_k}\right)\phi_{2k}^2\right]dx \tag{6.159}$$

$$\delta S_-: \ \delta\left(\frac{\bar{b}}{a}\right) = \frac{1}{2i\zeta a^2}\int_{-\infty}^{\infty} \delta q\psi_2^2 \, dx \quad ,$$

$$\beta_k\delta\zeta_k = \frac{1}{2i\zeta_k a_k'^2}\int_{-\infty}^{\infty} \delta q\psi_{2k}^2 \, dx \quad ,$$

$$\delta\beta_k = \frac{1}{2i\zeta_k a_k'^2}\int_{-\infty}^{\infty} \delta q\left[\left(\frac{\partial\psi_2^2}{\partial\zeta}\right)_k - \left(\frac{a_k''}{a_k'} + \frac{1}{\zeta_k}\right)\psi_{2k}^2\right]dx \ . \tag{6.160}$$

Therefore, it is appropriate to define the following dual sets[2] of functions;

$$\Phi(x,t;\zeta) = \phi_2^2 \ , \ \zeta \text{ real} \ ; \ \Phi_k(x,t) = \phi_2^2(\zeta_k) \quad , \ \chi_k(x,t) = \left(\frac{\partial\phi_2^2}{\partial\zeta}\right)_k \ , \ k = 1, \ldots, N \tag{6.161}$$

[2]Because $\partial\phi_2^2/\partial\zeta|_{\zeta_k}$ does not tend to zero as $x\to+\infty$, this set cannot be used as a basis for an arbitrary function $f(x)$.

$$\Psi(x,t;\zeta) = \psi_2^2 \ , \ \zeta \text{ real} \ ; \ \Psi_k(x,t) = \psi_2^2(\zeta_k) \ , \ \tau_k(x,t) = \left(\frac{\partial\psi_2^2}{\partial\zeta}\right)_k \ , \ k = 1, \ldots, N \ . \tag{6.162}$$

The set (6.161) of functions satisfies the equation

$$L_S\Phi = \zeta^2\Phi \ , \quad L_S = -\frac{1}{4}\frac{\partial^2}{\partial x^2} - q + \frac{1}{2}\int_{-\infty}^{x} dy\, q_y \ . \tag{6.163}$$

The adjoint set to(6.161) is

$$\Phi^A(x,t;\zeta) = \frac{\partial\psi_2^2}{\partial x} \ , \ \Phi_k^A(x,t) = \left(\frac{\partial\psi_2^2}{\partial x}\right) , \quad \chi_k^A(x,t) = \left(\frac{\partial^2\psi_2^2}{\partial\zeta\partial x}\right)_k \ , \tag{6.164}$$

and satisfies the equation

$$L_S^A\Phi^A = \zeta^2\Phi^A + \frac{1}{2}q_x\psi_2^2(R) \ , \ L_S^A = -\frac{1}{4}\frac{\partial^2}{\partial x^2} - q + \frac{1}{2}q_x\int_x^R dy \tag{6.165}$$

where in the application of (6.156) we consider the limit $R\to+\infty$.

The set (6.162) satisfies the equation

$$M_S\Psi = \zeta^2\Psi \ , \quad M_S = -\frac{1}{4}\frac{\partial^2}{\partial x^2} - q - \frac{1}{2}\int_x^{\infty} dy\, q_y \ . \tag{6.166}$$

The adjoint set to (6.162) is

$$\Psi^A(x,t,\zeta) = \frac{\partial\phi_2^2}{\partial x} \ , \ \Psi_k^A(x,t) = \left(\frac{\partial\phi_2^2}{\partial x}\right)_k \ , \ \tau_k^A(x,t) = \left(\frac{\partial^2\phi_2^2}{\partial x\partial\zeta}\right)_k \ , \tag{6.167}$$

and these elements satisfy

$$M_S^A\Psi^A = \zeta^2\Psi^A + \frac{1}{2}q_x\phi_2^2(-R) \ , \ M_S^A = -\frac{1}{4}\frac{\partial^2}{\partial x^2} - q - \frac{1}{2}q_x\int_{-R}^{x} dy \ , \tag{6.168}$$

where again we consider the limit $R\to+\infty$.

What is new here and different from before (Sect.6.4) is that the adjoint and the dual sets are no longer the same. Furthermore, the integral in the adjoint operator cannot be applied loosely since there is no longer an $r(x,t)$ or a $q(x,t)$ under the integral sign to ensure integrability. Nevertheless these definitions work and the reader may be quite interested in the details of the proof of the orthogonality relations given in Appendix C.

Using these relations, we may write the expansion of $q(x,t)$ and $\delta q(x,t)$ in terms of these bases[3]

$$q(x,t) = -\frac{2}{i\pi}\int_{-\infty}^{\infty}\zeta\,\frac{\bar{b}}{a}\,\Phi d\zeta + 4\sum_{k=1}^{N}\beta_k\zeta_k\Phi_k \tag{6.169}$$

[3]We note the expression for $q(x,t)$ contains no component of $\chi_k$ in (6.169) or $\tau_k$ in (6.170); namely, we can use (6.161) and (6.162) as bases for functions $f(x)$ for which $\int_{-\infty}^{\infty} f\phi_{kx}^2 = \int_{-\infty}^{\infty} f\psi_{kx}^2 = 0$, $k = 1, \ldots, N$. That this is true for $f(x) = q(x,t)$ may be seen from integrating (6C.1).

$$q(x,t) = \frac{2}{i\pi}\int_{-\infty}^{\infty} \zeta \frac{b}{a} \Psi d\zeta - 4\sum_{k=1}^{N} \gamma_k \zeta_k \Psi_k \tag{6.170}$$

$$\delta q(x,t) = \frac{1}{\pi}\int_{-\infty}^{\infty} \delta\left(\frac{b}{a}\right)\Phi^A d\zeta + 2i\sum_{k=1}^{N}\left(\delta\gamma_k \Phi_k^A + \gamma_k \delta\zeta_k \chi_k^A\right) , \tag{6.171}$$

$$\delta q(x,t) = -\frac{1}{\pi}\int_{-\infty}^{\infty} \delta\left(\frac{\bar{b}}{a}\right)\Psi^A d\zeta + 2i\sum_{k=1}^{N}\left(\delta\beta_k \Psi_k^A + \beta_k \delta\zeta_k \tau_k^A\right) . \tag{6.172}$$

The simplectic structure may be found by using the orthogonality relations. Consider

$$\int_{-\infty}^{\infty} \left(\frac{1}{2}\, \delta q \wedge \int_{-\infty}^{x} dy \delta q\right) dx$$

where we use (6.171) for the first $\delta q$ in the integrand and the x integral of (6.172) for the second. For simplicity, let q be real. Then we find that

$$\int_{-\infty}^{\infty} \left(\frac{1}{2}\, \delta q \wedge \int_{-\infty}^{x} dy \delta q\right) dx = \int_{0}^{\infty} - \frac{2\xi}{\pi}\, \delta \ln(1 - |R|^2) \wedge \delta \mathrm{Arg}\, b(\xi) + \sum_{k=1}^{N} \delta(-2\eta_k^2) \wedge \delta \ln b_k \tag{6.173}$$

where $R(\xi) = b/a$, the reflection coefficient. Hence, the inverse scattering transformation is a canonical transformation which carries $q(x,t)$ into the following action $[-(2\xi/\pi)\ln(1-|R|^2), -2\eta_k^2]$ and angle $|\mathrm{Arg}\, b(\xi), \ln b_k]$ coordinates. We note in particular, the simplicity of the derivation of (6.173) in comparison to the derivation of ZAKHAROV and FADDEEV [6.66].

From (6.169-172), the equations of motion which are integrable by means of the inverse scattering transformation may readily be found. Consider (6.170). Apply the operator $P_0(M_S)$ where $P_0$ is entire and find

$$\frac{\partial}{\partial x} P_0(M_S) q = \frac{2}{i\pi}\int_{-\infty}^{\infty} \zeta P_0(\zeta^2) \frac{b}{a} \psi_{2x}^2 d\zeta - 4\sum \zeta_k P_0(\zeta_k^2) \gamma_k \psi_{2kx}^2 \ .$$

By comparison with (6.171), we note that the equation

$$q_t + \frac{\partial}{\partial x} P_0(M_S) q = 0 \tag{6.174}$$

is equivalent to the scattering equations

$$\left(\frac{b}{a}\right)_t = -2i\zeta P_0(\zeta^2)\frac{b}{a}$$

$$\gamma_{kt} = -2i\zeta_k P_0(\zeta_k^2)\gamma_k$$

$$\zeta_{kt} = 0 \ . \tag{6.175}$$

The Hamiltonian structure may be found by discovering the functional H whose variation with respect to q is

$$\frac{\delta H}{\delta q} = -P_0(M_S) q \ , \tag{6.176}$$

for then (6.174) will read,

$$q_t = \frac{\partial}{\partial x}\frac{\delta H}{\delta q} \quad . \tag{6.177}$$

Again, the functional $\ln a(\zeta)$ provides the vital link. We can show from the variation of the scattering data that

$$\frac{\delta \ln a}{\delta q} = \frac{1}{2i\zeta}\frac{\phi_2\psi_2}{a} \quad . \tag{6.178}$$

But

$$\left(M_S - \zeta^2\right)\left(\frac{\phi_2\psi_2}{a} - 1\right) = \frac{q}{2} \quad . \tag{6.179}$$

Solving (6.179) for large $\zeta^2$, $\mathrm{Im}\{\zeta\} > 0$,

$$\frac{\phi_2\psi_2}{a} = 1 - \frac{1}{2}\sum_0^\infty \frac{1}{(\zeta^2)^{m+1}} M_S^m q \quad . \tag{6.180}$$

But

$$\ln a = -\sum \frac{1}{\zeta^n} C_n \quad ,$$

and thus, using (6.178,180), we see

$$\frac{\delta C_{2m+3}}{\delta q} = \frac{1}{4i} M_S^m q \quad . \tag{6.181}$$

Hence, if the dispersion relation is

$$\Omega(\zeta) = i\zeta P_0(\zeta^2) = i\zeta \sum_0 a_{2m+1}(\zeta^2)^m \quad , \tag{6.182}$$

then the corresponding Hamiltonian is

$$H_\Omega = -4i \sum a_{2m+1} C_{2m+3} \quad . \tag{6.183}$$

In scattering space, since we already have that

$$\ln a = \sum \ln\left(\frac{\zeta - i\eta_k}{\zeta + i\eta_k}\right) - \frac{\zeta}{2\pi i}\int_{-\infty}^{\infty} \frac{1}{\xi^2 - \zeta^2} \ln(1 - |R|^2)\,d\xi \tag{6.184}$$

the potential Hamiltonians are given by

$$C_{2m+1} = 2\sum \frac{(i\eta_k)^{2m+1}}{2m+1} - \frac{1}{\pi i}\int_0^\infty \xi^{2m} \ln(1 - |R|^2)\,d\xi \quad . \tag{6.185}$$

For example, if $\Omega(\zeta) = -4i\zeta^3$,

$$\begin{aligned} H &= \frac{1}{2}\int_{-\infty}^{\infty}\left(q_x^2 - 2q^3\right)dx \\ &= \frac{-32}{5}\sum \eta_k^5 - \frac{16}{\pi}\int_0^\infty \xi^4 \ln(1 - |R|^2)\,d\xi \quad . \end{aligned}$$

which corresponds to the Korteweg-deVries equation.

## 6.12 A Singular Perturbation Theory

The formulations of Sects.6.3,11 immediately allow us to examine the effects of perturbations. Suppose the equation we wish to study is a perturbation of one of the integrable flows (6.65)

$$\begin{bmatrix} r_t \\ -q_t \end{bmatrix} + 2\Omega(L^A)\begin{bmatrix} r \\ q \end{bmatrix} = \varepsilon\begin{bmatrix} r_t \\ -q_t \end{bmatrix}_{\text{pert.}} , \tag{6.186}$$

or (6.174)

$$q_t + \frac{\partial}{\partial x} P_0(M_S)q = \varepsilon(q_t)_{\text{pert.}} . \tag{6.187}$$

Then to calculate the rate of change of the scattering data $S_-$ in (6.21) or $S_-$ or $S_+$ in (6.185), we simply use the formulae (6.35-38) or (6.159,160). The integrable terms $2\Omega(L^A)\begin{bmatrix} r \\ q \end{bmatrix}$ and $\frac{\partial}{\partial x} P_0(M_S)q$ yield exact integrals in their respective cases whereas in general the perturbation term will not. For example, from (6.160) we will have

$$\zeta_{kt} = \frac{\varepsilon}{2i\zeta_k\beta_k a_k'^2}\int_{-\infty}^{\infty} (q_t)_{\text{pert.}}\, \psi_k^2 \, dx , \tag{6.188a}$$

$$\beta_{kt} + \left(\frac{a_k''}{a_k'} + \frac{1}{\zeta_k}\right)\beta_k\zeta_{kt} = + 2i\zeta_k P_0(\zeta_k^2)\beta_k + \frac{\varepsilon}{2i\zeta_k a_k'^2}\int_{-\infty}^{\infty} (q_t)_{\text{pert.}} \left(\frac{\partial\psi^2}{\partial\zeta}\right)_k dx , \tag{6.188b}$$

and

$$\left(\frac{b^*}{a}\right)_t = + 2i\zeta P_0(\zeta^2)b/a + \frac{\varepsilon}{2i\zeta a^2}\int_{-\infty}^{\infty} (q_t)_{\text{pert.}}\, \psi^2 \, dx . \tag{7.188c}$$

Now, straightaway we can determine the initial effects of the perturbation on the scattering data by solving (6.188) iteratively. However, we are usually interested in the cumulative effects of the perturbation over a long time $[t = O(\varepsilon^{-1})]$ during which order one changes can take place in those parameters such as the soliton amplitudes which in the unperturbed system would remain constant. What we are seeking then is to describe the slow changes in the action variables and angular frequencies (the derivatives of the angle phases) due to the perturbation. In order to determine these changes we solve (6.188) iteratively but allow the action variables of the unperturbed problem to vary slowly in time. The choice of the slow time behavior of the action variables is then made so as to eliminate secular terms (proportional to $\varepsilon t$) appearing in the asymptotic expansions for the scattering data. The question naturally arises as to whether this procedure is equivalent to the more naive approach of simply iterating directly on the equation itself. Namely suppose we are given the equation

$$q_t + 6qq_x + q_{xxx} = \varepsilon(q_t)_{\text{pert.}}$$

and simply expand $q = q_0 + \varepsilon(\delta q) + \ldots$ then we would obtain for the perturbation field

$$(\delta q)_t + \frac{\partial}{\partial x}\left[(\delta q)_{xx} + 6q_0\delta q\right] = (q_{0t})_{\text{pert.}} \quad . \tag{6.189}$$

Now from (6.160), we know that the corresponding change $\delta(\bar{b}/a)$ in the scattering data is given by (note for q real, $\bar{b} = b^*$)

$$\delta(\bar{b}/a) = \frac{1}{2i\zeta a^2}\int \delta q \psi_2^2 \, dx \tag{6.190}$$

plus the two relations for the corresponding rate of changes of the scattering data connected with the discrete spectrum. Inverting (6.190) means we can write $\delta q$ in terms of the adjoint basis $\Psi^A$ defined in (6.167) where all quantities are calculated in terms of the unperturbed $q_0(x,t)$. It is our claim that in this basis the left hand side of (6.189) separates and we find exactly the expressions for the time rate of change of $\delta(\bar{b}/a)$ as we would do if we simply substitute $\bar{b}/a = (\bar{b}/a)_0 + \varepsilon\delta(\bar{b}/a)$ into (6.188c) directly. To see this, we differentiate (6.190) with respect to time, keeping in mind that the unperturbed basis $\psi_2^2$ is time dependent. In fact, we will use that

$$(\psi_2^2)_t = (2q_x - 8i\zeta^3)\psi_2^2 + (4\zeta^2 - 2q)(\psi_2^2)_x \quad , \tag{6.191}$$

as well as (6.166) which may be rewritten

$$-\frac{1}{4}(\psi_2^2)_{xxx} - \frac{1}{2}q_x\psi_2^2 - q(\psi_2^2)_x = \zeta^2(\psi_2^2)_x \quad . \tag{6.192}$$

$$\begin{aligned}\delta\left(\frac{\bar{b}}{a}\right)_t &= \frac{1}{2i\zeta a^2}\int\Big\{\Big[-\frac{\partial}{\partial x}[(\delta q)_{xx} + 6q_0\delta q] + (q_t)_{\text{pert.}}\Big]\psi_2^2 \\ &\quad + \delta q\Big[(2q_{0x} - 8i\zeta^3)\psi_2^2 + (4\zeta^2 - 2q_0)(\psi_2^2)_x\Big]\Big\}dx \\ &= -8i\zeta^3\delta\left(\frac{\bar{b}}{a}\right) + \frac{1}{2i\zeta a^2}\int (q_t)_{\text{pert.}}\psi_2^2 \, dx + R\end{aligned} \tag{6.193}$$

where the remainder R, given by

$$R = \frac{1}{2i\zeta a^2}\int\Big\{-\frac{\partial}{\partial x}\Big[(\delta q)_{xx} + 6q_0\delta q\Big]\cdot\psi_2^2 + \delta q\Big[2q_{0x}\psi_2^2 + (4\zeta^2 - 2q_0)(\psi_2^2)_x\Big]\Big\}\, dx \quad ,$$

may be shown to be zero after integrating by parts and using (6.192).

Before we discuss concrete examples, several remarks are in order. First we list the advantages of the inverse scattering method for handling perturbations.

1) It is natural to analyze a mechanical system in terms of its action and angle variables or "normal modes" because in these coordinates the coupled system separates. In concrete terms, the inverse scattering approach tells us how to separate and therefore solve (6.189).

2) If one is simply looking for the slow changes which take place in the leading order approximations to the action variable over times $\varepsilon^{-1}$, one can directly use the conservations laws. This allows us to compute the slow rate of change of

the action variables. To leading order, the angle variables feel the effects of the perturbation through the slow changes of the action variable. We will illustrate this approach in each of the three examples.

3) The excitation of other normal modes due to the perturbation can be directly calculated. We will show how such a calculation leads to an understanding of the role which the 'shelf' plays in the perturbation of the Korteweg-deVries solitary wave.

The disadvantages of the method are:

1) It is difficult to apply when the unperturbed state is other than a multisoliton one (a criticism which applies even more seriously to the direct method).

2) It is difficult to calculate second-order effects and to investigate behavior on time scales longer than $\varepsilon^{-1}$. The reason for this is that in taking the limit $t \to \infty$, $\varepsilon t$ fixed, the behavior of the eigenfunctions and scattering data is nonuniform in $\zeta$. For example, for a pure multisoliton state, the reflection coefficient $R(\zeta)$ for the Schrödinger operator is identically zero. However, as soon as the perturbation is switched on, $R(0) = -1$.

3) It it difficult to handle the problem of soliton creation. We will make several further comments on this equation when we discuss the long-time behavior ($t \gg \varepsilon^{-1}$) of the Korteweg-deVries shelves.

On the other hand, the direct method has several serious disadvantages.

1) In general, it is impossible to separate (6.189). Usually, but not always, $q_0$ must be a function of a particular combination of x and t, the independent variables. In cases where it is not, such as in the investigation of the stability of the breather discussed below, where $q_0$ is periodic in time, the direct method requires one to average over a period of the fast time variable, a step which is not always justified (for example, when the frequency of the breather is order $\varepsilon$).

2) One must make an ansatz about the shape of the leading order approximation. For example, in the problem of a Korteweg-deVries solitary wave under perturbation, the earlier papers [6.67-71] assumed that the motion was adiabatic and took the form of a slowly varying solitary wave. This is not the case and as we discuss below, the leading order solution must contain a nonadiabatic part (a shelf) as well. Indeed the discovery of the shelf and the resolution of the dilemma facing the direct method was made using the inverse scattering perturbation method. Of course, once the answer is known, one can make the proper adjustments so that the problem can be tackled by the direct perturbation method.

3) It is often difficult to interpret the meaning of secular terms.

One advantage of the direct method is, of course, that it applies to systems in which the unperturbed problem is not integrable.

In summary, then, although the method and its variations described here are very powerful, there is no unique way of tackling a perturbation problem. Each method provides its own level of understanding and it is useful to analyze difficulties which arise with one method in the language of the other. We now discuss three concrete examples which not only serve to illustrate the method but which also, in their own right, are of general importance in applications. In particular we want to emphasize the value of using, in a judicious way, the conservation laws. Other examples and other important papers on perturbations are given in [6.57, 72-93].

The first example looks at the influence of external damping and forcing on the soliton of the nonlinear Schrödinger equation. Consider

$$q_t - iq_{xx} - 2iq^2q^* = -pq - E\,e^{i\Omega t} \quad , \tag{6.194}$$

which is (6.186) with $\Omega(\zeta) = -2i\zeta^2$, $\varepsilon(q_t)_{\text{pert.}} = -pq - E\,e^{i\Omega t}$ where p and E are small. Let the unperturbed state be

$$q(x,t) = 2\eta \operatorname{sech} 2\eta(x - \bar{x})\exp(-2i\xi x - 2i\bar{\sigma}) \quad ;$$

if no perturbations were present we would find

$$\eta_t = \xi_t = 0 \ , \quad \bar{\sigma}_t = 2(\xi^2 - \eta^2) \ , \quad \bar{x}_t = -4\xi \quad . \tag{6.195}$$

The expressions for the unperturbed eigenfunctions are as follows. Define

$$\sqrt{\gamma_k}\exp(i\zeta_k x) = \lambda_k \ ; \quad \sqrt{\gamma_k}\psi_j(\zeta_k) = u_{jk} \ , \quad k = 1,2, \ldots, N, \quad j = 1,2 \quad . \tag{6.196}$$

Then from the inverse scattering formulae [6.10,69], we find that

$$(I + B^*B)u_2^* = \lambda^*, \quad u_1 = -Bu_2^* \quad , \quad B = \left[\frac{\lambda_j\lambda_k^*}{\zeta_j - \zeta_k^*}\right] . \tag{6.197}$$

The derivatives of the squared eigenfunctions are

$$\gamma_k\left[\frac{\partial\psi_1^2}{\partial\zeta}\right]_k = 2ix(u_{1k})^2 + 2\sum_{\ell=1}^{N}\frac{\lambda_k\lambda_\ell^* u_{1k}u_{2\ell}^*}{(\zeta_\ell^* - \zeta_k)^2} \quad ,$$

$$\gamma_k\left[\frac{\partial\psi_2^2}{\partial\zeta}\right]_k = 2ix(u_{2k})^2 - 2\sum_{\ell=1}^{N}\frac{\lambda_k\lambda_\ell^* u_{2k}u_{1\ell}^*}{(\zeta_\ell^* - \zeta_k)^2} \quad , \quad k = 1, \ldots, N \tag{6.198}$$

and the eigenfunctions for real $\zeta$ are

$$\begin{bmatrix}\psi_1(\xi)\\ \psi_2(\xi)\end{bmatrix} = e^{i\xi x}\left\{\begin{bmatrix}0\\1\end{bmatrix} + \sum_{k=1}^{N}\frac{1}{\zeta_k^* - \xi}\begin{bmatrix}\lambda_k^* u_{2k}^*\\ -\lambda_k^* u_{1k}^*\end{bmatrix}\right\} \quad . \tag{6.199}$$

For the single soliton solution let $\zeta_1 = \xi_1 + i\eta_1$,

$$\gamma_1 = 2\eta_1\exp(2\bar{\theta}_1 + 2i\bar{\sigma}_1) \quad \left[\text{or} \quad \beta_1 = (\gamma_1 a_1'^2)^{-1} = -2\eta_1\exp(-2\bar{\theta}_1 - 2i\bar{\sigma}_1)\right] \quad ,$$

$a = (\zeta - \zeta_1)/(\zeta - \zeta_1^*)$. From (6.197,198) and (6.58), we find

$$q(x,t) = 2\eta_1 \operatorname{sech} 2\theta_1 \exp[-2i(\sigma_1 + \pi/4)] \tag{6.200}$$

where $\theta_1 = -\eta_1 x + \bar{\theta}_1$, $\sigma_1 = \xi_1 x + \bar{\sigma}_1$. In the case of a completely integrable system, the time parameter t enters the problem only through $\bar{\theta}_1$ and $\bar{\sigma}_1$ since $\zeta_{1t} = 0$. If

$\Omega(\zeta_1) = \Omega_r(\xi_1,\eta_1) + i\Omega_i(\xi_1,\eta_1)$, we have from (6.66) that

$$\bar{\theta}_{1t} = -\Omega_r \ , \quad \bar{\sigma}_{1t} = -\Omega_i \quad . \tag{6.201}$$

Using these results we find that

$$\eta_t = -\ 2p\eta + \frac{\pi E}{2} \text{ sech } \frac{\xi\pi}{2\eta} \sin\chi \ , \quad \xi_t = -\ \frac{\pi E\xi}{2\eta} \text{ sech } \frac{\xi\pi}{2\eta} \sin\chi \quad , \tag{6.202}$$

(where $\chi = \Omega t + 2\xi\bar{x} + 2\bar{\phi}$) from which we can show that $(\xi\eta)_t = -2p(\xi\eta)$ which implies that either $\xi$ or $\eta$ tends to zero. Namely if $\xi$ is too large, then there is a mismatch between the spatial phase of the soliton and the external field once the soliton decays due to damping. On the other hand if $\xi$ is not large initially then it will tend to zero and the spatial phase of the soliton becomes that of the external field. Let us take $\xi \equiv 0$. Then the equation for $\chi_t$ may be calculated to be

$$\chi_t = \Omega - 4\eta^2 \tag{6.203}$$

which together with

$$\eta_t = -\ 2p\eta + \pi E/2 \sin\chi \tag{6.204}$$

can be solved for $\eta$ and $\chi$. A phase plane analysis shows that the soliton *phase-locks* to the external field (i.e., $\chi_t = 0$) only when the frequency $\Omega$ lies in the window

$$0 < \Omega < \pi^2E^2/4p^2 \quad . \tag{6.205}$$

On the other hand if $\Omega > 0$ and is fixed and an external force is applied, the soliton will not respond and be entrained by the external field until the amplitude is greater than $2p/\pi\sqrt{\Omega}$. At this point the soliton will phase-lock to the external field and achieve an amplitude $\frac{1}{2}\sqrt{\Omega}$ entirely independent of the subsequent value of E or p as long as (6.205) is satisfied.

The equations (6.202) can be simply derived [6.75] by calculating

$$\frac{\partial}{\partial t}\int_{-\infty}^{\infty} qq^*dx \quad \text{and} \quad \frac{\partial}{\partial t}\int_{-\infty}^{\infty} (qq_x^* - q_xq^*)$$

directly from (6.194) using the unperturbed solution to calculate the terms on the right-hand side. In fact, the second equation will give us $\frac{\partial}{\partial t}(\xi\eta) = -2p(\xi\eta)$ directly.

In the first example, we only computed the changes in the soliton parameters and ignored the creation of the continuous spectrum. This step is not always justified as our second example will show. In what follows we will attempt to describe the propagation of a soliton in a shallow layer of water in which the depth begins to change slowly from some position $t = t_0$. The relevant equation is a perturbed form of (6.1)

$$q_t + 6qq_x + q_{xxx} = -\Gamma q \quad , \quad p = O(\varepsilon) \quad , \quad 0 < \varepsilon << 1 \quad , \tag{6.206}$$

in which t measures position, x measures time relative to the frame of the right going characteristic and $\Gamma$ is proportional to $D_t$, the change of depth. Note that if the depth decreases, $\Gamma < 0$, and $-\Gamma q$ acts as a negative damping. The unperturbed multisoliton state is given by the following equations:

$$\psi_j + \sum_{k=1}^{N} \frac{\gamma_k \exp i(\zeta_k+\zeta_j)x}{\zeta_k+\zeta_j} \psi_k = \exp(i\zeta_j x) \quad , \quad j = 1, \dots, N \quad , \tag{6.207a}$$

$$\left(\frac{\partial\psi}{\partial\zeta}\right)_{\zeta_j} = ix\psi_j + \sum_{k=1}^{N} \frac{\gamma_k\psi_k \exp i(\zeta_k+\zeta_j)x}{(\zeta_k+\zeta_j)^2} \quad , \quad j = 1, \dots , N \quad , \tag{6.207b}$$

$$\psi(\zeta) = e^{i\zeta x}\left(1 - \sum_{k=1}^{N} \frac{\gamma_k\psi_k \exp(i\zeta_k x)}{\zeta_k+\zeta}\right) \quad , \quad \zeta \text{ real} \quad . \tag{6.207c}$$

If $N = 1$, $\zeta_1 = i\eta$, $\gamma_1 = 2i\eta e^{2\bar{\theta}}$, then from (6.170),

$$q(x,t) = 2\eta^2 \operatorname{sech}\eta(x - \bar{x}) \quad , \quad \eta_t = 0 \quad , \quad \bar{x}_t = 4\eta^2 \quad . \tag{6.208}$$

We seek now to describe the change in the form of (6.208) after it has moved into a region where the depth is no longer constant. Using (6.207) and (6.188) we readily find that

$$\eta_t = -\frac{2}{3}\Gamma\eta \tag{6.209}$$

and, to leading order, that

$$\bar{x}_t = 4\eta^2 \quad . \tag{6.210}$$

Both expressions can be easily integrated to find the variation of the soliton amplitude and position with depth. What is of real interest here, however, is that (6.209) satisfies only the second of the three exact local relations for mass, energy and center of gravity, respectively,

$$\frac{\partial}{\partial t}\int_{-\infty}^{\infty} q\, dx = -\Gamma \int_{-\infty}^{\infty} q\, dx \quad , \tag{6.211a}$$

$$\frac{\partial}{\partial t}\int_{-\infty}^{\infty} q^2\, dx = -2\Gamma \int_{-\infty}^{\infty} q^2\, dx \quad , \tag{6.211b}$$

$$\frac{\partial}{\partial t}\int_{-\infty}^{\infty} xq\, dx = 3\int_{-\infty}^{\infty} q^2\, dx - \Gamma \int_{-\infty}^{\infty} xq\, dx \quad . \tag{6.211c}$$

Each expression integrates to give the global relations

$$M(t) = \int_{-\infty}^{\infty} q\, dx = M(t_0) \exp\left[-\int_{-\infty}^{\infty} \Gamma(s)ds\right] \quad , \tag{6.212a}$$

$$E(t) = \int_{-\infty}^{\infty} q^2\, dx = E(t_0) \exp\left[-2\int_{-\infty}^{\infty} \Gamma(s)ds\right] \quad , \tag{6.212b}$$

$$G(t) = \int_{-\infty}^{\infty} xq\, dx = \left\{G(t_0) + 3E(t_0)\int_{t_0}^{t} ds\left[\exp - \int_{t_0}^{s} \Gamma(r)dr\right]\right\}\exp\left[-\int_{t_0}^{t} \Gamma(s)ds\right] \quad . \tag{6.212c}$$

If $q(x,t)$ only consists of the soliton component, then $\frac{\partial}{\partial t}\int_{-\infty}^{\infty} q\, dx = \frac{\partial}{\partial t} 4\eta = \frac{-2}{3}\Gamma \int_{-\infty}^{\infty} q\, dx$. Thus if the depth decreases, the increasing soliton only absorbs two-thirds of the extra mass being created per unit distance t due to the decreasing depth. Where does the rest of the water go?

In order to find out we have to realize that as soon as a perturbation is applied to an integrable system, the system no longer separates into simple time evolutions for each of the individual modes as in (6.175). The modes become coupled and the perturbation may eventually cause all the normal modes of the unperturbed system to be excited, and some of them in such a way as to play a major role in the evolution of the system over long times. Such is the case here for when we calculate the reflection coefficient b/a from (6.188c) with $P_0 = -4\zeta^2$ and $b/a(\zeta,t = 0) = 0$, using (6.207c) and knowing to first order that $a = (\zeta - i\eta)/(\zeta + i\eta)$, we find that

$$\xi \frac{b(\xi,t)}{a(\xi,t)} = - \frac{2\Gamma\pi i}{3a \operatorname{sech} \frac{\xi\pi}{\eta}} \exp(-2i\xi x_0) \frac{\exp(-8i\xi\eta^2 t)-\exp(8i\xi^3 t)}{8i(\xi^2+\eta^2)} \tag{6.213}$$

where we have used $\bar{x} = 4\eta^2 t + x_0$ to first order. Note that the reflection coefficient b/a itself is singular like $1/\xi$ as $\xi \to 0$, but that $\xi b/a$ [which is the analogue of the Fourier transform; see (6.170)] grows linearly with t as the limit $\xi \to 0$. How do we interpret (6.213), for as it stands it has no limit as the fast time $t \to +\infty$ ? The answer is that we must look at the resulting behavior in physical space and calculate from (6.170) the contribution to $q(x,t)$ from the continuous spectrum which has been excited by the soliton. We call this contribution $q_c(x,t)$. We find

$$q_c(x,t) = \frac{4\Gamma}{3} \int_{-\infty}^{\infty} \frac{\xi d\xi}{a(\xi)\sinh \frac{\xi\pi}{\eta}} \left(1 - \frac{2\eta}{\eta-i\xi} \frac{e^{2\theta}}{1+e^{2\theta}}\right)^2 \exp 2i\xi(x - x_0)$$

$$\frac{\exp(-8i\xi\eta^2 t)-\exp(8i\xi^3 t)}{8i\xi(\xi^2+\eta^2)} \quad , \tag{6.214}$$

with $\theta = -\eta(x - \bar{x})$. From the Riemann-Lebesgue lemma, it is clear that the only nonvanishing contribution to (6.214) comes from the neighborhood of $\xi = 0$, whence we may write

$$q_c(x,t) = \frac{\Gamma}{6\pi\eta} \tanh^2\eta(x - \bar{x}) \int_{-\infty}^{\infty} \left[\frac{\sin 2\xi(x-x_0-4\eta^2 t)}{2\xi} - \frac{\sin 2\xi(x-x_0+4\xi^2 t)}{2\xi}\right] d(2\xi) \quad . \tag{6.215}$$

The first integral is $\pi \operatorname{sgn}(x - x_0 - 4\eta^2 t)$. The second is the integral of an Airy function whose value is $\pi$ for $x \gg x_0$ and $-\pi$ for $x \ll x_0$, the transition being achieved through a series of decreasing oscillations. Thus $q_c = 0$ outside the range $x_0 < x < \bar{x}$, where $\bar{x}$ is the soliton position and $x_0$ is that point which an unperturbed disturbance, triggered at the point at which the depth first changed, has reached. In the region $x_0 < x < \bar{x}$, $\bar{x} - x_0 \ll \varepsilon^{-1}$,

$$q_c(x,t) = -\Gamma/3\eta \ , \quad x_0 < x < \bar{x} \quad , \tag{6.216}$$

which represents a shelf of slowly varying amplitude $\Gamma$ reflecting the bottom topography. It is in this shelf that the extra water generated is stored. If we calculate the rate at which water is absorbed into the shelf we find

$$\frac{\partial}{\partial t}\int_{-\infty}^{\infty} q_c\, dx = \frac{\partial}{\partial t}\int_{x_0}^{\bar{x}} q_c\, dx = \bar{x}_t \cdot q_c = -\frac{1}{3}\Gamma \cdot 4\eta \tag{6.217}$$

which together with the rate at which extra water is absorbed by the soliton accounts for the total rate at which extra water is being produced. These results are to be found in [6.73,76,84].

However, the story still continues for if we are to describe the motion for times and distances of order $\varepsilon^{-1}$, we must account for the evolution of the shelf subsequent to its initial formation. This we do by appealing to (6.206) directly. This task is relatively simple because the shelf is small and slowly varying so that the evolution of the shelf is described by the balance $q_t = -\Gamma q$. At $\bar{t}$, when the solitary wave is at $\bar{x}(\bar{t}) = \int_{t_0}^{\bar{t}} 4\eta^2 dt$, the height of the shelf $q_c(x,\bar{t})$ at the point x is given by [6.83,84,93]

$$\begin{aligned} q_c(x,\bar{t}) &= -\Gamma/3\eta(t)\, \exp\int_{\bar{t}}^{t(x)} \Gamma(s)ds\ , \quad 0 < x < \bar{x} \\ &= 0\ , \quad \text{otherwise}\ , \end{aligned} \tag{6.218}$$

where $t = t(x)$ through the integration of $x_t = 4\eta^2$. With these results we can show that the local and global conservation laws are exactly satisfied. We find

$$\begin{aligned} M(\bar{t}) &= \int_{-\infty}^{\infty} q_s\, dx + \int_0^{\bar{x}} q_c\, dx \\ &= 4\eta_0 \exp\left[-\frac{2}{3}\int_{t_0}^{\bar{t}} \Gamma(s)ds\right] + 4\eta_0\left\{\left[\exp\left(-\int_{t_0}^{\bar{t}} \Gamma(s)ds\right)\right] - \exp\left[-\frac{2}{3}\int_{t_0}^{\bar{t}} \Gamma(s)ds\right]\right\} \\ &= 4\eta_0 \exp\left[-\int_{t_0}^{\bar{t}} \Gamma(s)ds\right]\ , \end{aligned} \tag{6.219}$$

which is exactly (6.212a). Similarly, to the first two orders in $\Gamma$, (6.212c) is satisfied.

We see, therefore, that the shelf plays a crucial role in the leading order behavior of the system. Although its amplitude is order $\varepsilon$, its mass content is order one.

It is useful to note that all these results can be obtained by the judicious use of the conservation laws directly [6.83-85]. Assume that $q(x,t)$ consists of a soliton $q_s = 2\eta^2 \operatorname{sech}^2\eta(x - \bar{x})$ and a shelf $q_c(x,t)$. First from (2.211b), we find $\eta_t = -\frac{2}{3}\Gamma\eta$. From (2.211c), we find $\bar{x}_t = 4\eta^2$; to leading order we may assume that the angle variable $\bar{x}$ changes adiabatically by reason only of the change in $\eta$. Next, use the local mass conservation law (2.211a), to find $(\partial/\partial t)\int_0^{\bar{x}} q_c\, dx = \bar{x}_t\, q_c(\bar{x}) = -4\eta\Gamma/3$ which implies that, at the moment of creation, $q_c(\bar{x}) = -\Gamma/3\eta$. Then use the argument of the last paragraph but one to find (6.210) and the global structure of the shelf. It is worth listing the results for several important examples.

Consider

$$q_t + 6q^r q_x + q_{xxx} = F \tag{6.220}$$

where $q_s = \alpha \operatorname{sech}^{2/r}\eta(x - \bar{x})$, $\alpha^r = (r+1)(r+2)\eta^2/3r^2$. Then (a), if $r = 1$, $F = \varepsilon q$,

$$\eta = \eta_0 \exp\left(\frac{2}{3}\varepsilon t\right), \quad \bar{x}_t = 4\eta^2, \quad q_c = \frac{\varepsilon \exp(\varepsilon t)}{3\eta_0\left(1 + \frac{\varepsilon x}{3\eta_0^2}\right)^{5/4}}, \quad 0 < x < \bar{x}. \tag{6.221a}$$

If (b), $r = 1$, $F = \varepsilon q_{xx}$,

$$\eta = \eta_0\left(1 + \frac{16\eta_0^2 \varepsilon t}{15}\right)^{-1/2}, \quad \bar{x}_t = 4\eta^2, \quad q_c = \frac{8\eta_0\varepsilon}{15}\exp\left(-\frac{2\varepsilon x}{15}\right), \quad 0 < x < \bar{x}. \tag{6.221b}$$

If (c), $r = 2$, $F = \varepsilon q$

$$\eta = \eta_0 \exp(2\varepsilon t), \quad \bar{x}_t = \eta^2, \quad q_c = \frac{\pi\varepsilon \exp(2\varepsilon t)}{\eta_0^2 + 4\varepsilon x}, \quad 0 < x < \bar{x}. \tag{6.221c}$$

We have now described, to leading order, the behavior of the system for x and t of order $\varepsilon^{-1}$. But the story does not end here because the shelf continues to evolve. Equation (6.185) with m = 1 gives

$$\int_{-\infty}^{\infty} q\, dx = 4\sum^{N} \eta_k + \frac{2}{\pi}\int_0^{\infty} \xi^{2m} \ln(1 - |R|^2)\, d\xi, \tag{6.222}$$

where N is the number of bound states corresponding to the potential q(x,t). Now if $\Gamma > 0$, the left-hand side is $4\eta_0 \exp(-\int^t \Gamma ds)$, the first term on the right-hand side is $4\eta_0 \exp(-\frac{2}{3}\int^t \Gamma ds)$, and the second term on the right-hand side, which corresponds to contributions from the continuous spectrum and which is always negative or zero, can make up the difference. However, if $\Gamma < 0$, in which case the shelf is one of elevation, it is clear that one needs more bound states in order to balance (6.222) (this point was first made by WRIGHT [6.94]). Therefore it is not correct to think of the shelf of elevation as corresponding to continuous spectrum. Rather it has the same spectral decomposition as a wide (of order $\int_{t_0}^t 4\eta^2 dt$, i.e., $\varepsilon^{-1}$) potential well of height $\Gamma$ (of order $\varepsilon$). This potential gives rise to a set of discrete eigenvalues $\{\eta_k\}_1^K$ with K of order $\varepsilon^{-1/2}$ which is densely distributed in an order $\varepsilon^{1/2}$ interval along the positive imaginary axis of the complex eigenvalue plane. In perturbation theory, evidence of this presence is indicated by the fact that $a(\zeta,t)$ develops nonuniformities in its expansion. KARPMAN and MASLOV [6.77] attempted to interpret the effect of these new bound states by inserting a single second soliton but this is not quite correct.

What happens next is that on the time scale $1/\varepsilon \ln 1/\varepsilon$, the nonlinear term $6qq_x$ becomes influential. Note from (6.221) that on this time and distance scale, the shelf is completely detached from the original soliton; i.e., if $\bar{x} \gg O(1/\varepsilon)$, $q_c$ is of the respective orders $(\varepsilon\bar{x})^{-5/4}$, $\exp(-2\varepsilon\bar{x}/15)$, $(\varepsilon\bar{x})^{-1}$ in the immediate rear of the soliton in examples (a), (b) and (c). The nonlinear term causes the derivative to increase until at $t = O(\frac{1}{\varepsilon}\ln\frac{1}{\varepsilon})$, it is order one. The subsequent evolution is thus again governed by the perturbed Korteweg-deVries equation with the $-\Gamma q$ term merely a perturbation. Beginning near the front of the shelf, a solitary wave train is formed which for intermediate times (long with respect to $\frac{1}{\varepsilon}\ln\frac{1}{\varepsilon}$ , but short with respect to the time for the rear of the train to propagate back to the end of the shelf) is similar to the solution found by GUREVICH and PITAEVSKII [6.95] to describe the solitary wave train resulting from a constant discontinuity in the initial profile. The leading pulse of this train has the form $q = 2\eta^2\,\mathrm{sech}^2\eta(x - 4\eta^2 t + 3/4\ln t + \text{constant})$ where the $\eta^2$ is given by the maximum amplitude of the discontinuity. Once the train has reached the rear end of the shelf, the individual pulses in the train become individual solitons. The problem here is more difficult as (a) the shelf is not constant in x and (b) the perturbation influences the evolution. Nevertheless, the picture presented here is qualitatively correct.

The case for $\Gamma > 0$ is less interesting. The soliton slowly decays and the shelf decays into a train of small amplitude, long, dispersive waves.

Our third example illustrates the use of perturbation theory to identify certain long wave, two- and three-dimensional instabilities of one-dimensional solitons and breathers. Consider (6.186) with $r = -q^*$ and $(\varepsilon q_t)_{\text{pert.}} = i\gamma q_{yy}$. Then if $\Omega = -2i\zeta^2$, (6.186) is $q_t - iq_{xx} - i\gamma q_{yy} - 2iq^2q^* = 0$, the nonlinear Schrödinger equation in two spatial dimensions. As mentioned already in the introduction, the one-dimensional soliton

$$q(x,t) = 2\eta\,\mathrm{sech}\,2\theta\,\exp[-2i(\sigma + \pi/4)] \quad , \tag{6.223}$$

with $\theta = -\eta x + \bar{\theta}$, $\sigma = \xi x + \bar{\sigma}$, $\bar{\theta}_t = -\Omega_r$, $\bar{\sigma}_t = -\Omega_i$, $\Omega(\xi + i\eta) = \Omega_r(\xi,\eta) + i\Omega_i(\xi,\eta)$, is unstable. We will explore this instability using a short cut version of our perturbation theory. As pointed out in each of the two previous examples, it is often sufficient to calculate the slow rate of change of the action variables alone assuming that, to leading order, angle variables only feel the effects of the perturbation through the slow change of the action variable. In such cases the quickest way to obtain the rates of change of the action variables is to use the conservation laws. For (6.186) with q given by (6.223) and $(\varepsilon q_t)_{\text{pert.}} = i\gamma q_{yy}$, with $|\gamma| \ll 1$, we find

$$\frac{d}{dt}\int_{-\infty}^{\infty} qq^*dx = \frac{d}{dt}\,4\eta = i\gamma\int_{-\infty}^{\infty}(q^*q_{yy} - qq^*_{yy})dx \quad , \tag{6.224}$$

$$\frac{d}{dt}\int_{-\infty}^{\infty}(qq^*_x - q^*q_x)dx = \frac{d}{dt}\,16i\xi\eta = 2i\gamma\int_{-\infty}^{\infty}(q^*_x q_{yy} + q_x q^*_{yy})dx \quad . \tag{6.225}$$

Now let $\eta = \eta_0 + \eta_1(y,t)$, $\xi = \xi_0 + \xi_1(y,t)$, which induces $\bar{\theta} = \bar{\theta}_0 + \bar{\theta}_1$, $\bar{\sigma} = \bar{\sigma}_0 + \bar{\sigma}_1$ and find

$$\eta_{1t} = 4\gamma\eta_0\bar{\sigma}_{1yy} + 4\gamma\bar{\theta}_0\xi_{1yy} \quad , \tag{6.226}$$

$$\xi_{1t} = -\frac{4}{3}\gamma\eta_0\,\bar{\theta}_{1yy} - \frac{4\gamma}{3}\,\eta_{1yy}\bar{\theta}_0 \quad . \tag{6.227}$$

Assuming $\bar{\theta}_t = -\Omega_r$, $\bar{\theta}_t = -\Omega_i$, we have

$$\bar{\theta}_{1t} = -\Omega_{r\eta}\eta_1 - \Omega_{r\xi}\xi_1 \quad , \tag{6.228}$$

$$\bar{\sigma}_{1t} = -\Omega_{i\eta}\eta_1 - \Omega_{i\xi}\xi_1 \quad , \tag{6.229}$$

where $\Omega_{r\eta} = \partial/_{\partial\eta}\Omega_r$. The second term on the right-hand side of both equations (6.226) and (6.227) is due to the fact that $\partial^2/\partial y^2\ (\xi x + \bar{\sigma})$ [and $\partial^2/\partial y^2\ (-\eta x + \bar{\theta})$] has the two terms $\xi_{yy}x$ and $\bar{\sigma}_{yy}$ $(-\eta_{yy}x, \bar{\theta}_{yy})$ contributing to the integrations in (6.224,225). However, they are not important as far as long wave instabilities are concerned. This is because it turns out that $(\partial/\partial t) = 0\sqrt{|\gamma|}$ and therefore $\eta_1, \xi_1$, are order $\sqrt{|\gamma|}$ less than $\bar{\sigma}_1$ and $\bar{\theta}_1$. Therefore, with a posteriori justification, we drop these terms and notice the coupling of the natural action-angle pairs $\eta$ with $\bar{\sigma}$, $\xi$ with $\bar{\theta}$. It is simple to show that

$$\left(\frac{\partial^2}{\partial t^2} + 4\gamma\Omega_{i\eta}\eta_0\frac{\partial^2}{\partial y^2}\right)\eta_1 = 4\gamma\left(-\Omega_r - \Omega_{i\xi}\eta_0\right)\frac{\partial^2\xi_1}{\partial y^2} \quad , \tag{6.230}$$

$$\left(\frac{\partial^2}{\partial y^2} - \frac{4}{3}\Omega_{r\xi}\eta_0\frac{\partial^2}{\partial y^2}\right)\xi_1 = \frac{4\gamma}{3}\left(-\Omega_r + \Omega_{r\xi}\eta_0\right)\frac{\partial^2\eta_1}{\partial y^2} \quad . \tag{6.231}$$

Setting $\eta_1, \xi_1, \propto \exp(iky + \nu t)$ we find (recall $\Omega$ is analytic and hence satisfies the Cauchy-Riemann equations $\Omega_{r\xi} = \Omega_{i\eta}$, $\Omega_{r\eta} = -\Omega_{i\xi}$),

$$\left(\nu^2 - 4\gamma\Omega_{i\eta}\eta_0 k^2\right)\left(\nu^2 + \frac{4\gamma}{3}\Omega_{r\xi}\eta_0 k^2\right) = \frac{16\gamma^2}{3}\left(-\Omega_r - \Omega_{i\xi}\eta_0\right)^2 k^4 \quad , \tag{6.232}$$

which means that, to leading order in $k^2$, either

$$\nu^2 = 4\gamma\Omega_{i\eta}\eta_0 k^2 \quad \text{or} \quad -4\gamma/3\Omega_{r\xi}\eta_0 k^2 \quad . \tag{6.233}$$

Recalling that $\Omega_{i\eta} = \Omega_{r\xi}$, we see that if $4\gamma\Omega_{i\eta}\eta_0 k^2 > 0$, $\eta$ and $\bar{\sigma}$ are unstable whereas $\xi$ and $\bar{\theta}$ are stable. This leads to the explosive instability. On the other hand, if $4\gamma/3\ \Omega_{r\xi}\eta_0 k^2 < 0$, then $\xi$ and $\bar{\theta}$ are unstable which case ZAKHAROV and RUBENCHIK [6.52] refer to as the snake-like instability as a twist develops in the soliton crest. For $\Omega = -2i\zeta^2$, the growth rates are $4\sqrt{\gamma}\eta_0$ if $\gamma > 0$ and $4\sqrt{\gamma}/3\ \eta_0 k$ for $\gamma < 0$, respectively. The two-dimensional nonlinear Schrödinger equation for deep water gravity waves has $\gamma < 0$ and thus it is the snake-like instability which occurs. In this context, the instability simply means that the underlying gravity wave train with wave vector $\underline{k}$ in the x direction is exciting its resonant partners [wave numbers $\underline{k} \pm \underline{K}$ where $2\underline{k} = \underline{k} + \underline{K} + \underline{k} - \underline{K}$, $\omega(2\underline{k}) = \omega(\underline{k} + \underline{K}) + \omega(\underline{k} - \underline{K})$] on the PHILLIPS [6.96] figure of eight resonance curve.

The next class of instabilities we investigate is long wave transverse perturbations of the equations discussed in Sect.6.10

$$q_t = \frac{\partial}{\partial x}\frac{\delta H}{\delta q} + \gamma u_{yy} \quad , \quad q = \frac{-u_x}{2} \quad . \tag{6.234}$$

For example if $H = \int_{\infty}^{\infty}\left(\frac{1}{4}q^4 + \frac{1}{2}q_x^2\right)$, $\frac{\partial}{\partial x}\frac{\delta H}{\delta q} = -q_{xxx} - 6q^2q_x$ and (6.234) would describe the propagation of pulses in a lattice with potential $a\Delta^2 + b\Delta^4$ in the x direction and linearly coupled in the transverse y direction. The soliton in this case is $q = 2\ \mathrm{sech}\ 2\theta$, $\theta = -\eta(x - \bar{x})$, $\bar{x}_t = V = -\Omega_r/\eta$, and using the conservation law

$$\frac{\partial}{\partial t}\int_{-\infty}^{\infty} q^2\, dx = 2\gamma \int_{-\infty}^{\infty} q u_{yy}\, dx \tag{6.235}$$

we find

$$\eta_{1t} = 4\gamma\eta_0\bar{x}_{1yy} \tag{6.236}$$

$$\bar{x}_{1t} = V_\eta \eta_1 \quad . \tag{6.237}$$

It is clear that there is instability if $\gamma V_\eta < 0$. For the modified Korteweg-deVries equation, $V = 4\eta^2$, and hence the instability criterion is $\gamma < 0$. For the sine-Gordon equation, $V = -1/(4\eta)^2$ and again instability occurs only when $\gamma < 0$. One can see that there is a simple reason for this. Write $q_t = -uxt/2$ in laboratory coordinates $X = x + t$, $T = x - t$ and find $q_t - \gamma u_{yy} = \frac{1}{2}u_{TT} - \frac{1}{2}u_{XX} - \gamma u_{yy}$. If $\gamma < 0$, the equation is of elliptic character in the y direction.

Next, we investigate the stability of solitons in the Korteweg-deVries family to transverse perturbations. Consider

$$q_t = -\frac{\partial}{\partial x}P(M_s)q + \gamma\psi_{yy} \quad , \quad q = \psi_x \quad . \tag{6.238}$$

If $P = -4M_s$ [with $M_s$ given by (6.166)] and $\gamma = -3/4$, then (6.238) is (6.11), the Kadomtsev-Petviashvili equation. Let $q = 2\eta^2\ \mathrm{sech}^2\eta(x - \bar{x})$, $\psi = 4\eta\tanh\eta(x - \bar{x}) + K$, and use the conservation of energy to find

$$\eta_t = 2\gamma\left(\frac{1}{4}\frac{K_{yy}}{K} - \eta_y\bar{x}_y - 1/3\eta\bar{x}_{yy}\right) \quad ,$$

$$\bar{x}_t = V(\eta) = P(-\eta^2) \quad . \tag{6.239}$$

Now let $\eta = \eta_0 + \eta_1(y,t)$, $\bar{x} = \bar{x}_0(t) + \bar{x}_1(y,t)$, linearize and find

$$\eta_{1t} = 2\gamma\left(\frac{1}{4}\frac{K_{yy}}{K} - 1/3\eta_0\bar{x}_{1yy}\right) \quad ,$$

$$\bar{x}_{1t} = V_\eta\eta_1 \quad . \tag{6.240}$$

The system that is unstable is

$$\gamma V_\eta > 0 \quad . \tag{6.241}$$

For (6.11), $\gamma < 0$, $V_\eta = 4\eta > 0$.

Finally, we present a new result. The one-dimensional breather or $0\pi$ pulse solution

$$u(x,t) = 4 \tan^{-1}(\eta/\xi \operatorname{sech} 2\theta \sin 2\sigma) \qquad (6.242)$$

of the sine-Gordon equation

$$u_{tt} - u_{xx} - Ku_{yy} + \sin u = 0 \qquad (6.243)$$

is *unstable*. In (6.242),

$$\theta_x = \frac{-\eta}{2}\left(1 + \frac{1}{4\zeta\zeta^*}\right), \quad \sigma_x = \frac{\xi}{2}\left(1 - \frac{1}{4\zeta\zeta^*}\right), \quad \zeta = \xi + i\eta \quad , \qquad (6.244a)$$

$$\theta_t = \frac{-\eta}{2}\left(1 - \frac{1}{4\zeta\zeta^*}\right), \quad \sigma_t = \frac{\xi}{2}\left(1 + \frac{1}{4\zeta\zeta^*}\right), \qquad (6.244b)$$

which if $\xi$ and $\eta$ are constant leads to

$$\theta = \frac{-\eta}{2\sqrt{\zeta\zeta^*}} \frac{x-vt}{\sqrt{1-V^2}} + \hat{\theta}_0 \quad , \quad \sigma = \frac{\xi}{2\sqrt{\zeta\zeta^*}} \frac{t-vx}{\sqrt{1-V^2}} + \hat{\sigma}_0 \quad , \quad V = \frac{1-4\zeta\zeta^*}{1+4\zeta\zeta^*} \quad . \qquad (6.245)$$

We will use the exact relations

$$\frac{dI_1}{dt} = K \int_{-\infty}^{\infty} u_{yy}u_t \, dx \quad , \quad \frac{dI_2}{dt} = K \int_{-\infty}^{\infty} u_{yy}u_x \, dx \quad , \qquad (6.246)$$

where $I_1 = \int_{-\infty}^{\infty} \frac{1}{2}\left(u_t^2 + u_x^2 + 4 \sin^2 u/2\right)dx = 16\eta\left[1 + \frac{1}{4} \zeta\zeta^*\right]$ and $I_2 = \int_{-\infty}^{\infty} u_x u_t \, dx = 16\eta\left[1 - \frac{1}{4} \zeta\zeta^*\right]$. Consider the one-dimensional solution $u_0$ given by (6.242) with $\zeta = \zeta_0 + i\eta_0$ and $\zeta_0\zeta_0^* = 1/4$ (the stationary breather), $\theta_{0t} = 0$, $\sigma_{0t} = \xi_0$. Now set $\eta = \eta_0 + \eta_1(y,t)$, $\xi = \xi_0 + \xi_1(y,t)$, $\theta = \theta_0(x,t) + \theta_1(x,y,t)$, $\theta = \sigma_0(x,t) + \sigma_1(x,y,t)$ and calculate the right-hand sides of (6.246) with $u = u_0$. We find

$$(\xi_0\eta_1 - \eta_0\xi_1)_t = \frac{\alpha^2 K}{2\eta_0} \sigma_{1yy}\left(\cos^2 2\sigma_0\right)J_1 \qquad (6.247)$$

$$(\xi_0\xi_1 + \eta_0\eta_1)_t = \frac{-\alpha^2 K}{2\eta_0} \theta_{1yy}\left(\sin^2 2\sigma_0\right)J_2 \qquad (6.248)$$

with $\alpha = \eta_0/\xi_0$, $D = y^2 + 2(1 + 2\alpha^2 \sin^2 2\sigma_0)y + 1$,

$$J_1 = \int_0^{\infty} \frac{(1+y)^2 dy}{D^2} = \frac{1}{1+\lambda}\left[1 + \frac{1}{\sqrt{\lambda^2-1}} \ln(\lambda + \sqrt{\lambda^2 - 1})\right] \quad , \quad \lambda = 1 + 2\alpha^2 \sin^2 2\sigma_0 \quad ,$$

$J_2 = \int_0^{\infty} (1 - y)^2 dy/D^2$, where we have justifiably neglected the terms in $\theta_1$ proportional to x.

From (6.244b),

$$\theta_{1t} = -4\eta_0(\eta_0\eta_1 + \xi_0\xi_1) \quad , \qquad (6.249)$$

$$\sigma_{1t} = -4\eta_0(\xi_0\eta_1 - \xi_1\eta_0) \quad . \qquad (6.250)$$

From (6.245,250), we note that the variations in phase $\theta$ and momentum $(\xi_0\xi_1 + \eta_0\eta_1)$ simply undergo oscillation. On the other hand, from (6.247,251) we have that

$$\hat{\sigma}_{1tt} = 2(\alpha k)^2\left(\cos^2 2\sigma_0\right)J_1\hat{\sigma}_1 \tag{6.251}$$

where we take K = 1 and $\sigma_1 = \hat{\sigma}(t)\exp(iky)$. Since $J_1$ is positive, the variation in phase $\sigma_1$ and the variation in energy $(\xi_0\eta_1 - \xi_1\eta_0)$ grow with time. There are two limits of interest. If the first, if we let $\alpha$ become small, $J_1 = 1$, then the sine-Gordon equation may be approximated by the nonlinear Schrödinger equation, $u = \exp(-it)\psi(x,t) + (*)$, and the average (over the period of $\sigma_0$) growth rate predicted by (6.252) is exactly that given in (6.233) with $\gamma = 1$, and $\Omega = -2i\zeta^2$. On the other hand as $\alpha$ becomes large, $\sigma_0$ becomes slowly varying in t ($2\sigma_0$ is approximately $t/\alpha$) and in this case we use a WKB approximation to show that

$$\hat{\sigma}_1 \propto \frac{\left[1+\alpha^2 \sin^2\left(\frac{t}{\alpha}\right)\right]^{\frac{1}{2}}}{\alpha k \cos(t/\alpha)} \exp(\pm\alpha k) \ln\left[\alpha \sin\frac{t}{\alpha} + \left(1 + \alpha^2 \sin^2\frac{t}{\alpha}\right)^{\frac{1}{2}}\right] \quad . \tag{6.252}$$

But when $\alpha$ is large, $\xi_0$ is small (recall $\xi_0^2 + \eta_0^2 = 1/4$) and the breather is about to form a kink-antikink pair. Indeed the limit $\alpha\to\infty$ in (6.242) gives

$$u(x,t) = 4 \tan^{-1} t \operatorname{sech} x \tag{6.253}$$

which corresponds to a kink-antikink pair separating at a logarithmic rate. A result consistent with the analysis above was also obtained by direct perturbation methods [6.91] for values of $\alpha$ up to order one.

## 6.13 Conclusion

Our aim in this article has been to emphasize the close interrelationship between the inverse scattering transform and the ideas of Fourier analysis and Hamiltonian mechanics. It should by now be apparent that it is the squares of eigenfunctions of the eigenvalue problems (6.16,155) which form the appropriate bases and play the dominant role, rather than the eigenfunctions themselves. Even in the linear limit (Remark 2, Sect.6.4), we see that the basis relaxes to $(e^{\pm i\zeta x})^2$. In those problems with singular dispersion relations (Sects.6.5,7-9), the quadratic products and squared eigenfunctions enter the partial differential equations explicitly as dependent variables and are then subject to physical interpretation.

The fact that the squared eigenfunctions were central had become clear to us in our first attempt [6.23] to generate the broadest class of integrable systems which where connected with the eigenvalue problem (6.16). In that paper, we added the time dependency of the eigenfunctions $v_1$, $v_2$ of (6.16) by assuming

$$v_{1t} = A(x,t,\zeta)v_1 + B(x,t,\zeta)v_2$$
$$v_{2t} = C(x,t,\zeta)v_1 - A(x,t,\zeta)v_2 \quad , \tag{6.254}$$

and obtaining the (A,B,C) equations

$$A_x = qC - rB \quad , \quad B_x + 2i\zeta B = q_t - 2qA \quad , \quad C_x - 2i\zeta C = r_t + 2Ar \quad , \tag{6.255}$$

by cross differentiating (6.16,219) and equating the coefficients of $v_1$, $v_2$ in the two expressions for $v_{1tx}$, $v_{1xt}$ and $v_{2tx}$, $v_{2xt}$. The connection with the squared eigenfunctions can be seen if one notices that the solutions to the homogeneous (A,B,C) equations (6.255) are simply the quadratic products of the v's, and thus the general expressions for $v_{1t}$ and $v_{2t}$ may be found this way. Once one goes beyond looking for polynomial solutions, however, it is more convenient to follow Sect.6.3.

Although the approach outlined in [6.23] and the previous paragraph does not present as complete a picture as we have developed in this paper, it has the one important and practical advantage that one can determine some of the integrable partial differentiable equations without having to know very much about the direct scattering problem. For example, one can readily find *polynomial* solutions of (6.220) in the form,

$$A = \sum_{n=0}^{N} A_n(x,t)\zeta^n \quad , \quad B = \sum_{n=0}^{N-1} B_n(x,t)\zeta^n \quad , \quad C = \sum_{n=0}^{N-1} C_n(x,t)\zeta^n \tag{6.256}$$

by substituting (6.256) in (6.255) and equating powers of $\zeta$. The $\zeta^0$ balance gives the evolution equations (these correspond to some of the equations in Class I). Of course, the reason for certain simplifications which seem to occur fortuitously is not clear; for example, the equation for $A_{nt}$ always can be integrated once. Nevertheless, the method and the computations are simple and it is recommended as the first thing to try if one simply wishes to determine if a given eigenvalue problem (such as $y'''' + (py)'' + qy = \lambda y$, for example) can provide a basis for solving partial differential equations.

In this review we have focused on characterizing and categorizing the evolution equations associated with (6.16,155). A completely parallel approach can be taken for other eigenvalue problems. For example, the class of evolution equations integrable by the $n^{th}$ order Zakharov-Shabat eigenvalue problem

$$V_x = [\zeta R_0 + P(x,t)]V \tag{6.257}$$

where $R_0$ is trace free and diagonal also admits simple and convenient characterization [6.57,97].

On the other hand, as we have mentioned in the introduction, a complete answer to the inverse problem of finding which eigenvalue problem, if any, is appropriate for solving a given evolution equation has not yet been given although some formidable progress has been made. It would be very useful if one could characterize identifiable features of the given evolution equation which mean that it is integrable.

The first step in the process is to write the evolution equation as the consistency condition

$$P_t - Q_x + [P,Q] = 0 \tag{6.258}$$

for the pair of equations

$$V_x = PV \tag{6.259}$$

$$V_t = QV \tag{6.260}$$

where P and Q are matrices whose entries depend on the unknown dependent variable (or variables) of the evolution equation (set of coupled evolution equations). For example if we are given (6.67) with $a_0 = a_1 = a_3 = 0$, $a_2 = -2i$, we write

$$P = X_1 + qX_2 + rX_3 \tag{6.261}$$

where $X_1$, $X_2$ and $X_3$ are to be determined. We substitute (6.261) in (6.258), using the expressions for $q_t$ and $r_t$. The goal now is to solve for Q. One way to do this is to solve iteratively. Write $P_t = (P_t)_0 + (P_t)_1 + (P_t)_2$ in order of decreasing derivatives. Now determine $Q_0$ from $Q_{0x} = (P_t)_0$; $Q_1$ from $(P_t)_1 + [P,Q_0] = Q_{1x}$ and close the system by stipulating $Q_2 = 0$. This will give a nonclosed Lie algrebra in $X_1$, $X_2$, $X_3$ and any matrix constants of integration introduced in solving for Q. The algebra can be arbitrarily closed by setting all brackets as linear combinations of $X_1$, $X_2$ and $X_3$ in such a way that all the Jacobi identities are satisfied. For most equations the algebra admits only the trivial solution ($X_i = 0$, $i = 1, 2, 3$ or $Q \propto P$ expressing the fact that the equation has a local conservation law). However, for an integrable equation there are an infinite number of solutions. In the example (6.67), we find, for arbitrary $\zeta$, the representation

$$X_1 = -i\zeta\begin{bmatrix}1 & 0\\ 0 & -1\end{bmatrix}, \quad X_2 = \begin{bmatrix}0 & 1\\ 0 & 0\end{bmatrix}, \quad X_3 = \begin{bmatrix}0 & 0\\ 1 & 0\end{bmatrix} \tag{6.262}$$

whence (6.257) is (6.16). For the massive Thirring model (6.7) or the derivative nonlinear Schrödinger equation

$$q_t = iq_{xx} - (q^2r)_x \quad , \quad r = \mp q^* \quad , \tag{6.263}$$

one representation for the solution of the algebra is

$$X_1 = -i\zeta^2\begin{bmatrix}1 & 0\\ 0 & -1\end{bmatrix}, \quad X_2 = \zeta\begin{bmatrix}0 & 1\\ 0 & 0\end{bmatrix}, \quad X_3 = \zeta\begin{bmatrix}0 & 0\\ 1 & 0\end{bmatrix}, \tag{6.264}$$

which means that

$$\begin{aligned} v_{1x} + i\zeta^2 v_1 &= \zeta q v_2 \\ v_{2x} - i\zeta^2 v_2 &= \zeta r v_1 \end{aligned} \tag{6.265}$$

is the appropriate eigenvalue problem. The free parameter $\zeta$ is crucial for it allows us to construct the solution to the evolution equation by constructing the IST mapping from the potential P

$$P(x,t) \to \Phi(x = +\infty, t; \zeta) \tag{6.266}$$

for each $\zeta$ to the scattering matrix $\Phi$. The time evolution of $\Phi(\infty,t;\zeta)$ is trivial.

What is it, then, about an evolution equation which renders it integrable? Clearly there must be a compatibility between the higher derivative terms which are less nonlinear and lower derivative ones which are more nonlinear. In the equations discussed in this chapter, these relations are expressed by the fact that the evolution of a quantity is expressed in terms of powers of an operator [cf. $L^A$ of (6.54)] which combines the $\partial/\partial x$ operator with a nonlinear operator in a very specific way. The answer to this question, when it comes, should also make it clear in a natural way that if an equation is integrable, then it is a member of a family of commuting flows, and that each member of the family is generated by using in turn each of the infinite number of conserved quantities as a Hamiltonian.

As a final remark, certain recent developments should be mentioned. The essence of the inverse scattering transform is that as the coefficient $q(x,t)$ contained in the operator $L = (-\partial^2/\partial x^2) - q(x,t)$ is deformed according to a certain class of nonlinear equations, the spectrum of L [defined as an operator on $L^2(R)$ for example] does not change. This is called an *isospectral deformation*. One might ask what other global properties of operators or, equivalently of systems of linear ordinary differential equations are preserved under deformation. Two are worthy of mention, one is contained in [6.98,99] and the other in [6.100-102]. We will illustrate each with the aid of the system

$$V_x = \begin{bmatrix} -i\zeta & q \\ q & i\zeta \end{bmatrix} V \quad , \tag{6.267}$$

$$V_t = \begin{bmatrix} -4i\zeta^3-2iq^2\zeta & 4\zeta^2q+2i\zeta q_x-q_{xx}+2q^3 \\ 4\zeta^2q-2i\zeta q_x-q_{xx}+2q^3 & 4i\zeta^3+2iq^2\zeta \end{bmatrix} V \quad , \tag{6.268}$$

whose compatability gives the modified Korteweg-deVries equation

$$q_t - 6q^2q_x + q_{xxx} = 0 \quad . \tag{6.269}$$

Inverse scattering works by concentrating on (6.267) over the interval $-\infty < x < \infty$ and defining scattering data which because q deforms according to (6.269) has a simple time evolution. Now suppose we wish to examine the stationary solutions $q(x,t) = q(x - ct)$ of (6.269). Then (6.268) becomes ($X = x - ct$, $T = t$)

$$V_T = \begin{bmatrix} -4i\zeta^3-2iq^2\zeta-i\zeta c & 4\zeta^2q+2i\zeta q_x+\nu \\ 4\zeta^2q-2i\zeta q_x+\nu & 4i\zeta^3+2iq^2\zeta+i\zeta c \end{bmatrix} V \tag{6.270}$$

where (6.269) integrates to

$$q_{xx} - 2q^3 - cq = -\nu \quad . \tag{6.271}$$

Since q is independent of T, let $V = \hat{V}\exp(\lambda T)$ whence (6.270) is

$$Q\hat{V} = \lambda\hat{V} \quad , \tag{6.272}$$

and $\hat{V}$ satisfies the X equation (6.267) as before. The method of KRICHEVER [6.98] and NOVIKOV [6.99] centers on (6.272) instead of (6.267). Note that the existence of $\hat{V}$ implies that

$$\det(Q - \lambda I) = 0 \tag{6.273}$$

which is an algebraic curve $F(\zeta,\lambda) = 0$ which defines a Riemann surface R of genus 2. It is easy to show that R is independent of X [Q(X) is similar to $Q(X_0)$ by virtue of the fact that it satisfies the equation $Q_X = \left[\begin{pmatrix} -i\zeta & q \\ q & i\zeta \end{pmatrix}, Q\right]$. Further the X dependence of the eigenvector $\hat{V}$ may be found from (6.267) which is now in the role of an auxiliary equation. It turns out that when the vector $\hat{V}(\zeta,\lambda)$ is normalized so as to have the behavior $\begin{pmatrix}1\\0\end{pmatrix}e^{-i\zeta X}, \begin{pmatrix}0\\1\end{pmatrix}e^{i\zeta X}$ at the two infinities on R, it is meromorphic on R-{∞} with poles which are independent of X. Thus the analogue of the spectrum in the inverse scattering transform is the Riemann surface. The analogue of the action variables are the poles which are independent of X.

The inverse problem is posed as follows. Given (1) a Riemann surface R of genus 2 with a point at ∞ (ii) a nonspecial divisor of poles $P_1 + P_2$, seek a function $\hat{V}[P(\zeta,\lambda),X]$ meromorphic for $P \in R$ except at $P = \infty$ such that $\hat{V} \sim \begin{pmatrix}1\\0\end{pmatrix}e^{-i\zeta x}, \begin{pmatrix}0\\1\end{pmatrix}e^{i\zeta x}$ at the two infinities with poles at $P_1$ and $P_2$ independent of X. Then $\hat{V}$ (6.272) satisfies (6.272) and (6.267) and q(x) can be constructed at any x.

Another kind of deformation which preserves certain global properties of the system (6.267,268) is associated with the self-similar solutions of (6.269). Let

$$q(x,t) = \frac{1}{(3t)^{1/3}} f\left[\frac{x}{(3t)^{1/3}}\right] , \quad \eta = \frac{x}{(3t)^{1/3}} \quad , \tag{6.274}$$

then $f(\eta)$ satisfies the second Painlevé equation

$$f'' = 2f^3 + \eta f - \nu \tag{6.275}$$

and with the transformation $V(x,t,\zeta) \to V(\eta,\xi)$, $\xi = \zeta(3t)^{1/3}$, we find

$$V_\eta = \begin{bmatrix} -i\xi & f(\eta) \\ f(\eta) & i\xi \end{bmatrix} V \tag{6.276}$$

$$\xi V_\xi = \begin{bmatrix} -4i\xi^3 - 2if^2\xi - i\eta\xi & 4\xi^2 f + 2i\xi f' + \nu \\ 4\xi^2 f - 2i\xi f' + \nu & 4i\xi^3 + 2if^2\xi + i\eta\xi \end{bmatrix} V \quad . \tag{6.277}$$

Now focus attention on (6.277). It is a system of ordinary differential equations with coefficients which depend polynomially on $\xi$ and which also contains as coefficients, $f(\eta)$ and its first two derivatives. Equation (6.277) has an irregular singular point of rank 3 at $\xi = \infty$ and a regular singular point at $\xi = 0$. Associated with each singular point is a monodromy matrix. For example if $\Phi(\xi)$ is the fundamental solution matrix of (6.277) near $\xi = 0$, then encircling $\xi = 0$ we have

$$\Phi(\xi e^{2\pi i}) = \Phi(\xi)J \quad . \tag{6.278}$$

Encircling the point at $\infty$ is a little more complicated as the neighborhood of infinity is divided by antiStokes lines into six sectors in which the fundamental solution matrices $\Psi_j(\xi)$, $j=1,2,\ldots 6$ have the asymptotic expansion $\begin{bmatrix} e^{-\theta} & 0 \\ 0 & e^{\theta} \end{bmatrix}$ with $\theta = 4i\xi^3/3 + i\eta\xi$. The connection matrices between $\Psi_{j+1}$ and $\Psi_j$ contain the Stokes multipliers, i.e., $\Psi_{j+1} = \Psi_j A_j$ where $A_j = \begin{bmatrix} 1 & b_j \\ a_j & 1 \end{bmatrix}$ where either $a_j$ or $b_j$ is zero. The matrices $A_j$, J and the connection matrix A between $\Psi_1$ and $\Phi(\Psi_1 = \Phi A)$ are called the monodromy data. We have the following remarkable result.

If $f(\eta)$ deforms according to (6.275), then all the monodromy data are constant in $\eta$. This is the analogue of the preservation of the spectrum in the inverse scattering transform and the preservation of the Riemann surface associated with the multiperiodic solutions of evolution equations.

The inverse problem then is as follows. Given matrices $\Psi_j, \Phi$ with prescribed asymptotic behavior at $\xi = \infty$ and 0, respectively, and given $A_j$, A, J, we can construct the common system of differential equations in both $\xi$ and $\eta$ which $\Psi_j$, $\Phi$ satisfy. This is the Riemann-Hilbert problem. Then the coefficient $f(\eta)$ in these equations [which is best identified from the second term in the asymptotic expansion of $\Psi_j(\xi)$ at $\xi = \infty$] can be computed and it satisfies (6.275). Details are given in [6.101], and in [6.102] a new type of solution, the multiphase similarity solution, of evolution equations is presented. The motivation for studying these problems was provided by the work of SATO et al. [6.100] whose results suggest that there is a close connection between exactly solvable models in statistical mechanics (such as the two-dimensional Ising model) and exactly integrable evolution equations.

Finally, I want to dedicate this article to my good colleagues Mark Ablowitz, David Kaup and Harvey Segur at Clarkson College and Hermann Flaschka at the University of Arizona and to thank them for the many stimulating hours of fun we have had in discovering these things together. The work was originally supported by the National Science Foundation under Grants No. MPS75-07568 and ATH75-06537a01. Later parts mentioned in Sects.12 and 13 were supported by the Office of Naval Research (N00014-76-C-0867) and the National Science Foundation (MCS75-07548A01).

## *Appendix A.* Orthogonality Relations

Define the inner product to be

$$\langle \underline{u}, \underline{v} \rangle = \int_{-\infty}^{\infty} u \cdot v \, dx \quad . \tag{6A.1}$$

If $L\underline{u} = \underline{u}$ and $L^A\underline{v} = \zeta\underline{v}$, then

$$\int_{-R}^{R} \underline{u}(\zeta) \cdot \underline{v}(\zeta')dx = \frac{1}{2i(\zeta-\zeta')} (u_2v_2 - u_1v_1)_{-R}^{R} \quad . \tag{6A.2}$$

We may use (6A.2) and (6.18) to find

$$\int_{-R}^{R} \Psi(\zeta)\cdot\Psi^A(\zeta')dx = \frac{a^2(\zeta)e^{-2i(\zeta-\zeta')R} - a^2(\zeta')e^{2i(\zeta-\zeta')R}}{2i(\zeta-\zeta')} \tag{6A.3}$$

From (6A.3), we can deduce

$$\begin{aligned}\left\langle \Psi(\zeta)\ ,\ \Psi^A(\zeta')\right\rangle &= -\pi a^2\delta(\zeta-\zeta')\ ,\quad \zeta,\zeta'\ \text{real}\ ,\\ &= 0\ ,\quad \text{otherwise}\ .\end{aligned} \tag{6A.4}$$

Note in particular that coefficient $a^2$ is the square of $a(\zeta,t)$. It is this feature which necessitates the introduction of the derivative states $\chi$ and $\tau$ in the respective bases. Differentiating (6A.3) with respect to $\zeta$ and $\zeta'$, we find,

$$\left\langle \frac{\partial}{\partial\zeta}\Psi(\zeta_k),\ \Psi^A(\zeta_j)\right\rangle = \left\langle \Psi(\zeta_k)\ ,\ \frac{\partial}{\partial\zeta}\Psi^A(\zeta_j)\right\rangle = -\frac{i}{2}a_k'^2\delta_{kj}\ , \tag{6A.5}$$

$$\left\langle \frac{\partial}{\partial\zeta}\Psi(\zeta_k),\ \frac{\partial}{\partial\zeta}\Psi^A(\zeta_j)\right\rangle = -\frac{i}{2}a_k'a_k''\delta_{kj}\ . \tag{6A.6}$$

Similar calculations yield

$$\begin{aligned}\left\langle \bar{\Psi}(\zeta),\ \bar{\Psi}^A(\zeta')\right\rangle &= \pi\bar{a}^2\delta(\zeta-\zeta'),\quad \zeta,\zeta'\ \text{real}\ ,\\ &= 0\ ,\ \text{otherwise}\ .\end{aligned} \tag{6A.7}$$

$$\left\langle \frac{\partial}{\partial\zeta}\bar{\Psi}(\zeta_k),\ \bar{\Psi}^A(\zeta_j)\right\rangle = \left\langle \bar{\Psi}(\zeta_k),\ \frac{\partial}{\partial\zeta}\bar{\Psi}^A(\zeta_j)\right\rangle = -\frac{i}{2}\bar{a}_k'^2\delta_{kj}\ , \tag{6A.8}$$

$$\left\langle \frac{\partial}{\partial\zeta}\bar{\Psi}(\zeta_k),\ \frac{\partial}{\partial\zeta}\bar{\Psi}^A(\zeta_j)\right\rangle = -\frac{i}{2}\bar{a}_k'\bar{a}_k''\delta_{kj}\ . \tag{6A.9}$$

All other inner products are zero.

*Appendix B.* Derivation of (6.146)

By using the relation between $\bar{\phi},\bar{\psi}$ and $\phi,\psi$ (see Sect.6.2) when $r = -q$, we may rewrite (6A.3) and its counterpart with $\Psi$ and $\Psi^A$ replaced by $\bar{\Psi}$ and $\bar{\Psi}^A$ in the form,

$$\int_{-R}^{R}\left(\phi_1^2+\phi_2^2\right)(\zeta')\left(\psi_2^2-\psi_1^2\right)(\zeta)dx = \frac{a^2(\zeta')e^{2i(\zeta-\zeta')R} - a^2(\zeta)e^{-2i(\zeta-\zeta')R}}{2i(\zeta-\zeta')}\ . \tag{6B.1}$$

From this we find,

$$\begin{aligned}\int_{-\infty}^{\infty}\left(\phi_1^2+\phi_2^2\right)(\zeta')\left(\psi_2^2-\psi_1^2\right)(\zeta)dx &= \pi a^2\delta(\zeta-\zeta')\ ,\quad \zeta,\zeta'\ \text{real}\\ &= 0\ ,\ \text{otherwise}\end{aligned}$$

$$\int_{-\infty}^{\infty}\frac{\partial}{\partial\zeta'}\left(\phi_1^2+\phi_2^2\right)(\zeta_k)\left(\psi_2^2-\psi_1^2\right)(\zeta_j) =$$

$$\int_{-\infty}^{\infty}\left(\phi_1^2+\phi_2^2\right)(\zeta_j)\frac{\partial}{\partial\zeta}\left(\psi_2^2-\psi_1^2\right)(\zeta_k) = \frac{i}{2}a_k'^2\delta_{kj}\ ,$$

$$\int_{-\infty}^{\infty} \frac{\partial}{\partial\zeta'}\left(\phi_1^2 + \phi_2^2\right)(\zeta_j)\,\frac{\partial}{\partial\zeta}\left(\psi_2^2 - \dot{\psi}_1^2\right)(\zeta_k) = \frac{i}{2}\,a_k' a_k'' \delta_{kj} \quad . \tag{6B.2}$$

In order to prove the invariance of the two-form

$$\int_{-\infty}^{\infty}\left(\frac{1}{2}\,\delta q \wedge \int_{-\infty}^{x} dy\,\delta q\right)dx \tag{6B.3}$$

we take the first $\delta q$ as given by (6.134) and the $\int_{-\infty}^{x} dy\delta q$ as the integral of (6.139)

$$\begin{aligned}\int_{\infty}^{x} dy\,\delta q = \frac{1}{\pi}\int_{-\infty}^{\infty} d\zeta\,\delta\left(\frac{\bar{b}}{a}\right)&\left(\frac{\phi_1^2+\phi_2^2}{2i\zeta} - \lim_{R\to\infty}\frac{e^{2i\zeta R}}{2i\zeta}\right)\\ - 2i\sum_1^N \delta\beta_k&\left(\frac{\phi_1^2+\phi_2^2}{2i\zeta} - \lim_{R\to\infty}\frac{e^{2i\zeta R}}{2i\zeta}\right)_{\zeta_k}\\ - 2i\sum_1^N \beta_k\delta\zeta_k\,\frac{\partial}{\partial\zeta}&\left(\frac{\phi_1^2+\phi_2^2}{2i\zeta} - \lim_{R\to\infty}\frac{e^{2i\zeta R}}{2i\zeta}\right)_{\zeta_k}\end{aligned} \tag{6B.4}$$

The terms arising from the value of $\phi_1^2 + \phi_2^2$ at $-\infty$ all give negligible contributions; the one in the integrand of the integral leads to a negligible contribution from the Riemann-Lebesgue lemma and the fact that the integrand vanishes at $\zeta = 0$; the ones in the summation vanish exponentially as $\mathrm{Im}\{\zeta_k\} > 0$. Thus, we find in taking the hook product and using (6B.2), (6B.4) becomes after a little calculation,

$$\int_0^{\infty}\frac{1}{2i\pi\zeta}\,\delta\ln b(\zeta)\wedge\delta\ln aa^* + \sum_{k=1}^N \delta\ln b_k \wedge \delta\ln\zeta_k$$

$$= \int_0^{\infty}\delta\frac{\ln aa^*}{-2\zeta\pi}\wedge\delta\mathrm{Arg}\,b(\zeta)d\zeta + \sum_{k=1}^N \delta(-\ln\zeta_k)\wedge\delta\ln b_k \quad . \tag{6B.5}$$

Hence, the conjugate variables in scattering space are $\left[-\frac{1}{2\zeta\pi}\ln aa^*,\ \mathrm{Arg}\,b(\zeta)\right]$, $\zeta$ real and positive, and $(-\ln\zeta_k,\ \ln b_k)$ with these pairs becoming action-angle variables for the special class of integrable Hamiltonians.

*Appendix C.* The Orthogonality Relations and the Preservation of the Two Forms Generated in Connection with the Schrödinger Equation

We make considerable use of the following relations. If $(v_1, v_2)$ and $(w_1, w_2)$ satisfy (6.156), then

$$(v_2w_2)_{xx} + 4\zeta^2(v_2w_2) + 2q(v_2w_2) = 2\left[v_1w_1 - i\zeta(v_1w_2 + v_2w_1)\right] \quad ,$$

$$\left[v_1w_1 - i\zeta(v_1w_2 + v_2w_1)\right]_x = -q(v_2w_2) \quad . \tag{6C.1}$$

In particular, if $v = w = \phi$, then

$$\phi_1^2 - 2i\zeta\phi_1\phi_2 = -q\phi_2^2 + \int_{-\infty}^{x} q_y\phi_2^2 dy \quad , \tag{6C.2}$$

and

$$-\frac{1}{4}(\phi_2^2)_{xx} - \zeta^2\phi_2^2 - q\phi_2^2 + \frac{1}{2}\int_{-\infty}^{x} q_y\phi_2^2 = L_S\phi_2^2 - \zeta^2\phi_2^2 = 0 \quad . \tag{6C.3}$$

We will now state and prove the orthogonality relations.

$$\int_{-\infty}^{\infty} \phi_2^2(\zeta')\psi_{2x}^2(\zeta)dx = 2i\zeta\pi a^2\delta(\zeta - \zeta') \quad , \quad \zeta,\zeta' \text{ real} \quad ,$$

$$= 0 \quad \text{otherwise} \quad ,$$

$$\int_{-\infty}^{\infty} \frac{\partial}{\partial\zeta}\psi_{2x}^2(\zeta_k)\phi_2^2(\zeta_j)dx = \int_{-\infty}^{\infty} \psi_{2x}^2(\zeta_k)\frac{\partial}{\partial\zeta}\phi_2^2(\zeta_j)dx = -\zeta_k a_k'^2\delta_{kj} \quad ,$$

$$\int_{-\infty}^{\infty} \frac{\partial}{\partial\zeta}\psi_{2x}^2(\zeta_k)\frac{\partial}{\partial\zeta}\phi_2^2(\zeta_j)dx = -a_k'(\zeta_k a_k'' + a_k')\delta_{kj} \quad . \tag{6C.4}$$

For the dual set, we obtain

$$\int_{-\infty}^{\infty} \phi_2^2(\zeta)\psi_2^2(\zeta')dx = -2i\zeta\pi a^2\delta(\zeta - \zeta') \quad , \quad \zeta,\zeta' \text{ real} \quad ,$$

$$= 0 \quad \text{otherwise} \quad ,$$

$$\int_{-\infty}^{\infty} \frac{\partial}{\partial\zeta}\phi_{2x}^2(\zeta_k)\psi_2^2(\zeta_j)dx = \int_{-\infty}^{\infty} \phi_{2x}^2(\zeta_k)\frac{\partial}{\partial\zeta}\psi_2^2(\zeta_j)dx = \zeta_k a_k'^2\delta_{jk} \quad ,$$

$$\int_{-\infty}^{\infty} \frac{\partial}{\partial\zeta}\phi_{2x}^2(\zeta_k)\frac{\partial}{\partial\zeta}\psi_2^2(\zeta_j)dx = a_k'(\zeta_k a_k'' + a_k')\delta_{jk} \quad . \tag{6C.5}$$

We will now prove (6C.4). From equations (6.163,165) which $\phi_2^2$ and $\psi_{2x}^2$ satisfy,

$$(\zeta'^2 - \zeta^2)\int_{-R}^{R} \phi_2^2(\zeta')\psi_{2x}^2(\zeta)dx$$

$$= \int_{-R}^{R} \left\{\psi_{2x}^2(\zeta)L_S\phi_2^2(\zeta') - \phi_2^2(\zeta')\left[L_S^A\psi_{2x}^2(\zeta) - \frac{1}{2}q_x\psi_2^2(R)\right]\right\}dx$$

$$= \int_{-R}^{R} \left\{\psi_{2x}^2(\zeta)\left[-\frac{1}{4}\phi_{2xx}^2(\zeta') - q\phi_2^2(\zeta') + \frac{1}{2}\int^{x} q_y\phi_2^2(\zeta')dx\right]\right.$$

$$\left. - \phi_2^2(\zeta')\left[-\frac{1}{4}\psi_{2xxx}^2(\zeta) - q\psi_{2x}^2(\zeta) - \frac{1}{2}q_x\psi_2^2(\xi)\right]\right\}dx \quad .$$

In this expression, the second and fifth terms cancel and there is further cancelation when we integrate the third term by parts, which leaves only a portion which may be integrated. Hence we have, on making further use of (6C.2),

$$(\zeta'^2 - \zeta^2)\int_{-R}^{R} \phi_2^2(\zeta')\psi_{2x}^2(\zeta)dx$$

$$= \frac{1}{4}\left[\phi_2^2(\zeta')\psi_{2xx}^2(\zeta) - \phi_{2x}^2(\zeta')\psi_{2x}^2(\zeta)\right]_{-R}^{R}$$

$$+\frac{1}{2}\left\{\psi_2^2(\zeta)\left[\phi_1^2(\zeta') - 2i\zeta'\phi_1(\zeta')\phi_2(\zeta')\right]\right\}_{x=R}$$

$$= -\zeta(\zeta + \zeta')a^2(\zeta')e^{2i(\zeta-\zeta')R} + \zeta(\zeta + \zeta')a^2(\zeta)e^{-2i(\zeta-\zeta')R}$$

$$-\zeta(\zeta - \zeta')b^2(\zeta')e^{2i(\zeta+\zeta')R} + \zeta(\zeta - \zeta')\bar{b}^2(\zeta)e^{2i(\zeta+\zeta')R}$$

$$-2(\zeta^2 - \zeta'^2)a(\zeta')b(\zeta')e^{2i\zeta R} \quad .$$

The third and fourth terms cancel, for as $R \to \infty$ they act (when divided by $\zeta^2 - \zeta'^2$) like Dirac delta functions $\delta(\zeta + \zeta')$ and $\bar{b}(\zeta) = b(-\zeta)$. The last term also cancels by use of the Riemann-Lebesgue lemma. It also vanishes when $\mathrm{Im}\{\zeta\} > 0$. Then we have,

$$\int_{-R}^{R} \phi_2^2(\zeta')\psi_{2x}^2(\zeta)dx = \frac{\zeta}{\zeta-\zeta'}\, a^2(\zeta')e^{2i(\zeta-\zeta')R} - \frac{\zeta'}{\zeta-\zeta'}\, a^2(\zeta)e^{-2i(\zeta-\zeta')R} + O(1) \tag{6C.6}$$

from which we readily deduce (6C.4). The relations (6C.5) are found in a similar manner.

We use the orthogonality relations to find the symplectic structure. Consider taking the hook product of (6.171) with the integral of (6.172). We obtain after use of the Riemann-Lebesgue lemma and the relations $b\bar{b} = a\bar{a} - 1$, $\bar{b}(\zeta) = b(-\zeta)$, $\bar{a}(\zeta) = a(-\zeta)$, $\gamma_k = b_k/a'_k$, $\beta_k = -1/b_k a'_k$, that

$$\int_{-\infty}^{\infty}\left(\frac{1}{2}\,\delta q \wedge \int_{-\infty}^{x} dy\,\delta q\right)dx = \int_{0}^{\infty}\frac{2i\zeta}{\pi}\,\delta \ln\left(1 - \frac{b\bar{b}}{a\bar{a}}\right)\wedge \delta \ln b(\zeta)d\zeta + \sum_{k=1}^{N} \delta(2\zeta_k^2) \wedge \delta \ln b_k \quad . \tag{6C.7}$$

Now, if $q(x,t)$ is real, $\bar{b}(\zeta) = b^*(\zeta)$, $\bar{a}(\zeta) = a^*(\zeta)$ for real $\zeta$ and $\zeta_k = i\eta_k$. Hence the expression (6C.7) becomes

$$\int_{0}^{\infty} -\frac{2\zeta}{\pi}\,\delta \ln(1 - |R|^2) \wedge \delta \mathrm{Arg}\, b(\zeta)d\zeta + \sum_{k=1}^{N} \delta(-2\eta_k^2) \wedge \delta \ln b_k$$

where $R(\zeta) = b/a$. Hence we have the canonically conjugate variables in scattering space and when H, the Hamiltonian belongs to the class of integrable systems, we have the action-angle variables.

References


6.1 C.S. Gardner, J.M. Greene, M.D. Kruskal, R.M. Miura: Phys. Rev. Lett. *19*, 1095 (1967)

6.2 C.S. Gardner, J.M. Greene, M.D. Kruskal, R.M. Miura: Comm. Pure App. Math. *27* 97 (1974)

6.3 N.J. Zabusky, M.D. Kruskal: Phys. Rev. Lett. *15*, 240 (1965)

6.4 M.J. Ablowitz, H. Segur: Stud. Appl. Math. *45*, 13-44 (1977)

6.5 D.J. Benney: J. Math. Phys. *57*, 52 (1966)

6.6 H. Washimi, T. Taniuti: Phys. Rev. Lett. *17*, 996 (1966)
6.7 S. Leibovich: J. Fluid Mech. *42*, 803-822 (1970)
6.8 N.J. Zabusky: Comp. Phys. Commun. *5*, 1 (1973)
6.9 E. Fermi, J. Pasta, S. Ulam: Los Alamos Report LA 1940 (May 1955); Reproduced in "Nonlinear Wave Motion", in *Lectures in Applied Mathematics*, Vol.15 (A.M.S., Providence, Rhode Island 1974) p.1430
6.10 V.E. Zakharov, A.B. Shabat: Sov. Phys. JETP *34*, 62 (1972)
6.11 P.D. Lax: Comm. Pure Appl. Math. *21*, 467 (1968)
6.12 V.L. Ginzburg, L.P. Pitaevskii: Sov. Phys. JETP *34*, 1240 (1958)
6.13 L.P. Pitaevskii: Sov. Phys. JETP *35*, 408 (1958)
6.14 P.L. Kelly: Phys. Rev. Lett. *15*, 1005 (1965)
6.15 V.I. Baspalov, A.G. Litvak, V.I. Talanov: Second All-Union Symposium on Non-linear Optics Collection of Papers (Nauka, Moscow 1968) (Russian)
6.16 D.J. Benney, A.C. Newell: J. Mathematical Phys. *46*, 133-139 (1967)
6.17 A.C. Newell: "Nonlinear Wave Motion", in *Lectures in Applied Mathematics*, Vol.15 (A.M.S., Providence, Rhode Island 1974) p.157
6.18 T.B. Benjamin, J.E. Feir: J. Fluid Mech. *27*, 417 (1966)
6.19 B.M. Lake, H.C. Yuen, H. Rungaldier, W.E. Ferguson: J. Fluid Mech. *83*, 49-74 (1977)
6.20 M. Wadati: J. Phys. Soc. Jpn. *32*, 1681 (1972)
6.21 M.J. Ablowitz, D.J. Kaup, A.C. Newell, H. Segur: Phys. Rev. Lett. *30*, 1462 (1973)
6.22 M.J. Ablowitz, D.J. Kaup, A.C. Newell, H. Segur: Phys. Rev. Lett. *31*, 125 (1973)
6.23 M.J. Ablowitz, D.J. Kaup, A.C. Newell, H. Segur: Stud. Appl. Math. *53*, 249 (1974)
6.24 G.L. Lamb: Phys. Rev. A*9*, 422 (1974)
6.25 L.D. Faddeev, L.A. Takhtajan: "Essentially Nonlinear One-Dimensional Model of Classical Fourier Theory; preprint (1974)
6.26 R. Miura: J. Math. Phys. *9*, 1202 (1968)
6.27 A.C. Newell: In *Bäcklund Transformations*, ed. by R.M. Miura, Lecture Notes in Mathematics, Vol.515 (Springer, Berlin, Heidelberg, New York 1976)
6.28 M.D. Kruskal: "Nonlinear Wave Motion", in *Lectures in Applied Mathematics*, Vol.15 (A.M.S., Providence, Rhode Island 1974) p.61
6.29 M.J. Ablowitz, D.J. Kaup, A.C. Newell: J. Math. Phys. *15*, 1852 (1974)
6.30 S.L. McCall, E.L. Hahn: Phys. Rev. Lett. *18*, 908 (1967); Phys. Rev. *103*, 183 (1969)
6.31 D.J. Kaup, A.C. Newell: Lett. Nuovo Cimento *20*, 325 (1977)
6.32 H. Flaschka: Theor. Phys. *51*, 703 (1974); Phys. Rev. B*9*, 1924 (1974)
6.33 J. Moser (ed.): *Dynamical Systems, Theory and Applications*, Lecture Notes in Physics, Vol.38 (Springer, Berlin, Heidelberg, New York 1975)
6.34 F. Calogero: Preprint (1975)
6.35 M.J. Ablowitz, J. Ladik: J. Math. Phys. *16*, 598 (1975)
6.36 M.J. Ablowitz, J. Ladik: Stud. Appl. Math. *57*, 1-12 (1977)
6.37 V.E. Zakharov, S.V. Manakov: JETP Lett. *18*, 242 (1973), also preprint of sequel article
6.38 K. Kenyon: Proc. Roy. Soc. A*299*, 141 (1966)
6.39 M.S. Longuet-Higgins, A.E. Gill: Proc. Roy. Soc. A*299*, 120 (1966)
6.40 A.C. Newell: J. Fluid Mech. *35*, 255 (1967)
6.41 A.C. Newell: J. Atmos. Sci. *29*, 64 (1972)
6.42 O.M. Phillips: *The Dynamics of the Upper Ocean* (Cambridge University Press, Cambridge 1966)
6.43 R.C. Davidson: *Methods in Nonlinear Plasma Theory* (Academic Press, New York 1972)
6.44 D.J. Kaup: Stud. Appl. Math. *54*, 165 (1975)
6.45 V.E. Zakharov, S.V. Manakov: Sov. Phys. JETP *42*, 842 (1976)
6.46 A.C. Newell: SIAM J. Appl. Math. *35*, 650-664 (1978)
6.47 D.J. Benney: Stud. Appl. Math. *56*, 81 (1977)
6.48 V.E. Zakharov, A.R. Shabat: Funkts. Anal. Prilozen. *8*, 43 (1974)
6.49 M.J. Ablowitz, R. Haberman: Phys. Rev. Lett. *35*, 1195 (1975)
6.50 F. Calogero: Nuovo Cimento *16*, 423 (1976)

6.51 V.E. Zakharov: Sov. Phys. JETP *35*, 908 (1972)
6.52 V.E. Zakharov, A.M. Rubenchik: Sov. Phys. JETP *38*, 494 (1974)
6.53 V.E. Zakharov, V.S. Synakh: Sov. Phys. JETP *41*, 485 (1976)
6.54 J.W. Miles: J. Fluid Mech. *79*, 157 (1977)
6.55 J.W. Miles: J. Fluid Mech. *79*, 171 (1977)
6.56 A.C. Newell, L.G. Redekopp: Phys. Rev. Lett. *38*, 377 (1977)
6.57 A.C. Newell: In *Proc. of Symposium on Nonlinear Evolution Equations Solvable by the Inverse Spectral Transform*, Rome 1977, ed. by F. Calogero (Pitman, London 1978)
6.58 B.B. Kadomtsev, V.I. Petviashvili: Sov. Phys. Dokl. *15*, 539 (1970)
6.59 D.J. Kaup, A.C. Newell: Adv. Math. *31*, 67-100 (1979)
6.60 H.D. Walhquist, F.B. Estabrook: J. Math. Phys. *16*, 1 (1975); see also [6.57]
6.61 R.M. Miura (ed.): *Bäcklund Transformations*, Lecture Notes in Mathematics, Vol.515 (Springer, Berlin, Heidelberg, New York 1976)
6.62 H.D. Morris: J. Math. Phys. *17*, 1867 (1976)
6.63 H. Flaschka, A.C. Newell: In *Dynamical Systems, Theory and Applications*, ed. by J. Moser, Lectures Notes in Physics, Vol.38 (Springer, Berlin, Heidelberg, New York 1975) p.355
6.64 D.J. Kaup: J. Math. Anal. Appl. *54*, 349 (1976)
6.65 D.J. Kaup, A.C. Newell: SIAM J. Appl. Math. *34*, 37-54 (1978)
6.66 V.E. Zakharov, L.D. Faddeev: Funk. Anal. Priloz. *5*, 18 (1971)
6.67 R. Grimshaw: J. Fluid Mech. *42*, 639-656 (1970); *46*, 611-622 (1971)
6.68 R.S. Johnson: J. Fluid Mech. *60*, 813-824 (1973); Proc. Camb. Philos. Soc. *73*, 183 (1973)
6.69 T. Kakutani: J. Phys. Soc. Jpn. *30*, 272 (1971)
6.70 S. Leibovich, J.D. Randall: J. Fluid Mech. *58*, 481-493 (1973)
6.71 E. Ott, R.N. Sudan: Phys. Fluids *13*, 1432 (1970)
6.72 D.J. Kaup: SIAM J. Appl. Math. *31*, 121-133 (1976)
6.73 D.J. Kaup, A.C. Newell: Proc. Roy. Soc. A*361*, 413-446 (1978)
6.74 A.C. Newell: Rocky Mount. J. Math. *8*, 25-52 (1978)
6.75 D.J. Kaup, A.C. Newell: Phys. Rev. B*18*, 5162-5167 (1978)
6.76 V.I. Karpman, E.M. Maslov: Phys. Lett. *60*A, 307-308 (1977)
6.77 V.I. Karpman, E.M. Maslov: Sov. Phys. JETP *48*, 252 (1978)
6.78 A.R. Bishop: *Solitons in Action* ed. by K. Longren, A.C. Scott (Academic Press, New York 1978) pp.61-68
6.79 M.B. Fogel, S.E. Trullinger, A.R. Bishop: Phys. Rev. Lett. *36*, 1411-1414 (1976); Phys. Rev. B*15*, 1578-1592 (1977)
6.80 J.P. Keener, D.W. McLaughlin: Phys. Rev. A*16*, 777-790 (1977); J. Math. Phys. *18*, 2008-2013 (1977)
6.81 D.W. McLaughlin, A.C. Scott: Appl. Phys. Lett. *30*, 545-547 (1977)
6.82 D.W. McLaughlin, A.C. Scott: *Solitons in Action*, ed. by K. Lonngren, A.C. Scott (Academic Press, New York 1978) pp.201-256
6.83 C.J. Knickerbocker, A.C. Newell: J. Fluid Mech. (1980) to appear
6.84 A.C. Newell: "Soliton Perturbations and Nonlinear Focussing", in *Solitons and Condensed Matter Physics*, ed. by A.R. Bishop, T. Schneider, Springer Series in Solid-State Sciences, Vol.8 (Springer, Berlin, Heidelberg, New York 1978) pp.52-68
6.85 C.J. Knickerbocker, A.C. Newell: Phys. Lett. *75*A, 326 (1980)
6.86 A.C. Newell: J. Math. Phys. *18*, 922 (1977)
6.87 A.C. Newell: J. Math. Phys. *19*, 1126-1133 (1978)
6.88 D.J. Kaup: "Solitons as Particles and the Effects of Perturbations", in *Significance of Nonlinearity in the Natural Sciences* (Plenum Press, New York 1977) pp.97-119
6.89 D.J. Kaup, A. Reimann, A. Bers: Rev. Mod. Phys. *51*, 275 (1979)
6.90 D.J. Kaup: Phys. Rev. A*16*, 704 (1977)
6.91 Y. Kodama: Ph.D. Thesis, Clarkson (1979)
6.92 Y. Kodama, T. Tanuiti: J. Phys. Soc. Jpn. *45*, 298-310, 311-314 (1970)
6.93 J.W. Miles: J. Fluid Mech. *91*, 181-190 (1979)
6.94 J. Wright: Private communication
6.95 A.V. Gurevich, L.P. Petaevskii: Sov. Phys. JETP *38*, 291 (1973)
6.96 O.M. Phillips: J. Fluid Mech. *9*, 193 (1960)

6.97 A.C. Newell: Proc. Roy. Soc. A*365*, 283-311 (1979)
6.98 I.M. Krichever: Funkts. Anal. Prilozen. *11*, 15-31 (1977)
6.99 S.P. Novikov: Rocky Mount. J. Math. *8*, 83-94 (1978)
6.100 S. Sato, T. Miwa, M. Jimbo: A series of papers entitled "Holonomic Quantum Fields": I, Publ. RIMS, Kyoto Univ. *14*, 223-267 (1977); II, ibid. *15*, 201-278 (1979); III, IV, V, RIMS Preprints 260 (1978), 263 (1978), and 267 (1978). The paper we refer to most often is III. See also a series of short notes: "Studies on Honomic Quantum Fields, I-XV", in Proc. Jpn. Acad.
6.101 H. Flaschka, A.C. Newell: "Monodromy and Spectrum Preserving Deformations, Part I", in Commun. Math. Phys. (1980) to appear
6.102 H. Flaschka, A.C. Newell: "Multiphase Similarity Solutions of Integrable Evolution Equations", in *Nonlinear Partial Differential Equations in Engineering and Applied Science*, ed. by R. Stemberg (Marcel Dekker, New York 1980)

# 7. The Inverse Scattering Method

V. E. Zakharov

The present article is devoted to a systematic exposition of different methods of obtaining equations which are integrable by the inverse scattering method. The exposition begins with elementary methods and concludes with the method of dressing operator families. Many results (this refers both to the elementary part and in particular to the method of dressing) are original and published for the first time.

## 7.1 Preliminary Remarks

Since 1967 when GARDNER, GREEN, KRUSKAL and MIURA (GGKM) [7.1] integrated the Korteweg-de Vries (KdV) equation thereby discovering the inverse scattering method, numerous attempts have been made to extend the range of application of this method. The importance of this problem can be explained both by the desire to understand more thoroughly the mathematical nature of the operations which constitute the method and by the hopes of applying this method in mechanics and theoretical physics. How wide is the class of equations which are integrable by the inverse scattering method?

A number of possibilities in the search for integrable equations have already been given in [7.1]. However, they were only realised in 1973 in work carried out by ABLOWITZ, KAUP, NEWELL and SEGUR (AKNS) [7.2]. Before this in 1968 LAX's work [7.3] appeared which put forward a simplified argument for the basic result of [7.1] and at the same time suggested the first method of searching for integrable equations. It was by this method that the first equations other than the KdV equation, integrable by the inverse scattering method [7.4-6] were discovered.

Both the above-mentioned methods have the advantage of being elementary and in essence are contained in the calculation of the conditions for the conservation of the eigenvalues of a certain spectral problem with a known method of varying the eigenfunctions. However, after obtaining the integrable equations, their actual integration or at least the calculation of particular exact solutions requires the development of a technique for solving the corresponding inverse spectral problem, which generally speaking can be difficult. This difficulty was partly overcome in [7.7] which demonstrated the possibility of constructing integrable equations together with a clear indication of the method of calculating their exact solutions. The method used in [7.7] was based on the following idea: linear operators with variable coefficients can be obtained by means of transformation operators from operators with constant coefficients (the usual method of solving inverse spectral problems); if two such constant operators which have a common spectrum are simultaneously transformed the condition of compatibility will assume the form of a nonlinear equation on the coefficients. This is the sought-after integrable equation. The procedure of transforming an equation with constant coefficients into an equation with variable coefficients can, in the language of theoretical physics, be called "dressing" and the entire method the dressing method.

A great advantage of the dressing method is its possible extension to the case of many variables, partially avoding the unpleasant question of the formulation and solubility of the corresponding inverse spectral problem. If we are considering its application to physics then the simplest exact solutions of such multidimensional equations are extremely interesting and informative.

In the present systematic exposition of different methods of obtaining equations which are integrable by the inverse scattering method, the degree of rigour will probably not satisfy a pure mathematician, and hardly anywhere is it stated in which functional spaces the equations which are discovered should be examined. In justification it must be said here that the central and invariant features of the work are algebraic in character. The analytical aspects are strongly influenced both by the concrete choice of the equation and by the choice for it of the boundary value problem. When selecting examples illustrating one or another variant of the method preference has been given to equations which have (at the present time) a physical meaning.

## 7.2 The Method of Finding "L-A" Pairs

Let $\psi(z)$ $(-\infty < z < \infty)$ be the complex vector functions $(\psi = \psi_1, \ldots, \psi_N)$ and $\hat{L}$ be the differential operator

$$\hat{L} = \ell_0 \frac{\partial^n}{\partial z^n} + u_1 \frac{\partial^{n-1}}{\partial z^{n-1}} + \ldots + u_n \quad ; \tag{7.1}$$

here $\ell_0$ is a constant nondegenerate matrix and $u_i(z)$ are variable matrices which have the constant matrices $\ell_i$ as limits at infinity, where

$$||u_i(z) - \ell_i|| < c(\mu)\, e^{-\mu|z|} \quad \text{as} \quad |z| \to \infty \quad \text{for any} \quad \mu > 0 \quad . \tag{7.2}$$

Thus when $z \to \pm\infty$ $\hat{L} \to \hat{L}_0$, where

$$\hat{L}_0 = \ell_0 \frac{\partial^n}{\partial z^n} + \ell_1 \frac{\partial^{n-1}}{\partial z^{n-1}} + \ldots + \ell_n$$

and the differential equation

$$\hat{L}\psi = \lambda\psi \tag{7.3}$$

degenerates into the equation with constant coefficients

$$\hat{L}_0\psi = \lambda\psi \quad . \tag{7.4}$$

Equation (7.4) gives a Riemann surface

$$\det\left\{\sum_{k=0}^{n} \ell_k \zeta^{n-k} - \lambda I\right\} = 0 \tag{7.5}$$

and has for each general point on it a set of fundamental solutions

$$\psi_i(\lambda,z) = \psi_i \exp(\zeta_i z) \qquad i = 1, \ldots, nN \quad . \tag{7.6}$$

Any solution of system (7.3) has the asymptotic character

$$\psi \to \sum_{i=1}^{nN} X_i(\lambda)\psi_i(\lambda,z) \qquad z \to \pm\infty \tag{7.7}$$

where

$$X_i^+(\lambda) = \sum_{j=1}^{nN} S_{ij}(\lambda)X_j^-(\lambda) \quad ; \tag{7.8}$$

the matrix $S_{ij}(\lambda)$, analytic by assumption (7.2) on the Riemann surface (7.5) will be called the full scattering matrix of the operator $\hat{L}$. $S_{ij}(\lambda)$ has an essential singularity at the point of infinity.

If $S_{ij}(\lambda)$ is the full scattering matrix for the operator $\hat{L}$ then it is obviously the full scattering matrix for the operator

$$\tilde{L} = e^{\gamma(z)}\, \hat{L}\, e^{-\gamma(z)} \tag{7.9}$$

as well, where $\gamma(z)$ is an arbitrary matrix commuting with $\ell_0$. From the whole class of $\tilde{L}$ we shall select the canonical operator which can be determined by the condition

$$u_1 = \ell_1 + [\ell_0,Q] \tag{7.10}$$

where $Q(z)$ is a certain matrix function of z. Let $S_{ij}(\lambda)$ be the full scattering matrix for a certain canonical operator $\hat{L}$. We shall say that for the operator $\hat{L}$ the inverse scattering method is determined identically if the operator $\hat{L}$ can be reconstructed uniquely from $S_{ij}(\lambda)$. We must point out that the problem of reconstructing the operator from an arbitrary matrix $S_{ij}(\lambda)$ is strongly overdetermined. This can be seen for instance from a calculation of the number y of functional coefficients of the matrix $S_{ij}(\lambda)$, which is $n^2N^2$, whereas y for the operator $\hat{L}$ taking into account (7.10) is not more than $nN^2 - N$. Consequently the operator $\hat{L}$ can be reconstructed from a set of $q \leqq nN^2 - N$ rational relations $T_i(\lambda)$ from the elements of the matrix $S_{ij}(\lambda)$. When (7.2) is replaced by a weaker condition for the vanishing of $u_i(z)$ as $z \to \pm\infty$, the elements of $T_{ij}(\lambda)$ are determined only for a certain set of points on the Riemann surface (7.5) — at the limit of the continuous and discrete spectrum of the operator $\hat{L}$. A detailed discussion of these questions is beyond the scope of the present chapter.

Let the coefficients of the canonical operator $\hat{L}$ with a given scattering problem further depend on the parameter t, in such a way that $\partial \ell_k/\partial t = 0$. We shall require that for each fixed value of $\lambda$ the semigroup of displacements with respect to t, generated by the differential equation

$$\frac{\partial\psi}{\partial t} + \hat{A}\psi = 0 \quad , \tag{7.11}$$

where

$$\hat{A} = a_0 \frac{\partial^m}{\partial z^m} + v_1 \frac{\partial^{m-1}}{\partial z^{m-1}} + \ldots + v_m$$

($a_0$ is a constant matrix) operates invariantly on the linear manifold of all the solutions of equation (7.3). This requirement is equivalent to satisfying the relation

$$\frac{\partial\hat{L}}{\partial t} = [\hat{L},\hat{A}] \quad . \tag{7.12}$$

The commutator $[\hat{L},\hat{A}]$ is in general an operator of order n + m, and condition (7.12) is equivalent to n + m + 1 equations, the first three of which have the forms

$$\begin{aligned} &[\ell_0,a_0] = 0 \quad , \\ &[\ell_0,v_1] - [a_0,u_1] = 0 \quad , \\ &[\ell_0,v_2] - [a_0,u_2] + [u_1,v_1] + n\ell_0 v_{1z} - m a_0 u_{1z} = 0 \end{aligned} \tag{7.13}$$

as $z \to \pm\infty$, $\hat{A} \to \hat{A}_0$, it being obvious that $[\hat{L}_0,\hat{A}_0] = 0$,

$$\hat{A}_0 = a_0 \frac{\partial^m}{\partial z^m} + a_1 \frac{\partial^{m-1}}{\partial z^{m-1}} + \ldots \quad . \tag{7.14}$$

The relation (7.14) can be regarded as an equation for determining $\hat{A}_0$. This equation with any $\hat{L}_0$ has a solution — it is sufficient to take for example as $\hat{A}_0$

an abitrary operator with constant numerical coefficients. However, we have the following result.

*Theorem* 1. A unique operator $\hat{A}$ exists which has as its limit as $z \to \pm\infty$ the operator $\hat{A}_0$.

*Proof*. Equations (7.13) determine the coefficients $v_k$ recursively—for k and k + 1 the equations have the form

$$[\ell_0, v_k] = f_k \quad , \tag{7.15}$$

$$[\ell_0, v_{k+1}] + [u_1, v_k] + n\ell_0 v_{kz} = f_{k+1} \quad . \tag{7.16}$$

Here $f_k$, $f_{k+1}$ depend only on $v_i$ $(i < k)$. Equation (7.15) determines $v_k$ exactly up to the q-dimensional subspace R of the matrices which commute with $\ell_0$. Multiplying (7.16) by $r_i \in R$, and calculating the trace, we see that

$$\mathrm{Tr}\left\{r_i\left[\ell_0, v_{k+1}\right]\right\} = \mathrm{Tr}\left\{v_{k+1}\left[r_i, \ell_0\right]\right\} = 0 \quad .$$

By virtue of the nondegeneracy of $\ell_0$ we obtain a system of q linear differential equations which are independent of the coefficients of the matrix $v_{k+1}$. Similar equations occur for the $a_k$, and the final result is found by satisfying (7.12).

From (7.13) it is easy to deduce that if the operator $\hat{A}_0$ is canonical then so is $\hat{A}$.

From (7.11) it follows that as $z \to \infty$,

$$\psi_i(\lambda, z, t) = \exp\left[- \int_0^t \hat{A}_0(\zeta_i)dt + \zeta_i z\right]\psi_i \tag{7.17}$$

(the coefficients of $\hat{A}_0$ may depend on time). Hence for the matrix S we obtain

$$S_{ij}(\lambda, t) = \exp\left[\int_0^t \hat{A}_0(\zeta_i)dt\right]S_{ij}(\lambda) \exp\left[- \int_0^t \hat{A}_0(\zeta_j)dt\right] \quad . \tag{7.18}$$

After determining the operator $\hat{A}$ the remaining n of the equations following from (7.12) represent a nonlinear system of evolution equations for $u_i(t)$. It can be written symbolically

$$\frac{\partial u_i}{\partial t} = \mathscr{F}(u_i) \tag{7.19}$$

where $\mathscr{F}(u_i)$ is a certain nonlinear operator. From (7.18) it follows that the Cauchy problem for this system can be solved by the following scheme:

$$u_i\Big|_{t=0} \xrightarrow{\mathrm{I}} S(\lambda, 0) \xrightarrow{\mathrm{II}} S(\lambda, t) \xrightarrow{\mathrm{III}} u_i(z, t) \quad . \tag{7.20}$$

At the first stage of the scheme the matrix $S(\lambda,0)$ must be found from the initial data $u_i(z,0)$ [or more accurately from the set of data $T_i(\lambda)$ sufficient for the reconstruction of $u_i(z)$], then at the second stage (7.18) must be applied and at the third stage the inverse scattering problem for the reconstruction of $u_i(z,t)$ must be solved. This is what the inverse scattering method consists of.

From what has been said above it is clear that (7.19) is reconstructed from the pair of asymptotic "bare" operators $\hat{L}_0$ and $\hat{A}_0$.

We introduce a number of very simple examples.

1) $$\hat{L}_0 = J\frac{\partial}{\partial z} \;;\quad \hat{A}_0 = I\frac{\partial}{\partial z} \;;\quad [I,J] = 0$$

$$\hat{L} = J\frac{\partial}{\partial z} + [J,Q] \;;\quad \hat{A} = I\frac{\partial}{\partial z} + [I,Q] \;.$$

From (7.12) it now follows (see [7.8-10])

$$\frac{\partial}{\partial t}[J,Q] = IQ_zJ - JQ_zI + \big[[J,Q],[I,Q]\big] \;. \tag{7.21}$$

2) $$\hat{L}_0 = \frac{\partial^2}{\partial z^2} \;;\quad \hat{A}_0 = 4\frac{\partial^3}{\partial z^3}$$

$$\hat{L} = \frac{\partial^2}{\partial z^2} - u \;;\quad \hat{A} = 4\frac{\partial^3}{\partial z^3} - 3\left(u\frac{\partial}{\partial z} + \frac{\partial}{\partial z}u\right)$$

$$u_t - 3(uu_z + u_zu) - u_{zzz} = 0 \;. \tag{7.22}$$

3) $$\hat{L}_0 = \frac{\partial^3}{\partial z^3} + s^2\frac{\partial}{\partial z} \;;\quad \hat{A}_0 = -i\beta\frac{\partial^2}{\partial z^2}$$

$$\hat{L} = \frac{\partial^3}{\partial z^3} - \frac{3}{4}\left(u\frac{\partial}{\partial z} + \frac{\partial}{\partial z}u\right) + \frac{3}{2}w + s^2\frac{\partial}{\partial z}$$

$$\hat{A} = -i\beta\left(\frac{\partial^2}{\partial z^2} - u\right)$$

$$u_t = 2\beta w_z$$

$$w_t = \frac{2}{3}\beta\left(-\frac{1}{4}u_{zzz} - s^2u_z + \frac{3}{4}uu_z + \frac{3}{4}u_zu\right) \;. \tag{7.23}$$

The physical meaning of all these equations will be discussed in Sect.7.5.

4) $$\hat{L}_0 = \begin{bmatrix} 1 & 0 \\ 0 & -1 \end{bmatrix}\frac{\partial}{\partial z} \qquad \hat{L} = \hat{L}_0 + \begin{bmatrix} 0 & r \\ q & 0 \end{bmatrix}$$

a) $$\hat{A}_0 = \alpha\begin{bmatrix} 1 & 0 \\ 0 & -1 \end{bmatrix}\frac{\partial^2}{\partial z^2}$$

$$\hat{A} = \hat{A}_0 + \frac{1}{2}\left(\begin{bmatrix} 0 & r \\ q & 0 \end{bmatrix} \frac{\partial}{\partial z} + \frac{\partial}{\partial z} \begin{bmatrix} 0 & r \\ q & 0 \end{bmatrix}\right) + \frac{1}{2} \begin{bmatrix} rq & 0 \\ 0 & -qr \end{bmatrix}$$

$$r_t = \frac{\alpha}{2} (r_{zz} + rqr)$$

$$q_t = - \frac{\alpha}{2} (q_{zz} + qrq) \qquad (7.24)$$

b) $\hat{A}_0 = \begin{bmatrix} 1 & 0 \\ 0 & 1 \end{bmatrix} \frac{\partial^3}{\partial z^3}$

$$\hat{A} = \hat{A}_0 - \frac{3}{4}\left(\begin{bmatrix} rq & -r_z \\ q_z & qr \end{bmatrix} \frac{\partial}{\partial z} + \frac{\partial}{\partial z} \begin{bmatrix} rq & -r_z \\ q_z & qr \end{bmatrix}\right)$$

$$r_t = - \frac{1}{2} r_{zzz} - \frac{3}{4} (r_z qr + rqr_z)$$

$$q_t = - \frac{1}{2} q_{zzz} - \frac{3}{4} (q_z rq + qrq_z) \quad . \qquad (7.25)$$

5) $\hat{L}_0 = \begin{bmatrix} 1 & 0 \\ 0 & 1 \end{bmatrix} \frac{\partial^2}{\partial z^2}$ ; $\hat{A}_0 = \frac{1}{\beta}\begin{bmatrix} 1 & 0 \\ 0 & -1 \end{bmatrix} \frac{\partial}{\partial z}$ ;

$\hat{L} = \hat{L}_0 + u$ ; $\hat{A} = \hat{A}_0 + \frac{1}{\beta} v$ ;

$\beta u_t = \frac{1}{2}\{I,u_z\} + [u,v]$ ; $v_z = - \frac{1}{2} [I,u]$ (see [7.11]) . (7.26)

## 7.3 Elementary Multidimensional Generalisation

The simplest way of realising multidimensional generalisations of the procedure examined above is to look at a single dimensional operator $\hat{L}$ with a multidimensional operator $\hat{A}$ of the type (see [7.12,13])

$$\hat{A} = \sum_{i=1}^{N} f_i(\hat{L}) \frac{\partial}{\partial x_i} + \hat{A}_1 \qquad (7.27)$$

where $f_i(\hat{L})$ are certain fixed polynomials of the operator $\hat{L}$. Since $[\hat{L},f_i(\hat{L})] = 0$ the basic equation (7.12) will now assume the form

$$\frac{\partial L}{\partial t} + [\hat{L},\hat{A}_1] + \sum_{i=1}^{N} f_i(\hat{L}) \frac{\partial \hat{L}}{\partial x_i} = 0 \quad . \qquad (7.28)$$

Equation (7.28) differs from (7.12) only because of the presence of a free term $f_i(\hat{L})\partial\hat{L}/\partial x_i$ and obviously because the strength of theorem 1 clearly determines $\hat{L}$ and $\hat{A}$ from the limiting values $\hat{L}_0$ and $\hat{A}_0$. As an example we shall examine

$$\hat{L} = i\begin{bmatrix}1 & 0\\ 0 & -1\end{bmatrix}\frac{\partial}{\partial z} + \begin{bmatrix}0 & u\\ -u^* & 0\end{bmatrix}$$

$$\hat{A} = \frac{\partial}{\partial x}\hat{L} + \hat{L}\frac{\partial}{\partial x} + \hat{A}_1$$

$$\hat{A}_1 = \frac{1}{2}\begin{bmatrix}w & 0\\ 0 & -w\end{bmatrix} .$$

Condition (7.28) leads to

$$\frac{\partial u}{\partial t} = i\,\frac{\partial^2 u}{\partial x \partial z} + iwu$$

$$\frac{\partial w}{\partial z} = -\,2\,\frac{\partial}{\partial x}|u|^2 \quad . \tag{7.29}$$

Unfortunately (7.29) and other equations of the type (7.28) do not as yet have any physical application.[1] More interesting multidimensional systems arise if the operator $\hat{L}$ itself becomes multidimensional. Let for example both operators $\hat{L}$ and $\hat{A}$ contain derivatives with respect to the variables x and z. Writing down (7.13) we can convince ourselves with simple examples that due to the appearance of mixed derivatives with respect to x and z the number of conditions on the coefficients of the operators $\hat{L}$ and $\hat{A}$ exceeds the number of these coefficients which makes it literally impossible to repeat the above stated procedure. The sole exception is the case where operator $\hat{L}$ contains only the first derivative with respect to the variable x (inclusion of the first derivative with respect to x in the operator $\hat{A}$ is equivalent to a redefinition of the variable t). In this case the operator $\hat{L}$ is replaced by M

$$\hat{M} = \alpha\,\frac{\partial}{\partial x} + \hat{L} \quad , \tag{7.30}$$

while (7.12) is replaced by

$$\frac{\partial L}{\partial t} - \alpha\,\frac{\partial A}{\partial x} = [\hat{L},\hat{A}] \quad . \tag{7.31}$$

[1]With the exception of [7.14].

Equation (7.31) written in terms of the derivatives differs from system (7.13) only by the addition of the derivatives $v_{ix}$ in the appropriate lines. Consequently only some of the first coefficients $v_i$ can be determined from algebraic equations. Partial differential equations arise for the remainder. Theorem 1 can now be reformulated as follows: the actions of the operators $\hat{A}_0$ and $\hat{M}_0 = \alpha(\partial/\partial x) + \hat{L}_0$ clearly define a system of equations for the coefficients of the operators $\hat{L}$ and $\hat{A}$. We now quote examples generalising those of Sect.7.2 to the two-dimensional case.

1) $\hat{L}_0 = J\frac{\partial}{\partial z}$ ; $\hat{A}_0 = I\frac{\partial}{\partial z}$

$$\frac{\partial}{\partial t}[J,Q] - \frac{\partial}{\partial x}[I,Q] = IQ_zJ - JQ_zI + \Big[[J,Q],[I,Q]\Big] . \quad (7.32)$$

2) The operators $\hat{L}$ and $\hat{A}$ from example (7.22) give with the help of (7.31) the system

$$u_t = 2\beta w_z$$

$$w_t = \frac{2}{3}\beta\left(-\frac{1}{4}u_{zzz} - s^2u_z + \frac{3}{4}uu_z + \frac{3}{4}u_zu\right) + \alpha u_x \quad . \quad (7.33)$$

System (7.33) is at the same time a two-dimensional generalisation of both (7.22, 23).

3) $\hat{L}_0 = \begin{bmatrix} \ell+1 & 0 \\ 0 & \ell \end{bmatrix}\frac{\partial}{\partial z}$ ; $\hat{L} = \hat{L}_0 + \begin{bmatrix} 0 & r \\ q & 0 \end{bmatrix}$ ;

$$\hat{A}_0 = \frac{1}{\beta}\begin{bmatrix} a+1 & 0 \\ 0 & a \end{bmatrix}\frac{\partial^2}{\partial z^2} \quad \text{(r and q are assumed to be scalar)}$$

$$\hat{A} = \hat{A}_0 + \frac{1}{2\beta}\left(\begin{bmatrix} 0 & r \\ q & 0 \end{bmatrix}\frac{\partial}{\partial z} + \frac{\partial}{\partial z}\begin{bmatrix} 0 & r \\ q & 0 \end{bmatrix}\right) + \frac{1}{\beta}\begin{bmatrix} r_1 & \phi_1 \\ \phi_2 & r_2 \end{bmatrix}$$

$$\beta r_t = \hat{D}_1 r + (r_1 - r_2)r$$

$$-\beta q_t = \hat{D}_1 r + (r_1 - r_2)q$$

$$\hat{D}_2(r_1 - r_2) + 2\hat{D}_1 rq = 0$$

$$\hat{D}_1 = \alpha^2\frac{\partial^2}{\partial x^2} + 2(\ell - a)\alpha\frac{\partial^2}{\partial x\partial z} + (\ell^2 - 2\ell a - a)\frac{\partial^2}{\partial z^2}$$

$$\hat{D}_2 = \alpha^2\frac{\partial^2}{\partial x^2} + \alpha(2\ell + 1)\frac{\partial^2}{\partial x\partial z} + \ell(\ell + 1)\frac{\partial^2}{\partial z^2} \, . \quad (7.34)$$

Obviously we have obtained a two-dimensional generalisation of (7.24).

Let us examine the spectral problem

$$\hat{M}\psi = \alpha \frac{\partial \psi}{\partial x} + \hat{L}\psi = \lambda\psi \quad . \tag{7.35}$$

It can easily be seen that this problem is degenerate with respect to $\lambda$. In fact the transformation $\psi = \exp(\lambda x/\alpha)\chi$ reduces (7.35) to

$$\alpha \frac{\partial \chi}{\partial x} + \hat{L}\chi = 0 \quad . \tag{7.36}$$

We shall examine (7.36) as $z \to \pm\infty$. It has a solution of the form

$$\chi_i(\lambda,z,x) = \chi_i \exp(\zeta_i z - \lambda x/\alpha) \quad ,$$

where $(\zeta,\lambda)$ are situated on the Riemann surface of (7.5). As $z \to \pm\infty$ a complete solution is expanded into the integral

$$\chi^{\pm}(z,x) = \int \sum_{i=1}^{nN} X_i^{\pm}(\lambda)\chi_i \exp(\zeta_i z - \lambda x/\alpha)d\lambda \quad .$$

The quantities $X_i^{\pm}(\lambda)$ are linked by a linear operator relation

$$X_i^{+}(\lambda) = \int \sum_{j=1}^{nN} S_{ij}(\lambda,\lambda')X_j^{-}(\lambda')d\lambda' \quad .$$

The operator matrix $S_{ij}(\lambda,\lambda')$ is now the complete scattering matrix and the inverse problem consists of reconstructing the coefficients of the operator $\hat{L}$ from the matrix $S_{ij}(\lambda,\lambda')$. That this reconstruction is possible has been finally proved in a number of cases [7.15]. The dependence of the operator $S_{ij}(\lambda,\lambda')$ on time t is given by

$$S_{ij}(\lambda,\lambda',t) = \exp\left[\int_0^t A(\zeta_i)dt\right]S_{ij}(\lambda,\lambda',0) \exp\left[-\int_0^t A(\zeta_j')dt\right] \quad . \tag{7.37}$$

If the coefficients of the operators $\hat{L}$ and $\hat{A}$ do not depend on x, then S $S_{ij}(\lambda,\lambda',t) = S_{ij}(\lambda)\delta(\lambda - \lambda')$ and we arrive at (7.18).

We also mention that (7.31) can conveniently be regarded as the condition for the consistency of (7.11) and

$$\hat{M}\psi = \alpha \frac{\partial \psi}{\partial x} + \hat{L}\psi = 0 \quad . \tag{7.38}$$

## 7.4 Dressing "$\hat{L}$-$\hat{A}$" Pairs

Hitherto we have only dealt with finding the equations which are in principle integrable by the inverse scattering method. In this section we shall describe a method of constructing their exact solutions without making use of the inverse scattering method. As before we start with the "bare" or "undressed" operators $\hat{L}_0$ and $\hat{A}_0$. Let $\hat{K}^{\pm}$ be a Volterra integral operator on the right determined on the same function space as $\hat{L}_0$.

$$\hat{K}^+\psi = \int_z^{\infty} K^+(z,z')\psi(z')dz' \quad . \tag{7.39}$$

On the strength of a known property of Volterra operators, the operator $1 + \hat{K}^+$ is invertible. We shall transform operator $\hat{L}_0$ by means of $1 + \hat{K}^+$, i.e., we shall examine the operator

$$\hat{L} = (1 + \hat{K}^+)\hat{L}_0(1 + \hat{K}^+)^{-1} \quad . \tag{7.40}$$

Generally speaking the operator $\hat{L}$ consists of two parts, a differential operator with variable coefficients depending on $\hat{K}^+$ and an integral Volterra operator on the right (the kernel of the operator $\hat{K}^+$ is assumed to be differentiable for a sufficient number of times). However, we wish to find operators $\hat{K}^+$ such that the integral part of $\hat{L}$ becomes zero. For this we shall examine the Fredholm operator

$$\hat{F}\psi = \int_{-\infty}^{\infty} F(z,z')\psi(z')dz' \quad ,$$

also possessing a sufficiently well-behaved kernel, and factorise this operator, i.e., we shall represent it as the product of two Volterra operators on opposite sides

$$1 + \hat{F} = (1 + \hat{K}^+)^{-1}(1 + \hat{K}^-) \tag{7.41}$$

$$\hat{K}^-\psi = \int_{-\infty}^{z} K^-(z,z')\psi(z')dz' \quad .$$

The operators $\hat{K}^{\pm}$ can be called the Volterra factors of the operator $\hat{F}$. Of all the operators $\hat{F}$ we shall concern ourselves with those that commute with the differential operators. Let, for example, $\hat{F}$ commute with the operator $\hat{M}_0 = \alpha\partial/\partial x + \hat{L}_0$, i.e.,

$$\hat{F}\hat{M}_0 - \hat{M}_0\hat{F} = 0 \quad . \tag{7.42}$$

Applying relation (7.42) to all functions $\psi$ and integrating the expression $\hat{F}\hat{M}_0\psi$ by parts we obtain a differential equation for the kernel of the operator $\hat{F}$

$$\alpha \frac{\partial F}{\partial x} + \sum_{k=0}^{n} \left[ \ell_k \frac{\partial^{n-k}}{\partial z^{n-k}} F + (-1)^{k+1} \frac{\partial^{n-k}}{\partial z'^{n-k}} F\ell_k \right] = 0 \tag{7.43}$$

which is equivalent to the commutability condition (7.42). Equation (7.43) can be written symbolically as

$$\alpha \frac{\partial F}{\partial x} + \hat{L}_0 F - F\hat{L}_0^+ = 0 \quad . \tag{7.44}$$

Expression $F\hat{L}_0^+$ signifies that the operator $\hat{L}_0^+$ is the conjugate of $\hat{L}_0$, differentiation is with respect to the variable z', and its matrix coefficients multiply the kernel F on the right.

It is important to examine the operators $\hat{F}$ which commute with $\hat{M}_0$ because of (7.44).

*Theorem* 2. If the operator $(1 + \hat{F})$ commutes with the differential operator $\hat{M}_0$, and is invertible then both its Volterra factors $(1 + \hat{K}^{\pm})$ transform $\hat{M}_0$ into one and the same operator $\hat{M}$.

*Proof*.

$$\begin{aligned} \hat{M} &= (1 + \hat{K}^+)\hat{M}_0(1 + \hat{K}^+)^{-1} \\ &= (1 + \hat{K}^-)(1 + \hat{F})^{-1}\hat{M}_0(1 + \hat{F})(1 + \hat{K}^-)^{-1} \\ &= (1 + \hat{K}^-)\hat{M}_0(1 + \hat{K}^-)^{-1} \quad . \end{aligned} \tag{7.45}$$

From (7.45) it follows that since both the operators $(1 + \hat{K}^{\pm})$ transform $\hat{M}_0$ into the same operator this operator must be purely differential, its integral Volterra components being zero.

The coefficients of the operator $\hat{M}$ can be calculated from the relation

$$\hat{M}(1 + \hat{K}^+) = (1 + \hat{K}^+)\hat{M}_0 \quad . \tag{7.46}$$

Writing (7.46) in its explicit form, it is easy to calculate the coefficients of the operator $\hat{M}$ explicitly

$$\hat{M} = \alpha \frac{\partial}{\partial x} + \hat{L} \quad , \qquad \text{where} \quad u_1(z) = [\ell_0,\xi_0] + \ell_1 \quad ,$$

$$\begin{aligned} u_2(z) &= (n - 1)\ell_0 \frac{d\xi_0}{dz} + \frac{1}{2}\left\{\frac{d\xi_0}{dz}, \ell_0\right\} + \frac{1}{2}[\ell_0,\xi_1] \\ &\quad + u_1\xi_0 + [\ell_1,\xi_0] \quad . \end{aligned} \tag{7.47}$$

In these formulae

$$\xi_i(z) = \left(\frac{\partial}{\partial z} - \frac{\partial}{\partial z'}\right)^i K(z,z')\Big|_{z=z'} \quad . \tag{7.48}$$

Thus the n coefficients of the operator $\hat{L}$ are expressed in terms of n quantities $\xi_i(z)$. In the case when the operator $\hat{L}_0$ is canonical, $\hat{L}$ is also canonical. This procedure for the explicit construction of the operators $\hat{L}$, $\hat{M}$ from the operators $\hat{L}_0$, $\hat{M}_0$ can be called the dressing of the operators $\hat{L}_0$, $\hat{M}_0$. The dressing of the operator $\partial/\partial t + \hat{A}_0$ can be carried out in a similar way. As a result of this dressing the operator $\partial/\partial t + \hat{A}$ arises, where the coefficients $v_i$ of the operator $\hat{A}$ are expressed in terms of $\xi_i$ by formulae similar to (7.47).

Let the operator $\hat{F}$ now commute simultaneously with operator $\hat{M}_0$ and $\partial/\partial t + \hat{A}_0$, i.e., as well as (7.43) its kernel also satisfies the equation

$$\frac{\partial F}{\partial t} + \hat{A}_0 F - F\hat{A}_0^+ = 0 \quad . \tag{7.49}$$

Then with the help of operator $\hat{K}^+$ the simultaneous dressing of both bare operators is effected. We shall now examine a certain function $\psi_0$ for which

$$\hat{M}_0\psi_0 = \alpha\,\frac{\partial\psi_0}{\partial x} + \hat{L}_0\psi_0 = 0 \quad ,$$

$$\frac{\partial\psi_0}{\partial t} + \hat{A}_0\psi_0 = 0 \quad . \tag{7.50}$$

From formula (7.34) it immediately follows that function

$$\psi = (1 + \hat{K}^+)\psi_0$$

satisfies (7.11,38).

Thus $u_i(z,x,t)$ and $v_i(z,x,t)$, determined from formula (7.47) with the help of operator $\hat{K}^+$ automatically become the solution of (7.31). In the particular case where $\alpha = 0$ and $\hat{M} = \hat{L}$ (7.44) is replaced by

$$\hat{L}_0 F - F\hat{L}_0^+ = 0 \tag{7.51}$$

and we obtain the solution to (7.12). If instead of operator $\partial/\partial t + \hat{A}_0$ we choose the operator $\sum f_i(\hat{L}_0)\partial/\partial x_i + \hat{A}_0$, the consistent solution to (7.51) and to

$$\sum f_i(\hat{L}_0)\,\frac{\partial F}{\partial x_i} + \hat{A}_0 F - F\hat{A}_0^+ = 0 \tag{7.52}$$

will give the solution to an equation of type (7.28). We further observe that all these equations can be examined directly as equations for the quantities $\xi_i(x,z,t)$. In this case the problem of searching for the operator $\hat{A}$ corresponding to a given operator $\hat{L}$ becomes superfluous.

Let us move on to the question of finding the kernel of the operator $\hat{K}^+$ from a given F. Multiplying (7.41) by $(1 + \hat{K}^+)$ and examining the result in the region $z' > z$ we obtain the relation

$$F(z,z') + K^+(z,z') + \int_z^\infty K^+(z,z'')F(z'',z)dz'' = 0 \quad . \tag{7.53}$$

Equation (7.53) can be considered for each $z = z_0$ as an integral Fredholm equation of the second kind for finding the function $K^+(z_0,z')$ [with a given $K^+$ it can also be considered as a Volterra equation for determining $F(z,z')$]. The following scheme has now arisen for constructing exact solutions to (7.31,28,12)

$$F(z,z',x,t) \overset{\text{I}}{\rightarrow} K^+(z,z',x,t) \overset{\text{II}}{\rightarrow} \xi_i(z,x,t) \overset{\text{III}}{\rightarrow} u_i(z,x,t),v_i(z,x,t) \quad . \tag{7.54}$$

At the first stage an arbitrary solution of the system of two equations for F is examined, then (7.53) is solved and the dressing of the bare operators and the calculation of the quantities $u_i$ and $v_i$ are carried out using the known $K^+$. Individual partial solutions can be found by separating the variables. Let (7.53,49) be satisfied. We shall look for their solution in the form

$$F = F_1(z,x,t)F_2(z',x,t) \quad . \tag{7.55}$$

It is easy to see that $F_1$ and $F_2$ must satisfy equations

$$\alpha \frac{\partial F_1}{\partial x} + L_0 F_1 = c_1 F_1$$

$$\frac{\partial F_1}{\partial t} + A_0 F_1 = c_2 F_1 \tag{7.56}$$

$$\alpha \frac{\partial F_2}{\partial x} - F_2 L_0 = -F_2 c_1$$

$$\frac{\partial F_2}{\partial t} - F_2 A_0 = -F_2 c_2 \quad . \tag{7.57}$$

Here $c_1$ and $c_2$ are as yet arbitrary matrix functions of x,t which must be chosen subject to the conditions for the consistency of the system (7.56,57). It is easy to see that for this

$$\frac{\partial c_1}{\partial t} - \alpha \frac{\partial c_2}{\partial t} = [c_1,c_2] \tag{7.58}$$

is necessary. The solution to (7.53) will be sought in the form

$$K^+ = K(z,x,t)F_2(z',x,t) \quad . \tag{7.59}$$

After substituting in (7.53) we will find that

$$K(z,x,t) = - F_1(z,x,t)\left[1 + \int_z^\infty F_2(z,x,t)F_1(z,x,t)dz\right]^{-1} \quad . \tag{7.60}$$

More general solutions can be sought in the form

$$\begin{aligned} F &= \sum_{k=1}^{n} F_1^{(k)}(z,x,t)F_2^{(k)}(z',x,t) \\ K^+ &= \sum_{k=1}^{n} K^{(k)}(z,x,t)F_2^{(k)}(z',x,t) \quad . \end{aligned} \tag{7.61}$$

All the $F_1^{(k)}$, $F_2^{(k)}$ satisfy (7.56,57) and the $K^{(k)}$ are determined by a linear algebraic system of equations with constant coefficients.

All the solutions which involve separation of variables and the reduction of (7.53) to a system of algebraic equations will be called N-soliton solutions.

The condition for the absence of the integral part in the "dressed" operators can be considered as the equation for the kernel of the operator $K^+$. It is easy to see that these equations have the form

$$\alpha \frac{\partial K^+}{\partial x} + \hat{L}K^+ - K^+\hat{L}_0 = 0 \tag{7.62}$$

$$\frac{\partial K}{\partial t} + \hat{A}K^+ - K^+\hat{A}_0 = 0 \quad . \tag{7.63}$$

The dressing method makes it possible at least in principle to solve the Cauchy problem for the equations in question via the following scheme [7.15]

$$u_i\big|_{t=0} \xrightarrow{\text{I}} \xi_i\big|_{t=0} \xrightarrow{\text{II}} K^+\big|_{t=0} \xrightarrow{\text{III}} F\big|_{t=0} \xrightarrow{\text{IV}} F\big|_{t=t} \xrightarrow{\text{V}} K^+\big|_{t=t} \xrightarrow{\text{VI}} \xi_i\big|_{t=t} \xrightarrow{\text{VIII}} u_i\big|_{t=t} \quad . \tag{7.64}$$

At the second stage of the scheme it is necessary to find the kernel $K^+(z,z',x,t)$ at the initial time $t = 0$. This can be done by solving the Cauchy-Goursat problem for (7.62) — the quantities $\xi_i(z,x,t)$ are here considered as data on the characteristic $z' = z$. At the third stage of the scheme $F(z,z')\big|_{t=0}$ must be found. Equation (7.53) may be used for this and we will consider it as a Volterra equation for F. Scheme (7.64) does not differ further from scheme (7.54).

As a very simple example we shall assume $\alpha = 0$; $\hat{L}_0 = \partial^2/\partial z^2$. Then (7.43) assumes the form

$$\frac{\partial^2 F}{\partial z^2} - \frac{\partial^2 F}{\partial z'^2} = 0 \quad . \tag{7.65}$$

In particular $F = F(z + z')$ may be chosen. In this case (7.53) becomes the well-known Marchenko equation for solving the inverse scattering problem for the Schrödinger operator. The question to what extent and in what sense (7.53) solves the inverse scattering problem for the operators $\hat{M}$ and $\partial/\partial t + \hat{A}$ has still not been investigated to any great extent. The common consistent solution to (7.43,49) can be written in the form

$$F(z,z',x,t) = \int \exp\left[i(\lambda z + \lambda' z') + A_0(i\lambda)t + \frac{1}{\alpha_0} L_0(i\lambda)x\right]$$

$$\times \tilde{F}(\lambda,\lambda')\exp\left[- A_0(i\lambda' t) - \frac{1}{\alpha} L_0(i\lambda')x\right]d\lambda d\lambda' \quad . \qquad (7.66)$$

Here $\tilde{F}(\lambda,\lambda')$ is an arbitrary matrix function, $A_0(i\lambda)$ and $L_0(i\lambda)$ represent the operators $\hat{A}_0$ and $\hat{L}_0$. The form of (7.66) given here makes use of the fact that $[\hat{A}_0,\hat{L}_0] = 0$.

We should point out that nowhere does the above-mentioned dressing scheme make use either of the fact that the operators $\hat{A}_0$ and $\hat{L}_0$ commute or of the fact that their coefficients are constant. In principle it would be possible to dress arbitrary operators $\hat{A}$ and $\hat{L}$ with variable coefficients dependent on x, t and z. In this case, however, (7.43,49) may turn out to be inconsistent. The realisation of condition (7.31) is enough for their consistency. In particular the coefficients of the new "bare" operators $\hat{L}$ and $\hat{A}$ can be one of the partial solutions to (7.31) which possibly do not vanish as $z \to \pm\infty$.

## 7.5 The Problem of Reduction and the Physical Interpretation of Examples

Although the inverse scattering method began with equations which have a transparent and sufficiently general physical meaning it is normally not easy to find a rational physical interpretation for the general equation obtained with the help of $\hat{L}$-$\hat{A}$ pairs. Experience shows that these equations most often acquire a physical meaning when additional restrictions, which we shall call reductions, are imposed on the form of the solutions.

The simplest form of reduction consists of fixing the algebra to which the coefficients of the operators $\hat{L}$ and $\hat{A}$ belong. Since when calculating the integrable equations only linear operations and multiplication (not necessarily commutative) took place these coefficients can belong to any associative algebra including non-matrix algebra (for example the algebra of operators in Hilbert space). Further reductions may conveniently be chosen on the examples of (7.21-32).

These equations are nontrivial starting with dimension N = 3. However, even in this very simple case they do not permit a physical interpretation (at the present time) with a general form for the matrices I, J and Q.

Let I, J and Q belong to an algebra which has defined on it an additive involution $A \to \bar{A}$ where

$$\bar{A}_1 \pm \bar{A}_2 = \overline{A_1 \pm A_2}, \ \bar{\bar{A}} = A \quad .$$

Also let the matrices be invariant with respect to this involution ($\bar{I} = I$, $\bar{J} = J$), then (7.21-32) allow a reduction

$$\bar{Q} = Q \quad . \tag{7.67}$$

This means that if we take initial conditions satisfying (7.67) then this condition is satisfied for all t (or x).

Let the algebra be fixed as that of complex $N \times N$ matrices, I and J being diagonal matrices. We shall define the involution by the formula

$$\bar{Q} = C^{-1}Q^{\dagger}C \tag{7.68}$$

where C is a diagonal unitary matrix ($CC^{\dagger} = 1$). The sign $^{\dagger}$ denotes Hermitian conjugation). In the general case $C_{kj} = \exp(i\phi_k)\delta_{kj}$ and (7.68) implies that

$$\bar{Q}_{kj} = \exp[i(\phi_j - \phi_k)]Q^*_{jk} \quad . \tag{7.69}$$

From the condition $\bar{\bar{Q}} = Q$ we obtain $\phi_k = 0, \pi$.

In the particular case when it is assumed that C = 1, $Q_{kj} = Q^*_{jk}$; i.e., we choose Q to be a Hermitian matrix. Let the matrix $J = \mathrm{diag}\, a_i$ ($a_{i+1} > a_i$); $I = \mathrm{diag}\, b_i$. We introduce the set of quantities

$$u_{ik} = \frac{i\, Q_{ik}}{\sqrt{a_k - a_i}} \quad i < k \quad .$$

Then (7.32) is equivalent to the Hamiltonian system

$$\frac{\partial u_{ik}}{\partial t} = i \frac{\delta}{\delta u^*_{ik}} \quad . \tag{7.70}$$

Here

$$H = \frac{i}{2} \sum_{i<k} \left[u^*_{ik}(\underline{v}_{ik} \cdot \nabla u_{ik}) - u_{ik}(\underline{v}_{ik} \cdot \nabla u_{ik})\right]$$

$$+ \sum_{i<k<j} \varepsilon_{ijk}(u_{ik}u_{kj}u^*_{ij} + u^*_{ik}u^*_{kj}u_{ij}) \tag{7.71}$$

where $\nabla$ is the gradient in the xz plane and $\underline{v}_{ik}$ are two-dimensional vectors

$$(v_{ik})_x = -\frac{b_i-b_k}{a_i-a_k} \quad ; \quad (v_{ik})_z = \frac{a_i b_k - a_k b_i}{a_i - a_k} \tag{7.72}$$

$$\varepsilon_{ikj} = \frac{a_i b_k - a_k b_i + a_k b_j - a_j b_k + a_j b_i - a_i b_j}{\sqrt{(a_k-a_i)(a_j-a_k)(a_j-a_i)}} \quad . \tag{7.73}$$

It obvious that system (7.70) only begins to be nontrivial when N = 3. In this case it has the form

$$\begin{aligned} &\frac{\partial u_0}{\partial t} + (\underline{v}_0 \cdot \nabla u_0) = i\varepsilon u_1 u_2 \\ &\frac{\partial u_1}{\partial t} + (\underline{v}_1 \cdot \nabla u_1) = i\varepsilon u_0 u_2^* \\ &\frac{\partial u_2}{\partial t} + (\underline{v}_2 \cdot \nabla u_2) = i\varepsilon u_0 u_1^* \quad , \end{aligned} \tag{7.74}$$

where

$$u_0 = u_{13} \quad ; \quad u_1 = u_{12} \quad ; \quad u_2 = u_{23}$$

$$\underline{v}_0 = \underline{v}_{13} \quad ; \quad \underline{v}_1 = \underline{v}_{12} \quad ; \quad \underline{v}_2 = \underline{v}_{23} \quad ; \quad \varepsilon = \varepsilon_{123} \quad .$$

System (7.74) is a fundamental system of nonlinear optical equations describing the breakdown of a pumping wave $u_0$ into "secondary waves" $u_1$ and $u_2$ and the reverse process of the merging of the secondary waves into the pumping wave (see [7.15,17]), $\underline{v}_0$, $\underline{v}_1$ and $\underline{v}_2$ are the group velocities of the waves. When deducing (7.74) from concrete physical equations (for example from nonlinear Maxwell equations) it may happen that these vectors are not coplanar. However, by transforming to a moving system of coordinates they can be made so. Thus (7.74) describes a general three-dimensional situation.

We now choose the matrix C in the form C = diag(-1,1,1). Here we once again arrive at the system (7.74). Now, however,

$$u_0 = u_{12} \quad ; \quad u_1 = u_{13} \quad ; \quad u_2 = u_{23} \quad ;$$

$$\underline{v}_0 = \underline{v}_{12} \quad ; \quad \underline{v}_1 = \underline{v}_{13} \quad ; \quad v_2 = \underline{v}_{23} \quad .$$

The wave $u_{12}$ is now the pumping wave and $u_{13}$ and $u_{23}$ are the secondary waves. Similarly when selecting C = diag(1,1,-1) wave $u_{23}$ becomes the pumping wave.

If we select C = diag(1,-1,1) then we arrive at the new system

$$\frac{\partial u_0}{\partial t} + (\underline{v}_0 \cdot \nabla u_0) = i\varepsilon u_1^* u_2^*$$

$$\frac{\partial u_1}{\partial t} + (\underline{v}_1 \cdot \nabla u_1) = i\varepsilon u_0^* u_2^* \tag{7.75}$$

$$\frac{\partial u_2}{\partial t} + (\underline{v}_2 \cdot \nabla u_2) = i\varepsilon u_0^* u_1^* \quad .$$

System (7.75) describes the nonlinear phase of "explosive instability" in a medium in which waves with negative energy can propagate. Now

$$u_0 = u_{13} \quad ; \quad u_1 = u_{12}^* \quad ; u_2 = u_{23}^* \quad . \tag{7.76}$$

We now raise the question of possible further reduction of the system described. This can be done by imposing further restriction on the choice of the matrices I and J. Thus we shall assume that in system (7.74) $a_2 = 0$, $(b_1 - b_2)/a_1 = (b_3 - b_2)/a_3$. Then $\underline{v}_1 = \underline{v}_2$ and it can be assumed that $u_1 = u_2$. System (7.75) assumes the form

$$\frac{\partial u_0}{\partial t} + (\underline{v}_0, \nabla u_0) = i\varepsilon u_1^2$$

$$\frac{\partial u_1}{\partial t} + (\underline{v}_1, \nabla u_1) = i\varepsilon u_0 u_1^* \quad . \tag{7.77}$$

System (7.77 is also wellknown in nonlinear optics in connection with the problem of the generation of a second harmonic in a nonlinear medium, see [7.17].

A further reduction can be realised by moving from the algebra of complex matrices to the algebra of real matrices. This can be realised by assuming that in the equations all the $u_{ij}$ are purely imaginary

$$u_{ij} = iw_{ij}$$

$$\frac{\partial w_0}{\partial t} + (\underline{v}_0 \cdot \nabla w_0) = -\varepsilon w_1 w_2$$

$$\frac{\partial w_1}{\partial t} + (\underline{v}_1 \cdot \nabla w_1) = \varepsilon w_0 w_2 \tag{7.78}$$

$$\frac{\partial w_2}{\partial t} + (\underline{v}_2 \cdot \nabla w_2) = \varepsilon w_0 w_1 \quad .$$

An analogous system arises from (7.75). These real systems correspond to the case of "exact resonance" in nonlinear optics. System (7.78) is remarkable because in the homogeneous case when $\nabla w_i = 0$ it coincides with Euler's equations for the free rotation of a solid body. This coincidence is not accidental. As MANAKOV [7.18] has shown (7.21) also contains the equations for the free movement of an N-dimensional solid body. In actual fact these equations have the form [7.19]

$$\dot{M} = [M,\Omega] \quad , \tag{7.79}$$

where M and $\Omega$ are real antisymmetric matrices, $M = I_0\Omega + \Omega I_0$, where $I_0$ is a positive definite symmetric matrix (the inertia tensor). We transform matrix $I_0$ to its principal axes and assume that $I_0 = \mathrm{diag}\, i_k$. Putting $J = I^2$, $I = I_0$, $M = [J,Q]$, $= [I,Q]$ and discarding the derivatives with respect to x and z we can see that the system of equations (7.21) coincides with (7.79).

Of the complex multidimensional system (7.70) with N = 4 [7.20] a logical physical interpretation has as yet been found only for one system, with the additional condition $a_1 - a_2 = a_3 - a_4$; $b_1 - b_2 = b_3 - b_4$. This system arises as a result of the further reduction $u_{13} = -u_{24}$, $u_{12} = u_{34}$ and leads to the problem of the interaction of four waves.

$$\frac{\partial u_{12}}{\partial t} + (\underline{v}_{12} \cdot \nabla u_{12}) = iq(u_{14}u_{13}^* + u_{13}u_{23}^*)$$

$$\frac{\partial u_{13}}{\partial t} + (\underline{v}_{13} \cdot \nabla u_{13}) = iq(u_{14}u_{12}^* + u_{13}u_{12}^*)$$

$$\frac{\partial u_{23}}{\partial t} + (\underline{v}_{23} \cdot \nabla u_{23}) = iqu_{13}u_{12}^* \tag{7.80}$$

$$\frac{\partial u_{14}}{\partial t} + (\underline{v}_{14} \cdot \nabla u_{14}) = iqu_{12}u_{23}^*$$

$$(\underline{v}_{12} - \underline{v}_{14}) \cdot (\underline{v}_{13} - \underline{v}_{23}) + (\underline{v}_{13} - \underline{v}_{14}) \cdot (\underline{v}_{12} - \underline{v}_{23}) = 0$$

$$q = \varepsilon_{123} = \varepsilon_{124} = -\varepsilon_{234} = -\varepsilon_{134} \quad .$$

System (7.80) describes the breakdown of the pumping wave $u_{13}$ into secondary waves $u_{12}$ and $u_{23}$ in the presence of a noninteracting (anti-Stokes) wave $u_{14}$ which in its turn can be broken down into the pumping $u_{13}$ and a secondary Stokes wave $u_{12}$. Equations (7.80) are met with in problems in nonlinear wave physics [7.21]. A full description of the reduction of (7.21,32) has not yet been given.

Equation (7.22) in scalar algebra is the famous Korteweg-de Vries (KdV) equation with which the inverse scattering method began. Its physical interpretation requires no explanation. We shall only point out that in commutative matrix algebra

$$u = \begin{bmatrix} u_0 & u_1 \\ 0 & u_0 \end{bmatrix} \quad , \tag{7.81}$$

(7.22) splits into two equations

$$u_{0t} = 6u_0u_{0z} + u_{0zzz}$$

$$u_{1t} = 6u_1u_{0z} + 6u_0u_{1z} + u_{1zzz} \quad , \tag{7.82}$$

of which the first is the normal KdV equation and the second is a KdV equation linearised about a given solution $u_0$. It frequently happens that the use of triangular matrices of the type (7.81) makes possible a study of linearised integrable systems.

System (7.23) with different choices of $\beta$ and s leads in scalar algebra to one of four canonical forms

$$u_{tt} \pm u_{xx} \pm u_{xxxx} + (u^2)_{xx} = 0 \quad . \tag{7.83}$$

The physical meaning of all the variants of this system is sufficiently clear.

System (7.24) is comprehensible if $\alpha = i$ and the reduction $q = \pm r^{\dagger}$ occurs. Then it leads to one equation

$$2r_t = i(r_{zz} \pm rr^{\dagger}r) \quad . \tag{7.84}$$

In scalar algebra $r^{\dagger} = r^*$, and this is the nonlinear Schrödinger equation which frequently arises in nonlinear wave physics in connection with problems of the self-focusing type [7.17]. In matrix algebra of the type

$$r = \begin{bmatrix} 0 & \dots & r_1 \\ \dots & \dots & \dots \\ 0 & \dots & r_n \end{bmatrix} \quad ,$$

it leads ([7.18]) to the system of nonlinear Schrödinger equations

$$-2ir_{nt} = r_{nzz} \pm \sum_{i=1}^{n} |r_i|^2 r_n \quad . \tag{7.85}$$

Equation (7.85) has a curious continuum limit

$$-2ir_t(t,z\xi) = \frac{\partial^2}{\partial z^2} r(t,z,\xi) \pm \int |r(t,z,\xi')|^2 d\xi' r(t,z,\xi) \tag{7.86}$$

where $\xi$ is a vector parameter—integration is carried out over a certain region in the space of this parameter. The system of equations of type (7.85) with $n = 2$ describes the self-interaction of electromagnetic waves with different polarisations and is encountered in nonlinear optics. It would be of great interest to examine (7.24) in the algebra of Heisenberg operators given by the commutation or anticommutation relations

$$[\psi(z),\psi(z')] = \delta(z - z') \tag{7.87}$$

$$\{\psi(z),\psi(z')\} = \delta(z - z') \quad . \tag{7.88}$$

In these instances (7.84) would describe a one-dimensional Bose or Fermi gas with a point interaction. In spite of the fact that appropriate theories have advanced a long way (the spectrum of the Hamiltonian operator and the statistical sum are well known) the use of the inverse scattering method could here be useful.

Equation (7.75) has been examined only in scalar algebra $q = \pm r^*$. In this there occurs the so-called "modified" KdV (or MKdV) equation

$$r_t + \frac{1}{2} r_{zzz} \pm \frac{3}{2}|r|^2 r_z = 0 \quad . \tag{7.89}$$

Both these equations are universal and are used to describe waves in media with weak dispersion in cases where cubic nonlinearity predominates.

System (7.33) is equivalent to equation

$$\frac{\partial^2 u}{\partial t \partial z} = \frac{\partial^2}{\partial z^2}\left(-\nu^2 u + \frac{1}{4} u_{zz} + \frac{3}{4} u^2\right) + \frac{3}{4} \beta^2 \frac{\partial^2 u}{y^2} \quad , \tag{7.90}$$

apparently first examined by KADOMTSEV and PETVIASHVILI [7.22]. This equation generalises the KdV equation when there is weak dependence on the transverse coordinate y. We shall examine only one of the numerous variants of (7.34). When $a = 0$ this system is reduced to the form ($\beta = i$, $q = \pm r^*$, $\alpha = 1$, the algebra is scalar)

$$ir_t = \frac{\partial^2 r}{\partial \xi^2} + ur \tag{7.91}$$

$$\frac{\partial u}{\partial \eta} = \pm \frac{\partial}{\partial \xi} |r|^2 \quad .$$

Here

$$u = r_1 - r_2 \qquad \frac{\partial}{\partial \xi} = \frac{\partial}{\partial x} + \ell \frac{\partial}{\partial z}$$

$$\frac{\partial}{\partial \eta} = \frac{\partial}{\partial x} + (\ell + 1) \frac{\partial}{\partial z} \quad .$$

System (7.91) is met with in plasma physics. When $\eta = t$ it describes the interaction of near-sonic Langmuir solitons [7.23,24].

## 7.6 Two Dimensional Instability of Solitons [7.25]

We shall move on to examine physical problems which can be solved by using the method presented in Sect.7.4. We shall restrict ourselves to very simple examples encompassed by the "single soliton" exact solution (7.60).

Let us examine (7.90. Equations (7.43,49) have the form

$$\frac{\partial F}{\partial t} = \frac{\partial^3 F}{\partial z^3} + \frac{\partial^3 F}{\partial z'^3} - \nu^2 \left(\frac{\partial F}{\partial z} + \frac{\partial F}{\partial z'}\right) \tag{7.92}$$

$$\beta \frac{\partial F}{\partial y} + \frac{\partial^2 F}{\partial z^2} - \frac{\partial^2 F}{\partial z'^2} = 0 \quad . \tag{7.93}$$

These equations have a simple common solution

$$F = F_0 = 2\nu\, e^{-\nu(z+z')} \quad . \tag{7.94}$$

From (7.53) we now obtain

$$K = K_0 = \frac{2\nu\, e^{-\nu(z+z')}}{1+e^{-2\nu z}} \quad . \tag{7.95}$$

The dressing formulae (7.47) give in this instance

$$u = 2\frac{d}{dz} K(z,z,x,t) \quad . \tag{7.96}$$

From (7.95,96) we have

$$u = u_0 = \frac{2\nu^2}{\cosh^2 \nu x} \quad . \tag{7.97}$$

The solution (7.97) is a soliton—an isolated undistorted wave in its own rest frame in which (7.90) is written. There is great physical interest in the problem of the instability of the soliton (7.97) with respect to transverse disturbances. This problem was partially solved by KADOMTSEV and PETVIASHVILI [7.22] who showed that if it is assumed that

$$u = u_0 + \delta u \quad ; \quad \delta u \cong e^{i\Omega t + ipy} \quad ,$$

then for small values of p

$$\Omega^2 = \beta^2 \nu^2 p^2 + \dots \quad . \tag{7.98}$$

Thus the soliton is stabke if $\beta^2 > 0$ and unstable if $\beta^2 < 0$. In the stable case (7.90) describes weakly nonlinear waves in a medium with the dispersion relation

$$\omega_k^2 = c^2 k^2 (1 - \lambda^2 k^2 + \dots) \quad , \tag{7.99}$$

while in the unstable case the dispersion relation is

$$\omega_k^2 = c^2 k^2 (1 + \lambda^2 k^2 + \dots) \quad . \tag{7.100}$$

Here $\lambda$ is the dispersion length of the medium. If $\ell_{\parallel}$ is the transverse dimension of the soliton and $\ell_{\perp}\sim 1/p$ is the dimension of the disturbance, then the inequality

$$\frac{1}{p} \gg \ell_{\parallel} \gg \lambda \tag{7.101}$$

is satisfied. The inverse scattering method makes possible a complete solution of the problem of the instability of the soliton because we can find the function $\Omega^2(p)$. For this we shall examine (7.90) in the algebra of triangular matrices of the type (7.81) already mentioned. The matrices F and K also belong to this algebra. For $u_1$ we obtain the linearisation of (7.90) while (7.53) acquires the form

$$F_1(z,z',t,y) + K_1(z,z',t,y) + \int_z^\infty K_1(z,z'',t,y)F_0(z'',z',t,y)dz''$$

$$+ \int_z^\infty K_0(z,z'',t,y)F_1(z'',z',t,y)dz'' = 0 \quad . \tag{7.102}$$

Equation (7.102) is easily solved. Assuming that

$$F_1 = \phi(y,t)\ e^{-\eta z - kz'} \quad ,$$

we find that

$$K_1(z,z,y,t) = \phi(y,t)\ e^{-(\eta+k)z}\left(-1 + \frac{2\nu}{(\nu+k)(1+e^{2\nu z})}\right)$$

$$\times \left(1 + \frac{1}{1+e^{2\nu z}}\right) \quad . \tag{7.103}$$

From the conditions for the vanishing of $u_1(z,y,t)$ as $z \to \pm\infty$ and

$$u_1(z,y,t) = 2\frac{d}{dz}K_1(z,z,y,t)$$

we find that $k = \nu$ and $\mathrm{Re}(\nu - \eta) > 0$. Now from (7.92,93) we obtain

$$\phi(y,t) = \phi_0\ e^{i\Omega t + ipy} \quad ; \quad \Omega^2 = \beta^2 p^2(\nu^2 - i\beta p) \quad . \tag{7.104}$$

The quantity $\beta$ must be chosen to satisfy the conditions $\mathrm{Im}\,\Omega \geqq 0$, as $|p| \to \infty$. Where $\beta^2 > 0$, formula (7.104) describes the spectrum of the damped oscillations of a soliton; where $\beta^2 < 0$ it describes the growth of the instability of the soliton. As $p^2 \to 0$ Kadomtsev's and Petviashili's result (7.98) follows from (7.104).

Let us return to scalar algebra and examine a solution of the type (7.55) assuming that

$$F = \psi(z,y,t)\psi^*(z',y,t) \quad . \tag{7.105}$$

In the scalar case the choice of $c_1$ and $c_2$ does not affect the final form of the solution and it can be assumed that $c_1 = c_2 = 0$.

$$\psi(z,y,t) = \int a(\eta)\exp[\eta(\nu^2 - \eta^2)t + i(\eta^2 - \nu^2)y - \eta x]d\eta \tag{7.106}$$

where $a(\eta)$ is an arbitrary function. The solution to (7.90) is now given by the formula

$$u(z,y,t) = -2\frac{d}{dz}\frac{|\psi(,y,t)|^2}{1 + \int\limits_z^\infty |\psi(z',y,t)|^2 dz'} \quad . \tag{7.107}$$

In particular if

$$\psi(z,y,t) = \int\limits_0^\nu a(\eta)\exp[\eta(\nu^2 - \eta^2)t + i(\eta^2 - \nu^2)y - \eta x]d\eta + \sqrt{2\nu}\, e^{-\nu x} \quad , \tag{7.108}$$

we obtain a solution tending to the soliton (7.97) as to $\to -\infty$. This solution describes the development of the instability of a soliton, varying with different choices of $a(\eta)$. If $a(\eta) = a\delta(\eta - \eta_0)(\eta_0 < \nu)$, then as a result of the development of the instability a soliton arises with a lower amplitude $\eta_0$. For a general choice of $a(\eta)$ the initial soliton disappears altogether. The energy contained in it transfers into an oscillating background uniformly diminishing with respect to x and t. Similar solutions describing the dampled nonlinear oscillations of a soliton can also be obtained in the stable case $\beta^2 > 0$.

## 7.7 Exact Solutions of Equations in Nonlinear Optics [7.26]

We shall now examine the exact solution of system (7.74) one of the basic systems of equations of nonlinear optics. We shall first observe without resorting to proof that the principal integral equation (7.53) can be replaced by

$$\begin{aligned} &F(z,z') + K(z,z') + \int\limits_z^\infty K(z,z'')CF(z'',z')dz'' \\ &- \int\limits_{-\infty}^z K(z,z'')(1 - C)F(z'',z')dz'' = 0 \quad , \end{aligned} \tag{7.109}$$

where C is an arbitrary matrix commuting with all of the $\ell_i$. We shall further observe that system (7.74) has trivial exact solutions when only one of the quantities $u_i$ differs from zero, for example $u_0 = u_1 = 0$; $u_2 = u(x - v_{2x}t, z - v_{2z}t)$. We shall introduce the dressing procedure on the basis of this exact solution of system (7.74). Here the matrix function F satisfies the two equations

$$\frac{\partial F}{\partial x} + I\frac{\partial F}{\partial z} + \frac{\partial F}{\partial z'} I + [I,P(z,x,t)]F - F[I,P(z',x,t)] = 0 \qquad (7.110)$$

$$\frac{\partial F}{\partial x} + J\frac{\partial F}{\partial z} + \frac{\partial F}{\partial z'} J + [J,P(z,x,t)]F - F[J,P(z',x,t)] = 0 \quad , \qquad (7.111)$$

where the matrix P has the form

$$P = - i\sqrt{a_2 - a_3}\begin{bmatrix} 0 & 0 & 0 \\ 0 & 0 & u \\ 0 & u^* & 0 \end{bmatrix} \qquad (7.112)$$

and $u(\xi,\eta)$ is an arbitrary matrix function.

We shall apply the following restriction on the form of the matrix F:

$$F = \begin{bmatrix} 0 & F_{12} & F_{13} \\ F_{21} & 0 & 0 \\ F_{31} & 0 & 0 \end{bmatrix} \quad , \qquad (7.113)$$

and we shall choose the matrix C in the form

$$C = \begin{bmatrix} 0 & 0 & 0 \\ 0 & 1 & 0 \\ 0 & 0 & 1 \end{bmatrix} \quad . \qquad (7.114)$$

It can be shown by looking at all possible u and all possible solutions of the system (7.110), (7.111) of the form (7.113) for which the integrals in (7.109) combine with C chosen in the form (7.114), that we obtain a general solution of the system (7.34). Here $u_0$, $u_1 \to 0$, $u_2 \to u$ as $z \to \pm\,\infty$.

System (7.74) is physically most interesting if the $u_i$ depend on three spatial coordinates. However, changing to a system of coordinates where the group velocity of one of the waves (for example $u_0$) is equal to zero, the system can be reduced to an effectively two-dimensional one for which it is sufficient to combine the coordinate plane with the plane of the vectors $v_2$ and $v_3$. Here the derivative along the transverse coordinate to this plane disappear. Let this coordinate by y. We shall assume that $b_1 = b_3 = 0$, then

$$v_{1x} = \frac{b}{a_1 - a_2} \quad ; \quad v_{1z} = \frac{a_1 b}{a_1 - a_2} \quad ;$$

$$v_{2x} = \frac{b}{a_2 - a_3} \quad ; \quad v_{2z} = \frac{a_2 b}{a_2 - a_3} \quad .$$

The numbers $a_1$, $a_2$ and $a_3$ have been normalised so that $\varepsilon = 1$. The reduction of the function F

$$F^{\dagger}(z,z') = F(z',z) \qquad (7.115)$$

corresponds to the choice of reduction (7.67). Formulae (7.47) give

$$q_0 = - i\sqrt{a_1 - a_3}\,u_0 = K_{13}(z,z,x,t) \quad ,$$

$$q_1 = i\sqrt{a_1 - a_2}\,u_1 = K_{12}(z,z,x,t) \quad , \tag{7.116}$$

$$q_2 = i\sqrt{a_2 - a_3}\,u_2 = K_{23}(z,z,x,t) + u \quad .$$

We shall examine the simplest case $u = 0$. Then

$$F_{12} = F(z - a_1x)\phi(z' - a_2x - bt) \quad ;$$

$$F_{13} = F(z - a_1x)\Psi(z - a_3x) \quad ;$$

$$\int_{-\infty}^{\infty} |F(s)|^2 ds = J_1 \quad ; \quad \int_{-\infty}^{\infty} |\Phi(s)|^2 ds = J_2 \quad ; \tag{7.117}$$

$$\int_{-\infty}^{\infty} |\Psi(s)|^2 ds = J_3 \quad .$$

It is easy to verify that (7.117) with arbitrary complex $F$, $\Phi$ and $\Psi$ is the solution to the system (7.110,111).

From (7.60) we find that

$$q_0 = - \frac{F\Psi}{\Delta} \quad ; \quad q_1 = - \frac{F\Phi}{\Delta} \quad ;$$

$$q_2 = \frac{\Phi^*\Psi}{\Delta} \int_{-\infty}^{z-a_1x} |F(s)|^2 ds \quad , \tag{7.118}$$

where

$$\Delta = 1 + \int_{-\infty}^{z-a_1x} |F(s)|^2 ds \left[ \int_{z-a_2x-bt}^{\infty} |\phi(s)|^2 ds + \int_{z-a_2x}^{\infty} |\psi(s)|^2 ds \right] \quad .$$

Asymptotically as $t \to - \infty$ solution (7.118) breaks down into a packet of pumping waves $u_0$ and a packet of waves $u_1^-$ with integrated intensities

$$I_0^- = \int |u_0(x,z)|^2 dx\, dz = q \ln(1 + J_1J_0) \quad ,$$

$$I_1^- = \int |u_1^-(x,z)|^2 dx\, dz = q \ln\left(1 + \frac{J_1J_2}{1+J_1J_2}\right) \quad .$$

As $t \to + \infty$ all three packets are present with intensities

$$I_0^+ = q \ln \left(1 + \frac{J_1J_3}{1+J_1J_2}\right) \quad ; \quad I_2^+ = I_1^+ - I_1^- \quad ;$$

$$I_1^+ = q \ln(1 + J_1J_2) \quad ; \quad q = \frac{(a_1-a_2)(a_2-a_3)}{b^2(a_1-a_3)} \quad .$$

If $J_1, J_3 \gg 1; J_1 \gg J_2, J_3$ we have the problem of the interaction of an intense packet of pumping waves with a small wave packet $u_2$. As can be seen from the intensities almost complete breakdown of the pumping occurs. The asymptotic form of the packets $u_1$, and $u_2$ depends on the function $\phi$ which does not enter into the asymptotic form of the pumping $u_0^-$. This fact reflects the breakdown instability of the pumping wave.

The whole of the above-mentioned technique can also be applied to the problem of explosive instability (7.75). The corresponding solution has the form ($\nabla_0 = 0, \ \varepsilon = 1$):

$$q_0 = - \frac{F\psi}{\Delta} \quad ; \quad q_1 = \frac{F^*\phi}{\Delta} \quad ; \quad q_2 = \frac{\phi^*\psi^*}{\Delta} \int_{-\infty}^{z-a_1x} |F(s)|^2 ds$$

$$\Delta = 1 - \int_{-\infty}^{z-a_1x} |F(s)|^2 ds \left[\int_{z-a_2x-bt}^{-\infty} |\phi(s)|^2 ds + \int_{z-a_2x}^{\infty} |\psi(s)|^2 ds\right] \quad . \tag{7.119}$$

With the evolution of the solution (7.119) it is possible for the denominator to go to zero at a certain point in the final moment of time. This proves that the development of explosive instability for three-dimensional wave packets leads to a point "collapse".

## 7.8 "$\hat{L},\hat{A},\hat{B}$" Triad

The possibility of reconstructing the coefficients of a two-dimensional operator from the scattering matrix given at one point of the spectrum led MANAKOV [7.27] to the discovery of a new class of integrable systems which do not permit the introduction of $\hat{L},\hat{A}$ pairs.

Let a differential operator $\hat{L}$ with respect to the two variables z and x be given together with the associated equation

$$\hat{L}\psi = 0 \quad . \tag{7.120}$$

The solutions to (7.120) form a linear space $\tilde{\Psi}$. We shall require that the semigroup of displacements with respect to t, which are determinable by the differential equation (7.111)

$$\psi_t + \hat{A}\psi = 0 \quad ,$$

acts on $\tilde{\Psi}$ invariantly. For this it is necessary and sufficient that the relation

$$\frac{\partial L}{\partial t} - [\hat{L},\hat{A}] = \hat{B}\hat{L} \tag{7.121}$$

is satisfied, where $\hat{B}$ is a certain differential operator. It is obvious that the operator $\hat{A}$ is not determined uniquely here but only up to the transformation

$$\hat{A} \to \hat{A} + \hat{C}\hat{L} \quad ; \quad \hat{B} \to \hat{B} + [\hat{L},\hat{C}] \tag{7.122}$$

where $\hat{C}$ is an arbitrary differential operator. The relation (7.121) gives a certain system of equations soluble by the inverse scattering method using the scheme (7.20-64). We shall describe (without proof) a class of operators $\hat{L}$ leading to equations of the type (7.121). We shall choose $\hat{L}$ in the form

$$\hat{L} = \hat{L}^{(0)} \frac{\partial}{\partial x} + \hat{L}^{(1)} \tag{7.123}$$

where $\hat{L}^{(0)}$ and $\hat{L}^{(1)}$ are $N \times N$ matrix differential operators with respect to z.

$$L^{(0)} = \ell_0^{(0)} \frac{\partial^{m_1}}{\partial z^{m_1}} + u_1 \frac{\partial^{m_1-1}}{\partial z^{m_1-1}} + \dots \tag{7.124}$$

$$L^{(1)} = v_0 \frac{\partial^{m_2}}{\partial z^{m_2}} + v_1 \frac{\partial^{m_2-1}}{\partial z^{m_2-1}} \dots \quad . \tag{7.125}$$

The matrix $\ell_0^{(0)}$ is constant and the matrix $v_0$ is constant if $m_2 \geqq m_1$. In the case where $m_2 = m_1 - 1$, $v_0$ is a variable matrix. As $z \to \infty$ the operator L assumes the limiting value $\hat{L}_0$. As operators $\hat{A}$ and $\hat{B}$ we can choose arbitrary differential operators with respect to x and z, characterised by their limiting values $A \to A_0$, $\hat{B} \to \hat{B}_0$ as $z \to \infty$, where

$$-[\hat{L}_0,\hat{A}_0] = \hat{B}_0\hat{L}_0 \quad . \tag{7.126}$$

The operators $\hat{A}$ and $\hat{B}$ can conveniently be represented in the form

$$\begin{aligned} \hat{A} &= \hat{A}^{(0)} \frac{\partial^n}{\partial x^n} + \dots + \hat{A}^{(n)} \\ \hat{B} &= \hat{B}^{(0)} \frac{\partial^n}{\partial x^n} + \dots + \hat{B}^{(n)} \end{aligned} \tag{7.127}$$

where $\hat{A}^{(i)}$ and $\hat{B}^{(i)}$ are differential operators with respect to z. Substitution in (7.121) gives a system of equations for the coefficients of the operators $\hat{A}$, $\hat{B}$

and for the coefficients of the operator $\hat{L}$. Generally speaking the operators $\hat{A}$ and $\hat{B}$ are not uniquely determined from these equations; however, all the nonuniqueness is associated with the transformation (7.122) and does not alter the form of the equations which are sought.

We shall examine the case when $\hat{A}$ and $\hat{B}$ are differential operators with respect to z. Then from (7.121) we have

$$\frac{\partial\hat{L}^{(0)}}{\partial t} - [\hat{L}^{(0)},\hat{A}] = \hat{B}\hat{L}^{(0)}$$
$$\frac{\partial\hat{L}^{(1)}}{\partial t} - [\hat{L}^{(1)},\hat{A}] = \hat{L}^{(0)}\hat{A}_x + \hat{B}\hat{L}^{(1)} \quad . \tag{7.128}$$

We shall choose as the simplest example

$$\hat{L}^{(0)} = \frac{\partial}{\partial z} + u \quad ; \quad \hat{L}^{(1)} = v \quad ; \quad \hat{A} = I\frac{\partial}{\partial z} + w \quad ;$$

$\hat{B} = [I,u]$ and I is a constant matrix.

We have

$$\frac{\partial u}{\partial t} = \frac{\partial}{\partial z}(w - Iu) + [u,w] + [I,u]u \quad ,$$
$$\frac{\partial v}{\partial t} = -\frac{\partial v}{\partial z}I + [v,w] - [uv,I] \quad , \tag{7.129}$$
$$\frac{\partial w}{\partial x} = [I,v] \quad .$$

System (7.129) is interesting in two limiting cases when the dependence on one of the coordinates is absent. If $\partial/\partial z = 0$, $u = 0$

$$\frac{\partial v}{\partial t} = [v,w] \quad ; \quad \frac{\partial w}{\partial x} = [I,v] \quad . \tag{7.130}$$

If $\frac{\partial}{\partial x} = 0$, $w = v = 0$

$$\frac{\partial u}{\partial t} = -I\frac{\partial u}{\partial z} + [I,u]u \quad . \tag{7.131}$$

We shall now show how to apply the dressing method to (7.121). For this we shall fix the limiting values of the operators $\hat{L}^{(0)}$ and $\hat{L}^{(1)}$:

$$\hat{L}^{(0)} \to \hat{M}_0 \quad ; \quad \hat{L}^{(1)} \to \hat{M}_1 \qquad \text{as} \qquad z \to \pm\infty \quad ,$$

so that

$$\hat{L}_0 = \hat{M}_0\frac{\partial}{\partial x} + \hat{M}_1 \tag{7.132}$$

and examine the space $\tilde{\Psi}_0$ of the solutions of equation

$$\hat{L}_0\psi_0 = 0 \quad . \tag{7.133}$$

We shall assume that operators $\hat{F}$ and $\hat{K}$ act on $\tilde{\Psi}_0$ like differential operators. All differential operators are now examined with respect to the ideal of operator $\hat{L}_0$, i.e., with exactness up to the addition of an arbitrary operator of the form $\hat{C}\hat{L}_0$.

Let the integral operator $\hat{F}$ with a smooth kernel transform the functions $\tilde{\Psi}_0$ back into $\tilde{\Psi}_0$ $(F:\tilde{\Psi}_0 = \tilde{\Psi}_0)$; then it is obvious that

$$(\hat{L}_0\hat{F} - \hat{F}\hat{L}_0)|\psi_0> = \hat{L}_0\hat{F}|\psi_0> = 0 \qquad (\psi_0 \text{ is a function from } \tilde{\Psi}_0) \quad . \tag{7.134}$$

Let $1 + \ddot{K}^+$ be the right Volterra factor of the operator $\hat{F}$. We shall examine the operator $\hat{L}$ dressed according to (7.40); from (7.46) it follows that

$$[\hat{L}(1 + \hat{K}^+) - (1 + \hat{K}^+)\hat{L}_0]|\psi_0> = \hat{L}(1 + \hat{K}^+)|\psi_0> \quad . \tag{7.135}$$

Thus functions of the type $\psi = (1 + K^+)|\psi_0>$ are solutions to (7.120) which was also necessary.

We shall now write (7.134) in an explicit form. It is equivalent to the condition

$$\hat{L}_0\hat{F} = \hat{G}\hat{L}_0^\dagger$$

where $\hat{G}$ is a certain integral operator or to the two conditions on the kernel

$$M_0\left(\frac{\partial}{\partial z}\right)F(z,z') = G(z,z')M_0^\dagger\left(\frac{\partial}{\partial z'}\right)$$

$$M_0\left(\frac{\partial}{\partial z}\right)\frac{\partial F}{\partial x}(z,z') + M_1\left(\frac{\partial}{\partial z}\right)F(z,z') = G(z,z')M_1^\dagger\left(\frac{\partial}{\partial z'}\right) \quad . \tag{7.136}$$

The pair of equations (7.136) replaces (7.44).

In a similar way (7.49) is replaced by the equations

$$\frac{\partial F}{\partial t} + \hat{A}_0F - F\hat{A}_0^\dagger = R\hat{L}_0 \quad , \tag{7.137}$$

$$R = \sum_{k=0}^{n} G_k \frac{\partial^k}{\partial x^k} \quad ,$$

which can also easily be written in the form of a system of equations for the kernels $G_k$ and F.

We shall now introduce explicit dressing formulae for the operator $\hat{L}$. Let for example

$$\hat{M}_0 = \ell_0 \frac{\partial}{\partial z^m} \quad ; \quad \hat{M}_1 = 0 \quad ,$$

then

$$\hat{L}_0 = \ell_0 \frac{\partial^m}{\partial z^m} + u_1 \frac{\partial^{m-1}}{\partial z^{m-1}} + \ldots$$
(7.138)
$$\hat{L}_1 = v_1 \frac{\partial^{m-1}}{\partial z^{m-1}} + \ldots$$

$$u_1 = \ell_0 K(z,z,x,t) - Q(z,z,x,t)\ell_0$$
(7.139)
$$v_1 = \frac{\partial}{\partial x} K(z,z,x,t) \quad .$$

Here the kernel Q is the solution to the equation

$$\hat{L}_0 \hat{K} = Q\hat{M}_0^{\dagger} \quad .$$

Analogous formulae can easily be obtained by cumbersome calculations for the operator $\hat{A}$. In principle these formulae make it possible to investigate (7.121) to the same depth as the equations which allow an $\hat{L}$-$\hat{A}$ pair.

We should point out that only derivatives of the quantities $\xi_i(z,x,t)$ with respect to z but not the quantities themselves enter into the dressing of operator $\hat{M}_0$.

## 7.9 The Conservation of the Spectrum of Operator Families

Let two consistent differential equations

$$\frac{\partial\psi}{\partial x} + \hat{L}\psi = 0 \qquad \frac{\partial\psi}{\partial t} + \hat{A}\psi = 0 \tag{7.140}$$

and their consistency condition

$$\frac{\partial L}{\partial t} - \frac{\partial A}{\partial x} = [\hat{L},\hat{A}] \tag{7.141}$$

be given.

Here $\hat{L}$ and $\hat{A}$ are as usual canonical differential operators with respect to z. Let the coefficients of these operators be indepedent of z and be only functions of x and t. Then the Fourier transform with respect to z can be applied to equations (7.140). The consistency of these equations ensures the existence of a common spectrum of polynomial operator families with respect to λ.

$$\psi_x + P(x,t,\lambda)\psi = 0 \tag{7.142}$$

$$\psi_t + R(x,t,\lambda)\psi = 0 \tag{7.143}$$

$$P = \ell_0\lambda^n + u_1\lambda^{n-1} + \ldots \quad ; \quad R = a_0\lambda^m + v_1\lambda^{m-1} + \ldots \quad . \tag{7.144}$$

To be more accurate for each value of $\lambda$ the semigroup generated by (7.143) must act invariantly on the space generated by the solutions of (7.142).

The problem of determining the equations to which the matrices P and R must be subjected in order that any solution of system (7.142) is simultaneously a solution of system (7.143) and conversely can be taken to be unconnected with the origin of (7.142) and (7.143) from an $\hat{L}$-$\hat{A}$ pair of differential operators with respect to z. Differentiating (7.142) with respect to t and (7.143) with respect to x, calculating the results and substituting for the derivatives with respect to x and t using formulae (7.142,143) we obtain

$$(P_t - R_x - [P,R])\psi = 0 \quad .$$

Hence we obtain the equation

$$P_t - R_x = [P,R] \tag{7.145}$$

which formally coincides with (7.141).

It is natural to examine the problem in a more general situation by considering P and R as rational functions of $\lambda$. Breaking them down into partial fractions, we assume

$$P = \ell_0\lambda^n + u_1\lambda^{n-1} + \ldots + u_n + \sum p_i$$
$$p_i = \sum_{j=1}^{s_i} \frac{p_{ij}}{(\lambda-\lambda_i)^j} \quad ; \quad \lambda_i \neq \lambda_j \tag{7.146}$$

$$R = a_0\lambda^m + v_1\lambda^{m-1} + \ldots + v_m + \sum Q_i$$
$$Q_i = \sum_{j=1}^{r_i} \frac{Q_{ij}}{(\tilde{\lambda}-\tilde{\lambda}_i)^j} \quad ; \quad \tilde{\lambda}_i \neq \tilde{\lambda}_j \quad . \tag{7.147}$$

Here $\lambda_i$, $\tilde{\lambda}_i$ are complex constants. The conditions on the matrices $u_i$, $v_i$, $p_{ij}$, $Q_{ij}$ must be obtained directly by substituting (7.146) and (7.147) in (7.145).

It is obvious that for the existence of such equations the commutability of matrices $\ell_0$ and $a_0$ is essential. When this condition is satisfied the equations always exist but can have nonunique solutions. It is easy to verify that for the equations to have a unique solution for $u_i$, $v_i$, $p_{ij}$, $Q_{ij}$ the following two conditions are necessary and sufficient:

1) At least one of the numbers n and m is equal to zero.
2) None of the numbers $\lambda_i$ coincide with any of the numbers $\tilde{\lambda}_i$.

To prove this statement we must point out that when conditions 1 and 2 are fulfilled neither new powers of $\lambda$ nor new partial fractions which do not exist in P or R appear in the commutator [P,R]. Therefore the commutator can be broken down into the same powers and partial fractions as P and R. Thus all the equations are determined uniquely.

We shall introduce a number of examples.

1) Let

$$P = \frac{p_1}{\lambda - a} \quad ; \quad R = \frac{q_1}{\lambda + a} \quad .$$

From (7.145) it follows that

$$p_{1t} - q_{1x} = 0 \quad ; \quad a(p_{1t} + q_{1x}) = [p_1, q_1] \quad .$$

Or, assuming

$$p_1 = \phi_x \quad ; \quad q_1 = \phi_t$$

$$2a\phi_{xt} = [\phi_x, \phi_t] \quad , \tag{7.148}$$

and putting

$$\xi = x + t \quad ; \quad \eta = x - t \quad ,$$

we reduce system (7.148) to the form

$$a\left(\frac{\partial^2}{\partial \xi^2} - \frac{\partial^2}{\partial \eta^2}\right)\phi = [\phi_\xi, \phi_\eta] \quad ,$$

from which its relativistic invariance is apparent.

2) Let

$$p = \frac{v}{\lambda} \quad ; \quad R = I\lambda - w \quad .$$

After substitution in (7.145), system (7.130) arises. In the algebra of $2 \times 2$ matrices this system coincides with the Bloch-Bloembergen equations describing the propagation of light pulses in a two-level medium with infinite relaxation times. Assuming in these equations that

$$I = i\begin{bmatrix} 0 & 1 \\ 1 & 0 \end{bmatrix} \quad , \quad v = i\begin{bmatrix} 0 & e^{i\phi} \\ e^{-i\phi} & 0 \end{bmatrix} \quad , \quad w = i\begin{bmatrix} w & 0 \\ 0 & -w \end{bmatrix}$$

we arrive at the well-known "sine-Grodon" equation [7.2,28,29]

$$\frac{\partial^2 \phi}{\partial t \partial x} + 2 \sin 2\phi = 0 \quad . \tag{7.149}$$

3)

$$P = I\lambda + u + \frac{v}{\lambda} \quad ; \quad R = J\lambda + w \quad .$$

This example does not satisfy condition 1 since m = n = 1. Substituting in (7.145) we obtain

$$[I,w] = [J,u] \tag{7.150}$$

$$\frac{\partial u}{\partial t} - \frac{\partial w}{\partial x} = - [J,v] + [u,w] \tag{7.151}$$

$$\frac{\partial v}{\partial t} = [v,w] \quad . \tag{7.152}$$

In these equations nonuniqueness can be avoided by assuming that the matrices u, w arose (when v = 0) from the canonical operators, i.e.,

$$u = [I,Q] \quad ; \quad w = [J,Q] \quad . \tag{7.153}$$

Then when v = 0 this system coincides with (7.32) if $\partial/\partial z = 0$. Similar difficulties of nonuniqueness also occur when there are double zeros in the denominators of $R(x,t,\lambda)$ and $P(x,t,\lambda)$.

In these cases the rules of predetermination, in particular those on which the choice (7.153) is based, will be formulated in Sect.7.10. Let us now turn to (7.121) admitting an "$\hat{L},\hat{A},\hat{B}$ triad". They can be represented as the consistency conditions for the two equations

$$\hat{L}_0\psi_x + \hat{L}_1\psi = 0 \tag{7.154}$$

$$\psi_t + \hat{A}_\psi = 0 \quad . \tag{7.155}$$

The operators $\hat{L}_0$, $\hat{L}_1$ contain only derivatives with respect to z while the operator $\hat{A}$ contains derivatives with respect to both x and z. Let the coefficients of these operators be independent of z and the coefficients of operator $\hat{L}_0$ be independent of x as well (which conforms to the dressing rule of $\hat{L},\hat{A},\hat{B}$ triads). Completing the Fourier transform with respect to z, we get from (7.154) the equation

$$\psi_x = \frac{P_1(x,t,\lambda)}{P_2(\lambda)}\,\psi \tag{7.156}$$

where $P_1(\lambda,x,t)$ and $P_2 = \det \| e^{-\lambda z}\hat{L}_0 e^{\lambda z} \|$ are polynomials with respect to $\lambda$.

With the help of (7.156) it is easy to calculate the higher derivatives of $\psi$ with respect to x. It is obvious that

$$\psi^{(n)} = \frac{P_n(x,t,\lambda)}{[P_2(\lambda)]^n} \quad . \tag{7.157}$$

Using (7.157) to express the derivatives with respect to x in (7.155) we obtain (7.143) in which, however, the rational function R has a very special form. Its denominator is a power of the denominator of the function P. In the particular case of (7.128) when the operator $\hat{A}$ does not contain derivatives with respect to x, P is an arbitrary rational function and R is a polynomial. It is natural to assume that the general equation (7.145) for arbitrary rational operator families is also a result of discarding the derivatives with respect to z in a certain more complex equation, but its calculation goes beyond the bounds of the present work. The question can still be raised as to the conservation with respect to t of the spectrum of an operator family containing higher derivatives with respect to x. Let for example the following system be consistent for all $\lambda$:

$$\psi_{xx} = P(x,t,\lambda)\psi \quad , \tag{7.158}$$

$$\psi_t = R(x,t,\lambda)\psi_x + R_1(x,t,\lambda)\psi \quad . \tag{7.159}$$

Differentiating (7.158) with respect to t and (7.159) twice with respect to x, subtracting and expressing $\psi_t$, $\psi_{xx}$, $\psi_{xxx}$ from (7.158,159) we obtain by equating the coefficients of $\psi$, $\psi_x$ to zero:

$$[P,R] = R_{xx} + 2R_{1x} \quad , \tag{7.160}$$

$$P_t + [P,R_1] - 2R_xP - RP_x - R_{1xx} = 0 \quad . \tag{7.161}$$

If P, R, $R_1$ commute (are scalars or belong to a commuting matrix algebra) (7.160, 161) simplify to one equation

$$P_t - 2R_xP - RP_x + \frac{1}{2}R_{xxx} = 0 \quad . \tag{7.162}$$

Substituting P and R in the forms (7.146,147) into (7.162), it is easy to obtain the equations for the coefficients $u_i$, $P_{ij}$, $v_i$, $Q_{ij}$ which guarantee the consistency of (7.158,159). Here, however, it is possible not to place any limitations on the numbers $\lambda_i$ and $\tilde{\lambda}_i$; in every case a definite system of equations arises.

Let us quote a number of examples.

1)
$$\begin{aligned} P &= \lambda + u \quad , \\ R &= \lambda^m + v_1\lambda^{m-1} + \ldots + v_m \quad . \end{aligned} \tag{7.163}$$

Substituting (7.163) in (7.162) we obviously obtain equations determining an "$\hat{L},\hat{A}$" pair, $\hat{L}$ being the operator $(d^2/dx^2) - u$ and $\hat{A}$ being a scalar operator of order 2m + 1. In particular if m = 1, choosing R = 4$\lambda$ + v we obtain from (7.162)

$$v = -2u$$

$$u_t + 6uu_x - u_{xxx} = 0 \quad ,$$

the KdV equation. If R is a rational function equations occur which no longer possess an "$\hat{L}$-$\hat{A}$" pair. Thus for example assuming $R = v/\lambda$ we obtain the system

$$u_t = 2v_x$$

$$v_{xxx} = 4uv_x + 2vu_x \quad .$$

2)
$$P = \lambda^2 + u_1\lambda + u_2$$
$$R = \lambda + u \quad . \tag{7.164}$$

We now have a system [7.30] describing long waves in shallow water:

$$u_{1t} + \frac{3}{2} u_1 u_{1x} - u_{2x} = 0$$
$$u_{2t} + u_2 u_{1x} + \frac{1}{2} u_1 u_{2x} - \frac{1}{4} u_{1xxx} = 0 \quad . \tag{7.165}$$

3)
$$P = \lambda + u + \frac{v}{\lambda}$$

$$R = \lambda + w$$

$$w = -\frac{1}{2} u$$

$$u_t + \frac{3}{2} uu_x - \frac{1}{4} u_{xxx} = v_x$$
$$v_t + vu_x + \frac{1}{2} uv_x = 0 \quad . \tag{7.166}$$

In the more general case of the noncommuting equations (7.161) the situation is more complex. Rigid restriction are imposed on the form of the dependency of P, R, $R_1$ on $\lambda$. When we substitute arbitrary polynomials of $\lambda$ in (7.161) we obtain generally speaking, an overdetermined system of equations for the coefficients. Nonetheless, in certain cases, this overdetermined system has solutions. This follows, for instance, from the fact that an equation of the type (7.158) can be obtained from a system of two first-order equations (7.142).

From the formal point of view all the equations obtained from (7.158,159) can be regarded as particular cases of the equations obtained with the help of the general procedure (7.142,143), since (7.158) can be reduced to a system of equations of the first order (7.142,143) in which the dimension of the matrices is doubled in comparison with (7.158), the matrices P and R here having the form

$$P = \begin{bmatrix} 0 & 1 \\ P & 0 \end{bmatrix} \quad ; \quad R = \begin{bmatrix} R_1 & R \\ P + R_{1x} & R_x + R_1 \end{bmatrix} \tag{7.167}$$

and the vector

$$\psi = \begin{bmatrix} \psi \\ \psi_x \end{bmatrix} .$$

However, it is nontrivial that the matrix P chosen in the form (7.167) retains this form in time.

## 7.10 The "Dressing" of Operator Families [7.31]

We shall now describe the method of constructing solutions to the equations described in Sect.7.9 and also a method of predetermining the equations where necessary. As in Sect.7.2 we require for the construction of the equations (and their solutions) "bare" commuting operators with constant coefficients. For these we shall examine two commuting rational functions of $\lambda$, $P_0(\lambda)$ and $R_0(\lambda)$, $[P_0,R_0] = 0$. $P_0$ and $R_0$ are assumed to be square matrices of order N.

We shall examine an arbitrary contour G in the complex plane of $\lambda$ and ascribe to it a matrix function $S(\lambda)$. We shall construct a singular integral equation for the matrix function $\tilde{\psi}(\lambda)$:

$$\tilde{\psi}(\lambda) = S(\lambda) + \int_G \frac{\tilde{\psi}(\lambda')}{\lambda-\lambda'+i0} d\lambda' S(\lambda) \quad . \tag{7.168}$$

We shall now impose on the contour G and the function $S(\lambda)$ a single requirement — that (7.168) be uniquely soluble (this can always be achieved by selecting $S(\lambda)$ sufficiently small for example).

Now let $S(\lambda)$ further depend on the two parameters x and t:

$$S(\lambda,x,t) = \exp[P_0(\lambda)x + R_0(\lambda)t]S(\lambda)\exp[- P_0(\lambda)x - R_0(\lambda)t] \quad . \tag{7.169}$$

Hence it follows

$$S_x = [P_0(\lambda),S] \tag{7.170}$$

$$S_t = [R_0(\lambda),S] \quad . \tag{7.171}$$

We shall now show that the function $\tilde{\psi}$ satisfies the two equations

$$\tilde{\psi}_x = P(\lambda,x,t)\tilde{\psi} - \tilde{\psi}P_0(\lambda) \tag{7.172}$$

$$\tilde{\psi}_t = R(\lambda,x,t)\tilde{\psi} - \tilde{\psi}R_0(\lambda) \tag{7.173}$$

where P, R are certain matrices explicitly expressed in terms of $\tilde{\psi}$, $P_0$, $R_0$. We shall denote

$$\chi = \tilde{\psi}_x - P\tilde{\psi} + \tilde{\psi}P_0 \quad . \tag{7.174}$$

We differentiate (7.168) with respect to x and express $\tilde{\psi}_x$ in terms of $\chi$. After elementary operations using (7.168,170) we obtain

$$\chi(\lambda) = gS(\lambda) + \int \frac{\chi(\lambda')}{\lambda-\lambda'+i0}\, d\lambda' S(\lambda) \quad , \tag{7.175}$$

where

$$g(\lambda,x,t) = [P_0(\lambda) - P(\lambda,x,t)] + \int \frac{P(\lambda')-P(\lambda)}{\lambda-\lambda'+i0}\, \tilde{\psi}(\lambda')d\lambda'$$

$$+ \int \frac{\tilde{\psi}(\lambda')}{\lambda-\lambda'+i0} [P_0(\lambda) - P_0(\lambda')]d\lambda' \quad . \tag{7.176}$$

We shall choose the matrix P so that $g \equiv 0$. From the fact that (7.168) has a unique solution it now follows that $\chi \equiv 0$, i.e. (7.172) is satisfied. Similarly the matrix $R(\lambda)$ can be selected so that (7.173) is satisfied.

Introducing

$$\tilde{\psi} = \psi e^{-P_0(\lambda)x-R_0(\lambda)t} \tag{7.177}$$

we can convince ourselves that $\psi$ satisfies (7.142,143) and consequently the matrices P and R satisfy (7.145).

We shall examine in greater detail the condition $g \equiv 0$.

$$P(\lambda,x,t) - P_0(\lambda) = \int \psi(\lambda') \frac{P_0(\lambda)-P_0(\lambda')}{\lambda-\lambda'}\, d\lambda'$$

$$- \int \frac{P(\lambda,x,t)-P(\lambda',x,t)}{\lambda-\lambda'}\, \psi(\lambda')d\lambda' \quad . \tag{7.178}$$

Let $P_0(\lambda)$ be broken down into partial fractions. We shall seek $P(\lambda,x,t)$ in the form of a rational function possessing the same partial fractions as $P_0(\lambda)$. The equality (7.178) is satisfied at each pole of $P_0(\lambda)$ (including the case where $\lambda = \infty$), therefore it can be considered that

$$\begin{aligned} P_0(\ ) &= \frac{p_{01}}{\lambda-\lambda_0} + \dots + \frac{p_{0k}}{(\lambda-\lambda_0)^k} \quad , \\ P(\lambda) &= \frac{p_1}{\lambda-\lambda_0} + \dots + \frac{p_k}{(\lambda-\lambda_0)^k} \quad . \end{aligned} \tag{7.179}$$

Substituting (7.179) in (7.178) we obtain a system of equations for $p_k$:

$$p_k - p_{0k} = p_k I_1 - I_1 p_{0k} \quad ,$$

$$p_{k-1} - p_{0,k-1} = p_{k-1}I_1 - I_1 p_{0,k-1} + p_k I_2 - I_2 p_{0k} \quad , \tag{7.180}$$

$$\ldots \quad .$$

Here

$$I_k(x,t) = \int \frac{\psi(\lambda',x,t)}{(\lambda'-\lambda_0)^k} \, d\lambda' \quad . \tag{7.181}$$

The system of equations (7.180) can be solved recurrently.

$$p_k = (1 - I_1)p_{0k}(1 - I_1)^{-1}$$

$$p_{k-1} = (1 - I_1)p_{0,k-1}(1 - I_1)^{-1} + (p_k I_2 - I_2 p_{0k})(1 - I_1)^{-1}) \tag{7.182}$$

$$\ldots \quad .$$

Similarly the coefficients of the polynomial part of P can be found. Assuming

$$P_0(\lambda) = \ell_0 \lambda^n + \ell_1 \lambda^{n-1} + \ldots$$

$$P(\lambda) = \ell_0 \lambda^n + u_1 \lambda^{n-1} + \ldots \tag{7.183}$$

and substituting (7.183) in (7.178) we obtain the recurrence relations

$$u_1 = \ell_1 + [J_0, \ell_0] \tag{7.184}$$

$$u_2 = \ell_2 + J_0 \ell_1 - u_1 J_0 + [J_1, \ell_0] \tag{7.185}$$

$$\ldots \quad .$$

Here

$$J_i = \int \lambda^i \psi(\lambda,x,t) d\lambda \quad . \tag{7.186}$$

The coefficients of the matrix $R(\lambda,x,t)$ can also be determined in terms of $R_0(\lambda)$, $I_i$ and $J_i$ in a similar way. This process of constructing the variable matrices P and R from the constant matrices $P_0$ and $R_0$ can be called the process of "dressing" the "bare" or "undressed" matrices. The dressing process eliminates the nonuniqueness in the construction of integrable equations when P and R have poles that coincide. All their coefficients at an $n^{th}$ order pole are expressed in terms of n quantities $I_k$ and in terms of the commuting quantities $p_{0k}$, $R_{0k}$. In particular if P and R have a simple pole then at the pole position

$$p_1 = (1 - I_1)p_{01}(1 - I_1)^{-1}$$
$$R_1 = (1 - I_1)R_{01}(1 - I_1)^{-1} \quad , \tag{7.187}$$

i.e., the commuting matrices $p_1$, $R_1$ are similar by the same transformation to the constant matrices $p_{01}$ and $R_{01}$ [this follows from (7.145)!]. Similarly the non-uniqueness in the polynomial parts of P and R is removed. From (7.184) it follows that where $p_1 = 0$, P and R are canonical.

The dressing method enables us to find the solution to the Cauchy problem with respect to t of equations described in Sect.7.9 by the following scheme:

$$P\Big|_{t=0} \to \psi(\lambda,x)\Big|_{t=0} \to S(\lambda,x)\Big|_{t=0} \to S(\lambda,x,t) \to \psi(\lambda,x,t) \to P(\lambda,x,t) \quad . \tag{7.188}$$

Similarly the Cauchy problem with respect to x can be solved. Moreover, by making an arbitrary choice of G and $S(\lambda)$ we can construct various exact solutions of the equation under consideration. In particular the contour can be chosen in the form of a set of discrete points $\mu_i$ (not coinciding with the poles of P and R!) and replaces the integral (7.168) by the sum

$$\tilde{\psi}_n = S_n + \sum_{m \neq n} \frac{\tilde{\psi}_n}{\mu_n - \mu_m} S_n \quad . \tag{7.189}$$

This is a linear algebraic system of equations. Its solution can be found in an explicit form. Now it is obvious that

$$S_n(x,t) = \exp[P_0(\mu_n)x + R_0(\mu_n)t]S_n \times \exp[-P_0(\mu_n)x - R_0(\mu_n)t] \tag{7.190}$$

$$I_k = \sum_n \frac{\tilde{\psi}_n(x,t)}{(\mu_n - \lambda_0)^k} \quad ; \quad J_k = \sum_n \mu_n^k \tilde{\psi}_n \quad . \tag{7.191}$$

In the simplest case where there are only two points in all $\mu_1$, $\mu_2$, $\mu_1 - \mu_2 = \mu$, the solution of system (7.189) assumes the form

$$\psi_1 = \mu(\mu S_1 + S_2^2)(\mu^2 + S_1 S_2)^{-1}$$
$$\psi_2 = \mu(\mu S_2 - S_1^2)(\mu^2 + S_2 S_1)^{-1} \quad . \tag{7.192}$$

Formulae (7.192) encompass a large class of physically important solutions to the equations of Sect.7.9 — analogues of the single soliton solutions.

Let the contour G represent the real axis $-\infty < \lambda < \infty$. Further let $\tilde{\psi}(\lambda)$ and $S(\lambda)$ have absolutely integrable Fourier transforms. Then (7.168) is equivalent to the equation

$$K(z - z') + F(z - z') + \int_{z}^{\infty} K(z - z'')F(z'' - z')dz'' = 0 \qquad (7.193)$$

where

$$K(\zeta) = -i \int_{-\infty}^{\infty} \tilde{\psi}(\lambda)e^{i\lambda\zeta}d\lambda$$

$$F(\zeta) = i \int_{-\infty}^{\infty} S(\lambda)e^{i\lambda\zeta}d\lambda \quad . \qquad (7.194)$$

Let F and K be the kernels of certain integral operators and the operator $\hat{K}$ be a Volterra operator on the right. Equation (7.193) is obviously connected with the problem of the factorisation of the operator $\hat{F}$, whose kernel depends only on the difference of the arguments, in the form of a product of Volterra operators. Another point of view pertaining to (7.168) is that it solves the Riemann problem concerning the conjunction of matrix functions which are analytic on opposite sides of the contour G. Let us introduce

$$\phi^{\pm}(\lambda) = \frac{1}{2\pi i} \int \frac{\tilde{\psi}(\lambda')}{\lambda-\lambda' \ i0} d\lambda' \quad . \qquad (7.195)$$

Then as we know

$$\tilde{\psi}(\lambda) = \phi^{-}(\lambda) - \phi^{+}(\lambda) \quad ,$$

and it now follows from (7.168) that

$$\phi^{-} = S + \phi^{+}(1 + 2\pi i S) \qquad (7.196)$$

is a linear relation connecting the analytical matrices $\phi^{\pm}(\lambda)$ on the contour G. The theory of the Riemann matrix function in connection with the equations from Sect.7.5 has been developed in detail in [7.31] which shows that there are certain restrictions on the forms of $P_0(\lambda)$ and $R_0(\lambda)$. Equation (7.168) can be considered as a Gel'fand-Levitan equation which solves the inverse scattering problem for the operators (7.172,173). It is clear that for equations permitting an $\hat{L}$-$\hat{A}$ pair or an $\hat{L}$-$\hat{A}$-$\hat{B}$ triad, the dressing method given above is a variant of the method described in Sect.7.4.

## References

7.1 C.S. Gardner, J.M. Greene, M.D. Kruskal, R.M. Miura: Phys. Rev. Lett. *19*, 1095 (1967)
7.2 M.J. Ablowitz, D.J. Kaup, A.C. Newell, H. Segur: Stud. Appl. Math. *53*, 249 (1974)
7.3 P.D. Lax: Comm. Pure Appl. Math. *21*, 467 (1968)
7.4 V.E. Zakharov, A.B. Shabat: Sov. Phys. JETP *61*, 118 (1971)

7.5 V.E. Zakharov: Sov. Phys. JETP *65*, 219 (1973)
7.6 M. Wadati: J. Phys. Soc. Jpn. *34*, 1289 (1973)
7.7 V.E. Zakharov, A.B. Shabat: Funct. Anal. Priloz. *8*, 43 (1974)
7.8 V.E. Zakharov, S.V. Manakov: JETP Lett. *18*, 413 (1973)
7.9 V.E. Zakharov, S.V. Manakov: Sov. Phys. JETP *69*, 1654 (1975)
7.10 D.J. Kaup: Stud. Appl. Math. *55*, 9 (1976)
7.11 F. Calogero, A. Degasperis: Lett. Nuovo Cimento, *16*, 425 (1976)
7.12 F. Calogero: Lett. Nuovo Cimento *14*, 433 (1975)
7.13 V.E. Zakharov, S.V. Manakov: Theor. Math. Phys. *27*, 283 (1976)
7.14 R. Hirota, I. Satsuma: J. Phys. Soc. Jpn. *40*, 611 (1976)
7.15 L.P. Nitzhnik: *The Non-stationary Inverse Scattering Problem* (Nauka Dumka, Kiev 1973)
7.16 A.B. Shabat: Dokl. Akad. Nauk SSSR *211*, 1310 (1973); see also V.E. Zakharov: "The inverse scattering method", in *The Theory of Elasticity in a Medium with Microstructure*, ed. by I.A. Kunina (Nauka, Moscow 1975)
7.17 N. Bloembergen: *Nonlinear Optics, a Lecture Note* (W.A. Benjamin, New York-Amsterdam 1965)
7.18 S.V. Manakov: Funct. Anal. Priloz. *10*, 93 (1976)
7.19 V.I. Arnold: *Mathematical Methods of Classical Mechanics* (Nauka, Moscow 1974) (in Russian)
7.20 This paper was not published. The result was published in V.E. Zakharov, S.V. Manakov: "Soliton theory". Sov. Sci. Rev. Sect.A *1*, 133 (1979)
7.21 S.V. Manakov: Sov. Phys. JETP *65*, 219 (1973)
7.22 B.B. Kadomtsev, V.I. Petviashvili: Dokl. Akad. Nauk. *192*, 753 (1970)
7.23 V.E. Zakharov: Sov. Phys. JETP *62*, 1945 (1972)
7.24 N. Yaiima, M. Oikawa: "Formation and Interaction of Sonic-Langmuir Solitons" (Inverse scattering method), preprint, Research Inst. of Appl. Mech. Phys., Japan (1976)
7.25 V.E. Zakharov: JETP Lett. *22*, 364 (1975)
7.26 V.E. Zakharov: Dokl. Akad. Nauk. SSSR *229*, 1314 (1976)
7.27 S.V. Manakov: Usp. Mat. Nauk.*31*(5), 245 (1976)
7.28 L.A. Takhtajan: Sov. Phys. JETP *66*, 476 (1974)
7.29 V.E. Zakharov, L.A. Takhtajan, L.D. Faddeev: Dokl. Akad. Nauk. SSSR *219*, 1334 (1974)
7.30 D.J. Kaup: "A Higher Order Water Wave Equation and the Method for Solving It", Preprint, New York (1975)
7.31 V.E. Zakharov, A.B. Shabat: Funct. Anal. Priloz. *13*(3), 13 (1979)

# 8. Generalized Matrix Form of the Inverse Scattering Method

M. Wadati

This is a review on the matrix generalization of the inverse scattering method. First, the inverse scattering problem for $n \times n$ Schrödinger equation is discussed. Second, the inverse scattering method is extended into $n \times n$ matrix form. Nonlinear evolution equations which are solvable by the extension are presented. In addition, it is pointed out that the same generalization is possible for discrete cases (lattice problems).

## 8.1 Historical Remarks

In first reviewing a short history of the inverse scattering method (sometimes ISM for short), I want to make clear the motivation and the purpose of the matrix generalization of the ISM.

In 1967, GARDNER et al. [8.1] invented the method to solve the Korteweg-de Vries (KdV) equation. They considered a Schrödinger equation whose potential is the solution of the KdV equation and showed that exact solutions can be obtained by clever use of the solution of the inverse scattering problem for the Schrödinger equation. This method was soon expressed in a general form by LAX [8.2]. He observed that the "unitary equivalence" of the associated eigenvalue operator is the essence of the method. In 1972, ZAKHAROV and SHABAT [8.3] solved the nonlinear Schrödinger equation by the ISM. They used a $2 \times 2$ Dirac type operator with derivatives of first order. Their work was very important in the developments of the ISM. It then became clear that the ISM is not a special device merely for solving the KdV equation. Subsequently I was able to show [8.4] that the modified KdV equation is solvable by the Zakharov-Shabat formalism. Another important aspect of Zakharov and Shabat's paper is that the associated eigenvalue operator is not Hermitian. Therefore Lax's unitary equivalence is not the unique way of introducing the ISM. Observing this fact, ABLOWITZ et al. (AKNS) [8.5] showed that the sine-Gordon equation in the light-cone frame is also solvable and further succeeded in formulating the ISM in a simple form.

With the experience of studying Zakharov and Shabat's paper and solving the modified KdV equation I noticed that the ISM may be formulated by using a matrix form of the Schrödinger equation. Around the same time, LAMB [8.6] observed a similar result related to the study of coherent pulse propagation in nonlinear optics. Also I thought that the presentation of a general eigenvalue problem for which the inverse scattering problem is solvable clarifies the applicability of the ISM. Then the matrix generalization of the ISM was proposed [8.7]. Very recently I found that the same generalization is possible for lattice problems.

There is no doubt that the ISM is one of the biggest discoveries in modern mathematical physics. Here, I make two comments. First, the ISM has been a unique method whereby initial value problems of nonlinear evolution equations can be solved exactly. Second, the method leads almost naturally to the conclusion that a soliton system is a completely integrable system.

This work is devoted to a summary of the generalized matrix form of the ISM. In Sect.8.2, the inverse scattering problem for an $m \times m$ Schrödinger equation is discussed. In Sect.8.3, the ISM is generalized to matrix form and solvable equations associated with an $m \times m$ Schrödinger equation are presented. In Sect.8.4, the extension to lattice problems is proposed. The last section contains some concluding remarks.

## 8.2 The Inverse Scattering Problem

Consider the eigenvalue problem for an $m \times m$ Schrödinger operator on the entire axis $(-\infty < x < \infty)$;

$$L\psi(x,k) = \lambda\psi(x,k) \quad , \quad \lambda = k^2 \quad , \tag{8.1}$$

where

$$L = -(\partial^2/\partial x^2)I + U(x) \quad , \tag{8.2}$$

$$I = (\delta_{ij}), \quad U(x) = [u_{ij}(x)] \quad , \quad i,j = 1,2, \ldots m \quad ,$$

and

$$\tilde{\psi}(x,k) = [\psi_1(x,k) \quad , \quad \psi_2(x,k), \ldots, \psi_m(x,k)] \quad . \tag{8.3}$$

Every set of m solutions of the system (8.1) can be represented as an $m \times m$ matrix $\Psi(x,k)$ satisfying the equation

$$-\Psi''(x,k) + U(x,)\Psi(x,k) = k^2\Psi(x,k) \quad . \tag{8.4}$$

We shall make two assumptions for the potential $U(x)$;
(a) the potential $U(x)$ is Hermitian, i.e., $U^*(x) = U(x)$,

(b) the potential U(x) is continuous and decreases rapidly enough as $|x| \to \infty$.

Comments on these assumptions will be given in the end of this section. Later on I shall quote a theorem and lemmas without proofs. The proofs are found in an original paper [8.7].

### 8.2.1 Jost Functions

For real k, we introduce Jost functions F(x,k) and G(x,k) bound by the following boundary conditions at infinity:

$$\lim_{x \to \infty} F(x,k)\exp(-ikx) = I \quad ,$$
$$\lim_{x \to -\infty} G(x,k)\exp(ikx) = I \quad . \tag{8.5}$$

Under the condition (b) for the potential we can show that F(x,k) and G(x,k) are analytic in the upper half plane $\mathrm{Im}\{k\} \geqq 0$.

### 8.2.2 A Fundamental System of Solutions

For real k, the pairs F(x,k), F(x,-k) and G(x,k), G(x,-k) are the fundamental systems of solutions of (8.4). In fact, we have

$$W\{F^*(x,\bar{k}),\, F(x,k)\} = -W\{G^*(x,\bar{k}),\, G(x,k)\} = 2ikI \quad ,$$
$$W\{F^*(x,\bar{k}),\, F(x,-k)\} = W\{G^*(x,\bar{k}),\, G(x,-k)\} = 0 \quad , \tag{8.6}$$

where

$$W\{Y^*(x),Z(x)\} = Y^*(x)Z'(x) - Y^{*\prime}(x)Z(x) \tag{8.7}$$

is the Wronskian of matrix function Y and G and a bar denotes the complex conjugate. Then, we can express G(x,k) as a linear combination of F(x,k) and F(x,-k), and F(x,k) as a linear combination of G(x,k) and G(x,-k).

$$G(x,k) = F(x,-k)A(k) + F(x,k)B(k) \quad , \tag{8.8a}$$

$$F(x,k) = G(x,-k)C(k) + G(x,k)D(k) \quad . \tag{8.8b}$$

The compatibility of relations (8.8) requires

$$B(k)D(k) + A(-k)C(k) = D(k)B(k) + C(-k)A(k) = I \quad ,$$
$$A(k)D(k) + B(-k)C(k) = C(k)B(k) + D(-k)A(k) = 0 \quad . \tag{8.9}$$

The coefficient matrices A(k), B(k), C(k) and D(k) can be expressed in terms of the Wronskians of the solutions F(x,k), G(x,k) and their complex conjugate transposes.

$$A(k) = -W\{F^*(x,-k),\ G(x,k)\}/2ik,\quad B(k) = W\{F^*(x,k),\ G(x,k)\}/2ik \quad ,$$
$$C(k) = W\{G^*(x,-k),\ F(x,k)\}/2ik,\quad D(k) = -W\{G^*(x,k),\ F(x,k)\}/2ik \quad , \tag{8.10}$$

for real k. From (8.10), we find that

$$A(k) = C^*(-k) \quad , \quad B(k) = -D^*(k) \quad . \tag{8.11}$$

For complex $k(\mathrm{Im}\{k\} \geqq 0)$, we define the coefficient matrices as follows:

$$A(k) = -W\{F^*(x,-\bar{k}),\ G(x,k)\}/2ik,\quad B(k) = W\{F^*(x,\bar{k}),\ G(x,k)\}/2ik \quad ,$$
$$C(k) = W\{G^*(x,-\bar{k}),\ F(x,k)\}/2ik,\quad D(k) = -W\{G^*(x,\bar{k}),\ F(x,k)\}/2ik \quad , \tag{8.12}$$

which reduce to (8.10) on the real axis. Since $F(x,k)$, $G(x,k)$, $F^*(x,-\bar{k})$ and $G^*(x,-\bar{k})$ are analytic in the upper half plane, $A(k)$ and $C(k)$ are analytic in the upper half plane. From (8.12), we find that for complex k ($\mathrm{Im}\{k\} \geqq 0$)

$$A(k) = C^*(-\bar{k}) \quad , \quad B(k) = -D^*(\bar{k}) \quad . \tag{8.13}$$

I present two lemmas which will be used in later discussions.

*Lemma 1*: For real k, $\det\{A(k)\} \neq 0$.

*Lemma 2*: For $|k| \to \infty$

$$A(k) = I + O(1/|k|),\quad B(k) = O(1/|k|) \quad ,$$
$$C(k) = I + O(1/|k|),\quad D(k) = O(1/|k|) \quad . \tag{8.14}$$

### 8.2.3 Bound States

The bound states are determined from the condition

$$\det\{A(k)\} = 0 \quad . \tag{8.15}$$

If, for a given $k_j$, the condition (8.15) is satisfied, then there exists a non-zero vector $\underline{a}$ so that $A(k)\underline{a} = 0$. In this case we can find a solution of the basic system (8.1) such that

$$\begin{aligned}\psi(x,k_j) &= G(x,k_j)\underline{a} \\ &= F(x,k_j)B(k_j)\underline{a} \quad .\end{aligned} \tag{8.16}$$

From the boundary conditions (8.5), we have

$$\begin{aligned}\psi(x,k_j) &\to \exp(-ik_jx)\underline{a} \quad \text{for} \quad x \to -\infty \\ &\to \exp(ik_jx)\underline{a} \quad \text{for} \quad x \to \infty \quad .\end{aligned} \tag{8.17}$$

Then we conclude that for $k = k_j$ ($\mathrm{Im}\{k_j\} > 0$), the system (8.1) has a solution which is quadratically integrable on the entire axis. Since the potential $U(x)$ is Hermitian, the energy eigenvalue $k_j^2$ is real, and then $k_j = i\kappa_j$ ($\kappa_j$ is a real positive number). The bound states therefore correspond exactly to the points $\mathrm{Im}\{k\} > 0$, $\mathrm{Re}\{k\} = 0$, where $\det\{A(k)\} = 0$ and where, therefore $A^{-1}(k)$ has a pole.

By virtue of Lemma 1, $\det\{A(k)\}$ does not vanish on the real axis. Further, the asymptotic form (8.14) indicates that $\det\{A(k)\}$ differs from zero for sufficiently large k. Hence $\det\{A(k)\}$ can have only a finite number of zeros, which implies that the number of bound states is finite. Now we find an important theorem.

*Theorem 1*: All the singularities of the matrix $A^{-1}(k)$ in the upper half plane $\mathrm{Im}\{k\} > 0$ are simple poles. Around each pole $k_j = i\kappa_j$ ($\kappa_j$ real positive, $j = 1,2, \ldots s$), the matrix $A^{-1}(k)$ has the form

$$A^{-1}(k) = N_j/(k - k_j) + \ldots \quad , \tag{8.18}$$

where the matrix $N_j \neq 0$.

## 8.2.4 The Gelfand-Levitan-Marchenko Equation

We represent the Jost function F(x,k) in the following form:

$$F(x,k) = \exp(ikx)I + \int_x^\infty K(x,y)\exp(iky)dy \quad . \tag{8.19}$$

Substitution of the expression (8.19) for F(x,k) into (8.4) yields

$$(d/dx)K(x,x) = -(1/2)U(x) \quad , \tag{8.20}$$

$$[-\partial^2/\partial x^2 + \partial^2/\partial y^2 + U(x)]K(x,y) = 0 \quad . \tag{8.21}$$

On the basis of Lemma 1, (8.8a) can be written in the form

$$G(x,k)A^{-1}(k) = F(x,-k) + F(x,k)R(k) \quad , \tag{8.22}$$

where

$$R(k) = B(k)A^{-1}(k) \quad . \tag{8.23}$$

From (8.9,11) it is easily found that $R(k) = R^*(-k)$. We call the matrix R(k) the reflection coefficient matrix.

We are in a position to derive the Gelfand-Levitan-Marchenko equation. Fourier transforming (8.22) and using (8.16,18,19), we arrive at

$$K(x,y) + H(x+y) + \int_x^\infty dz\, K(x,z)\, H(y+z) = 0 \quad , \quad x \leqq y \quad , \tag{8.24}$$

where

$$H(x) = \frac{1}{2\pi}\int_{-\infty}^{\infty} dk\, R(k)\, \exp(ikx) - i\sum_{j=1}^{s} R_j \exp(ik_j x) \quad , \tag{8.25a}$$

$$R_j = B(k_j)N_j \quad . \tag{8.25b}$$

We refer to (8.24) with (8.25) as the Gelfand-Levitan-Marchenko equation. The aggregate of quantities

$$\{R(k) : k_j^2 \quad , \quad R_j (j = 1,2, \ldots, s)\} \tag{8.26}$$

is called the scattering data of the problem (8.1). The Gelfand-Levitan-Marchenko equation indicates that when we know the scattering data, we can evaluate $K(x,y)$ and then construct the potential $U(x)$ from (8.20).

Thus we have proved that the inverse scattering problem can be solved for system (8.1), which gives the basis for the application of the ISM. However, in the discussions, we have assumed two conditions for the potential $U(x)$. Condition (b) is of no interest here and we do not pursue this subject. Condition (a) is important for practical reasons, because we often encounter cases where the potential $U(x)$ is not Hermitian. At the moment, it is an open question as to what are the most general conditions for which the inverse scattering problem can be solved. I have considered three cases:

I) $U(x)$ is diagonal (and can be complex),
II) $\tilde{U}(x) = JU(x)J$; $J$ a constant matrix, $J^2 = 1$,
III) $U^*(x) = JU(x)J$; $J$ a constant matrix, $J^2 = 1$.

The first crucial point in the discussions is the definition of the Wronskian. Although we need extra assumptions, similar arguments seem to be valid. Case I) was studied for a $2\times 2$ matrix related to (8.41). I want to add that the periodic boundary case should be studied.

## 8.3 The Inverse Scattering Method and Solvable Equations

We are given a nonlinear evolution equation of the form

$$U_t = S\{U\} \tag{8.27}$$

where $S$ is in general a nonlinear operator and the matrix function $U(x,t)$ is an $m\times m$ matrix. Assuming the operator $L$ is linear with respect to $U$, we write (8.27) in the operator form

$$L_t = [B,L] = BL-LB \quad , \tag{8.28}$$

by choosing an appropriate $m\times m$ matrix operator $B$. Instead of solving (8.28), we introduce the eigenvalue problem

$$L\psi = \lambda\psi \quad , \tag{8.29}$$

whose eigenfunction varies with time as

$$\psi_t = B\psi \quad . \tag{8.30}$$

It is readily seen from the time evolution properties of $L$ and $\psi$ that the eigenvalue $\lambda$ does not depend on time

$$\lambda_t = 0 \quad . \tag{8.31}$$

We observe that a nonlinear equation (8.27), which is equivalent to (8.28), is also equivalent to the set of equations (8.29-31). Study of the inverse scattering problem for (8.29) indicates that the sought after matrix function U(x,t) can be reconstructed from knowledge of the scattering data only. The time dependence of the scattering data is determined from (8.30) in the limit $x \to \pm\infty$. This procedure for finding a solution of the system (8.27) is the extension of the ISM to matrix form. It is to be remarked that

I) if $L^* = L$, then $B^* = -B$,
II) if $L^* = JLJ(J^* = J,\ J^2 = 1)$, then $B^* = -JBJ$,
III) if $\tilde{L} = JLJ(\tilde{J} = J,\ J^2 = 1)$, then $\tilde{B} = -JBJ$.

By choosing the operator L in the form of an $m \times m$ Schrödinger operator we can write down nonlinear evolution euqations which may be solved by our extension of the ISM. Before doing it I want to point out that the AKNS formalism is included in our system. In fact, the system

$$\partial\psi_1/\partial x + i\zeta\psi_1 = q(x,t)\psi_2 \quad , \tag{8.32a}$$

$$\partial\psi_2/\partial x - i\zeta\psi_2 = r(x,t)\psi_1 \quad , \tag{8.32b}$$

can be rewritten in the form

$$\left[-\begin{pmatrix}1 & 0\\ 0 & 1\end{pmatrix}\frac{\partial^2}{\partial x^2} + \begin{pmatrix}rq & q_x\\ r_x & rq\end{pmatrix}\right]\begin{pmatrix}\psi_1\\ \psi_2\end{pmatrix} = \zeta^2\begin{pmatrix}\psi_1\\ \psi_2\end{pmatrix} \quad . \tag{8.33}$$

Moreover we find that this system belongs to a class L = JLJ and that, when r = a + bq (a,b constants), the system can be diagonalized. Explicit applications to the AKNS system are given in [8.7].

Here I want to discuss the KdV and nonlinear Schrödinger equations in matrix form.

### 8.3.1 The Korteweg-de Vries Equation in Matrix Form

If we choose

$$B = B_2\partial^3 + B_1\partial + \partial B_1 \quad , \quad B_2 = \text{constant and } \partial = \partial/\partial x \quad , \tag{8.34}$$

(8.28) with (8.2) and (8.34) yields

$$U_t = \frac{1}{4} B_2 U_{xxx} - \frac{3}{4} B_2 (U^2)_x \tag{8.35}$$

with the conditions $B_2U = UB_2$ and $3B_2U = -4B_1$. Equation (8.35) is the KdV equation in matrix form. The simplest choice is

$$U(x,t) = u(x,t) \quad , \quad B_2 = -4 \cdot \, , \tag{8.36}$$

then we have from (8.35)

$$u_t - 6uu_x + u_{xxx} = 0 \quad , \tag{8.37}$$

which is the well-known KdV equation. If we choose the components of the $2 \times 2$ matrix $U(x,t)$ as follows

$$\begin{aligned} &U_{11}(x,t) = -u^2 - iu_x \quad , \quad U_{22}(x,t) = U_{11}^*(x,t) \quad , \\ &U_{12}(x,t) = U_{21}(x,t) = 0 \quad , \quad u \ \text{ real} \end{aligned} \tag{8.38}$$

and set $B_2 = -4I$, then we have from (8.35)

$$u_t + 6u^2u_x + u_{xxx} = 0 \quad , \tag{8.39}$$

which is the modified KdV equation. More generally, if we replace $U_{11}(x,t)$ in (8.38) by

$$U_{11}(x,t) = -(\alpha u + \beta u^2) - i\sqrt{\beta}u_x \quad , \tag{8.40}$$

we obtain

$$u_t + 6\alpha uu_x + 6\beta u^2 u_x + u_{xxx} = 0 \quad , \quad (\beta > 0) \quad . \tag{8.41}$$

This equation contains the KdV equation and the modified KdV equation, and is interesting itself as a model for wave propagation in a nonlinear lattice [8.8].

It is possible to derive various forms of multi-component KdV equation from (8.35). However, I do not cite them since their physical applications are, at present, not clear.

### 8.3.2 The Nonlinear Schrödinger Equation in Matrix Form

Let us take an $m \times m$ matrix operator B such that

$$B = 2iJ\partial^2 - 2iV\partial - iJU \quad , \tag{8.42}$$

where the components of constant matrix J are given by

$$J_{ij} = (1 \ \text{ for } \ i = j = 1 \ ; \ -1 \ \text{ for } i = j > 2 \ ; \ \text{ otherwise zero}) \quad ,$$

and we assume the matrix function V has the properties $V = V^*$ and $JV = -VJ$. Substitution of (8.2) and (8.42) into (8.28) yields

$$V_t = iJV_{xx} + 2iJV^3 \quad , \tag{8.43}$$

$$U = -V^2 + JV_x \quad . \tag{8.44}$$

If we choose the components of the matrix V so that

$$V_{ij} = (q_j \quad \text{for} \quad i = 1,\ j = 2, \ldots, m \ ; \quad \bar{q}_i \quad \text{for } j = 1 \ ,$$

$$i = 2, \ldots, m \ ; \quad \text{otherwise zero}) \ , \tag{8.45}$$

then (8.43) gives

$$i(q_k)_t + (q_k)_{xx} + 2\left(\sum_{n=1}^{m} \bar{q}_m q_m\right) q_k = 0 \ . \tag{8.46}$$

This is the m-component nonlinear Schrödinger equation. The cases $m = 1$ and $m = 2$ were studied by ZAKHAROV and SHABAT [8.3], and MANAKOV [8.9], respectively, in relation to nonlinear optics.

I have not worked out how widely this formalism applies in the case of $m > 3$. Very recently, CALOGERO and DEGASPERIS [8.10] have made some progress in this subject.

## 8.4 Extension to Lattice Problems

In this section I shall propose that matrix generalization is also possible for the discrete case (lattice problems). Let us consider the eigenvalue problem

$$L(n)\psi(n) = \psi(n + 1) + S(n)\psi(n) + T(n)\psi(n - 1) = \lambda\psi(n) \ , \tag{8.47}$$

$$n = -\infty, \ldots, -1, 0, 1, \ldots, +\infty \ ,$$

where $\psi(n)$ is an m-componente column vector and $S(n)$ and $T(n)$ are $m \times m$ matrices. Hereafter we assume that $S(n) \to 0$ and $T(n) \to I$ as $|n| \to \infty$. Then (8.47) is a discrete version of the $m \times m$ Schrödinger equation. The inverse scattering problem for the system (8.47) was studied by CASE and KAC [8.11] and FLASCHKA [8.12] in the case of $m = 1$. The general case in matrix form is to be studied.

Nonlinear differential-difference equations solvable via the ISM associated with the system (8.47) are, for instance, the following.

### 8.4.1 Volterra Systems

Consider a set of equations;

$$\psi(n + 1) + T(n)\psi(n - 1) = \lambda\psi(n) \ , \tag{8.48}$$

$$\dot{\psi}(n) = \psi(n + 2) + W(n)\psi(n) \ . \tag{8.49}$$

Here the dot indicates the time derivative. We impose the boundary condition $W(n) \to cI$; c constant, as $|n| \to \infty$. The condition $\lambda_t = 0$ implies

$$\dot{T}(n) = T(n + 1)T(n) - T(n)T(n - 1) \ , \tag{8.50}$$

with

$$W(n) = T(n+1) + T(n) + (c-2)I \quad . \tag{8.51}$$

Equation (8.50) contains a family of Volterra systems. The simplest choice

$$T(n) = N_n \quad , \tag{8.52}$$

yields

$$\dot{N}_n = N_n(N_{n+1} - N_{n-1}) \quad . \tag{8.53}$$

This equation has been proposed by several authors. Especially, ZAKHAROV et al. [8.13] who introduced (8.53) to describe the spectrum of Langmuir waves in plasma physics.

If we take for T(n) the $2 \times 2$ matrix whose components are

$$\begin{aligned} T_{11}(n) &= 1 + i(M_n - M_{n-1}) + M_n M_{n-1} = T_{22}^*(n) \quad , \\ T_{12}(n) &= T_{21}(n) = 0 \quad , \quad M_n \ \text{real} \quad , \end{aligned} \tag{8.54}$$

then (8.50) yields

$$\dot{M}_n = (1 + M_n^2)(M_{n+1} - M_{n-1}) \quad . \tag{8.55}$$

More generally if we replace $T_{11}(n)$ in (8.54) by

$$T_{11}(n) = 1 + \frac{1}{2}(\alpha - i\sqrt{4\beta - \alpha^2})Q_{n-1} + \frac{1}{2}(\alpha + i\sqrt{4\beta - \alpha^2})Q_n + \beta Q_n Q_{n-1} \quad ; \tag{8.56}$$

$\alpha$, $\beta$ and $Q_n$ real

we have from (8.50)

$$\dot{Q}_n = (1 + \alpha Q_n + \beta Q_n^2)(Q_{n+1} - Q_{n-1}) \tag{8.57}$$

under the condition $\beta = 0$ or $4\beta - \alpha^2 > 0$. The latter condition assures the positive definiteness of $(1 + \alpha Q_n + \beta Q_n^2)$. Equation (8.57) obviously encompasses (8.53) and (8.55).

### 8.4.2 The Toda Lattice Equation and a Nonlinear Self-Dual Network Equation

Consider a set of equations:

$$\psi(n+1) + S(n)\psi(n) + T(n)\psi(n-1) = \lambda\psi(n) \quad , \tag{8.58}$$

$$\dot{\psi}(n) = -T(n)\psi(n-1) \quad . \tag{8.59}$$

The condition $\lambda_t = 0$ yields

$$\dot{S}(n) = T(n+1) - T(n) \quad , \tag{8.60a}$$

$$\dot{T}(n) = S(n)T(n) - T(n)S(n-1) \quad . \tag{8.60b}$$

If we set

$$S(n) = -P_n \quad , \quad T(n) = \exp[-(Q_n - Q_{n-1})] \quad , \tag{8.61}$$

we have from (8.58) the TODA lattice equation [8.14]

$$P_n = \dot{Q}_n \quad , \quad \dot{P}_n = \exp[-(Q_n - Q_{n-1})] - \exp[-(Q_{n+1} - Q_n)] \quad . \tag{8.62}$$

In the case of $2\times 2$ matrices, if we choose

$$\begin{aligned} &S_{11}(n) = i[I_{n+1}(1 + iV_n) - I_n(1 - iV_n)] = S^*_{22}(n) \quad , \\ &T_{11}(n) = (1 + I_n^2)(1 + iV_{n-1})(1 - iV_n) = T^*_{22}(n) \quad , \\ &S_{12}(n) = S_{21}(n) = T_{12}(n) = T_{21}(n) = 0 \quad ; \\ &I_n \, , \; V_n \; \text{ real} \end{aligned} \tag{8.63}$$

then (8.60) give a nonlinear self-dual network equation [8.15]

$$\dot{I}_n = (1 + I_n^2)(V_{n-1} - V_n) \, , \quad \dot{V}_n = (1 + V_n^2)(I_n - I_{n+1}) \quad . \tag{8.64}$$

### 8.4.3 Discrete Nonlinear Schrödinger Equation

Consider a set of equations

$$\psi(n + 1) + S(n)\psi(n) + T(n)\psi(n - 1) = \lambda\psi(n) \quad , \tag{8.65}$$

$$\dot{\psi}(n) = - iJ\psi(n + 2) + C(n)\psi(n + 1) + D(n)\psi(n) \quad , \tag{8.66}$$

where

$$S(n) = \begin{bmatrix} 0 & -i\left(\bar{R}_n - \bar{R}_{n-1}\right) \\ -i\left(R_n - R_{n-1}\right) & 0 \end{bmatrix} \quad , \quad T(n) = \left(1 + |R_{n-1}|^2\right)I \quad ,$$

$$J = \begin{pmatrix} 1 & 0 \\ 0 & -1 \end{pmatrix} \, , \quad C(n) = \begin{bmatrix} 0 & -\left(\bar{R}_{n+1} + \bar{R}_{n-1}\right) \\ \left(R_{n+1} + R_{n-1}\right) & 0 \end{bmatrix} \quad ,$$

$$D(n) = \begin{bmatrix} i\left(1 - R_{n-1}\bar{R}_n + \bar{R}_{n-1}R_n\right) & 0 \\ 0 & -i\left(1 + R_{n-1}\bar{R}_n - \bar{R}_{n-1}R_n\right) \end{bmatrix} \quad . \tag{8.67}$$

The condition $\lambda_t = 0$ gives a discrete nonlinear Schrödinger equation [8.16]

$$i\dot{R}_n + R_{n+1} + R_{n-1} - 2R_n + |R_n|^2(R_{n+1} + R_{n-1}) = 0 \quad . \tag{8.68}$$

Thus we have obtained a matrix generalization for the discrete case. I want to emphasize that there are strong similarities between the continuous and discrete cases. Mathematical methods which are valid and useful in the continuous case work also in the discrete case [8.17], and it seems that we have solvable nonlinear difference-differential equations corresponding to solvable nonlinear differential equations.

Let us confirm the comment taking (8.41,57) as an example. GARDNER et al. [8.1, 18] introduced the ISM by linearizing the Miura transformation. Essentially they have done the following. By substitution of

$$u = i\psi_x/(\sqrt{\beta}\psi) - \alpha/2\beta \quad , \tag{8.69}$$

into the (generalized) Miura transformation (8.40), we obtain the Schrödinger equation

$$-\psi_{xx} + U_{11}\psi = \lambda\psi \ , \quad \lambda = \alpha^2/4\beta \quad . \tag{8.70}$$

On the other hand, substitution of

$$Q_{n-1} = i\phi(n)/[\sqrt{\beta}\phi(n-1)] - (\alpha + i\sqrt{4\beta - \alpha^2})/2\beta \quad , \tag{8.71}$$

into (8.56) yields

$$\phi(n+1) + T_{11}(n)\phi(n-1) = \lambda\phi(n) \quad , \quad \lambda = \sqrt{4 - \alpha^2/\beta} \quad . \tag{8.72}$$

In every step we have similar equations.

## 8.5 Concluding Remarks

In this work I have summarized the present status of the study of the generalized matrix form of the ISM. I do not claim this formalism is the best. Each formulation has its own advantages and usefulness and sometimes makes up the other's deficiencies.

Recnet studies have shown that there are many varieties of the eigenvalue problems associated with solvable nonlinear differential equations. The operator L has a third-order derivative for the Boussinesq equation [8.19], a $3 \times 3$ Dirac-type operator for the equation of three wave interactions [8.20] and the long wave – short wave equation [8.21], and eigenvalue dependent potentials for the water wave equations [8.22] and the sine-Gordon equation in laboratory frame [8.23]. Moreover, we have the eigenvalue problem with two independent variables for the Kadomtsev-Petviashvili equation [8.24]. We do not know what is the widest formulation of the ISM. However, we are surely making a great contribution to mathematical physics. For a long time, Fourier transformation, i.e., decomposition into sinusoidal waves, governed our way of thinking. Now we have started a real understanding of nonlinear waves by the use of the ISM and the concept of the soliton.

References


8.1 C.S. Gardner, J.M. Greene, M.D. Kruskal, R.M Miura: Phys. Rev. Lett. *19*, 1095 (1967); Commun. Pure Appl. Math. *27*, 97 (1974)
See also M. Wadati, M. Toda: J. Phys. Soc. Jpn. *32*, 1403 (1972)

8.2 P.D. Lax: Commun. Pure Appl. Math. *21*, 467 (1968)

8.3 V.E. Zakharov, A.B. Shabat: Soviet Phys. JETP *34*, 62 (1972)

8.4 M. Wadati: J. Phys. Soc. Jpn. *32*, 1681 (1972); *34*, 1289 (1973)

8.5 M.J. Ablowitz, D.J, Kaup, A.C. Newell, H. Segur: Phys. Rev. Lett. *30*, 1262 (1973); *31*, 125 (1973); Stud. Appl. Math. *53*, 249 (1974)

8.6 G.L. Lamb, Jr.: Phys. Rev. A*9*, 422 (1974)

8.7 M. Wadati, T. Kamijo: Prog. Theor. Phys. *52*, 397 (1974)

8.8 M. Wadati: J. Phys. Soc. Jpn. *38*, 673, 681 (1975)

8.9 S.V. Manakov: Soviet Phys. JETP *38*, 248 (1974)

8.10 F. Calogero, A. Degasperis: in this volume

8.11 K.M. Case, M. Kac: J. Math. Phys. *14*, 594 (1973)
K.M. Case: J. Math. Phys. *14*, 916 (1973)

8.12 H. Flaschka: Prog. Theor. Phys. *51*, 703 (1974)

8.13 V. Zakharov, S. Musher, A. Rubenshik: JETP Lett. *19*, 151 (1974)

8.14 M. Toda: Phys. Rep. *18*C, 1 (1975)

8.15 R. Hirota: J. Phys. Soc. Jpn. *34*, 289 (1973)

8.16 M.J. Ablowitz, J.F. Ladik: J. Math. Phys. *17*, 1011 (1976)

8.17 M. Wadati: Suppl. Prog. Theor. Phys. *59*, 36 (1976)

8.18 R.M. Miura, C.S. Gardner, M.D. Kruskal: J. Math. Phys. *9*, 1204 (1968)

8.19 V.E. Zakharov: Zh. Eksp. Theor. Fiz. *65*, 219 (1973)

8.20 V.E. Zakharov, S.V. Manakov: JETP Lett. *18*, 243 (1973)
D.J, Kaup: Stud. Appl. Math. *55*, 9 (1976)

8.21 A.C. Newell: Preprint

8.22 D.J. Kaup: Prog. Theor. Phys. *54*, 396 (1975)

8.23 V.E. Zakharov, L.A. Takhtadzhyan, L.D. Faddeev: Sov. Phys. Dokl. *19*, 824 (1975)
D.J. Kaup: Stud. Appl. Math. *54*, 165 (1975)

8.24 V. Dryuma: JETP Lett. *19*, 387 (1974)

# 9. Nonlinear Evolution Equations Solvable by the Inverse Spectral Transform Associated with the Matrix Schrödinger Equation

F. Calogero, and A. Degasperis

With 1 Figure

The class of nonlinear evolution equations (NEEs) that can be solved via the inverse spectral transform (IST) associated with the (linear) matrix Schrödinger eigenvalue problem can be extended beyond that described by M. Wadati in the preceding chapter. The most remarkable feature of the novel solvable NEEs thus obtained is the behavior of their solitons, which, while possessing all the stability properties that characterize solitons, generally do not move with constant speed, behaving instead as particles in an external field of force (whose features depend generally on the parameters of each soliton, i.e. its "shape" and initial "polarization"). In this chapter we describe this extended class of NEEs, the solution of the (Cauchy) initial-value problem by the IST, and various properties of their solutions, including a large class of Bäcklund transformations (BTs) and several related results (nonlinear superposition principle, multisoliton ladder, conserved quantities, generalized resolvent formula, nonlinear operator identities); we also analyze the simplest novel NEE belonging to this class, and the behavior of its solitons (or rather "boomerons").

Our presentation will be terse (especially the part on BTs and related topics); for a more detailed treatment of our method, based on generalized Wronskian relations, and for a description of its relationship to other approaches, we refer to papers we have published elsewhere [9.1-3].

## 9.1 Direct and Inverse Matrix Schrödinger Problem; Notation

The main results have been discussed by M. Wadati in the preceding chapter; here we mereley report the main equations, to establish the notation that will be used in the rest of this chapter.

*Basic notation.* We use generally upper case characters for matrices ($N \times N$); an exception to this convention is the notation $\sigma_n$, $n = 1, 2, \ldots, N^2 - 1$, for $N^2 - 1$ Hermitian matrices that, together with the matrix $\sigma_0 = I$, constitute a basis for $N \times N$ matrices. Greek indices run over the values $0, 1, 2, \ldots, N^2 - 1$; Latin indices

Since this chapter was first written, our view has changed; it is now our feeling that we should drop the adjective "inverse" in the title, and accordingly in the text.

over 1, 2, ..., $N^2 - 1$; the convention that repeated indices are summed upon is always understood. We also use Latin indices to label vector components and matrix elements; thus for instance $(v,v) = v_n^* v_n \equiv \sum_{n=1}^N v_n^* v_n$. The notation for commutators and anticommutators used throughout reads: $[A,B] = AB - BA$, $\{A,B\} = AB + BA$. We also use occasionally the dyadic notation $vv^\dagger$ for matrices: of ourse $W = vv^\dagger$ implies $W_{\ell m} = v_\ell v_m^*$.

*The direct problem.* The matrix Schrödinger problem is characterized by the linear differential equation

$$\psi_{xx} = [Q - k^2]\psi \quad . \tag{9.1}$$

In this equation the $N \times N$ matrix $Q$ depends on the real variable $x$, and the vector $\psi$ depends on $x$ and $k$; both may depend parametrically on other variables (see below).

We always assume the "potential" matrix $Q(x)$ to vanish asymptotically (sufficiently fast [9.1]). The continuous part of the spectrum of the eigenvalue problem (9.1) is then characterized by the asymptotic boundary conditions

$$\Psi(x,k) \to \exp(-ikx) + R(k)\exp(ikx) \quad \text{as} \quad x \to +\infty \quad , \tag{9.2a}$$

$$\Psi(x,k) \to T(k)\exp(-ikx) \quad \text{as} \quad x \to -\infty \quad ; \tag{9.2b}$$

here $\Psi$ is a $N \times N$ matrix whose columns are solutions of (9.1) and $k$ is real (positive by convention). Clearly if $Q$, and therefore $\Psi$, depend parametrically on other variables, so do the (matrix) "reflection coefficient" $R$ and "transmission coefficient" $T$. The discrete part of the spectrum of (9.1) consists of a finite number of eigenvalues $k^{(j)} = ip^{(j)}$, whose corresponding eigenfunctions satisfy the Schrödinger equation

$$\psi_{xx}^{(j)} = \left[Q + p^{(j)^2}\right]\psi^{(j)} \tag{9.3}$$

and the normalization condition

$$\int_{-\infty}^{+\infty} dx \left(\psi^{(j)}, \psi^{(j)}\right) = 1 \quad . \tag{9.4}$$

In the following we shall omit the superscript labelling different discrete eigenstates, whenever this can be done without causing confusion.

If the potential matrix $Q$ is Hermitian (as we generally assume), the $p$'s are real (and, by convention, positive); the asymptotic behavior of the discrete spectrum eigenfunctions is

$$\psi(x) \to c\exp(-px) \quad \text{as} \quad x \to +\infty \quad , \tag{9.5a}$$

$$\psi(x) \to c_-\exp(px) \quad \text{as} \quad x \to -\infty \quad ; \tag{9.5b}$$

the matrix (dyadic notation)

$$C = cc^\dagger \quad , \tag{9.6}$$

that plays an important role in the following, is related to the residue of $R(k)$ at $k = ip$ by the formula

$$\lim_{k \to ip} [(k - ip)R(k)] = iC \quad , \tag{9.7}$$

provided the potential $Q(x)$ vanishes asymptotically sufficiently fast [9.1] (and the discrete eigenvalue is nondegenerate; this is hereafter assumed, for simplicity).

Clearly if $Q(x)$ is given, the matrices $R(k)$ and $T(k)$, and the parameters $p^{(j)}$, $c^{(j)}$ and $c_-^{(j)}$ characterizing the discrete part of the spectrum (if any), are uniquely determined through the equations given above. Such a determination constitutes the direct one-dimensional matrix Schrödinger problem.

*The inverse problem.* The input data that are sufficient to determine uniquely the "potential" $Q(x)$ are the reflection coefficient $R(k)$, the discrete eigenvalues $p^{(j)}$ (if any) and the corresponding matrices $C^{(j)}$. The formulae read

$$M(x) = \sum_j C^{(j)} \exp[-p^{(j)}x] + (2\pi)^{-1} \int_{-\infty}^{+\infty} dk\, R(k) \exp(ikx) \quad , \tag{9.8}$$

$$K(x,x') + M(x + x') + \int_x^{+\infty} dx''\, K(x,x'')M(x' + x'') = 0 \quad , \quad x \leqq x' \quad , \tag{9.9}$$

$$Q(x) = -2 \frac{d}{dx} K(x,x) \quad . \tag{9.10}$$

The values of $R(k)$ for negative $k$ that enter in the definition (9.8) of the kernel $M(x)$ can be obtained by analytic continuation from the values of $R(k)$ for real positive $k$; for Hermitian $Q(x)$ they can also be obtained from the formula $R(-k) = R^{\dagger}(k)$.

## 9.2 Generalized Wronskian Relations; Basic Formulae

All the results described in the following sections derive from the following basic formulae:

$$2ikf(-4k^2)[MR(k) - R'(k)M]$$
$$= \int_{-\infty}^{+\infty} dx\bar{\Psi}'(x,k)\{f(\Lambda)[MQ(x) - Q'(x)M]\}\Psi(x,k) \quad , \tag{9.11}$$

$$(2ik)^2 g(-4k^2)[NR(k) + R'(k)N]$$
$$= \int_{-\infty}^{+\infty} dx\bar{\Psi}'(x,k)\{g(\Lambda)\Gamma N\}\Psi(x,k) \quad . \tag{9.12}$$

Here $f(z)$ and $g(z)$ are two arbitrary (entire) functions; $M$ and $N$ are two constant (i.e., x-independent) matrices; $R(k)$ and $R'(k)$ are the reflection coefficients for the potentials $Q(x)$ and $Q'(x)$, respectively. The two operators $\Lambda$ and $\Gamma$, that act on whatever stands to their right within the curly brackets, are defined by the following formulae, which specify their action on a generic matrix $F(x)$ (vanishing as $x \to +\infty$ faster than $1/x$):

$$\Lambda F(x) = F_{xx}(x) - 2[Q'(x)F(x) + F(x)Q(x)] + \Gamma \int_x^{+\infty} dx' F(x') \quad , \tag{9.13}$$

$$\Gamma F(x) = Q'_x(x)F(x) + F(x)Q_x(x)$$
$$+ \int_x^{+\infty} dx' [Q'(x)Q'(x')F(x') - Q'(x)F(x')Q(x') - Q'(x')F(x')Q(x)$$
$$+ F(x')Q(x')Q(x)] \quad . \tag{9.14}$$

These formulae obtain in the framework of the approach based on generalized Wronskian relations [9.1]; they display a property of the Schrödinger eigenvalue problem described in the previous section. Their usefulness resides in the possibility they imply of finding a class of nonlinear relations connecting two potentials (that differ by a finite amount or infinitesimally, see below), such that the relations connecting the corresponding reflection coefficients are particularly simple.

Analogous relations involving the transmission coefficients and/or the discrete spectrum parameters can also be given [9.1]; however, they are not essential for the derivation of the results of main interest here (the formulae for the discrete spectrum parameters, some of which are indeed essential for the results given below, can all be obtained from those for the reflection coefficients, provided consideration be restricted to potentials that vanish asymptotically faster than exponentially, since then all discrete spectrum parameters are obtainable via (9.7); and the formulae thus obtained have undoubtedly a more general validity).

## 9.3 Nonlinear Evolution Equations Solvable by the Inverse Spectral Transform; Solitons

*The limit $Q' \to Q$.* Introduce a parametric dependence on a new variable, say $y$, setting

$$Q(x) = Q(x,y) \quad , \quad Q'(x) = Q(x,y+\Delta y) \quad , \tag{9.15}$$

which of course also implies

$$R(k) = R(k,y) \quad , \quad R'(k) = R(k,y+\Delta y) \quad ; \tag{9.16}$$

insert this ansatz in the equations of the previous section, and take the limit $\Delta y \to 0$, so that $Q'(x) = Q(x) = Q(x,y)$ and the operators $\Lambda$ and $\Gamma$ go over into the operators L and G defined as follows [F(x) is a generic matrix vanishing as $x \to +\infty$; we omit explicit indication of the y dependence]:

$$LF(x) = F_{xx}(x) - 2\{Q(x),F(x)\} + G \int_x^{+\infty} dx'F(x') \quad , \tag{9.17}$$

$$GF(x) = \{Q_x(x),F(x)\} + \left[Q(x), \int_x^{+\infty} dx'[Q(x'),F(x')]\right] \quad . \tag{9.18}$$

Moreover setting, $M = 1/\Delta y$ and $M = \sigma_n$ in (9.11), and $N = \sigma_\mu$ in (9.12), one gets three formulae:

$$2ikf(-4k^2)R_y(k,y) = \int_{-\infty}^{+\infty} dx\bar{\Psi}(x,k,y)\{f(L)Q_y(x,y)\}\Psi(x,k,y) \quad , \tag{9.19}$$

$$2ikf_n(-4k^2)[\sigma_n,R(k,y)] = \int_{-\infty}^{+\infty} dx\bar{\Psi}(x,k,y)\Big\{f_n(L)[\sigma_n,Q(x,y)]\Big\}\Psi(x,k,y) \quad , \tag{9.20}$$

$$(2ik)^2 g_\mu(-4k^2)\Big\{\sigma_\mu,R(k,y)\Big\} = \int_{-\infty}^{+\infty} dx\bar{\Psi}(x,k,y)\Big\{g_\mu(L)G\sigma_\mu\Big\}\Psi(x,k,y) \quad . \tag{9.21}$$

Here the functions f, $f_n$ and $g_\mu$ are arbitrary (entire); they could of course depend parametrically on other variables, such as y (they must be independent of x). The derivation we have just outlined implies that (9.20,21) hold even if the repeated indices n and $\mu$ were not summed upon (respectively from 1, and from 0, to $N^2 - 1$); no additional generality is implied by this, however, due to the arbitrariness of the functions $f_n$ and $g_\mu$.

*NEEs solvable by IST.* Assume Q, and therefore also R, to depend parametrically on several variables; call one of these t ("time"), and denote all the others by an M-dimensional vector $\underline{y}$

$$Q \equiv Q(x,\underline{y},t) \quad , \quad R \equiv R(k,\underline{y},t) \quad . \tag{9.22}$$

Imagine (9.19) written with y replaced by t, or by any one of the components of $\underline{y}$; in each case with an arbitrary, and different, function f. From such equations, and from (9.20,21), the following result is immediately implied: if the potential matrix $Q(x,\underline{y},t)$ *satisfies the nonlinear equation*

$$\begin{aligned} f_0(L,\underline{y},t)Q_t &= 2\beta_0(L,\underline{y},t)Q_x + \alpha_n(L,y,t)[\sigma_n,Q] \\ &+ \beta_n(L,\underline{y},t)G\sigma_n + \underline{\gamma}(L,\underline{y},t)\cdot\frac{\partial}{\partial \underline{y}} Q \quad , \end{aligned} \tag{9.23}$$

*the corresponding reflection coefficient $R(k,\underline{y},t)$ satisfies the linear partial differential equation*

$$\begin{aligned} f_0(-4k^2,\underline{y},t)R_t &= \alpha_n(-4k^2,\underline{y},t)[\sigma_n,R] \\ &+ 2ik\beta_\nu(-4k^2,\underline{y},t)\{\sigma_\nu,R\} + \underline{\gamma}(-4k^2,\underline{y},t)\cdot\frac{\partial}{\partial \underline{y}} R \quad . \end{aligned} \tag{9.24}$$

In the following we restrict for simplicity our attention to the case $f_0 = 1$, so that (9.23) has more explicitly the structure of a NEE; and moreover we assume the (entire, but otherwise arbitrary) functions $\beta_n$ and $\alpha_n$ to be independent of $\underline{y}$ and t, and the functions $\beta_0$ and $\underline{\gamma}$ to be independent of $\underline{y}$. Thus we focus attention on the NEE

$$Q_t(x,\underline{y},t) = 2\beta_0(L,t)Q_x(x,\underline{y},t) + \alpha_n(L)[\sigma_n,Q(x,\underline{y},t)] + \beta_n(L)G\sigma_n + \underline{\gamma}(L,t)\cdot\frac{\partial}{\partial\underline{y}}Q(x,\underline{y},t) \quad , \tag{9.25}$$

indicating in particular how the initial-value problem, characterized by the initial condition

$$Q(x,\underline{y},t_0) = Q_0(x,\underline{y}) \quad , \tag{9.26}$$

can be solved by IST. Here $Q_0(x,\underline{y})$ is a given (matrix) function, of course vanishing asymptotically in x.

Let $R_0(k,\underline{y}) = R(k,\underline{y},t_0)$ be the reflection coefficient corresponding to $Q_0(x,\underline{y})$; it can of course be determined by solving the direct Schrödinger problem. But we also know that, if $Q(x,\underline{y},t)$ evolves according to (9.25) the corresponding reflection coefficient evolves according to the linear equation

$$R_t(k,\underline{y},t) = 4ik\beta_0(-4k^2,t)R(k,\underline{y},t) + \alpha_n(-4k^2)[\sigma_n,R(k,\underline{y},t)] + 2ik\beta_n(-4k^2)\{\sigma_n,R(k,\underline{y},t)\} + \underline{\gamma}(-4k^2,t)\cdot\frac{\partial}{\partial\underline{y}}R(k,\underline{y},t) \quad , \tag{9.27}$$

which can be integrated in closed form, yielding, together with the boundary condition given above, the explicit solution

$$R(k,\underline{y},t) = \exp\left[4ik\int_{t_0}^{t} dt'\beta_0(-4k^2,t')\right] \cdot \exp\left\{(t-t_0)\left[\alpha_n(-4k^2) + 2ik\beta_n(-4k^2)\right]\sigma_n\right\}\cdot R_0\left[k,\underline{y}+\int_{t_0}^{t} dt'\underline{\gamma}(-4k,t')\right] \cdot \exp\left\{(t-t_0)\left[-\alpha_n(-4k^2) + 2ik\beta_n(-4k^2)\right]\sigma_n\right\} \quad . \tag{9.28}$$

From the reflection coefficient R at time t, given explicitly by this formula, one can reconstruct the potential Q at time t, solving the inverse problem (see Sect. 9.1; the discrete spectrum parameters, which are also required inputs in the inverse problem, are discussed below); the potential $Q(x,\underline{y},t)$ thus obtained is clearly the solution of the NEE (9.25), characterized by the initial condition (9.26). It should be emphasized that this procedure yields the solution of the Cauchy problem for (9.25), which is nonlinear because the operators L and G depend on Q [see their definitions (9.17,18)], through the solution of the direct and inverse spectral problems, both of which involve only linear equations, i.e., the Schrödinger equation (9.1) for the direct problem and the linear integral equation (9.9) for the inverse problem. Moreover, it allows us to analyze quite explicitly the time evolution of any solution $Q(x,\underline{y},t)$ of the NEE (9.25) in terms of two parts, one associated with the continuous spectrum of the associated spectral problem, and another (to be discussed below) associated with the discrete spectrum.

The subclass of NEEs that is obtained by setting, in (9.25), $\beta_0(z,t) = \beta_0(z)$ and $\alpha_n = \beta_n = \underline{\gamma} = 0$, constitutes a compact way of writing the NEEs treated by Wadati in the preceding chapter.

Special NEEs belonging to the class (9.25) are discussed below; here we merely note that, if the basis matrices $\sigma_n$ are Hermitian, sufficient conditions to guarantee that $Q(x,\underline{y},t)$ remains Hermitian throughout its time evolution if it is Hermitian at the initial time $t_0$, are that the functions $\beta_\nu$ and $\underline{\gamma}$ be real while the $\alpha_n$'s are pure imaginary [9.1]

$$\alpha_n = -\alpha_n^* \quad , \quad \beta_\nu = \beta_\nu^* \quad , \quad \underline{\gamma} = \underline{\gamma}^* \; . \tag{9.29}$$

*Evolution of the discrete spectrum parameters.* The time evolution corresponding to the NEE (9.25) is characterized by the (nonlinear) partial differential equation

$$p_t(\underline{y},t) = \underline{\gamma}(4p^2,t) \cdot \frac{\partial}{\partial \underline{y}} p(\underline{y},t) \quad , \tag{9.30}$$

whose (implicit) solution is

$$p(\underline{y},t) = p_0\Big[\underline{y} + \int_{t_0}^{t} dt' \underline{\gamma}[4p^2(\underline{y},t),t']\Big] \quad , \tag{9.31}$$

where of course

$$p_0(\underline{y}) = p(\underline{y},t_0) \tag{9.32}$$

should be obtained from the initial potential $Q_0(x,\underline{y}) = Q(x,\underline{y},t_0)$. Note that, if the $\underline{y}$ variable is not present [i.e., $\underline{\gamma} = 0$ in (9.25) or, for that matter, (9.23)], the discrete eigenvalues remain fixed throughout the time evolution. If instead $\underline{\gamma} \neq 0$, the time evolution of $p(\underline{y},t)$ is far from trivial; indeed in this case even the number of discrete eigenvalues might change throughout the motion. A detailed analysis of this case will not be attempted here.

The other quantity whose time evolution must be ascertained is the matrix $C(\underline{y},t)$ [defined by (9.6) and entering in (9.8)]; it is characterized by the linear partial differential matrix equation [corresponding to the NEE (9.25) for Q]

$$\begin{aligned} C_t(\underline{y},t) = {} & \underline{\gamma}(4p^2,t) \cdot \frac{\partial}{\partial \underline{y}} C(\underline{y},t) + C(\underline{y},t)\underline{\gamma}_z(z,t)\Big|_{z=4p^2} \cdot \frac{\partial}{\partial \underline{y}} (4p^2) \\ & + \alpha_n(4p^2)[\sigma_n,C(\underline{y},t)] - 2p\beta_\nu(4p^2)\{\sigma_\nu,C(\underline{y},t)\} \quad , \end{aligned} \tag{9.33}$$

where of course $p \equiv p(\underline{y},t)$ is the solution of the equations given immediately above. An explicit solution of this equation is

$$C(\underline{y},t) = \exp\Big[-4p \int_{t_0}^{t} dt\beta_0(4p^2,t')\Big]$$

$$\left\{1 - \left[\int_{t_0}^{t} dt'\underline{\gamma}_z(z,t')\Big|_{z=4p^2}\right] 8p \cdot \frac{\partial}{\partial \underline{y}'} p_0(\underline{y}')\Big|_{\underline{y}'=\underline{y}} + \int_{t_0}^{t} dt'\underline{\gamma}(4p^2,t')\right\}^{-1}$$

$$\cdot \exp\left\{(t - t_0)\left[\alpha_n(4p^2) - 2p\beta_n(4p^2)\right]\sigma_n\right\} \cdot C_0\left[\underline{y} + \int_{t_0}^{t} dt'\underline{\gamma}(4p^2,t')\right]$$

$$\cdot \exp\left\{(t - t_0)\left[-\alpha_n(4p^2) - 2p\beta_n(4p^2)\right]\sigma_n\right\} \quad , \tag{9.34}$$

where, of course,

$$C_0(\underline{y}) = C(\underline{y},t_0) \tag{9.35}$$

should be obtained from the initial potential $Q_0(x,\underline{y}) = Q(x,\underline{y},t_0)$. Note that also in (9.34) $p \equiv p(\underline{y},t)$.

If $\underline{\gamma} = 0$, so that the $\underline{y}$ variable disappears, (9.34) simplifies substantially. In view of the importance of this case, we write out the relevant formula

$$C(t) = \exp\left[-4p \int_{t_0}^{t} dt'\beta_0(4p^2,t')\right]$$

$$\cdot \exp\left\{(t - t_0)\left[\alpha_n(4p^2) - 2p\beta_n(4p^2)\right]\sigma_n\right\} \cdot C(t_0)$$

$$\cdot \exp\left\{(t - t_0)\left[-\alpha_n(4p^2) - 2p\beta_n(4p^2)\right]\sigma_n\right\} \quad , \tag{9.36}$$

where now p is constant.

It should be emphasized that the time evolution of the parameters p and C characterizing each discrete eigenvalue is quite independent of the eventual presence of other discrete eigenvalues, or of the continuum spectrum. As is well known, it is this fact, coupled with the solvability through the IST procedure, that causes the remarkable behavior of "solitons", in particular their stability. It also motivates the interest in the single-soliton solution, since this not only provides a remarkable special solution of the class of NEEs being investigated, but indeed constitutes a component (which may, or may not, show up asymptotically; see below) of a broad class of solutions.

*Solitons.* The most straightforward way to obtain the single-soliton solution is to solve (9.9) in the special case with R = 0 and with only one discrete eigenvalue. The result reads

$$Q(x,\underline{y},t) = -2p^2\{\cosh[p(x - \xi)]\}^{-2} P \quad . \tag{9.37}$$

Here $p \equiv p(\underline{y},t)$ is the parameter characterizing the value of the discrete eigenvalue, $P \equiv P(\underline{y},t)$ is a matrix satisfying the projection operator restriction

$$P^2 = P \quad , \tag{9.38}$$

and it is related to the matrix $C(\underline{y},t)$ by the formula

$$C = 2p \exp(2p\xi) P \tag{9.39}$$

which, together with (9.38), also defines $\xi = \xi(\underline{y},t)$. Of course also in this equation $p \equiv p(\underline{y},t)$.

We thus see that, for the whole class of NEEs considered in this paper, the single-soliton solution has the usual $(\cosh)^{-2}$ shape in the x variable, being moreover characterized by a quantity $\xi$ which must clearly be interpreted as the *position* of the soliton (on the x axis), and by a matrix P [whose projection property, and normalization, are characterized by (9.38)], to which the soliton solution is proportional, and which shall occasionally be referred to as the *polarization* of the soliton. The $\underline{y}$ and t dependence of $\xi$ and P are determined by their relationship to p and C, (9.38,39), and the $\underline{y}$ and t dependence of p and C, given above.

Clearly if the $\underline{y}$ variable is actually present, the situation is rather complicated (and interesting). But even if $\underline{\gamma} = 0$ (as we assume for simplicity in the rest of this paper), so that the $\underline{y}$ dependence disappears and moreover p is t independent, the soliton solution (9.37) is quite interesting, its most remarkable feature being the fact that $\xi$ is generally not linear in t, i.e., the soliton generally moves with a speed that varies in time. Indeed a little matrix algebra yields

$$\xi(t) = \xi(t_0) - 2\int_{t_0}^{t} dt'\beta_0(4p^2,t') + (2p)^{-1} \ln[(c_0,E(t-t_0)c_0)/(c_0,c_0)] \quad , \quad (9.40)$$

with

$$E(t) = \exp\left\{-t\left[\alpha_n(4p^2) + 2p\beta_n(4p^2)\right]\sigma_n\right\} \cdot \exp\left\{t\left[\alpha_n(4p^2) - 2p\beta_n(4p^2)\right]\sigma_n\right\} \quad . \quad (9.41)$$

The corresponding formula for P is

$$P(t) = \left\{(c_0,c_0)/\left(c_0,E(t-t_0)c_0\right)\right\}\cdot\exp\left\{(t-t_0)\left[\alpha_n(4p^2) - 2p\beta_n(4p^2)\right]\sigma_n\right\} \cdot P(t_0)\exp\left\{-(t-t_0)\left[\alpha_n(4p^2) + 2p\beta_n(4p^2)\right]\sigma_n\right\} \quad . \quad (9.42)$$

Here of course

$$c_0 = c(t_0) \quad (9.43)$$

is determined by the initial conditions, and we have used the relationship

$$P(t) = c(t)c^{\dagger}(t)/\left(c(t),c(t)\right) \quad (9.44)$$

implied by (9.39), (9.38) and (9.6).

To discuss the behavior of the soliton it is convenient to introduce the spectral decomposition of the matrix E(t)

$$E(t) = \sum_{k=1}^{N} \exp[2p\zeta_k(t)]E_k(t) \quad , \quad (9.45)$$

where of course

$$E_k(t)E_e(t) = \delta_{ke}E_k(t) \quad . \quad (9.46)$$

Note that the quantities $\zeta_k(t)$, as well as the projection operators $E_k(t)$, depend on the functions $\alpha_n$ and $\beta_n$ that characterize the structure of the NEE under consideration [see (9.25)]; they depend on the initial conditions only through the value of p. The time evolution of $\xi(t)$ and $P(t)$ depends moreover on the initial vector $c_0 = c(t_0)$, through the quantities

$$e_k(t) = \left(c_0, E_k(t) c_0\right)/(c_0, c_0) \quad . \tag{9.47}$$

The explicit formulae are

$$\xi(t) = \xi(t_0) - 2 \int_{t_0}^{t} dt' \beta_0(4p^2, t') + (2p)^{-1} \ln\left\{ \sum_{k=1}^{N} e_k(t - t_0) \exp[2p\zeta_k(t - t_0)]\right\} , \tag{9.48}$$

$$\begin{aligned} P(t) = &\left\{ \sum_{k=1}^{N} e_k(t - t_0)\exp[2p\zeta_k(t - t_0)]\right\}^{-1} \\ &\cdot \exp\left\{(t - t_0)\left[\alpha_n(4p^2) - 2p\beta_n(4p^2)\right]\sigma_n\right\} \cdot P(t_0) \\ &\cdot \exp\left\{- (t - t_0)\left[\alpha_n(4p^2) + 2p\beta_n(4p^2)\right]\sigma_n\right\} \quad . \end{aligned} \tag{9.49}$$

Let us now consider two special cases: because the t dependence due to $\beta_0$ is relatively trivial, we set for simplicity $\beta_0 = 0$.

Assume first that $\beta_n(4p^2) = 0$. Then $E(t) = 1$, $\xi(t) = \xi(t_0)$; the soliton does not move [it would move with constant velocity if $\beta_0(4p^2,t) = \beta_0(4p^2) \neq 0$].

Assume next that

$$\left[\alpha_n(4p^2)\sigma_n \quad , \quad \beta_m(4p^2)\sigma_m\right] = 0 \quad ; \tag{9.50}$$

note that this case includes the possibility $\alpha_n(4p^2) = 0$, while we exclude now the case $\beta_n(4p^2) = 0$ treated above. Then

$$\zeta_k(t) = -2t\beta^{(k)} \quad , \quad k = 1, 2, \ldots, N \quad , \tag{9.51}$$

where $\beta^{(k)}$ are the eigenvalues of the matrix $\beta_n(4p^2)\sigma_n$. Thus in this case if the initial conditions are such that

$$e_k(0) = \delta_{k\ell} \quad , \tag{9.52}$$

the soliton moves with constant velocity,

$$\xi(t) = \xi(t_0) - 2(t - t_0)\beta^{(\ell)} \quad ; \tag{9.53}$$

while if more generally, $e_1(0) \neq 0$ and $e_N(0) \neq 0$ at large $|t|$

$$\begin{aligned} \xi(t) = &-2(t - t_0)\beta^{(1)} + \xi(t_0) + (2p)^{-1} \ln[e_1(0)] \\ &+ O\left\{\exp\left[-4tp\left(\beta^{(2)} - \beta^{(1)}\right)\right]\right\} \quad \text{as} \quad t \to +\infty \quad , \end{aligned} \tag{9.54a}$$

$$\begin{aligned} \xi(t) = &-2(t - t_0)\beta^{(N)} + \xi(t_0) + (2p)^{-1} \ln[e_N(0)] \\ &+ O\left\{\exp\left[4tp\left(\beta^{(N)} - \beta^{(N-1)}\right)\right]\right\} \quad \text{as} \quad t \to -\infty \quad . \end{aligned} \tag{9.54b}$$

Here we assuming the $\beta_n$'s to be real, so that the eigenvalues $\beta^{(k)}$ are also real and can be ordered according to the prescription $\beta^{(k)} < \beta^{(k+1)}$ (assuming no degeneracy). Thus in this case the soliton moves asymptotically with constant speed; if $\beta^{(N)} > 0 > \beta^{(1)}$, it moves towards the right as $t \to +\infty$, and it recedes also to the right as $t \to -\infty$. This behavior justifies the introduction of the term "boomeron" [9.3].

Returning to the general case, we note that, if the $\alpha_n$'s are imaginary and the $\beta_\nu$'s are real, as is required in order that Q may be Hermitian at all times, it is easy to show that, in a reference frame moving with speed $\beta_0(4p^2,t)$, if the soliton escapes to infinity, it can do so at most with a speed that is asymptotically constant (that solitons may indeed move asymptotically with constant speed, escaping to infinity, or that they may instead oscillate indefinitely, is demonstrated by the example discussed below).

This same approach can be used to obtain the (pure) multisoliton solutions. We write explicitly only the formula for the two-soliton case:

$$Q(x,\underline{y},t) = -W_x(x,\underline{y},t) \quad , \tag{9.55}$$

$$W(x,\underline{y},t) = -2(p_1+p_2)[1-\rho\tau_1\tau_2]^{-1} \cdot [\tau_1 P_1 + \tau_2 P_2 - \tau_1\tau_2\{P_1,P_2\}] \quad , \tag{9.56}$$

$$\tau_k = [P_k/(p_1+p_2)] \cdot \{1-\tanh[p_k(x-\xi_k)]\} \quad , \quad k = 1, 2 \quad . \tag{9.57}$$

In these equations $\xi_k = \xi_k(\underline{y},t)$ and $P_k = P_k(\underline{y},t)$ should be interpreted as the "position" and "polarization" of the $k^{th}$ soliton ($k = 1,2$), which is characterized by the positive parameter $p_k$; their time evolution is given by the explicit equations displayed and discussed above, of course with p replaced by $p_k$. Finally the scalar $\rho = \rho(\underline{y},t)$ is defined by either one of the following two (equivalent) equations:

$$\rho P_1 = P_1 P_2 P_1 \quad , \tag{9.58a}$$

$$\rho P_2 = P_2 P_1 P_2 \quad . \tag{9.58b}$$

## 9.4 The Boomeron Equation and Other Solvable Nonlinear Evolution Equations Related to it; Boomerons

The simplest new NEE of the class described above is obtained for $N = 2$, $\beta_0 = \underline{\gamma} = 0$, and $\alpha_n$, $\beta_n$ constant, $n = 1, 2, 3$. It is convenient to set $b_n = 2\beta_n$, $a_n = 2i\alpha_n$, and

$$U(x,t) + \sigma_n V_n(x,t) = \int_x^{+\infty} dx' Q(x',t) \quad ;$$

the $\sigma_n$'s are the 3 Pauli matrices. The "boomeron equation" is then conveniently written in terms of the 3-vectors $\underline{a}$, $\underline{b}$ and $\underline{V}(x,t)$, and of the scalar $U(x,t)$, as follows:

$$U_t(x,t) = \underline{b} \cdot \underline{V}_x(x,t) \tag{9.59a}$$

$$\underline{V}_{xt}(x,t) = U_{xx}(x,t)\underline{b} + \underline{a} \times \underline{V}_x(x,t) - 2\underline{V}_x(x,t) \times [\underline{V}(x,t) \times \underline{b}] \quad . \tag{9.59b}$$

Clearly the quantities $U(x,t)$ and $\underline{V}(x,t)$ satisfy the asymptotic conditions

$$U(+\infty,t) = U_x(\pm\infty,t) = 0 \quad , \quad \underline{V}(+\infty,t) = \underline{V}_x(\pm\infty,t) = 0 \quad . \tag{9.60}$$

So far no physical phenomenon has been modeled by this NEE (or rather system of coupled NEEs); in view of the remarkable features of the solutions described below this constitutes a very interesting goal. In this connection it may be of interest to note that the NEE (9.59) can be derived from the Lagrangian density

$$L = \frac{1}{2} U_t U_x + \frac{1}{2} \underline{V}_t \cdot \underline{V}_x - U_x \underline{b} \cdot \underline{V}_x + \frac{1}{2}(\underline{a} \times \underline{V}_x) \cdot \underline{V} + 2(\underline{V}_x \cdot \underline{V})(\underline{b} \cdot \underline{V}) \quad . \tag{9.61}$$

The solution of the initial-value problem for the NEE (9.59), in the framework of solutions obeying the boundary conditions (9.60), can be achieved by the IST technique described in the previous section; but now a more explicit description of the behavior of the soliton solution

$$U(x,t) = -p\{1 - \tanh[p(x - \xi(t))]\} \quad , \quad \underline{V}(x,t) = \hat{\underline{n}}(t) U(x,t) \quad , \tag{9.62}$$

can be given. Indeed it can be shown that its "position" $\xi(t)$ and "polarization" $\hat{\underline{n}}(t)$ (a unit vector; we always use the notation $\underline{v} \equiv v\hat{\underline{v}}$) evolve in time according to the equations

$$\hat{\underline{n}}_t(t) = \underline{a} \times \hat{\underline{n}}(t) + 2p\hat{\underline{n}}(t) \times [\hat{\underline{n}}(t) \times \underline{b}] \quad , \tag{9.63a}$$

$$\xi_t(t) = -\underline{b} \cdot \hat{\underline{n}}(t) \quad ; \tag{9.63b}$$

and these equations can be solved explicitly:

$$\xi(t) = \xi_0 + (2p)^{-1} \ln[n_+ E_+(t) + n_- E_-(t) + \bar{s} S(t) + \bar{c} C(t)] \quad , \tag{9.64}$$

$$\hat{\underline{n}}(t) = [\underline{n}_+ E_+(t) + \underline{n}_- E_-(t) + \underline{s} S(t) + \underline{c} C(t)]/[n_+ E_+(t) + n_- E_-(t) + \bar{s} S(t) + \bar{c} C(t)] \quad , \tag{9.65}$$

$$E_\pm(t) = \exp[\pm a\nu_-(t - t_0)] \quad , \quad S(t) = \sin[a\nu_+(t - t_0)] \quad ,$$
$$C(t) = \cos[a\nu_+(t - t_0)] \quad , \tag{9.66}$$

$$\nu_\pm = \left\{\left[\lambda^2(\hat{\underline{a}} \cdot \hat{\underline{b}})^2 + \frac{1}{4}(1 - \lambda^2)^2\right]^{\frac{1}{2}} \pm \frac{1}{2}(1 - \lambda^2)\right\}^{\frac{1}{2}} \quad , \tag{9.67}$$

$$\lambda = 2pb/a \quad , \tag{9.68}$$

$$\underline{n}_\pm = \frac{1}{2}\Big\{(\nu_+^2 - 1)\hat{\underline{n}}_0 + [(\hat{\underline{a}} \cdot \hat{\underline{n}}_0) \mp \eta\nu_+]\hat{\underline{a}} + \lambda[\lambda(\hat{\underline{b}} \cdot \hat{\underline{n}}_0) \mp \nu_-]\hat{\underline{b}}$$
$$\pm \nu_-(\hat{\underline{a}} \times \hat{\underline{n}}_0) \mp \eta\lambda\nu_+(\hat{\underline{b}} \times \hat{\underline{n}}_0) - \lambda(\hat{\underline{a}} \times \hat{\underline{b}})\Big\}/(\nu_+^2 + \nu_-^2) \quad , \quad \eta = \mathrm{sign}(\hat{\underline{a}} \cdot \hat{\underline{b}}) \quad , \tag{9.69}$$

$$\underline{s} = \eta[\nu_- \hat{\underline{a}} - \eta\lambda\nu_+ \hat{\underline{b}} + \eta\nu_+(\hat{\underline{a}} \times \hat{\underline{n}}_0) + \lambda\nu_-(\hat{\underline{b}} \times \hat{\underline{n}}_0)]/(\nu_+^2 + \nu_-^2) \quad , \quad \underline{c} = \underline{n}_0 - \underline{n}_+ - \underline{n}_- \quad , \tag{9.70}$$

$$\bar{s} = \eta[\nu_-(\hat{\underline{a}} \cdot \hat{\underline{n}}_0) - \eta\lambda\nu_+(\hat{\underline{b}} \cdot \hat{\underline{n}}_0)]/(\nu_+^2 + \nu_-^2) \quad , \quad \bar{c} = 1 - n_+ - n_- \quad , \tag{9.71}$$

The following important properties implied by these formulae should be noted: I) $\hat{\underline{n}}(t)$ is a unit vector for all t; II) the unit vectors $\hat{\underline{n}}_\pm = (\mp \eta\nu_+ \hat{\underline{a}} \mp \lambda\nu_- \hat{\underline{b}} - \lambda\hat{\underline{a}} \times \hat{\underline{b}})/$

$(1 + \nu_-^2)$ are independent of the initial polarization $\underline{n}_0$; III) iff $\hat{\underline{n}}_0 = \hat{\underline{n}}_+$ or $\hat{\underline{n}}_0 = \hat{\underline{n}}_-$ , the polarization remains constant, $\hat{\underline{n}}(t) = \hat{\underline{n}}_0$.

The speed $\xi_t(t)$ of the soliton, as well as its polarization, generally changes with time. Let us discuss briefly this remarkable phenomenon, excluding at first the special cases when $\underline{a} \times \underline{b} = \underline{0}$ or $\hat{\underline{a}} \cdot \hat{\underline{b}} = 0$ which are discussed separately below.

If the initial polarization $\hat{\underline{n}}_0$ coincides with one of the two unit vectors $\hat{\underline{n}}_+$ or $\hat{\underline{n}}_-$, the soliton moves (to the left or to the right; see below) with a constant speed of modulus

$$v = \frac{1}{2} a\nu_-/p = b\nu_-/\lambda \quad , \tag{9.72}$$

and the polarization remains constant

$$\hat{\underline{n}}(t) = \hat{\underline{n}}_0 \quad , \quad \xi(t) = \xi_0 \pm v(t - t_0) \quad \text{if} \quad \hat{\underline{n}}_0 = \hat{\underline{n}}_\pm \quad . \tag{9.73}$$

Otherwise both $\hat{\underline{n}}(t)$ and $\xi_t(t)$ vary with time; at no finite time does $\hat{\underline{n}}(t)$ coincide with $\hat{\underline{n}}_+$ or $\hat{\underline{n}}_-$, while asymptotically

$$\hat{\underline{n}}(\pm\infty) = \hat{\underline{n}}_\pm \quad , \quad \xi_t(\pm\infty) = \pm v \quad . \tag{9.74}$$

Note that this implies that, independently of the conditions assigned at time $t_0$ (provided $\hat{\underline{n}}_0 \neq \hat{\underline{n}}_\pm$), in the extreme future and in the remote past the soliton (or rather the *boomeron*) escapes towards and comes in from the right, with the (same) asymptotically constant speed v

$$\xi(t) - [\xi_0 + v|t - t_0|] = O[\exp(-a\nu_-|t|)] \quad \text{as} \quad t \to \pm\infty \quad . \tag{9.75}$$

The behavior of this localized solution of the NEE (9.59) should be noted: for given $\underline{a}$ and $\underline{b}$ [i.e., for a given NEE of type (9.59)], the positive constant p characterizes both the shape (specifically, the "amplitude" and the "width") of the boomeron and its asymptotic speed v; the initial position $\xi_0$ of the boomeron can of course be assigned arbitrarily [corresponding to the translation invariant nature of the NEE (9.59)]; its initial polarization $\hat{\underline{n}}_0$ characterizes its initial speed and also determines its subsequent behavior (both its speed and its polarization), but only for finite times. The asymptotic behavior of the boomeron is essentially independent of $\hat{\underline{n}}_0$, except for whether $\hat{\underline{n}}_0$ coincides or not with $\hat{\underline{n}}_+$ or $\hat{\underline{n}}_-$; indeed, the asymptotic speed (as $t \to \pm\infty$) always has the magnitude v, although its sign may be different depending whether $\hat{\underline{n}}_0$ coincides with $\hat{\underline{n}}_+$ or $\hat{\underline{n}}_-$ or differs from them (even by a small amount). Note finally that only if $\hat{\underline{n}}_0 = \hat{\underline{n}}_-$ or $\hat{\underline{n}}_0 = \hat{\underline{n}}_+$ can the soliton escape to, or come in from the left asymptotically.

*In the special case* $\hat{\underline{a}} \times \hat{\underline{b}} = 0$, the boomeron is still described by (9.62,63), but the behavior of $\xi(t)$ and $\hat{\underline{n}}(t)$ is now given by the simpler expressions

$$\xi(t) = \xi_0 + (2p)^{-1} \ln\left\{\cosh[2pb(t - t_0)] - (\hat{\underline{b}} \cdot \hat{\underline{n}}_0)\sinh[2pb(t - t_0)]\right\} \quad , \tag{9.76}$$

$$\hat{\underline{n}}(t) = \left\{\hat{\underline{b}}\left[(\hat{\underline{b}} \cdot \hat{\underline{n}}_0)\cosh[2pb(t - t_0)] - \sinh[2pb(t - t_0)]\right]\right.$$

$$+ (\hat{\underline{b}} \times \hat{\underline{n}}_0)\sin[a(t - t_0)] + [\hat{\underline{n}}_0 - \hat{\underline{b}}(\hat{\underline{b}} \cdot \hat{\underline{n}}_0)]\cos[a(t - t_0)]\Big\}$$

$$/\Big\{\cosh[2pb(t - t_0)] - (\hat{\underline{b}} \cdot \hat{\underline{n}}_0)\sinh[2pb(t - t_0)]\Big\} \quad . \tag{9.77}$$

These expressions can be obtained from those given above for the general case, and imply a behavior of the boomeron analogous to that described above (with $\hat{\underline{n}}_\pm = \mp\hat{\underline{b}}$): if $\hat{\underline{n}}_0 = -\hat{\underline{b}}$ or $\hat{\underline{n}}_0 = \hat{\underline{b}}$, $\hat{\underline{n}}(t) = \hat{\underline{n}}_0$, $\xi(t) = \xi_0 \pm b(t - t_0)$; if $\hat{\underline{n}}_0 \neq \pm\hat{\underline{b}}$, $\hat{\underline{n}}(t) \neq \hat{\underline{b}}$ and $\hat{\underline{n}}(t) \neq -\hat{\underline{b}}$ for all (finite) t, $\hat{\underline{n}}(\pm\infty) = \mp\hat{\underline{b}}$, $\xi_t(\pm\infty) = \pm b$, and for large $|t|$ $\xi(t) = \xi_0 + b[t - t_0] + O[\exp(-2pb|t|)]$. In this case, moreover, the following elegant theorem holds: *the boomeron coordinate $\xi(t)$ coincides with that of a (nonrelativistic) particle of unit mass, with initial position $\xi(t_0) = \xi_0$ and initial speed $\xi_t(t_0) = -(\hat{\underline{b}} \cdot \hat{\underline{n}}_0)$, moving in the external potential $\Phi(\xi - \xi_0)$, with*

$$\Phi(x) = \frac{1}{2} b^2[1 - (\hat{\underline{b}} \cdot \hat{\underline{n}}_0)^2]\exp(-4px) \quad . \tag{9.78}$$

If the initial velocity is negative [namely, if the initial polarization of the boomeron is such that $(\hat{\underline{b}} \cdot \hat{\underline{n}}_0) > 0$], the particle starts towards the left, moving with a decreasing velocity until it reaches the turning point $\bar{\xi} = \xi_0 + (4p)^{-1} \ln[1 - (\hat{\underline{b}} \cdot \hat{\underline{n}}_0)^2]$; then it comes back, escaping finally to the right with the asymptotically constant speed b. If instead the boomeron starts towards the right $[(\hat{\underline{b}} \cdot \hat{\underline{n}}_0) < 0]$, it continues with an increasing velocity that tends asymptotically to the same value b. Note that this analogy applies also to the special cases $\hat{\underline{n}}_0 = -\hat{\underline{b}}$ or $\hat{\underline{n}}_0 = \hat{\underline{b}}$ (the only ones for which the boomeron does not boomerang, moving instead with constant velocity), for in these cases the potential $\Phi(x)$ of (9.78) vanishes. And note finally that the potential $\Phi(x)$ depends on the parameters of the boomeron (its initial polarization $\hat{\underline{n}}_0$ and its shape parameter p), and on $\underline{b}$, but it is independent of the value of a, which influences only the time evolution of the boomeron polarization at finite times.

*In the other special case, $\hat{\underline{a}} \cdot \hat{\underline{b}} = 0$,* the boomeron is still described by (9.62, 63), but the behavior of $\xi(t)$ and $\hat{\underline{n}}(t)$ is given by the simpler expressions

$$\xi(t) = \xi_0 + (2p)^{-1} \ln\Big\{1 - \lambda\beta S_1(t) + \frac{1}{2}\lambda(\lambda + \gamma)S_2(t)\Big\} \quad , \tag{9.79}$$

$$\hat{\underline{n}}(t) = \Big\{\hat{\underline{n}}_0 - [\lambda\hat{\underline{b}} - (\hat{\underline{a}} \times \hat{\underline{n}}_0)]S_1(t) - \frac{1}{2}[\hat{\underline{n}}_0 - \alpha\hat{\underline{a}} - \lambda^2\beta\hat{\underline{b}} + \lambda(\hat{\underline{a}} \times \hat{\underline{b}})]S_2(t)\Big\}$$

$$/\Big\{1 - \lambda\beta S_1(t) + \frac{1}{2}\lambda(\lambda + \gamma)S_2(t)\Big\} \quad , \tag{9.80}$$

$$S_1(t) = (\lambda^2 - 1)^{-\frac{1}{2}}\sinh[a(\lambda^2 - 1)^{\frac{1}{2}}(t - t_0)] \quad ,$$

$$S_2(t) = 4(\lambda^2 - 1)^{-1} \sinh^2\Big[\frac{a}{2}(\lambda^2 - 1)^{\frac{1}{2}}(t - t_0)\Big] \quad \text{if} \quad \lambda > 1 \quad ,$$

$$S_1(t) = a(t - t_0) \quad , \quad S_2(t) = [a(t - t_0)]^2 \quad \text{if} \quad \lambda = 1 \quad ,$$

$$S_1(t) = (1 - \lambda^2)^{-\frac{1}{2}}\sin[a(1 - \lambda^2)^{\frac{1}{2}}(t - t_0)] \quad ,$$

$$S_2(t) = 4(1-\lambda^2)^{-1}\sin^2\left[\frac{a}{2}(1-\lambda^2)^{\frac{1}{2}}(t-t_0)\right] \quad \text{if} \quad \lambda<1 \quad , \tag{9.81}$$

$$\alpha = (\hat{\underline{a}}\cdot\hat{\underline{n}}_0) \quad , \quad \beta = (\hat{\underline{b}}\cdot\hat{\underline{n}}_0) \quad , \quad \gamma = (\hat{\underline{a}}\times\hat{\underline{b}}\cdot\hat{\underline{n}}_0) = \pm(1-\alpha^2-\beta^2)^{\frac{1}{2}} \quad , \tag{9.82}$$

with $\lambda$ defined by (9.68). Thus if the initial polarization $\hat{\underline{n}}_0$ coincides with either one of the unit vectors [$\theta(x) = 1$ if $x>0$, $\theta(0) = \frac{1}{2}$, $\theta(x) = 0$ if $x<0$]

$$\hat{\underline{n}}_\pm = -\left\{\pm\lambda(1-\lambda^2)^{\frac{1}{2}}\theta(1-\lambda)\hat{\underline{a}}\pm(\lambda^2-1)^{\frac{1}{2}}\theta(\lambda-1)\hat{\underline{b}} + [\theta(\lambda-1)+\lambda^2\theta(1-\lambda)]\hat{\underline{a}}\times\hat{\underline{b}}\right\}/\lambda \quad , \tag{9.83}$$

it does not change with time, $\hat{\underline{n}}(t) = \hat{\underline{n}}_0$, and the soliton does not move at all [i.e., $\xi(t) = \xi_0$] if $\lambda \leq 1$, while if $\lambda>1$ it moves with the constant speed $\pm v$ [i.e., $\xi(t) = \xi_0 \pm v(t-t_0)$ if $\hat{\underline{n}}_0 = \hat{\underline{n}}_\pm$] where now

$$v = (\lambda^2-1)^{\frac{1}{2}}b/\lambda \quad . \tag{9.84}$$

If instead the initial polarization $\hat{\underline{n}}_0$ coincides neither with $\hat{\underline{n}}_+$ nor with $\hat{\underline{n}}_-$, the boomeron polarization $\hat{\underline{n}}(t)$ does change with time: it never coincides with $\hat{\underline{n}}_+$ or $\hat{\underline{n}}_-$ but, if $\lambda>1$, it tends asymptotically to the values $\hat{\underline{n}}_\pm$ [i.e., $\hat{\underline{n}}(\pm\infty) = \hat{\underline{n}}_\pm$], while if $\lambda<1$ it precedes periodically [with period $(2\pi/a)(1-\lambda^2)^{-\frac{1}{2}}$].

As for the behavior of the boomeron coordinate $\xi(t)$, it is best understood noting that *it coincides with that of a particle of unit mass, with initial position* $\xi(t_0) = \xi_0$ *and initial speed* $\xi_t(t_0) = -\beta b$, *moving in the external potential* $\tilde{\Phi}(\xi-\xi_0)$, *with*

$$\tilde{\Phi}(x) = \frac{1}{2}(b/\lambda)^2\exp(-2px)\left\{\left[\left([(\alpha^2+\gamma^2)\lambda+\gamma]^2+\alpha^2\right)/(\alpha^2+\gamma^2)\right]\exp(-2px) - 2(1+\lambda\gamma)\right\} \quad . \tag{9.85}$$

This potential vanishes as $x\to+\infty$; it has a single (negative) minimum at $x = \bar{x} \equiv (2p)^{-1}\ln\left\{\left(\left[(\alpha^2+\gamma^2)\lambda+\gamma\right]^2+\alpha^2\right)/\left[(\alpha^2+\gamma^2)(1+\lambda\gamma)\right]\right\}$ if $\lambda\gamma>-1$; and it diverges to positive infinity as $x\to-\infty$ (unless $\alpha = 1+\lambda\gamma = 0$, when it vanishes identically; this happens if $\hat{\underline{n}}_0 = \hat{\underline{n}}_+$ or $\hat{\underline{n}}_0 = \hat{\underline{n}}_-$). Note that the total energy of this particle has the simple expression

$$E = \tilde{\Phi}(0) + \frac{1}{2}\beta^2 b^2 = \frac{1}{2}v^2\,\mathrm{sign}(\lambda-1) \tag{9.86}$$

with $v$ defined by (9.84). Thus it is positive if $\lambda>1$, in which case the boomeron escapes to, or comes in from the right as $t\to\pm\infty$, moving asymptotically with the constant speed $v$ of (9.84). The total energy $E$ is instead negative if $\lambda<1$, so that in this case the boomeron behaves as a particle trapped in the potential, oscillating indefinitely around the equilibrium position $\xi_0+\bar{x}$ (see above). This oscillatory behavior is of course also given, in this case with $\lambda<1$, by the explicit equation (9.79); it is periodic, with period $(2\pi/a)(1-\lambda^2)^{-\frac{1}{2}}$. The marginal case $\lambda = 1$ corresponds to the motion of a zero-energy particle in the potential (9.85),

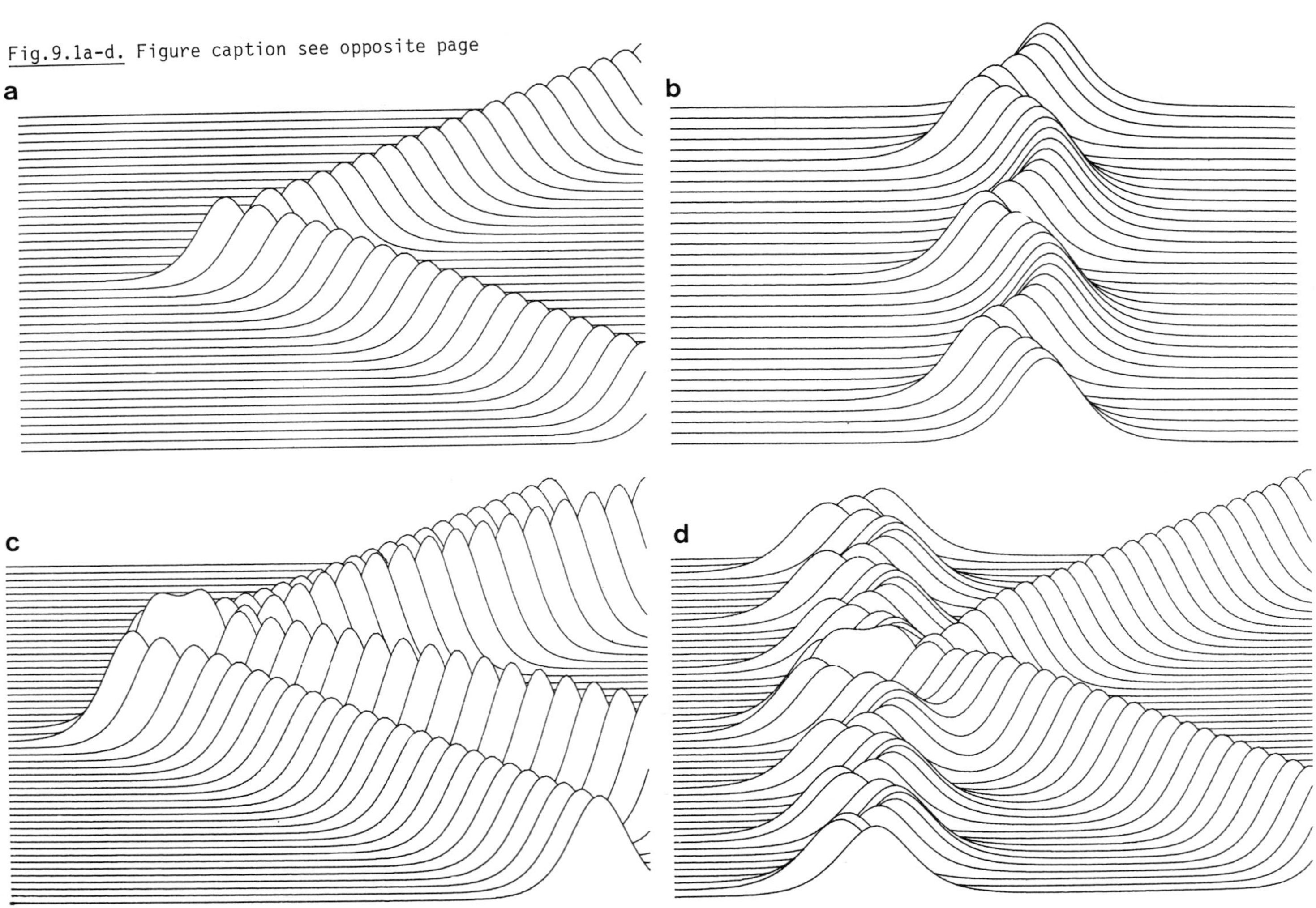

Fig.9.1a-d. Figure caption see opposite page

e

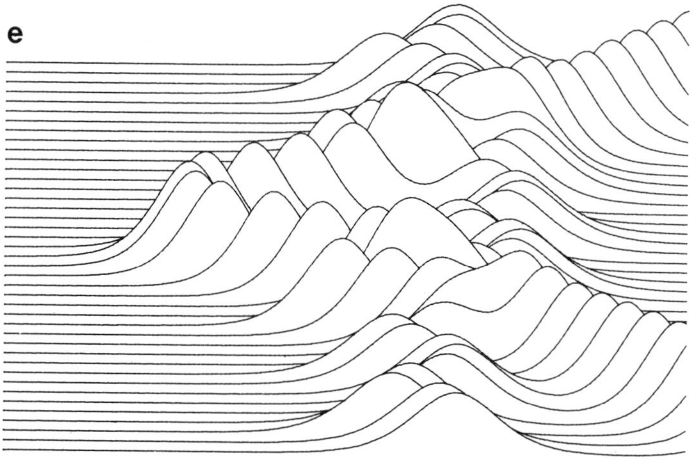

Fig.9.1a-e. *Boomeron Behavior*. $U_x(x,t)$ is displayed as a function of x at equally spaced times. All five cases refer to (9.59) with $\underline{a}\cdot\underline{b}=0$. Parts (a) and (b) have been computed from (9.62), (9.79) and (9.81). Parts (c-e) have been computed using (9.56) in addition. All the figures were computer produced by Dr. J.C. Eilbeck (Dept. of Mathematics, Heriot-Watt University, Edinburgh EH14 4AS)

that always escapes to (or comes in from) positive infinity, but with a vanishing asymptotic velocity.

Note that the analogy of the motion of the boomeron to that of a particle in the external potential (9.85) covers also the cases when the initial polarization coincides with $\underline{n}_+$ or $\underline{n}_-$, since the potential then vanishes identically (and moreover the initial speed of the boomeron $-\beta b$ vanishes if $\lambda < 1$).

It should be emphasized that, for a given NEE of type (9.59) with $\hat{\underline{a}}\cdot\hat{\underline{b}} = 0$, in a solution containing several solitons all the types of soliton behavior described above may be simultaneously present—the oscillatory one, the boomeranging one [$\xi(t) \to +\infty$ as $t \to \pm\infty$], the static one [$\xi(t) = \xi_0$], the uniform one [$\xi(t) = \xi_0 \pm v(t - t_0)$]—depending on the values of the shape parameter p [in particular, whether or not it exceeds the quantity a/(2b)] and of the initial polarization $\hat{\underline{n}}_0$ of each soliton; the latter two behaviors listed correspond, however, to a set of initial conditions having a lower dimensionality. Note moreover that, in order that the phenomenon of soliton trapping (oscillatory behavior) may occur, it is *required* that $\hat{\underline{a}}\cdot\hat{\underline{b}} = 0$, since this phenomenon is characteristic of this special case only.

The boomeron equation (9.59) yields, in special cases, other solvable NEEs. An interesting instance comes from the following remark: *if the vectors $\underline{a}$ and $\underline{b}$ are orthogonal, and the vector $\underline{V}$ is initially orthogonal to $\underline{a}$, it always remains orthogonal to $\underline{a}$*

$$\underline{a}\cdot\underline{b} = 0 \quad \text{and} \quad \underline{a}\cdot\underline{V}(x,t_0) = 0 \quad \text{imply} \quad \underline{a}\cdot\underline{V}(x,t) = 0 \quad . \tag{9.87}$$

Then in this case one can set

$$\hat{\underline{b}}\cdot\underline{V}(x,t/b) = -Y(x,t) \quad , \tag{9.88a}$$

$$(\hat{\underline{b}}\times\hat{\underline{a}})\cdot\underline{V}(x,t/b) = Z(x,t) - [a/(2b)] \quad , \tag{9.88b}$$

$$U_x(x,t/b) = -Y_t(x,t) - Z^2(x,t) + [a/(2b)]^2 \quad , \tag{9.88c}$$

thereby obtaining for Y and Z the simple system of coupled nonlinear partial differential equations

$$Y_{tt}(x,t) - Y_{xx}(x,t) = -[Z^2(x,t)]_t \quad , \tag{9.89a}$$

$$Z_{xt}(x,t) = 2Y_x(x,t)Z(x,t) \quad . \tag{9.89b}$$

These NEEs are therefore solvable by IST, through their connection with the boomeron equation in the special case of (9.87), provided the initial data

$$Z(x,t_0) = Z_0(x) \quad , \quad Y(x,t_0) = Y_0(x) \quad , \quad Y_t(x,t_0) = W(x) \tag{9.90}$$

are given, with

$$Z_{0x}(\pm\infty) = Y_0(+\infty) = Y_{0x}(\pm\infty) = W(+\infty) = W_x(\pm\infty) = 0 \quad ; \tag{9.91}$$

and the solutions display the rich phenomenology discussed above.

Note that the NEEs (9.89), besides being invariant under time and space translations, are invariant under the scale transformation

$$t \to \eta t \quad , \quad x \to \eta x \quad , \quad Y \to Y/\eta \quad , \quad Z \to Z/\eta \quad ; \tag{9.92}$$

they are moreover invariant under the discrete transformation

$$Z(x,t) \to -Z(x,t) \quad . \tag{9.93}$$

We finally note that, differentiating (9.89a) with respect to x and using (9.89b), one can obtain an intriguing nonlinear partial differential equation for the single field Z(x,t), namely

$$[\partial^2/\partial t^2 - \partial^2/\partial x^2](Z_{xt}/Z) + 2(Z^2)_{xt} = 0 \quad , \tag{9.94}$$

whose Cauchy problem is therefore also solvable by IST. The initial data to be provided are

$$Z(x,t_0) = Z_0(x) \quad , \quad Z_t(x,t_0) = Z_1(x) \quad , \quad Z_{tt}(x,t_0) = Z_2(x) \quad , \tag{9.95}$$

and it is sufficient (but not necessary) that they satisfy the requirements

$$Z_0(x) \neq 0 \quad , \quad Z_{0x}(\pm\infty) = Z_{1x}(\pm\infty) = Z_{2x}(\pm\infty) = 0 \quad . \tag{9.96}$$

## 9.5 Bäcklund Transformations

The basic formulae are obtained directly from (9.11,12), since the arbitrariness of f and g, and of M and N, imply that, *if the two matrix potentials* $Q(x)$ *and* $Q'(x)$ are related by the formula

$$a_\mu(\Lambda)[\sigma_\mu Q(x) - Q'(x)\sigma_\mu] + b_\mu(\Lambda)\Gamma\sigma_\mu = 0 \quad , \tag{9.97}$$

*the corresponding reflection coefficients* $R(k)$ *and* $R'(k)$ *are related by the equation*

$$R'(k) = \left\{\left[a_\mu(-4k^2) + 2ikb_\mu(-4k^2)\right]\sigma_\mu\right\}R(k)\left\{\left[a_\mu(-4k^2) - 2ikb_\mu(-4k^2)\right]\sigma_\mu\right\}^{-1} \quad . \tag{9.98}$$

The operators $\Lambda$ and $\Gamma$ in (9.97) are of course defined by (9.13,14), while the functions $a_\mu(z)$ and $b_\mu(z)$ are arbitrary [except for the requirement that they make some sense out of (9.97); a sufficient condition is that they be entire]. It should be emphasized that *the functions* $a_\mu$ *and* $b_\mu$ *might also depend on other variables such as* $t$; they must of course be independent of x.

The importance of the formulae (9.97) is that they display in closed form the explicit, if complicated, relationships between two potentials Q and Q', whose corresponding reflection coefficients are related by the simple linear formula (9.98).

Assume now the matrix Q(x,t) to satisfy the NEE (9.25) and the matrix Q'(x,t) to be related to it by (9.97) with $a_\mu$ and $b_\mu$ independent of t and $\underline{y}$. Then, provided for all values of z

$$[\alpha_m(z)a_n(z) + z\beta_m(z)b_n(z)][\sigma_m,\sigma_n] = 0 \quad , \tag{9.99a}$$

$$[\alpha_m(z)b_n(z) + \beta_m(z)a_n(z)][\sigma_m,\sigma_n] = 0 \quad , \tag{9.99b}$$

(9.98) implies that R'(k,t) is related to $R_0'(k) \equiv R'(k,t_0)$ by the same equation (9.28) that relates R(k,t) to $R_0(k) \equiv R(k,t_0)$. But such a time evolution for the reflection coefficient R'(k,t) corresponds to the time evolution (9.25) for the corresponding potential Q'(x,t). Conclusion: *if the matrix* $Q(x,t)$ *satisfies the NEE (9.25), and* $Q'(x,t)$ *is related to* $Q(x,t)$ *by (9.97) (with* $a_\mu$ *and* $b_\mu$ *independent of* $y$ *and* $t$*), and satisfying (9.99), then* $Q'(x,t)$ *satisfies the same NEE (9.25).*

We therefore term the transformation (9.97), with the conditions (9.99), a *Bäcklund transformation* (BT). The arbitrariness of the functions $a_\mu$ and $b_\mu$, that are constrained only by the limitations (9.99), should be emphasized; note in particular that $a_0$ and $b_0$ are not required to satisfy any constraint. Additional constraints may, however, be implied by the requirement that, given a matrix Q having some special property (for instance, hermiticity), the Q' related to it by the BT (9.97) have the same property.

Hereafter we limit our consideration to the subclass of BTs characterized by the condition $a_n = b_n = 0$; for notational convenience we moreover set $a_0 = f$, $b_0 = g$, so that the BT is now written

$$f(\Lambda)[Q(x,t) - Q'(x,t)] + g(\Lambda)\Gamma = 0 \quad , \tag{9.100}$$

and the corresponding formula for the reflection coefficients reads

$$R'(k,t) = R(k,t)\left\{[f(-4k^2) + 2ik\, g(-4k^2)]/[f(-4k^2) - 2ik\, g(-4k^2)]\right\} \quad . \tag{9.101}$$

This class of BTs is particularly interesting, because it holds for the whole class (9.25) [or, for that matter, (9.24)] of NEEs, there being now no restriction, related to the structure of the NEE (i.e., to the functions $\alpha_n$ and $\beta_\mu$), on the functions f and g that characterize the BT (9.100). Moreover, even though these BTs are generally only a small subclass of the BTs (9.97), they are nevertheless sufficiently general to yield a number of important results, that actually follow essentially from an even simpler subcase of (9.100), namely that corresponding to both f and g being constant. It is convenient to characterize this transformation by the constant $p = -\frac{1}{2} f/g$, so that (9.101) becomes

$$R'(k,t) = -[(k + ip)/(k - ip)]R(k,t) \tag{9.102}$$

and the BT reads

$$Q'(x,t) = Q(x,t) - (2p)^{-1}\Gamma \quad . \tag{9.103}$$

To discuss BTs and related results, it is convenient to work in terms of the integral of Q, rather than Q itself. We therefore introduce the matrix

$$W(x,t) = \int_x^{+\infty} dx' Q(x',t) \quad , \tag{9.104}$$

that clearly satisfies the boundary conditions

$$W(+\infty,t) = W_x(\pm\infty,t) = 0 \quad . \tag{9.105}$$

Indeed it is often convenient to write also the NEEs (9.25) in terms of W rather than Q; for instance it is in terms of W that the boomeron equation was written as a pure partial differential, rather than an integro-differential, nonlinear equation. Of course an analogous notation is used for $W'(x,t) = \int_x^{+\infty} dx' Q'(x',t)$.

In terms of W and W' the BT (9.103) becomes [9.1]

$$W'_x(x,t) + W_x(x,t) = -\frac{1}{2}[W'(x,t) - W(x,t)][4p + W'(x,t) - W(x,t)] \quad . \tag{9.106}$$

An instructive exercise is to solve this equation in the special case W = 0. A little matrix algebra yields then the single-soliton solution, namely

$$W'(x,t) = -2p\{1 - \tanh[p(x - \xi)]\}P \quad , \tag{9.107}$$

with

$$P^2 = P \quad . \tag{9.108}$$

The "constants of integration" $\xi$ and P depend of course on t; their explicit time evolution may be ascertained inserting (9.107) in the NEE (9.25) and solving the resulting equations, that involve only the variable t, recovering thereby for $\xi$ and P the expressions (9.48,49).

## 9.6 Nonlinear Superposition Principle

Clearly (9.102) implies that two BTs (9.106), characterized by the parameters $p_1$ and $p_2$, commute. There immediately follows [9.1] the validity of the *nonlinear superposition principle* characterized by the formula

$$(p_1 - p_2)[W_{12} - W_0] + \frac{1}{4}\left\{[W_{12} - W_0] \quad , \quad [W_1 - W_2]\right\} = - (p_1 + p_2)[W_1 - W_2] \quad , \tag{9.109}$$

where $W_0(x,t)$ is a (matrix) solution of an equation of the class (9.25) [rewritten of course in terms of W rather than Q; see (9.104)], $W_1(x,t)$ and $W_2(x,t)$ are the solutions of the same NEE related to $W_0$ by the BT (9.106) with parameters $p_1$ and $p_2$, respectively, and $W_{12}(x,t)$ is also a solution of the same NEE [precisely that related to $W_1$ by the BT (9.106) with $p = p_2$, or to $W_2$ by the same BT but with $p = p_1$].

This remarkable result may be used, starting from $W_0 = 0$ and two soliton solutions $W_1$ and $W_2$, to obtain directly the two-soliton solution [9.1]; and the process can be continued, adding one soliton at a time (soliton ladder), to obtain by purely algebraic steps multisoliton solutions [9.1].

## 9.7 Conserved Quantities

Presumably all the NEEs discussed in this chapter possess an infinite sequence of conserved quantities. The explicit form of these conserved quantities has been explicitly obtained for the following NEE:

$$Q_t(x,t) = [A(t),Q(x,t)] + \left\{B(t),Q_x(x,t)\right\} + \left[Q(x,t),[W(x,t),B(t)]\right] \quad . \tag{9.110}$$

Here A(t) and B(t) are arbitrary N-dimensional matrices, and the convenient notation (9.104) is used; the boomeron equation of Sect.9.4 is the special case of this NEE corresponding to N = 2 and to constant (and traceless) matrices A and B.

The conserved quantities $c_m$ are given by the formula [9.1]

$$c_m = \frac{1}{4}\int_{-\infty}^{+\infty} dx \; \mathrm{tr}\left[W^{(m)}(x,t)\right] \quad , \tag{9.111}$$

the matrices $W^{(m)}$ being defined by the recursion relations

$$W^{(m+1)}(x,t) = W_x^{(m)}(x,t) + \sum_{\ell=1}^{m-1} W^{(\ell)}(x,t)W^{(m-\ell)}(x,t) \tag{9.112}$$

with

$$W^{(1)}(x,t) = W_x(x,t) = -Q(x,t) \quad . \tag{9.113}$$

However, only the odd numbered $c_{2n+1}$ yield nontrivial conserved quantities, since all the even numbered $c_{2n}$ vanish. The first three conserved quantities read

$$c_1 = -\frac{1}{4}\,\mathrm{tr}\left\{\int_{-\infty}^{+\infty} dx\; Q(x,t)\right\} ,$$

$$c_3 = \frac{1}{4}\,\mathrm{tr}\left\{\int_{-\infty}^{+\infty} dx[Q(x,t)]^2\right\} ,$$

$$c_5 = -\frac{1}{4}\,\mathrm{tr}\left\{\int_{-\infty}^{+\infty} dx\left(2[Q(x,t)]^3 + [Q_x(x,t)]^2\right)\right\} .$$

## 9.8 Generalized Resolvent Formula

Consider the linear partial differential equation

$$q_t(x,t) = i\omega(\partial/\partial x,t)q(x,t) \tag{9.114}$$

where $\omega(z,t)$ is entire in z. Then the solution of the Cauchy problem can be written concisely through the "resolvent formula"

$$q(x,t) = \exp\left\{i\int_{t_0}^{t} dt'\omega(\partial/\partial x,t')\right\}q(x,t_0) \;. \tag{9.115}$$

It is easy to derive an analogous (albeit much more implicit) expression for the class of NEEs (9.25), comparing (9.97,98) (which hold even if f and g depend on t) with (9.28) [9.1]. The formula reads

$$a_\mu(\Lambda,t_0,t)[\sigma_\mu Q(x,t_0) - Q(x,t)\sigma_\mu] + b_\mu(\Lambda,t_0,t)\Gamma\sigma_\mu = 0 \tag{9.116}$$

with $a_\mu$ and $b_\mu$ defined by

$$\left[a_\mu(z^2,t_0,t) + zb_\mu(z^2,t_0,t)\right]\sigma_\mu$$
$$= \exp\left[z\int_{t_0}^{t} dt'\beta_0(z^2,t')\right]\exp\left\{(t-t_0)\left[\alpha_n(z^2) + z\beta_n(z^2)\right]\sigma_n\right\} , \tag{9.117}$$

and $\Lambda$ and $\Gamma$ defined by (9.13,14) with $Q'(x) = Q(x,t)$ and $Q(x) = Q(x,t_0)$.

It is also possible, in a similar manner, to relate the solution, evaluated at time t', of a NEE of the class (9.25), to the solution, evaluated at time t, of another NEE of the same class; such transformations have been termed generalized Bäcklund Transformations [9.1].

## 9.9 Nonlinear Operator Identities

The linear partial differential equation (9.114) has, for $\omega(z,t) = -iz$ the solution $q(x,t) = f(x + t)$, with f an arbitrary function. Insertion of this solution in (9.115) [with $\omega(z,t) = -iz$] yields the well-known operator identity

$$f(x + y) = \exp\left(y \frac{d}{dx}\right) f(x) \quad . \qquad (9.118)$$

In a similar manner, but starting from (9.116,117) rather than (9.115) [and of course (9.25) in place of (9.114)], it is possible to obtain a whole class of intriguing nonlinear operator identities [9.1].

## References


9.1 F. Calogero, A. Degasperis: Nuovo Cimento (in press). The present paper follows closely the treatment given in this reference, where additional details (including the proofs of all the results quoted here) can be found

9.2 F. Calogero, A. Degasperis: Lett. Nuovo Cimento *16*, 181 (1976)

9.3 F. Calogero, A. Degasperis: Lett. Nuovo Cimento *16*, 425; 434 (1976)

9.4 F. Calogero: "Generalized Wronskian Relations: a Novel Approach to Bargmann-Equivalent and Phase-Equivalent Potentials", in *Studies in Mathematical Physics* (Essays in Honor of Valentine Bargmann), ed. by E.H. Lieb, B. Simon, A.S. Wightman (Princeton University Press, Princeton, N.J. 1976); Lett. Nuovo Cimento *14*, 443 (1975); Nuovo Cimento *31*B, 229 (1976); Lett. Nuovo Cimento *14*, 537 (1975)
F. Calogero, A. Degasperis: Nuovo Cimento *32*B, 201 (1976)

# 10. A Method of Solving the Periodic Problem for the KDV Equation and Its Generalization

S. P. Novikov

## 10.1 One-Dimensional Systems Admitting a Lax Representation; Their Stationary Solutions

We consider a translation invariant evolution equation of the form

$$\dot{u} = K[u] \quad , \quad \left(u' = \frac{\partial u}{\partial x} \quad , \quad \dot{u} = \frac{\partial u}{\partial x}\right) \tag{10.1}$$

where u is a vector-function of a continuous variable x or of a discrete variable n; the equation (10.1) is invariant with respect to translations in x (or n).

*Definition 10.1.* We say that (10.1) admits a Lax representation if there exist two operators L and A, acting on functions f(x) [or f(n)], the coefficients of which are expressed in terms of u, such that (10.1) is equivalent to $\dot{L} = [L,A]$. We shall always take the class of operators L and A to be translation invariant.

In general, L and A can be given in the form of matrices using some basis $\{f_j(x)\}$ or $\{f_j(n)\}$ for the space of functions. In what follows we shall assume that the evolution equation (10.1) and the operator L associated with it are *local*. In the continuous case that means that the right-hand site K[u] depends only on a finite number of derivatives $u(x), \dots, u^{(q)}(x)$ at a given point x, and the associated L is a differential operator whose coefficients depend only on a finite number of derivatives $u(x), \dots u^{(q)}(x)$ at this point. In the discrete case 'local' means that K[u] for any n and the coefficients of L are functions of u on an interval of definite length $[n-\Delta, n+\Delta]$. The length of this interval is called the order of the operator L. This requirement makes sense only when the variable n can take an infinite sequence of values.

A survey of the equations admitting a Lax representation can be found in [10.1]. Of the known systems the only ones that are not local are the systems of MOSER [10.2] and Calogero of particles on the line with a certain two-body interaction potential where the associated operator L has all its matrix elements nonzero. In the local case ('Jacobi matrices') the nonzero elements are concentrated in a certain strip of finite width around the main diagonal.

From the requirement of being local and the translation invariance it follows that (10.1) defines a flow (or dynamical system) on any translation invariant class of functions (rapidly decreasing, periodic with any period, almost periodic with any group of periods, and others). The operator L acts on all such classes of functions; its eigenvalues, in principle, can give integrals of (10.1). Our aim is to explain some basic methods for studying the periodic and quasi-periodic case, developed by the author, Dubrovin, Matveev and Its (see, for example [10.3]; and also by LAX, McKEAN and VAN MOERBEKE, see [10.1]); these methods are explained in more detail in the survey [10.1].

In all the cases where our method is successful, there is an infinite number of local translation invariant evolutionary systems of the form (10.1)

$$\dot{u} = K_m[u] \quad , \quad m = 1,2, \ldots \tag{10.2}$$

whose flows commute with each other; the original equation (10.1) is obtained for m = 1. All the systems (10.2) have Lax representations $\dot{L} = [L,A_m]$ with the same L. The best-known example is the KdV equation, where

$$L = \frac{d^2}{dx^2} + u(x) \ , \quad A = -4\frac{d^3}{dx^3} + 3\left(u\frac{d}{dx} + \frac{d}{dx}u\right) + \lambda\frac{d}{dx} \ . \tag{10.3}$$

The KdV equation has 'higher analogues' where the operator $A_m$ has order 2m + 1. Of the discrete systems of local type the best known are the Toda lattice and the discrete KdV equation. These systems, and also the well-known 'sine-Gordon' equation, the nonlinear Schrödinger equation and others, have 'higher analogues' (10.2), which admit Lax representations with the same associated operator L.

Although, as a rule, only the original evolution equation (10.1) is of physical interest, the existence of the higher analogues turns out to be very important in constructing solutions. It is natural to reason as follows: we wish to find solutions of the original equation (10.1). For this we consider stationary solutions of any one of its higher analogues

$$\dot{u} = \sum_{q=1}^{m} c_q K_q[u] = 0 \ . \tag{10.4}$$

Since all the systems (10.2) commute, the set of stationary solutions of (10.4) is invariant under the flow (10.1) on any translation invariant class of functions. If we can explicitly solve the stationary equation (10.4) for the higher analogues of (10.1) and find on this set of functions the dynamics of the original system (10.1), then we obtain a class of exact solutions of the original system.

As $m \to \infty$, the class of solutions of (10.4) increases.

*Definition 10.2.* We say the collection of evolutionary systems (10.2) is *complete for the periodic problem* if the totality of periodic solutions of the equations

(10.4) for all possible constants $\{c_q\}$ is dense in the space of all continuous periodic functions of x.

An analogous concept of completeness can be introduced for other translation invariant classes of functions. It is fairly clear (although this is not strictly proved in the work being discussed) that for the KdV equation the collection of higher analogues of the KdV equations is complete for the periodic problem. However, it is not complete for the class of rapidly decreasing functions, for which the method of the inverse scattering problem was first discovered by GARDNER et al. [10.4].

It turns out that the existence of complete collections of evolutionary systems associated with an operator L makes it possible not only to find solutions of the system (10.1), but also to give an effective solution of the inverse problem for L.

This idea for studying equation (10.1) and the operator L requires the solution of the following problems:

1) How can we solve the stationary equations (10.4)? Will these finite-dimensional systems be completely integrable?
2) What are the spectral properties of operators L whose coefficients satisfy the stationary equations (10.4)?

## 10.2 Finite-Zoned Linear Operators

We consider an operator L of order k with periodic coeffcients of period T. We choose a basis $(\phi_1, \ldots, \phi_n)$ of solutions of the equation $L\phi = E\phi$ at the point $x = x_0$, for example, as follows:
$\phi_1(x_0) = 1$, $\phi_1^{(q)}(x_0) = 0$, $1 \leqq 1 \leqq k - 1$, $\phi_j^{(j-1)}(x_0) = 1$, $\phi_j^{(q)}(x_0) = 0$, $q \neq j - 1$.

A basis $\phi_j(n, n_0, E)$ can also be defined in the discrete case, if a solution of $L\phi = E\phi$ is determined by the initial values at k neighboring points $n_0$, $n_0 + 1$, $\ldots$, $n_0 + k-1$. One usually associates the following concepts with the operator L:

a) the translation (monodromy) matrix $\hat{T}\phi_j(x, x_0, E) = \phi_i(x + T, x_0, E) = \sum_{i=1}^{k} \alpha_{ij}(x_0,E)\phi_i(x, x_0, E)$, $\hat{T} = (\alpha_{ij})$.

b) the Bloch eigenfunctions (or Floquet functions), such that $\hat{T}\psi_j = \mu_j(E)\psi_j$, where $\mu_j(E)$ are the eigenvalues of the matrix $\hat{T}$, independent of $x_0$.

If $L = -d^2/dx^2 + u(x)$, then $\phi(x, x_0, E)$ is usually called $C(x,x_0,E)$, and $\phi_2(x,x_0,E)$ is called $S(x,x_0,E)$. If $\mu_j(E) = \exp(\pm ipT)$, where $\mu_1 = \exp(ipT)$, $\mu_2 = \exp(-ipT)$, then p(E) is called the 'quasimomentum' and the dependence $E = E(p)$ is called the 'dispersion relation'.

We shall normalize the Bloch function by requiring $\psi_\pm = 1$ for $x = x_0$; the function $\psi_\pm(x,x_0,E)$ is bounded in modulus wherever p(E) is real. Such segments on the

E axis are called 'permitted zones', stability zones', or the 'spectrum' of the operator L on $L_2(-\infty,\infty)$; the complementary regions are called 'lacunas'. For two solutions $L\phi = E\phi$, $L\psi = E\psi$, th Wronskian $W(\phi,\psi) = \phi\psi' - \psi\phi'$ is preserved: $W' = 0$. Hence the matrix T is unimodular.

We have the simple

*Lemma 10.1*. For the Schrödinger operator $L = -d^2/dx^2 + u(x)$, the Bloch eigenfunctions $\psi_\pm(x,E)$ are single valued and meromorphic in E on the Riemann surface $\Gamma$ which is a 2-sheeted covering of the E plane branched at all the ends of the permitted zones $E_i$. The poles of the function $\psi_\pm(x,x_0,E)$ are independent of x and are situated on $\Gamma$ at points $Q_j(x_0) = (\gamma_j,\pm)$, one over each lacuna of finite length (or at the ends of a lacuna) only on one of the sheets $\pm$. The zeros of $\psi_\pm$ are situated at points $Q_q(x)$ on $\Gamma$. As $E \to \infty$ we have asymptotically $\psi_\pm(x,x_0,E) \sim \exp[\pm i\sqrt{E}(x-x_0)]$.

This assertion is essentially trivial; we state it because formerly, before the work on the KdV equation, this important connection with Riemann surfaces did not play a serious role in the theory of the Sturm-Liouville (Schrödinger) equation with periodic potential u(x). The Riemann surface $\Gamma$ has a natural projection onto the E plane $\pi: \Gamma \to C$. We shall denote the function $\psi_\pm(x,x_0,E)$ by $\psi(x,x_0,P)$, where P is a point of $\Gamma$. The collection of points $P_j, j = \pm$ from the pre-image $\pi^{-1}(E) = (P_+,P_-)$ gives a basis of solutions $\psi_+,\psi_-$ for a given energy E. From the condition $\hat{T}\psi = \mu(E)\psi$ we get: the logarithmic derivative $d\ln\psi/dx = i\chi(x,E)$ does not depend on $x_0$ and is periodic with period T.

All these properties lead to natural generalizations of the Bloch function to the almost periodic case and to more general operators L.

*Definition 10.3*. Let L be an operator of order k with almost periodic coefficients (in general complex, and possibly, meromorphic in x). We say L is *admissible* if for all complex E the equation $L\psi = E\psi$ has a 'Bloch' solution $\psi(x,P)$ meromorphic on some k-sheeted Riemann surface $\Gamma$ over the E-plane $\pi: \Gamma \to C$, where P is a point of and the poles of $\psi$ do not depend on x. It is required that the logarithmic derivative $d\ln\psi/dx$ should be almost periodic with the same group of periods as L; furthermore the collection $\psi(x,P_j)$ must give a basis of solutions of the equation $L\psi = E\psi$ if $\pi(P_j) = E$ and E is not a branch point. As $E \to \infty$, on each sheet we must have asymptotically for $\psi$: $\psi_j \sim \exp[k_j(x-x_0)]$, where $k_j^{-1}$ is a local parameter for $\Gamma$ on that sheet $[k_\pm = \sqrt{E}, L = -d^2/dx^2 + u(x)]$. We call the Riemann surface $\Gamma$ the *spectrum* of the admissible operator L.

*Definition 10.4*. We say the admissible operator L is *finite-zoned* if its spectrum (the surface $\Gamma$) has finite genus. In this case the number of poles of the Bloch function $\psi(x,P)$ is equal to the genus.

One introduces analogous concepts for discrete (local) operators L. The significance of this concept of finite-zoned operator lies in the following assertion.

*Theorem 10.1.* If the translation invariant (local) evolutionary equation (10.1) admits a Lax representation with local translation invariant operators L, A, then every periodic and quasi-periodic real and complex (possible meromorphic in x) solution of the stationary equation K[u] = 0 defines a finite-zoned operator L.

For the Schrödinger operator $L = -d^2/dx^2 + u(x)$, and also for all the L's that have arisen in the theory of nonlinear systems of KdV type, a converse theorem has been proved. Moreover, in all these cases the stationary equation (10.4) turns out to be a completely integrable Hamiltonian system, where solution is quasi-periodic and given in terms of $\theta$ functions associated with the surface $\Gamma$ (see the survey [10.1]).

In the periodic case the translation matrix $\hat{T}$ for the operator L (of order k) is a function of $x_0$, E, and satisfies the equation in $x_0$

$$\frac{\partial T}{\partial x_0} = [Q,\hat{T}] \quad , \tag{10.5}$$

where the matrix $Q(x_0,E)$ is polynomial in E and is expressed in terms of $u(x_0)$, $u'(x_0)$, ... . In the case $L = -d^2/dx^2 + u$, Q has the form $Q = \begin{pmatrix} 0 & u-E \\ 1 & 0 \end{pmatrix}$. If (10.4) holds, the associated operators L and A commute, [A,L] = 0. In the basis $\phi_j(x,x_0,E)$ of solutions of $L\phi_j = E\phi_j$, A acts as the matrix

$$A\phi_j = \sum_i \lambda_{ij}\phi_i \quad , \qquad \Lambda = (\lambda_{ij}) = \sum c_q\Lambda_q \quad .$$

For the nonstationary equation $\dot{L} = [L,A]$ the matrix $\Lambda$ is defined by

$$\dot{\phi}_j - A\phi_j = \sum_i \lambda_{ij}\phi_i \quad . \tag{10.6}$$

We have the equation

$$\frac{\partial\hat{T}}{\partial t} = [\Lambda,T] \quad . \tag{10.7}$$

In the stationary case $[\Lambda,\hat{T}] = 0$.

From this we deduce: the Bloch function $\psi_j(x,x_0,E)$ which is defined as an eigenvector of the matrix $\hat{T}$ in the periodic case, is also an eigenvector for the matrix $\Lambda$. The matrix $\Lambda$ itself is polynomial in E (if L has finite order) and depends only on E, $u(x_0)$, $u'(x_0)$, ... . In view of this the Bloch function $\psi$ (as eigenvector of $\Lambda$, local in $x_0$) is also defined in the nonperiodic case.

As an eigenvector of the matrix $\Lambda$ which is polynomial in E, the Bloch function $\psi$ is meromorphic on the Riemann surface $\Gamma$ over the E plane

$$\det[y - \Lambda(E)] = P(y,E) = 0 \tag{10.8}$$

and has the correct asymptotic as $E \to \infty$.

From these considerations we deduce that the operator L is finite-zoned, which completes the proof of theorem 10.1.

If $L = -d^2/dx^2 + u$, then $\mathrm{tr}\Lambda = 0$ and $P(y,E) = y^2 - R(E)$ where $R(E) = \det\Lambda$. In the general, nonstationary case, we have (10.7) and (10.6). From the requirement that both of these hold we get

$$\frac{\partial\Lambda}{\partial x_0} - \frac{\partial Q}{\partial t} = [Q,\Lambda] \ . \tag{10.9}$$

In the stationary case we have

$$\frac{\partial\hat{T}}{\partial t} = \frac{\partial Q}{\partial t} = 0 \quad , \quad \frac{\partial\Lambda}{\partial x_0} = [Q,\Lambda] \ . \tag{10.10}$$

This gives a representation of Lax type for the stationary equations (10.4) on matrices depending polynomially on E. Integrals of (10.4) are obtained as the coefficients of the characteristic polynomial

$$\det[y - \Lambda(E)] = P(y,E) = 0 \quad ,$$

where the coefficients of $\Lambda$ are expressed in terms of $u(x_0), u'(x_0), \ldots$ . For the KdV equation and all its higher analogues, where $L = -d^2/dx^2 + u$, the Riemann surface $\Gamma$ is hyperelliptic $y^2 = R_{2n+1}(E) = \det\Lambda$. In the basis $\phi_1 = C(x,x_0,E)$, $\phi_2 = S(x,x_0,E)$ the matrices $\Lambda$, $\hat{T}$ are real if the potential u is real. If $i\chi_\pm(x,E) = d\ln\chi_\pm/dx$, then $i\chi' + \chi^2 = u-E$, $\chi = \chi_R + i\chi_I$, where $-\chi_I = \frac{1}{2}(\ln\chi_R)'$. This $\chi_R$ coincides with the Wronskian

$$2\chi_R = \psi_+'\psi_- - \psi_-'\psi_+ = W(\psi_+,\psi_-) \quad .$$

We have the equalities

$$\psi_\pm = C(x,x_0,E) + i\chi_\pm(x_0,E)S(x,x_0,E)$$

$$\psi_\pm = \frac{\chi_R(x,E)}{\chi_R(x_0,E)} \exp\left(i\int_{x_0}^{x} \chi_R dx\right)$$

$$\hat{T} = (\alpha_{ij}),\ \chi_R(x_0,E) = \frac{\sqrt{1-a^2}}{\alpha_{21}} \quad , \quad a = \frac{\alpha_{11}+\alpha_{22}}{2} = \frac{\mathrm{tr}\,\hat{T}}{2} \quad .$$

For the time dynamics, by virtue of any of the higher KdV equations, it follows that from (10.5,7)

$$\dot{\chi}_R = (\lambda_{21}\chi_R)' \quad . \tag{10.11}$$

If L is finite-zoned, then $\chi_R$ has the form

$$\chi_R(x,E) = \frac{\sqrt{\Pi(E-E_i)}}{P_n(x,E)} \quad , \tag{10.12}$$

where $E_i$ are the boundaries of the lacunas (branch points of $\Gamma$) and $P_n(x,E) = \Pi_{j=1}^{n}[E - \gamma_j(x)]$ is a polynomial of degree n in E.

The poles of the function $\psi_\pm(x,x_0,E)$ lie on $\Gamma$ over the points $\gamma_j(x_0)$ and the zeros over the points $\gamma_j(x)$. From (10.11) it follows taht the integral $p(E) = \frac{1}{T}\int_{x_0}^{x_0+T} \chi_R\, dx$ is conserved for any E; comparing (10.11) and (10.12) we get the t dynamics of the parameters $\gamma_j$

$$\dot{\gamma}_j = (\lambda_{21})_{E=\gamma_j} \sqrt{R(E)}/\Pi_{k\neq j} (\gamma_j - \gamma_k) \quad . \tag{10.13}$$

The dependence on x for $\gamma_j$ has the form

$$\gamma_j' = \pm\sqrt{R(E)}/\Pi_{k\neq j} (\gamma_j - \gamma_k) \quad . \tag{10.14}$$

All these equations are linearized by the so-called 'Abel substitution' which we shall discuss later. From (10.13,14) it is clear that they are written for the collection of points $(Q_j = (\gamma_j,\pm)$ on the Riemann surface $\Gamma$.

## 10.3 Hamiltonian Formalism for KdV in Stationary and Nonstationary Problems

All the higher KdV equations have the form

$$\dot{u} = \left(\frac{\delta I_n}{\delta u} + c_1 \frac{\delta I_{n-1}}{\delta u} + \ldots + c_n \frac{\delta I_0}{\delta u}\right)' \tag{10.15}$$

where the $I_n$ are obtained as the coefficients of the asymptotic expansion $p(E) = 1/T\chi_R dx = \sqrt{E}(1 + \sum_{n\geq -1}^{\infty} I_n/E^{n+2})$; from the Riccati equation for $\chi$ follows the expression $I_n = \int P_n(u,u', \ldots, u^{(n)}dx$, where the $P_n$ are polynomials, $I_{-1}$ = const. $\int u dx$, $I_0$ = const. $\int u^2/2 dx$.

According to GARDNER, ZAKHAROV and FADEEV [10.5a,b], all the equations (10.15) are Hamiltonian systems with an infinite number of degrees of freedom. More precisely, let $I = \int P(u,u', \ldots, u^{(m)})dx$ be any functional having a meaning on any translation invariant class of functions (for example, rapidly decreasing, periodic or almost periodic in x) for which the integral I converges. We consider the equation

$$u = \left(\frac{\delta I}{\delta u}\right)' \quad . \tag{10.16}$$

This flow on the chosen function space defines a Hamiltonian system, where d/dx is the skew-symmetric operator of the symplectic structure. The Poisson bracket of two functionals $\mathcal{J}_\alpha = \int P_\alpha(u,u', \ldots)dx$, $\alpha = 1,2$, by definition has the form

$$[\mathcal{J}_1, \mathcal{J}_2] = \int \frac{\delta\mathcal{J}_1}{\delta u}\left(\frac{\delta\mathcal{J}_2}{\delta u}\right)' dx \quad . \tag{10.17}$$

We shall examine further only functionals $\mathcal{J} = \int P dx$ where P does not depend explicitly on x. Commutativity of two functionals, $[\mathcal{J}_1, \mathcal{J}_2] = 0$, means that there is a formal identity

$$\frac{\delta\mathcal{J}_1}{\delta u}\left(\frac{\delta\mathcal{J}_2}{\delta u}\right)' = [Q_{12}(u,u',\ldots)]' \quad . \tag{10.18}$$

However, we shall examine the flow (10.16) formally on any translation invariant class of smooth functions, where the integral I, possibly, has no meaning. In this general case too, having the local identity (10.18) means that the flows $(\mathcal{J}_1)$ and $(\mathcal{J}_2)$ commute, where the $(\mathcal{J}_\alpha)$ are written down in the form of equation

$$\dot{u} = \left(\frac{\delta\mathcal{J}_\alpha}{\delta u}\right)' \quad .$$

GARDNER, FADDEEV and ZAKHAROV [10.5a,b] have shown that all the higher KdV's, generated by the integrals $I_n$, commute pairwise: $[I_n, I_m] = 0$.

As already mentioned in Sect.10.1, we consider the KdV flow on the set of stationary solutions of (10.4)

$$0 = \frac{\delta I_n}{\delta u} + c_1\frac{\delta I_{n-1}}{\delta u} + \ldots + c_n\frac{\delta I_0}{\delta u} + c_{n+1}\frac{\delta I_{-1}}{\delta u} \quad , \tag{10.19}$$

where $I_1 = \gamma\int(u'^2/2 + u^3)dx$, $I_2 = \gamma\int(u''^2/2 - 5u^2u''/2 + 5u^4/2)dx$.

From the general viewpoint we have two flows $(\mathcal{J}_\alpha)$, $\alpha = 1,2$ of the form (10.16) for which (10.18) holds, considered on any translationally invariant class of functions. The actual integrals $\mathcal{J}_\alpha$ may have no meaning. On the set of fixed points of the flow $(\mathcal{J}_1)$, we consider the flow $(\mathcal{J}_2)$ as a finite dimensional dynamical system. From (10.18) it follows that $Q_{12}$ gives us an integral of (10.20) for the fixed points of the flow $(\mathcal{J}_1)$

$$\frac{\delta\mathcal{J}_1}{\delta u} - d = 0 \ , \quad Q_{12}^{(d)\prime} = \left(Q_{12} - d\,\frac{\delta\mathcal{J}_2}{\delta u}\right)' = \left(\frac{\delta\mathcal{J}_1}{\delta u} - d\right)\left(\frac{\delta\mathcal{J}_2}{\delta u}\right)' \quad . \tag{10.20}$$

Taking the functional $\mathcal{J}_1 = \int P_1 dx$ nondegenerate, we introduce in the usual way phase coordinates and $(q_1, \ldots, q_n, p_1, \ldots, p_n)$, where $P_1 = P_1(u, \ldots, u^{(n)})$.

Let us express $Q_{12}^{(d)}$ in terms of (q,p). We have *Theorem 10.2.* $Q_{12}^{(d)}$ is the finite dimensional Hamiltonian of the flow $(\mathcal{J}_2)$, restricted to the set of stationary solutions of ($\mathcal{J}_1$). In the trivial case $\mathcal{J}_2 = \int u^2/2ds$, the flow ($\mathcal{J}_2$) is the group of translations in x, and $Q_{12}^{(d)}$ is the Hamiltonian of (10.20) in x.

In the case of the higher KdV's we have $\mathcal{J}_1 = \sum_{q\geq -1}^{n} I_q c_{n-1}$, where $c_0 = 1$, $c_{-1} = d$ and $\mathcal{J}_2^{(\ell)} = I_\ell$ for $Q \leq \ell \leq n-1$. All the flows $\mathcal{J}_2^{(\ell)}$ commute, and hence all the Hamiltonians $Q_{12(\ell)}^{(d)}$ for all $Q \leq \ell \leq n-1$ commute pairwise (have zero Poisson bracket) in the finite dimensional space (q,p). Hence in particular the system (10.20) is completely integrable. The integrals $Q_{12(\ell)}^{(d)}$ have a simple connection with another collection of integrals—the coefficients of the characteristic polynomial $\det\Lambda = \sum_{i=0}^{2n} (E - E_i)$, which is expressed in terms of the collection of constants $c_q$ and the derivatives $u, u', \ldots, u^{(2n-1)}$, as indicated in Sect.10.2. We have the formula

$$Q_{12(\ell)}^{(d)} = 2^{2\ell+1}(-1)^{n+\ell+1} \sum_{k_1 + \ldots + k_{2n+1} = n+\ell+2} \beta_{k_1} \ldots \beta_{k_{2n+1}} E_0^{k_1} \ldots E_{2n}^{k_{2n+1}}$$

$$\text{where } \beta_k = \frac{1}{2}\left(\frac{1}{2} - 1\right) \ldots \left(\frac{1}{2} - k + 1\right)\Big/k! \qquad (10.21)$$

The Hamiltonians $Q_{12(\ell)}^{(d)}$ are expressed in terms of the boundaries of the zones $E_i$ (10.21) $\det\Lambda = P(u,E) = \sum_{i>0} a_{2n+1-i}E^i$;

$$a_{2n+1-i} = \sum_{k+\ell=i+1} 2^{2i+3} c_{n-k+1} c_{n-\ell+1} - 8 \sum_{\substack{i+\ell\leq n \\ \ell>0}} 2^{2i} Q_{12(\ell-1)}^{(d)} c_{n-(i+\ell)} \quad . \qquad (10.22)$$

This formalism has a natural generalization to other systems admitting a Lax representation. Finishing the integration of the equations (10.19) requires the introduction of 'angular' variables on the common level surface of the commuting integrals. Convenient variables on this level surface are the $(\gamma_j,\pm)$ indicated in Sect.10.2. The connection between the collection $(\gamma_j)$ and the potential u and its derivatives, and thus also the phase variables of the problem (10.19), is obtained from (10.12)

$$\chi_R = \sqrt{\Pi(E - E_i)}/\Pi[E - \gamma_j(x)] \ , \quad \frac{1}{T}\int_0^T \chi_R \, dx \sim \sqrt{E}\left(1 + \sum_{n\geq -1} \frac{I_n}{E^{n+2}}\right) \ .$$

In particular, we have

$$u(x) = -2\sum_{j=1}^{n} \gamma_j(x) + \sum_{i=0}^{2n} E_i \quad . \qquad (10.23)$$

The angular variables are introduced using the 'Abel map' (see Sect.10.4). As will be explained in Sect.10.4, the complex level surface of all the commuting integrals of (10.19) is an Abelian variety (the 'Jacobi torus').

## 10.4 The Akhiezer Function and Its Applications

For the finite-zoned Schrödinger operator $L = -d^2/dx^2 + u$ we obtained the Bloch function $\psi_\pm$, meromorphic on the hyperelliptic Riemann surface $\Gamma$ of genus $g > 0$, two-sheeted over the E plane $\pi: \Gamma \to C$. As $E \to \infty$ we have asymptotically $\psi \sim \exp[\pm i\, E(x - x_0)]$. The function $\psi$ has a collection of poles $D = P_1 + \ldots + P_g$ independent of x, and a collection of zeros $Q_1(x), \ldots, Q_g(x)$ independent of $x_0$.

The normalization is such that $\psi = 1$ for $x = x_0$ (see Sect.10.2).

From the geometrical viewpoint, we have

a) a certain Riemann surface $\Gamma$, hyperelliptic and of finite genus g (an 'algebraic curve')

b) a distinguished branch point $P_\infty$ ('infinity')

c) the divisor of poles $D = P_1 + \ldots + P_g$ of degree g.

Let $w = k^{-1}$ be a local parameter near the point $P_\infty (w = 0)$. We have the function $\psi(x,P)$, meromorphic for $P \neq P_\infty$, depending on the parameter x; the collection of poles of this function does not depend on x, and $\psi = \exp[k(x - x_0)][1 + 0(k^{-1})]$ as $k \to \infty$ (or $P \to P_\infty$). Such a function is uniquely determined by the curve $\Gamma$, the point $P_\infty$ and the divisor of poles D. It is constructed as follows: the differential $\Omega = \psi^{-1}(d\psi/dw)dw$ is algebraic on $\Gamma$; it has first-order poles at the points $P_j$, with residue -1, and first-order poles with residue +1 at the points $Q_j(x)$. Moreover, $\Omega$ has the form $\Omega = (x - x_0)w^{-2}dw + v$, where v has no poles near the point $P_\infty$.

Let $(a_1, \ldots, a_g, b_1, \ldots, b_g)$ be the usual basis of cycles on $\Gamma$, where $a_i \circ a_j = b_i \circ b_j = 0$ and $a_i \circ b_j = \delta_{ij}$. Let $\omega_1, \ldots, \omega_g$ be the basis for the holomorphic differentials on $\Gamma$ such that $\oint_{a_j} \omega_k = 2\pi i \delta_{jk}$ and $B_{kj} = \oint_{b_k} \omega_j$. We choose a differential $\omega$ of the second kind with just one double pole at $P_\infty$, where $\oint_{a_j} \omega = 0$. We choose differentials $\Omega_j$ of the third kind, having their only poles at the pair of points $P_j$ and $Q_j$, with the same residues as $\Omega$. Let

$$\Omega = \sum_{j=1}^{g} \Omega_j - (x - x_0)\omega + \sum_{j=1}^{g} \delta_j \omega_j \quad . \tag{10.24}$$

We use the requirement that $\Omega = d\ln\psi$, where $\psi$ is a single-valued function. That means that $\oint_{a_j} \Omega = 2\pi i m_j$, $\oint_{b_j} \Omega = 2\pi i n_j$, where $m_j$, $n_j$ are integers. We apply these relations to the formula (10.24) and use the well-known identity

$$\oint_{b_k} \Omega_j = \int_{P_j}^{Q_j(x)} \omega_k \quad . \tag{10.25}$$

After this we get

$$(x - x_0)\mathcal{U}_j = (x - x_0)\oint_{b_j} \omega = \sum_{k=1}^{g} \int_{P_k}^{Q_k(x)} \omega_j + (\text{lattice}) \tag{10.26}$$

where the lattice in $C^g$ is defined by the Riemann matrix $(2\pi i\delta_{kj}, B_{kj})$. The quotient $C^g$/(lattice) is the Jacobi torus $\mathcal{J}(\Gamma)$.

Thus the parameters $\eta_k$ are 'angles', such that $\eta_k' = \mathcal{U}_k$ where $\eta_k = \sum_j \int_{P_j}^{Q_j} \omega_k$. Since the set of points $(P_j)$ is fixed we have 'Abel's map' $S^g(\Gamma) \to \mathcal{J}(\Gamma)$. The torus $\mathcal{J}(\Gamma)$ is the quotient space of $C^g$ by the lattice in the coordinates $\eta_k$. $S^g(\Gamma)$ is the symmetric power of the curve $\Gamma$, consisting of unordered collections of points $(P_1, \ldots, P_g)$. The functions on the variety $S^g(\Gamma)$ are generated by the symmetric functions of $P_1, \ldots, P_g$. In particular, according to (10.23), the potential is expressed by the formula

$$u(x) = -2 \sum_{j=1}^{g} \pi[Q_j(x)] + \sum E_i, \quad \pi: \Gamma \to C \quad . \tag{10.27}$$

From this already follows, in principle, an expression for the potential $u(x)$ in terms of $\theta$ functions in accordance with textbooks on Riemann surfaces and $\theta$ functions, for example [10.6]. However, one can get a much more convenient formula in terms of Riemann's $\theta$ function

$$\begin{aligned} u(x) &= -2 \frac{d^2}{dx^2} \ln\theta\left[\eta_1^0 + {}_1(x - x_0) - K_1, \ldots\right] + C \quad , \\ K_j &= \frac{1}{2}\left(\sum_{k=1}^{g} B_{kj} - 2\pi ij\right), \; \eta_j^0 = \sum_k \int_\infty^{P_k(x_0)} \omega_k, \quad C = C(\Gamma) \quad , \end{aligned} \tag{10.28}$$

and the Riemann $\theta$ function is defined by the series

$$\theta(\eta_1, \ldots, \eta_g) = \sum_{m_1 \ldots m_g} \exp\left\{\frac{1}{2} \sum_{jk} B_{jk} m_j m_k + \sum_k \eta_k m_k\right\} \quad . \tag{10.29}$$

The time development of the potential under KdV (or a higher KdV with index $q \geqq 1$) can be obtained from the asymptotics as $P \to P_\infty$

$$\psi \sim \exp[kx + k^3 t]\left[1 + 0\left(\frac{1}{k}\right)\right] \tag{10.30}$$

or

$$\psi \sim \exp\left[kx + P_{2q+1}(k)t\right]\left[1 + 0\left(\frac{1}{k}\right)\right]$$

where $P_{2q+1}$ is a polynomial in k of degree $2q + 1$.

From this follows the formula

$$- C + u(x,t) = - 2 \frac{d^2}{dx^2} \ln\theta(x\underline{\mathcal{U}} + t\underline{W}^{(q)} + \underline{\mathcal{U}}_0) \quad , \tag{10.31}$$

where $W_j^{(q)}$ are the periods over the cycles $b_j$ of certain differentials with poles at $P_\infty$, by analogy with the x dynamics. In fact, (10.13,14) in x and t were really written on the variety $S^g(\Gamma)$ and the angular parameters $\eta_k$ introduced on the torus $\mathcal{J}(\Gamma)$ by Abel's map. Therefore the dynamics of all the higher KdV's is given by straight-line spirals on the torus $\mathcal{J}(\Gamma)$. Comparing with what we did before, we see that for the completely integrable systems (10.19) the complex level surface of all the commuting integrals is an Abelian variety, birationally isomorphic to $\mathcal{J}(\Gamma)$. If the potential u is real and bounded for all x and t, then the point $(Q_1, \ldots, Q_g) \in S^g(\Gamma)$ moves according to (10.13,14) on the real torus $T^g \subset \mathcal{J}(\Gamma)$, which is the product of the cycles $a_j$, $T^g = a_1 \times \ldots \times a_g$, $Q_j = (\gamma_j, \pm) \in a_j$.

The cycles $a_j$ are chosen on $\Gamma$ as the pre-images of the lacunas under the projection $\pi$ on to the E plane. Various concrete calculations and applications can be found in the survey [10.1].

New applications of functions of Akhiezer type have arisen in recent work [10.7,8].

Following [10.7], we consider an arbitrary algebraic curve $\Gamma$ (not necessarily hyperelliptic) and a point on it $P_\infty$ (not necessarily a branch point).

We take a local parameter $w = k^{-1}$ near $P_\infty (w = 0)$ and a divisor of poles $D = P_1 + \ldots + P_g$ of degree g. Following the construction indicated above, we construct an 'Akhiezer function' $\psi(x,y,t,P)$ with poles at the divisor D and with asymptotics

$$\psi \sim \exp[kx + \alpha_m(k)y + \beta_n(k)t]$$

near $P_\infty$; $\alpha_m(k)$ and $\beta_n(k)$ are polynomials of degree m and n, respectively. The following facts are true:

a) we can find operators $L_m$ and $L_n$ in x of degrees m and n with constant leading terms, for which we have

$$\frac{\partial\psi}{\partial y} = L_m\psi, \quad \frac{\partial\psi}{\partial t} = L_n\psi \quad ; \tag{10.32}$$

b) we have the equation of compatibility

$$\left[\frac{\partial}{\partial y} - L_m, \frac{\partial}{\partial t} - L_n\right] = 0 \tag{10.33}$$

equivalent to the nonlinear Zakharov-Shabat equations on the coefficients of $L_m$, $L_n$.

c) If $\alpha_m = k^2$, $\beta_n = k^3 + \lambda k$, then we have

$$L_2 = -\frac{d^2}{dx^2} + u, \quad L_3 = -4\frac{d^3}{dx^3} + 3\left(u\frac{d}{dx} + \frac{d}{dx}u\right) + \lambda\frac{d}{dx} + w \quad . \tag{10.34}$$

The nonlinear equation (10.33) coincides with the '2-dimensional KdV' or 'KADOMTSEV-PETVIASHVILI equation' [10.9] for the transverse perturbation of solutions of KdV

$$\frac{3}{4}\frac{\partial u}{\partial y} = -\frac{\partial w}{\partial x}, \frac{\partial w}{\partial y} = \frac{\partial u}{\partial t} + \lambda\frac{\partial u}{\partial x} + \frac{1}{4}\left(u\frac{\partial u}{\partial x} + \frac{\partial^3 u}{\partial x^3}\right) \quad . \tag{10.35}$$

d) If $\alpha_m(k)$ is the principal part of the expansion in k of the algebraic function f(P) on the curve $\Gamma$ with its only pole at $P_\infty$, then the function $\psi$ has the form

$$\psi = \exp[fy]\psi_0(x,y,P) \quad ,$$

and the coefficients of $L_m$, $L_n$ do not depend on y. From this it follows that to each such function f with its only pole at $P_\infty$ (let t = 0) we can assign naturally an operator $L_f$ in x [one constructs an entire commutative algebra of operators, isomorphic to the whole ring of these functions $A(\Gamma,P_\infty)$ with poles at $P_\infty$]. This leads to an interesting classification of commutative rings of differential operators. The coefficients of these operators and the function $\psi$ are expressed in terms of the Riemann $\theta$ function.

In recent work, which we shall not discuss in more detail here, algebro-geometric ideas and Akhiezer-type functions are applied to the inverse problems for the two-dimensional Schrödinger equation with periodic coefficients (see [10.8]). These considerations can also be applied in higher dimensions $n > 2$.

Finally, we remark that most of the present text is based on the cycle of works of the author, Dubrovin, Matveev, and Its, carried out and, mainly, published in 1974 (see [10.10,3]) and also on the work of LAX [10.11a]. In writing Sect.10.3 I also used results of later work by LAX [10.11b], GEL'FAND and DIKII [10.12], and NOVIKOV and BOGOYAVL'ENSKII [10.13,14]. At the end of Sect.10.4 I discussed recent work by KRICHEVER [10.7]. Part of the results on the theory of the one-dimensional Schrödinger equation with periodic finite-zoned potential, using Abelian varieties, was also obtained in 1975 by McKEAN and VAN MOERBEKE [10.15]. A more detailed exposition of this theory, with references and history, can be found in [10.1].

## References


10.1 B.A. Dubrovin, S.P. Novikov, V.B. Matveev: Usp. Mat. Nauk *31*, 55 (1976)
10.2 J. Moser: Adv. Math. *16*, 354 (1975)
10.3 B.A. Dubrovin, S.P. Novikov: Dokl. Akad. Nauk SSSR *219*, 19 (1974); Zh. Eksp. Teor. Phys. *67*, 2131 (1974)
10.4 C.S. Gardner, J.M. Green, M.D. Kruskal, R.M. Miura: Phys. Rev. Lett. *19*, 1095 (1967)
10.5a C.S. Gardner: J. Math. Phys. *12*:8, 1548 (1971)
10.5b L.D. Faddeev, V.E. Zacharov: Funkts. Anal. *5*:4, 18 (1971)
10.6 A. Krazer: *Lehrbuch der Thetafunktionen* (Teubner, Leipzig 1903)
10.7 I.M. Krichever: Dokl. Akad. Nauk SSSR *227*, 2 (1976); Funkts. Anal. *10*, 75 (1976)

10.8 B.A. Dubrovin, I.M. Krichever, S.P. Novikov: Dokl. Akad. Nauk SSSR *229*:1, 15 (1976)
10.9 B.B. Kadomtsev, V.I. Petviashvili: Dokl. Akad. Nauk SSSR *192*:4, 753 (1970)
10.10 S.P. Novikov: Funkts. Anal. *8*, 54 (1974)
10.11a P.D. Lax: Lect. Appl. Math. *15*, 85 (1974);
10.11b Commun. Pure Appl. Math. *28*, 141 (1975)
10.12 I.M. Gel'fand, L.A. Dikii: Usp. Mat. Nauk *30*, 185 (1975); Funkts. Anal. *10*, 18 (1976)
10.13 O.I. Bogoyavl'enskii, S.P. Novikov: Funkts. Anal. *10*, 9 (1976)
10.14 O.I. Bogoyavl'enskii: Funkts. Anal. *10*, 2 (1976)
10.15 H. McKean, P. Van Moerbeke: Inventiones Math. *30*, 217 (1975)

# 11. A Hamiltonian Interpretation of the Inverse Scattering Method

L. D. Faddeev

In this chapter we shall discuss another aspect of the theory of nonlinear evolution equations which are integrable by the inverse scattering method. Those having applications of physical interest prove to be infinite-dimensional Hamiltonian systems. Their explicit solvability has the following interpretation in the language of Hamiltonian systems: a transform from the initial Cauchy data to the scattering data which underlies the inverse scattering method represents a nonlinear canonical transformation to variables of the action-angle type. This interpretation was originally suggested by ZAKHAROV and this author [11.1] for the case of the Korteweg-de Vries equation. Its most interesting applications are in the quantization problem for nonlinear equations. It also played an important heuristic role in clarifying the slow stochastization of an oscillator lattice [11.2] and in deriving the integrability of finite-dimensional systems of stationary points for the higher conservation laws [11.3]. In this connection we note that the N-dimensional rigid body equations have recently been shown to be integrable by MANAKOV [11.4].

## 11.1 The Hamiltonian Formulation

Let us recall several concepts from Hamiltonian dynamics necessary to clarify the above assertion. A rigorous exposition may be found in [11.5,6] for example.

Underlying the Hamiltonian formulation of mechanics are

I. A phase space $\Gamma_{2n}$ which is a 2n-dimensional manifold.

II. A closed nondegenerate 2-form $\Omega$ (a symplectic form) on $\Gamma_{2n}$

III. A Hamiltonian function $\mathscr{H}$ on $\Gamma_{2n}$

In local coordinates $\xi$ on $\Gamma_{2n}$ the form $\Omega$ appears as

$$\Omega = \sum_{\alpha,\beta=1}^{2n} \Omega_{\alpha\beta}(\xi) d\xi^{\alpha} \wedge d\xi^{\beta}$$

where the matrix $\Omega_{\alpha\beta}$ is antisymmetric, nondegenerate, has $\det\|\Omega_{\alpha\beta}\| \neq 0$ and satisfies the condition

$$\partial_\mu \Omega_{\alpha\beta} + \text{cyclic perm.} = 0 \quad .$$

The equations of motion have the form

$$\frac{d}{dt}\xi^\alpha = \dot{\xi}^\alpha = \sum_\beta \Omega^{\alpha\beta} \frac{\partial H}{\partial \xi^\beta}$$

where H is the Hamiltonian and the matrix $\Omega^{\alpha\beta}$ is the inverse of $\Omega_{\alpha\beta}$. The tensor field $\Omega^{\alpha\beta}$ is often considered to be more fundamental than $\Omega_{\alpha\beta}$ for it determines the Poisson bracket of functions on $\Gamma_{2n}$:

$$\{f(\xi),g(\xi)\} = \sum_{\alpha,\beta} \Omega^{\alpha\beta} \frac{\partial f}{\partial \xi^\alpha} \frac{\partial g}{\partial \xi^\beta} \quad .$$

Clearly the equations of motion can be put in the form

$$\dot{\xi}^\alpha = \{\xi^\alpha, H\} \quad .$$

A system is said to admit variables of the action-angle type if there exists a transformation $(p,q) \to \xi$ which carries the form $\Omega$ and the Hamiltonian H into

$$\Omega = \sum_{i=1}^n dp_i \wedge dq_i \quad ; \quad H = H(p) \quad .$$

The Hamiltonian equations appear in the new coordinates as

$$\dot{p} = 0 \quad ; \quad \dot{q} = \frac{\partial H}{\partial p}$$

and their solution is a trivial matter

$$p(t) = p(0) \quad ; \quad q(t) = q(0) + \frac{\partial H}{\partial p} t \quad . \tag{11.1}$$

If there exist global variables of the action-angle type, the Hamiltonian system is said to be completely integrable.

Complete integrability is a rare phenomenon in mechanics. Equations solvable by the inverse scattering method extend the set of examples of completely integrable systems by adding physically interesting infinite-dimensional systems. We speak of infinite dimensions because in what follows the variables $\xi$ will be represented by functions of $x \in \mathbb{R}^1$ with x taking the place of the index $\alpha$.

We now present the Hamiltonian formulation for a number of integrable systems.

*The Sine-Gordon Equation.* The phase space comprises pairs of functions $u(x)$, $\pi(x)$ which vanish as $|x| \to \infty$. The Hamiltonian H and the form $\Omega$ are

$$H = \int_{-\infty}^{\infty} \left(\frac{\gamma}{2}\pi^2 + \frac{1}{2\gamma}\left[u_x^2 + 2m^2(1 - \cos u)\right]\right)dx$$

$$\Omega = \int_{-\infty}^{\infty} \delta\pi(x)\wedge\delta u(x)dx \tag{11.2}$$

respectively. Here m and $\gamma$ are the mass and coupling constant. The equations of motion,

$$\dot{u} = \gamma\pi \ ; \quad \dot{\pi} = \frac{1}{\gamma}(u_{xx} - m^2 \sin u)$$

are equivalent to the second-order equation

$$u_{tt} - u_{xx} + m^2 \sin u = 0 \tag{11.3}$$

which has the colloquial name of "sine-Gordon". The coupling constant $\gamma$ which has slipped out of the equation and into the Poisson bracket

$$\{u_t(x),u(y)\} = \gamma\delta(x - y) \quad ,$$

emerges during quantization in the role of Planck's constant $\hbar$.

2. *The Nonlinear Schrödinger Equation.* Now the phase space consists of complex-valued functions $\psi(x)$ vanishing as $|x|\to\infty$. The Hamiltonian

$$H = \int_{-\infty}^{\infty} \left(\frac{1}{2m}\bar{\psi}_x\psi_x \pm \gamma|\psi|^4\right)dx$$

and the form

$$\Omega = \mathrm{Im}\int_{-\infty}^{\infty} \delta\bar{\psi}(x)\wedge\delta\psi(x)dx \quad ; \quad \{\bar{\psi}(x),\psi(y)\} = i\delta(x - y) \tag{11.4}$$

lead to the equation of motion

$$i\psi_t = -\frac{1}{2m}\psi_{xx} \pm 2\gamma|\psi|^2\psi \quad ,$$

which is called the nonlinear Schrödinger equation. Here m and $\gamma$ are the mass and coupling constant and the sign in front of $\gamma$ distinguishes repulsion (+) from attraction (-). The case $|\psi(x)| \to \rho$ as $|x| \to \infty$ is also of physical interest. The corresponding form $\Omega$ requires further definition which will not be discussed here.

Symplectic forms of type (11.2,4) are characteristic of field theory, but they do not exhaust all of the interesting examples. Let us convince ourselves of this.

3. *The Korteweg-de Vries Equation.* The phase space consists of real-valued functions $v(x)$ vanishing as $|x| \to \infty$. We have

$$H = \int_{-\infty}^{\infty}\left(\frac{1}{2}v_x^2 + v^3\right)dx \quad ; \quad \Omega = \int_{-\infty}^{\infty}\int_{-\infty}^{\infty} \delta v(x)\wedge\delta v(y)dx\,dy \quad ,$$

so that

$$\{v(x),v(y)\} = \delta'(x - y)$$

and the equation of motion

$$v_t = \{v,H\} = \frac{d}{dx}\frac{\delta H}{\delta v} = 6vv_x - v_{xxx} \tag{11.5}$$

actually coincides with the KdV equation. This form of the KdV equation is due to GARDNER who used it to clarify its Hamiltonian nature in [11.7].

4. *The Sine-Gordon Equation in Light Cone Coordinates.* In light cone coordinates

$$\xi = x + t \quad ; \quad \eta = x - t$$

(11.3) becomes

$$u_{\xi\eta} = m^2 \sin u \tag{11.6}$$

and its solution can be parametrized by the initial data $u(\eta) = u(\xi,\eta)\big|_{\xi}$, i.e., by a single function. Equation (11.6) is of Hamiltonian type and is generated by the Hamiltonian $P_+$ and the form $\Omega$

$$P_+ = \frac{m^2}{\gamma}\int_{-\infty}^{\infty}[1 - \cos u(\eta)]d\eta$$

$$\Omega = \frac{1}{\gamma}\int_{-\infty}^{\infty} \delta u_x(\eta)\wedge\delta u(\eta)d\eta \quad .$$

It is not hard to see that when restricted to solutions of the equation of motion the form $\Omega$ is equivalent to the form (11.2) and that $P_+$ is the displacement operator along $\xi$, $P_+ = H + P$, $P$ being the momentum of the field $u,\pi$

$$P = -\int_{-\infty}^{\infty} \pi u_x d_x \quad .$$

5. *Three Waves.* The phase space comprises triplets of complex-valued functions $\psi_\alpha(x)$, $\alpha = 1,2,3$ with the usual symplectic form

$$\Omega = \mathrm{Im}\sum_\alpha \delta\bar{\psi}_\alpha\wedge\delta\psi_\alpha \quad ,$$

so that

$$\{\bar{\psi}_\alpha(x),\psi_\beta(y)\} = i\delta_{\alpha\beta}\delta(x - y) \quad .$$

The Hamiltonian

$$H = \frac{1}{2i}\int_{-\infty}^{\infty}\left[\sum_\alpha v_\alpha(\bar{\psi}_{\alpha x}\psi_\alpha - \bar{\psi}_\alpha\psi_{\alpha x})\right]dx + \int_{-\infty}^{\infty} (\bar{\psi}_1\psi_2\psi_3 + \psi_1\bar{\psi}_2\bar{\psi}_3)\, dx$$

leads to the well-known equations for three waves

$$\psi_{1t} + v_1\psi_{1x} = i\psi_2\psi_3 \quad ; \quad \psi_{2t} + v_2\psi_{2x} = i\psi_1\bar{\psi}_3 \quad ; \quad \psi_{3t} + v_3\psi_{3x} = i\psi_1\bar{\psi}_2 \quad .$$

*6. The Toda Lattice.* Instead of the functions which we were dealing with in the preceding examples we now encounter an infinite sequence of variables $p_n$, $q_n$, $n = \dots -1,0,1, \dots$ . The form $\Omega$ is of the usual type

$$\Omega = \sum_n dp_n \wedge dq_n$$

and the Hamiltonian is taken to be

$$H = \frac{1}{2}\sum_n p_n^2 + \sum_n [\exp(q_n - q_{n-1}) - 1 - q_n + q_{n-1}] \quad .$$

The equation of motion

$$\dot{q}_n = p_n \quad ; \quad \dot{p}_n = \exp(q_n - q_{n-1}) - \exp(q_{n+1} - q_n)$$

were introduced by TODA [11.8] as a model for a lattice of interacting oscillators.

With this we end our list of Hamiltonian systems to which the inverse scattering method applies. The corresponding action-angle variables will be defined later on.

The idea that the KdV equation is completely integrable stems from the following observation: the formula [11.9]

$$r(\kappa,t) = \exp\{8i\kappa^3 t\} r(\kappa,0)$$

which gives the time dependence of the reflection coefficient for the Schrödinger equation with a potential subject to the KdV equation may be rewritten in the form

$$|r(\kappa,t)| = |r(\kappa,0)| \quad ; \quad \arg r(\kappa,t) = \arg r(\kappa,0) + 8\kappa^3 t \quad ,$$

which makes it clear that exactly one half of the scattering data is time dependent. Comparison with (11.1) makes the idea of complete integrability of the KdV equation transparent; its detailed realization was carried out in [11.1].

After [11.1] the complete integrability was established for the above-mentioned examples: the nonlinear Schrödinger was studied in [11.10], the sine-Gordon equation in [11.11], the three waves and the Toda lattice in [11.12]; see also [11.13].

## 11.2 Complete Integrability of the Nonlinear Schrödinger Equation

To illustrate the derivation of complete integrability we take the nonlinear Schrödinger equation, reserving but a few remarks for the other cases.

The auxiliary spectral problem has the form [11.14] ($m = \frac{1}{2}$, $\gamma = 1$)

$$L\phi = \kappa\phi \; ; \qquad L = \frac{1}{i}\begin{pmatrix}1 & 0\\ 0 & -1\end{pmatrix}\frac{d}{dx} + \begin{pmatrix}0 & \psi\\ \pm\bar{\psi} & 0\end{pmatrix} . \tag{11.7}$$

Recall the definition of the scattering data. Let $\mathcal{J}$ and $\mathcal{G}$ be the fundamental matrix solutions of problem (11.7) such that

$$\begin{aligned}\mathcal{J}(x,\kappa) &\sim \mathcal{E}(x,\kappa) \quad \text{as} \quad x \to \infty \; ;\\ \mathcal{J}(x,\kappa) &\sim \mathcal{E}(x,\kappa) \quad \text{as} \quad x \to \infty \; ;\end{aligned} \qquad \mathcal{E}(x,\kappa) = \begin{pmatrix}e^{i\kappa x} & 0\\ 0 & e^{-i\kappa x}\end{pmatrix} .$$

Then

$$\mathcal{J}(x,\kappa) = \mathcal{G}(x,\kappa)T(\kappa)$$

with a transition matrix T of the special form

$$T(\kappa) = \begin{pmatrix}a(\kappa) & b(\kappa)\\ \pm\bar{b}(\kappa) & \bar{a}(\kappa)\end{pmatrix}$$

where the transition coefficients $a(\kappa)$ and $b(\kappa)$ satisfy

$$|a|^2 \mp |b|^2 = 1 \quad ; \quad a(\kappa) = 1 + O\left(\frac{1}{|\kappa|}\right) , \quad b(\kappa) = O\left(\frac{1}{|\kappa|}\right) , \quad |\kappa| \to \infty .$$

Moreover, $a(\kappa)$ is the boundary value of a function which is analytic in the upper half-plane, so that

$$a(\kappa) = \exp\left\{-\frac{1}{\pi i}\int_{-\infty}^{\infty} \frac{\ln|a(\kappa')|}{\kappa-\kappa'+i0}d\kappa'\right\}\prod_{\ell}\frac{\kappa-\kappa_{\ell}}{\kappa-\bar{\kappa}_{\ell}} , \tag{11.8}$$

$k_\ell$ being the zeros of $a(\kappa)$ in the upper half-plane (existing only in the case of attraction). We assume that their total number N is finite and that $a(\kappa)$ has no real zeros; this requirement will be discussed later. For $\kappa = k_\ell$, $\mathrm{Im}\{k_\ell\} > 0$ the spectral problem has a solution with the asymptotic behavior

$$\chi_\ell(x) = \begin{pmatrix}0\\ \exp(ik_\ell x)\end{pmatrix} , \quad x \to \infty \; ; \quad \chi_\ell(x) = d_\ell\begin{pmatrix}\exp(ik_\ell x)\\ 0\end{pmatrix} , \quad x \to \infty ,$$

where $d_\ell$ is a complex coefficient.

The scattering data comprise the coefficient $b(\kappa)$ and the set of numbers $k_\ell$, $d_\ell$, $\ell = 1, \ldots, N$. The function $\psi(x)$ may be recovered from the scattering data by using the Gelfand-Levitan equation for a $2\times 2$ matrix kernel $K(x,y)$

$$K(x,y) + F(x+y) + \int_x^{\infty} K(x,z)F(z+y)dz = 0,\ x < y \quad , \tag{11.9}$$

where the matrix kernel F(x) has the form

$$F = \begin{pmatrix} 0 & F \\ -\bar{F} & 0 \end{pmatrix} \quad ; \quad F(x) = \frac{1}{2\pi}\int_{-\infty}^{\infty} r(\kappa)\exp(i\kappa x)d\kappa + \sum_{\ell} m_{\ell}\exp(ik_{\ell}x)$$

with

$$r(\kappa) = \frac{b(\kappa)}{a(\kappa)} \quad ; \quad m_{\ell} = d_{\ell}/\left[i \frac{d}{d\kappa} a(\kappa)\Big|_{\kappa=k_{\ell}}\right] \quad .$$

The equation is written down for the attractive case. For the sake of definiteness this is the only case which we shall consider.

The function $\psi(x)$ is given by the formula

$$\psi(x) = -2iK_{12}(x,x) \quad . \tag{11.10}$$

Our task is to calculate the transformations of the Hamiltonian H and the form $\Omega$ under the mapping

scattering data $\rightarrow [\bar{\psi}(x),\psi(x)]$ .

To calculate the Hamiltonian we use the following method: it turns out that H is a coefficient in the expansion of $\ln a(\kappa)$ for large $|\kappa|$

$$\ln a(\kappa) = \sum_{n=1} \frac{C_n}{\kappa^n} \quad .$$

There are two ways of calculating the coefficients $C_n$. From (11.8) we obtain

$$C_n = -\frac{1}{\pi i}\int_{-\infty}^{\infty} \ln|a(\kappa)|\kappa^{n-1}d\kappa - \sum_{\ell=1}^{N} \frac{1}{n}(k_{\ell}^n - \bar{k}_{\ell}^n) \quad .$$

On the other hand they can be expressed in terms of local functionals of $\bar{\psi}$ and $\psi$. To do this consider the matrix element $f_{11}(x,\kappa)$ of the matrix solution $\mathscr{J}(x,\kappa)$. It can be verified that for $\mathrm{Im}\{\kappa\} > 0$

$$f_{11}(x,\kappa) = e^{i\kappa x}[1+o(1)] \quad , \quad x \rightarrow \infty; \quad f_{11}(x,\kappa) = a(\kappa)e^{i\kappa x}[1+o(1)] \quad , \quad x \rightarrow -\infty \quad .$$

The basic differential equation implies that

$$\chi(x,\kappa) = \frac{d}{dx}\ln f_{11}(x,\kappa) - i\kappa$$

satisfies the equation

$$\chi = -i\psi\phi \quad ; \quad \phi_x = \frac{1}{i}(2\kappa\phi - \psi\phi^2 - \bar{\psi}) \tag{11.11}$$

and from the asymptotic behavior of $f_{11}(x,\kappa)$ we obtain

$$\ln a(\kappa) = -\int_{-\infty}^{\infty} \chi(x,\kappa)dx \quad .$$

We may look for a solution of (11.11) in the form

$$\chi(x,\kappa) = \sum_n \frac{\chi_n(x)}{\kappa^n} \quad .$$

As a result we have a method of calculating the coefficients $C_n$ in terms of $\bar{\psi}$ and $\psi$ and their derivatives. The first three coefficients are

$$C_1 = -\frac{i}{2}\int_{-\infty}^{\infty} \bar{\psi}\psi dx$$

$$C_2 = +\frac{1}{8}\int_{-\infty}^{\infty} (\bar{\psi}_x\psi - \bar{\psi}\psi_x)dx$$

$$C_3 = -\frac{i}{8}\int_{-\infty}^{\infty} (\bar{\psi}_x\psi - |\psi|^4)dx$$

and we can see that up to constant factors they coincide with the important observables, the number of particles, the momentum and the energy.

The resulting formula

$$H = -\frac{8}{\pi}\int_{-\infty}^{\infty} \kappa^2 \ln|a(\kappa)|d\kappa - \frac{8i}{3}\sum_{\ell} (k_\ell^3 - \bar{k}_\ell^3) = \int_{-\infty}^{\infty} (|\psi_x|^2 - |\psi|^4)dx$$

provides the desired expression for the Hamiltonian in terms of the scattering data.

Next we calculate the form $\Omega$. To this end we express the variations $\delta\bar{\psi}$ and $\delta\psi$ of the functions $\bar{\psi}$ and $\psi$ in terms of the variations of the scattering data. To do this we use the Gelfand-Levitan equation. Let us introduce its resolvent, i.e., a kernel $\Gamma_x(y,z)$ such that the solution of the equation

$$f(y) + g(y) + \int_x^{\infty} f(z)F(z + y)dz = 0 \quad , \quad y > x$$

is representable in the form

$$f(y) = -g(y) - \int_x^{\infty} g(z)\Gamma_x(y,z)dz \quad .$$

By comparing these formulae with the Gelfand-Levitan equation (11.9) we obtain

$$\Gamma_x(y,x) = TK^*(x,y)T \quad , \quad T = \begin{pmatrix} 1 & 0 \\ 0 & -1 \end{pmatrix} \quad . \tag{11.12}$$

The variation $\delta K$ of the kernel of the transformation operator satisfies the equation

$$K(x,y) + \delta F(x + y) + \int_x^{\infty} K(x,z)\delta F(z + y)dz + \int_x^{\infty} \delta K(x,z)F(z + y)dz = 0 \quad .$$

Using the resolvent we may write its solution as

$$\delta K(x,y) = -(I + K)\delta F(I + \Gamma_x)(x,y) \quad .$$

In particular this implies

$$\delta\psi(x) = \frac{i}{\pi}\int_{-\infty}^{\infty} [\delta r(\kappa) f_{11}^2(x,\kappa) - \delta\bar{r}(\kappa) f_{12}^2(x,\kappa)]d\kappa$$

$$+\sum_{\ell}\Big[\delta m_\ell f_{11}^2(x,k_\ell) - \delta\bar{m}_\ell f_{12}^2(x,\bar{k}_\ell) + m_\ell \delta k_\ell \frac{d}{d\kappa} f_{11}^2(x,\kappa)\Big|_{\kappa=k_\ell}$$

$$- \bar{m}_\ell \delta\bar{k}_\ell \frac{d}{d\kappa} f_{12}^2(x,\kappa)\Big|_{\kappa=\bar{k}_\ell}\Big]$$

due to (11.10), (11.12) and to the fact that there is a representation for the matrix solution $\mathcal{J}(x,\kappa)$ in terms of the transformation operator

$$\mathcal{J}(x,\kappa) = \mathcal{E}(x,\kappa) + \int_x^{\infty} K(x,y)\, \mathcal{E}(y,\kappa)dy \quad .$$

Inserting the expression for $\delta\psi$ into the form $\Omega$ [see (11.4)] we can integrate over x explicitly because the relations

$$u_1(x,\kappa)v_2(x,\kappa') + u_2(x,\kappa)v_1(x,\kappa')$$

$$= \frac{1}{i(\kappa-\kappa')} \frac{d}{dx} \{u_1(x,\kappa)v_2(x,\kappa') - u_2(x,\kappa)v_1(x,\kappa')\}$$

hold for all solutions of the spectral problem (11.7).
The resulting expression for the form $\Omega$ in terms of the scattering data and their variations is

$$\Omega = \frac{1}{\pi i}\int_{-\infty}^{\infty} |a(\kappa)|^2 \delta r(\kappa)\wedge\delta\bar{r}(\kappa)d\kappa + \frac{1}{2\pi^2} \text{V.p.}\int_{-\infty}^{\infty} d\kappa d\ell \, \frac{\delta F(\kappa)\wedge\delta F(\ell)}{\kappa-\ell}$$

$$+ \sum \frac{2}{\pi i}\int_{\infty}^{\infty} \delta F(\kappa)\wedge\left(\frac{\delta k_\ell}{\kappa-\kappa_\ell} - \frac{\delta\bar{k}_\ell}{\kappa-\bar{\kappa}_\ell}\right) + 2\sum_{\ell\neq j}\left(\frac{\delta k_\ell\wedge\delta k_j}{k_\ell-k_j} + \frac{\delta\bar{k}_\ell\wedge\delta\bar{k}_j}{\bar{k}_\ell-\bar{k}_j}\right)$$

$$- 4\sum_{\ell,j} \frac{\delta k_\ell\wedge\delta\bar{k}_j}{k_\ell-\bar{k}_j} + 2\sum_{\ell}\left(\frac{\delta m_\ell\wedge\delta k_\ell}{m_\ell} + \frac{\delta\bar{m}_\ell\wedge\delta\bar{k}_\ell}{\bar{m}_\ell}\right) \; ;$$

$$\delta F(\kappa) = a(\kappa)\bar{b}(\kappa)\delta r(\kappa) + \bar{a}(\kappa)b(\kappa)\delta\bar{r}(\kappa) = -2\delta \ln|a(\kappa)| \quad .$$

The following combinations of the scattering data

$$P(\kappa) = -\frac{2}{\pi}\ln|a(\kappa)| \quad ; \quad Q(\kappa) = \arg b(\kappa) \quad ; \quad -\infty < \kappa < \infty \quad ;$$

$$\xi_\ell = 4\,\mathrm{Re}\{k_\ell\} \;;\quad \eta_\ell = -\ln|d_\ell|$$

$$\alpha_\ell = 4\,\mathrm{Im}\{k_\ell\} \;;\quad \beta_\ell = \arg d_\ell \quad ; \quad \ell = 1, \ldots, N$$

contain all the information concerning the scattering data and may be taken as the canonical variables. The form $\Omega$ now becomes

$$\Omega = \int_{-\infty}^{\infty} \delta P(\kappa)\wedge\delta Q(\kappa)d\kappa + \sum_{\ell=1}^{N} (\delta\xi_\ell\wedge\delta\eta_\ell + \delta\alpha_\ell\wedge\delta\beta_\ell) \quad .$$

We see that the observables N, P, H can be expressed in terms of the momentum-type variables only.

$$N = \int_{-\infty}^{\infty} |\psi|^2 dx = \int_{-\infty}^{\infty} P(\kappa)d\kappa + \sum_\ell \alpha_\ell \;,$$

$$P = \frac{1}{2i}\int_{-\infty}^{\infty} (\bar{\psi}\psi_x - \bar{\psi}_x\psi)dx = \int_{-\infty}^{\infty} 2\kappa P(\kappa)d\kappa + \sum_\ell \frac{1}{2}\xi_\ell\alpha_\ell \quad ,$$

$$H = \int_{-\infty}^{\infty} (|\psi_x|^2 - |\psi|^4)dx = \int_{-\infty}^{\infty} 4\kappa^2 P(\kappa)d\kappa + \sum_\ell \left(\frac{1}{4}\xi_\ell^2\alpha_\ell - \alpha_\ell^3/12\right) \quad .$$

The following simple formulae for the evolution of the scattering data

$$\frac{d}{dt} b(\kappa,t) = 4i\kappa^3 t b(\kappa,t) \quad ; \quad \frac{d}{dt} m_\ell(t) = 4ik_\ell^2 t m_\ell(t)$$

represent the Hamiltonian equations of motion in terms of the scattering data.

We have shown that the transition to the scattering data for the nonlinear Schrödinger equation is a canonical transformation to variables of the action-angle type. In the next section we discuss some applications of this result. We conclude this section with a number of formulae concerning the action-angle type variables for some other examples of integrable equations mentioned in Sect.11.1

*1) The Korteweg-de Vries Equation.* The auxiliary spectral problem [11.9]

$$-y'' + v(x)y = \kappa^2 y \quad ; \quad v(x) \to 0 \quad , \quad |x| \to \infty$$

has solutions $f(x,\kappa)$, $g(x,\kappa)$ with the asymptotic behavior

$$f(x,\kappa) = e^{i\kappa x}[1 + o(1)] \quad , \quad x\to\infty; \quad g(x,\kappa) = e^{-i\kappa x}[1 + o(1)] \quad , \quad x\to-\infty \quad .$$

The transition coefficients are defined by

$$f(x,\kappa) = a(\kappa)g(x,-\kappa) + b(\kappa)g(x,\kappa)$$

$$f(x,ik_\ell) = d_\ell g(x,ik_\ell) \quad ; \quad a(ik_\ell) = 0 \quad , \quad \ell = 1, \ldots, N \ .$$

The variables of action-angle type are expressed in terms of the scattering data by the formulae [11.1]

$$P(\kappa) = \frac{4\kappa}{\pi} \ln|a(\kappa)| \quad ; \quad Q(\kappa) = \arg b(\kappa) \quad ; \quad 0 \leqq \kappa < \infty \quad ;$$

$$\xi_\ell = k_\ell^2 \quad ; \quad \eta_\ell = 2 \ln d_\ell \quad ; \quad \ell = 1, \ldots, N \quad ;$$

$$\Omega = \int_0^\infty \delta P(\kappa) \wedge \delta Q(\kappa) d\kappa + \sum_\ell d\xi_\ell \wedge d\eta_\ell \quad ;$$

$$H = 8 \int_0^\infty \kappa^3 P(\kappa) d\kappa + \frac{32}{5} \sum_\ell \xi_\ell^{5/2} \quad .$$

*2) The Sine-Gordon Equation.* We shall consider only the light cone coordinate form of this equation. The auxiliary spectral problem is the same as for the nonlinear Schrödinger equation with $\psi(x) = iu_\eta(\eta)$ assuming purely imaginary values [11.15,16]. The scattering data for this problem satisfy the following additional relations:

$$a(\kappa) = \overline{a(-\kappa)} \quad ; \quad b(\kappa) = -b(-\kappa)$$

and the zeros of $a(\kappa)$ are symmetric with respect to the imaginary axis. The transition coefficients $d_\ell$ are real for purely imaginary $k_\ell$. The variables of action-angle type are [11.11] $(m = 1)$

$$P(\kappa) = -\frac{8}{\pi\gamma\kappa} \ln|a(\kappa)| \quad ; \quad Q(\kappa) = \arg b(\kappa) \quad , \quad Q < \kappa < \infty$$

$$\xi_a = \frac{1}{\gamma} \ln \lambda_a \quad ; \quad \eta_a = 8 \ln|c_a| \quad ; \quad a = 1, \ldots, A \quad ;$$

$$\xi_b = \frac{4}{\gamma} \ln|\mu_b| \quad ; \quad \eta_b = \frac{4}{\gamma} \ln|d_b| \quad ; \quad \alpha_b = \arg \mu_b \quad ; \quad \beta_b = -\frac{16}{\gamma} \arg d_b \quad ;$$

where $i\lambda_a$ denote the purely imaginary zeros, $\mu_b$ the complex ones with $0 < \arg \mu_b < \pi/2$; $c_a$ and $d_b$ the corresponding transition coefficients. The form $\Omega$ is

$$\Omega = \int_0^\infty P(\kappa) \wedge \delta Q(\kappa) d\kappa + \sum_a d\xi_a \wedge d\eta_a + \sum_b (d\xi_b \wedge d\eta_b + d\alpha_b \wedge d\beta_b) \quad .$$

The Hamiltonian H and the momentum P are

$$H = \int_0^\infty \left(2\kappa + \frac{1}{8\kappa}\right) P(\kappa) d\kappa + \frac{1}{\gamma} \sum_a \left(\frac{1}{\lambda_a} + 16\lambda_a\right) + \sum_b \frac{\mu_b - \bar{\mu}_b}{i\gamma} \left(\frac{1}{|\mu_b|^2} + 16\right) \quad ;$$

$$P = \int_0^\infty \left(2\kappa - \frac{1}{8\kappa}\right) P(\kappa) d\kappa + \frac{1}{\gamma} \sum_a \left(\frac{1}{\lambda_a} - 16\lambda_a\right) + \sum_b \frac{\mu_b - \bar{\mu}_b}{i\gamma} \left(\frac{1}{|\mu_b|^2} - 16\right) \quad .$$

*3) The Toda Lattice.* The auxiliary spectral problem is formulated in terms of an infinite matrix [11.17,18].

$$L\phi(n) = \alpha_{n-1}\phi(n-1) + \alpha_n\phi(n+1) + \beta_n\phi(n) = \lambda\phi(n) \quad ;$$

$$\alpha_n = \exp(q_n - q_{n-1}) \quad ; \quad \beta_n = -\frac{1}{2}p_n \quad .$$

The solutions $f(n,\zeta)$ and $g(n,\zeta)$ are defined by the asymptotic behavior

$$f(n,\zeta) = \zeta^n[1+O(1)] \quad , \quad n\to\infty \; ; \quad g(n,\zeta) = \zeta^{-n}[1+O(1)] \quad , \quad n\to-\infty \quad ;$$

$$2\lambda = \zeta + \frac{1}{\zeta} \quad ; \quad \zeta = e^{i\phi} \quad , \quad 0 \leqq \phi \leqq 2\pi \quad .$$

The transition coefficients are defined by

$$f(n,\zeta) = a(\zeta)g(n,1/\zeta) + b(\zeta)g(n,\zeta) \quad ;$$

$$f(n,\zeta_\ell) = d_\ell g(n,\zeta_\ell) \quad ; \quad a(\zeta_\ell) = 0 \quad ; \quad -1 < \zeta_\ell < 1 \quad , \quad \ell = 1, \ldots, N \quad .$$

The variables of the action-angle type are given by the formulae [11.12,13]

$$P(\phi) = \frac{1}{\pi}\sin\phi\,\ln|a(e^{i\phi})| \quad ; \quad Q(\phi) = \arg b(e^{i\phi}) \quad ;$$

$$\xi_\ell = \zeta_\ell + \frac{1}{\zeta_\ell} \quad ; \quad \eta_\ell = \ln d_\ell \quad ;$$

$$\Omega = \int_0^\pi \delta P(\phi)\wedge\delta Q(\phi)d\phi + \sum_\ell d\xi_\ell\wedge d\eta_\ell \quad .$$

The Hamiltonian may be expressed in terms of the momentum-type variables only

$$H = \int_0^\pi 2\sin\phi P(\phi)d\phi + \sum_\ell\left[\frac{1}{2}\left(\zeta_\ell^2 + \frac{1}{\zeta_\ell^2}\right) + \ln\zeta_\ell^2\right] \quad .$$

With this we end our list of action-angle type variables and refer to the paper [11.19] for the three wave system.

## 11.3 Applications to the Quantization Problem

As we have mentioned above, the Hamiltonian formulation of integrable equations is most useful when applied to their quantization. From this point of view the most interesting are the nonlinear Schrödinger and sine-Gordon equations. In the quantum case the first one describes a system of indefinitely many particles which interact pairwise via a $\delta$-like potential. The second one provides a nontrivial relativistic example of a self-interacting scalar field.

Writing the Hamiltonian in terms of action-type variable makes it possible to describe the spectrum. In fact, when quantized these variables become commuting operators. Their spectrum is determined by the topology of the underlying phase space. The eigenvalues of the quantized Hamiltonian may be obtained by inserting the eigenvalues of the action-type variables into the classical Hamiltonian functional.

Let us give some illustrative examples. To begin with, consider the nonlinear Schrödinger equation. By performing a simple change of variables we write the observables N, P, and H in the form

$$N = \int_{-\infty}^{\infty} \rho(p)dp + \sum_{\ell} n_{\ell} \quad ; \quad \rho(p) = \frac{1}{2} P\left(\frac{p}{2}\right) \quad ; \quad n_{\ell} = \alpha_{\ell} \quad ;$$

$$P = \int_{-\infty}^{\infty} p\rho(p)dp + \sum_{\ell} p_{\ell} \quad ; \quad p_{\ell} = \frac{1}{2}\xi_{\ell}\alpha_{\ell} \quad ;$$

$$H = \int_{-\infty}^{\infty} p^2\rho(p)dp + \sum_{\ell}\left(\frac{p_{\ell}^2}{n_{\ell}} - \frac{n_{\ell}^3}{12}\right) \quad .$$

The quantity $\phi(p)$, canonically conjugated to the positive quantity $\rho(p)$ takes on values on the circle $0 \leq \phi(p) \leq 2\pi$. This implies that the complex quantities

$$a^*(p) = \sqrt{\rho(p)}\ e^{-i\phi(p)} \quad ; \quad a(p) = \sqrt{\rho(p)}\ e^{i\phi(p)}$$

are well defined and have Poisson brackets

$$\{a^*(p), a(p')\} = i\delta(p - p') \quad .$$

When quantized, the operators $\hat{a}^*(p)$ and $\hat{a}(p)$ assume the meaning of the usual creation and annihilation operators acting in the Fock space. In particular, the spectrum of the operator

$$\hat{\rho}(p) = \hat{a}^*(p)\hat{a}(p)$$

consists of the eigenvalues

$$\rho'(p) = \sum_{i=1}^{M} \delta(p - p_i) \quad . \tag{11.13}$$

Here the set of real numbers $\{p_i\}$ and an integer M are interpreted as the number of particles and their momenta. The contribution of $\hat{\rho}(p)$ to the spectrum of $\hat{N}$, $\hat{P}$ and $\hat{H}$ is

$$N' = M \quad ; \quad P' = \sum_{i=1}^{M} p_i \quad ; \quad H = \sum_{i=1}^{M} p_i^2$$

respectively and represents the energy and momentum of M nonrelativistic particles of mass $\frac{1}{2}$ .

The quantity $n_\ell$ is also conjugated with a phase-like variable, so that the spectrum of the corresponding operator consists of integers. At the same time the quantity $p_\ell$ is conjugated with the quantity $q_\ell$ which assumes arbitrary real values. It follows that its spectrum must also contain all real numbers. The resulting contribution of the soliton variables to the spectrum of the quantized Hamiltonian and momentum is of the form

$$P' = \sum_{\ell=1}^{N} p \quad ; \quad H = \sum_{\ell=1}^{N} \left( \frac{p_\ell^2}{n_\ell} - \frac{n_\ell^3}{12} \right) ,$$

i.e., it represents the momentum and energy of a system of particles with mass $n_\ell/2$ and internal energy $-n_\ell^3/12$. For $n = 2,3, \ldots$ such particles can be interpreted as bound states of n fundamental particles of mass $\frac{1}{2}$. The term $-n^3/12$ is their binding energy. In the case $n = 1$ we obtain particles of mass $\frac{1}{2}$, i.e., the same as described above. Their energy assumes the form $p^2$ if we substitute

$$H \to H + N/12 ,$$

which can be justified by the ordering of the quantum operators $\hat{\psi}^*$, $\hat{\psi}$. The binding energy of n particles is then

$$\varepsilon_n = \frac{n^3 - n}{12} .$$

The intriguing coincidence of the spectral contributions from the continuous spectrum and from the soliton variables for $n = 1$ is not yet fully understood. It is likely to be clarified by a correct description of the topology of the phase space in terms of action-angle variables. Thus a large set of zeros with small imaginary parts would be close to the continuous spectrum. This would make soliton and continuous variables interchangeable.

Traditionally, one studies the spectrum of the operator $\hat{H}$ in a representation in which the operators $\hat{\psi}^*(x)$, $\hat{\psi}(x)$ have the following commutation relations

$$[\hat{\psi}^*(x), \hat{\psi}(y)] = \delta(x - y)$$

(we take Planck's constant $\hbar$ equal to 1).
The operators

$$\hat{N} = \int_{-\infty}^{\infty} \hat{\psi}^* \hat{\psi} dx \quad ; \quad \hat{H} = \int_{-\infty}^{\infty} (\hat{\psi}_x^* \hat{\psi}_x - \hat{\psi}^* \hat{\psi}^* \hat{\psi} \hat{\psi}) dx$$

commute, the spectrum of $\hat{N}$ is integral and in a subspace with a given value of $N' = M$ the operator $\hat{H}$ is nothing but an M-particle Schrödinger operator

$$H_M = - \sum_{i=1}^{M} \frac{\partial^2}{\partial x_i^2} + \sum_{i<j}^{M} \delta(x_i - x_j)$$

acting in the subspace of symmetric functions. The spectrum of $H_M$ has been studied in detail in [11.20] and consists of the branches

$$p_1^2 + \dots + p_M^2 \quad ; \quad p_1^2 + \dots + p_{M-2}^2 + \frac{p^2}{2} - \frac{2^3-2}{12} \quad ; \dots$$

$$\frac{p_{m_1}^2}{m_1} - \varepsilon_{m_1} + \dots + \frac{p_{m_k}^2}{m_k} - \varepsilon_{m_k} \quad , \quad m_1 + \dots + m_k = M \quad ; \dots \frac{p_M^2}{M} - \varepsilon_M \quad .$$

We observe that the two derivations of the spectrum of the quantized Hamiltonian for the nonlinear Schrödinger equation lead to identical results. From general considerations it follows that both approaches should give the same result in the leading order with respect to $\hbar$, but quantizations in different canonical variables need not coincide exactly. In any case the above experiment shows that the quantization in action-angle variables reveals directly the spectrum of the Hamiltonian in a straightforward and simple manner. One is therefore tempted to apply the described procedure in a case in which no exact quantum solution is yet known. The sine-Gordon equation being such a case, we now proceed to study it.

Redefining the action-type variables in a suitable way we may write the Hamiltonian $\hat{H}$ and the momentum $\hat{P}$ in the form

$$P = \int_{-\infty}^{\infty} p\rho(p)dp + \sum_a^A p_a + \sum_b^B p_b \quad ;$$

$$H = \int_{-\infty}^{\infty} \sqrt{p^2 + 1}\rho(p)dp + \sum_a^A \sqrt{p_a^2 + M^2} + \sum_b^B \sqrt{p_b^2 + M^2(\alpha_b)} \quad ;$$

$$\rho(p)dp = P(\kappa)d\kappa \quad ; \quad p = \frac{1}{4}\left(\kappa + \sqrt{\kappa^2 + 1}\right) \quad ; \quad p_a = \frac{1}{\gamma}\left(\frac{1}{\lambda_a} - 16\lambda_a\right) \quad ;$$

$$p_b = \frac{\mu_b - \bar{\mu}_b}{i\gamma}\left(\frac{1}{|\mu_b|^2} - 16\right) \quad ; \quad M = \frac{8}{\gamma} \quad ; \quad M(\alpha) = \frac{16}{\gamma}\sin\alpha \quad .$$

The quantity $\rho(p)$ is conjugated with a phase-type quantity, so that the spectrum of the corresponding quantum operator consists of the values (11.13). The spectra of the quantities $\hat{p}_a$ and $\hat{p}_b$ fill the entire axis. Finally, the quantity $\alpha$ is conjugated with the quantity $\beta$ which has values in the interval $0 \leqq \beta \leqq 32\pi/\gamma$, so that the spectrum of $\alpha$ consists of the values $\alpha_n = \gamma n/16$ in the interval $0 < \alpha < \pi/2$. The result is that the spectrum of $\hat{H}$ is composed of particles with masses $m = 1$, $M = 8/\gamma$ and $M_n = 16\gamma^{-1}\sin(\gamma n/16)$, $0 < n < 8\pi/\gamma$. This answer is certainly exact only to leading order with respect to $\hbar$ (or $\gamma$ as we have mentioned above). In particular it seems that the particles with masses $m = 1$ and $M_1 = 16\gamma^{-1}\sin\gamma/16$ are actually identical which parallels the corresponding result for the nonlinear Schrödinger equation.

We see that the exact classical solution of the sine-Gordon equation enables us to predict a rich spectrum of particles for the corresponding quantum problem. The inverse scattering method with its Hamiltonian interpretation made it possible to go beyond the limits of perturbation theory, the very ideology of the latter permitting the sine-Gordon equation to describe only one sort of interacting particle, namely the fundamental particle of mass 1. Following this route it has been shown for the first time that localized solutions of nonlinear classical equations in field theory correspond to particles in the quantum theory [11.21,22]. A soliton mechanism for the description of the mass spectrum is now becoming increasingly popular. In particular, the mass spectrum of the sine-Gordon equation has been obtained in [11.23] by an alternative method.

A review of this trend in quantum field theory would, however, lead far from the basic theme of this book, which is devoted to the inverse scattering method. We conclude therefore simply by referring to a number of original works [11.22-26] and to reviews [11.27-29].

## References

11.1 V.E. Zakharov, L.D. Faddeev: Funct. Anal. Prilozen. *5*:4, 18(1971)
11.2 V.E. Zakharov: Sov. Phys. JETP *65*, 219 (1973)
11.3 S.P. Novikov: Funct. Anal. Prilozen. *8*:3, 54 (1974)
11.4 S.V. Manakov: Funct. Anal. Prilozen. *10*:4, 93 (1976)
11.5 V.I. Arnold: *Mathematical Methods of Classical Mechanics* (in Russian) (Nauka, Moscow 1974)
11.6 H. Goldstein: *Classical Mechanics* (Addison-Wesley, London 1950)
11.7 C.S. Gardner: J. Math. Phys. *12*, 1548 (1971)
11.8 M. Toda: Phys. Rep. *18*C, 1 (1974)
11.9 C.S. Gardner, J.M. Greene, M.D. Kruskal, R.M. Miura: Phys. Rev. Lett. *19*, 1905 (1967)
11.10 V.E. Zakharov, S.V. Manakov: Theor. Math. Phys. *19*, 332 (1974)
11.11 L.D. Faddeev, L.A. Takhtajan: Theor. Math. Phys. *21*, 160 (1974)
11.12 S.V. Manakov: Thesis, L.D. Landau Institute for Theoretical Physics (1974)
11.13 D.W. McLaughlin: J. Math. Phys. *16*, 96 (1975)
11.14 V.E. Zakharov, A.B. Shabat: Sov. Phys. JETP *61*, 118 (1971)
11.15 M. Ablowitz, D. Kaup, A. Newell, H. Segur: Phys. Rev. Lett. *31*, 125 (1973)
11.16 L.A. Takhtajan: Sov. Phys. JETP *66*, 476 (1974)
11.17 H. Flaschka: Prog. Theor. Phys. *51*, 703 (1974)
11.18 S.V. Manakov: Sov. Phys. JETP *76*, 543 (1974)
11.19 S.V. Manakov: Theor. Math. Phys. *28*, 172 (1976)
11.20 F.A. Beresin, G.P. Pokhill, V.M. Finkelberg: Vestn. Moscow State Univ. *1*, 21 (1964)
11.21 L.D. Faddeev, L.A. Takhtajan: Uspekhi Math. Nauk *29*, 249 (1974)
11.22 V.E. Korepin, L.D. Faddeev: Theor. Math. Phys. *25*, 147 (1975)
11.23 R. Dashen, B. Hasslacher, A. Neveu: Phys. Rev. *D10*, 1449 (1974)
11.24 R. Dashen, B. Hasslacher, A. Neveu: Phys. Rev. *D11*, 3424 (1975)
11.25 J. Goldstone, R. Jackiw: Phys. Rev. *D11*, 1486 (1975)
11.26 J. Gervais, R. Sakita: Phys. Rev. *D12*, 1038 (1975)
11.27 R. Rajaraman: Phys. Reports *21C*, 229 (1975)
11.28 L.D. Faddeev: preprint, Institute for Advanced Study, Princeton (1975)
11.29 L.D. Faddeev, V.E. Korepin: Phys. Rep. *42*C, 1 (1978)

# 12. Quantum Solitons in Statistical Physics

A. H. Luther

Quantum solitons are common in many fundamental model of field theories, statistical mechanics, and solid state physics in one dimension. This article reviews the interrelationships between these models, their formulation, and the operators which describe them. The common denominator is the quantum sine-Gordon model, which is reduced to solvable form. The solution for the eigenvalue spectrum is discussed, and the equivalence of solitons, particles, and spin waves, is emphasized. The principle of universality, or operator democracy, is proposed to explain these many equivalences.

## 12.1 Preliminary Remarks

The scalar interacting fermion problem in one dimension has played a major role in the recent progress of theoretical physics. It is a source problem, that is, it is a central problem not only in many different fields, but also it is a building element in more complicated models. The problem has many names. With the parameters chosen to satisfy Lorentz invariance, it is the massive THIRRING model [12.1]. Parameterized to be consistent with Fermi liquid theory, it is the (massive) LUTTINGER model [12.2]. On a lattice it is called the spin-½ x-y-z model [12.3,4], and as a transfer matrix, it is related to the two-dimensional Ising model and eight-vertex model [12.3]. In a deeply unifying manner, all of these are equivalent to the quantum sine-Gordon equation, [12.5], the prototype nonlinear equation in one dimension, and quantum solitons become another name for fermions, or spin waves.

These many models are all related to one famous problem of statistical mechanics, the eight-vertex model [12.3]. An exact calculation of the eigenvalue spectrum for the spin-½ x-y-z chain was performed by JOHNSON, KRINSKY, and McCOY [12.4], extending the methods developed by BAXTER to solve the eight-vertex model [12.3]. There are simple physical ideas which connect the continuum field theories to the spin chain [12.6], and the purpose of this chapter is to review some of these models, the relations between them, and the calculation of their eigenvalue spectrum.

It is not possible here to discuss all of these models in any great depth. Although an adequate literature concerning some of them exists [12.1,2], some material should await further distillation. Perhaps additional progress will then provide the opportunity for a broader perspective. Indeed much remains to be done. While the eigenvalue spectrum for the massive fermion problem is well understood, there are remarkably few results yet for the correlation functions. Technical problems also remain in proving the relations between the different models in some ranges of the parameters [12.7].

There is considerable interest in the recent proposals [12.8] for the S matrix of the massive fermion problem. Certainly a microscopic derivation of this object would be of great interest. Obviously much attention will be directed to these areas, and the understanding of the eigenvalue spectrum which has emerged tends to suggest optimism that a complete solution will be possible in the near future.

The intent of this article is to indicate the connection between the fermion problem in one dimension and the spin-½ x-y-z spin chain problem; and its direction for proceeding is from the eigenvalue spectrum for the spin chain to the fermion spectrum. A complete mathematical discussion of the relationships between the various models cannot be presented here, but a simple, heuristic, view about them is discussed, which should provide a helpful perspective.

Even within the restricted field of quantum solitons in field theory, a substantial literature already exists, some of which is referenced in Chap.1. The canonical formalism is described at the classical level by Fadeev in Chap.11, and its semi-classical quantization has been reviewed by FADDEEV and KOREPIN [12.9]. They emphasize that quantum solitons result from the quantization of action-angle-type variables. Fine examples are the sine-Gordon equation (also studied by BULLOUGH and DODD [12.10]) and the nonlinear Schrödinger equation (also studied by KAUP [12.11]). No attempt is made here to touch upon this canonical formulation of field theory or to relate the methods of statistical mechanics to those of field theory—except to interpolate the remark, that the answers are the same whenever comparison is possible.

Even within the restricted field of statistical mechanics, the substantial literature on the problem of direct interest here defies adequate review within the confines of this article. Only the particular chain of theoretical development which has led to the calculation of the eigenvalue spectrum is discussed, and many promising applications to physical systems in the realms of phase transition theory, quasi-one-dimensional conductors, and one-dimensional magnetic chains, must be omitted (brief reference to one of these is, however, made in Chap.3).

Even within the confines remaining, an adequate summary may not be possible without losing the reader in masses of technical detail, or boring the expert with well-known or inadequate arguments. These risks are compensated by the possibility of stimulating interest in a strikingly universal problem—one that crosses interdisciplinary boundaries on a scale of hitherto unreached proportions. Surely an understanding from all viewpoints, and their unification, would represent a satisfactory compensation.

The new technical viewpoint discussed here involves the words "continuum limit", representing the replacement of continuum mechanics by discrete mechanics. After solving the discrete problem, the resurrection of the continuum mechanics is accomplished by letting the spacing between points tend to zero — and at the same time, redefining operators in the problem in an appropriate way. In field theory, this procedure is called "renormalization", a program which can be carried through explicitly for the one-dimensional fermion problem. Somewhere, underlying this continuum limit, lies the universal problem of common interest.

Of course, the appeal of this universal problem is of more than interdisciplinary interest. There are many seemingly unrelated problems which reduce to one, and a solution of one then solves them all. Given several general characteristics of the one-dimensional fermion problem, such as its scalar symmetry, Lorentz invariance, and locality, it is furthermore possible to argue that it is the only possible problem. It is to this unique problem and the characterization of its multifaceted uniqueness, that attention is now directed.

## 12.2 Quantization and Quantum Solitons

The classical sine-Gordon equation, or any equation of motion for that matter, can exhibit a qualitatively different behavior at the quantum level. A classical equation is quantized, according to the traditional methods, from the canonical action-angle variables and the equations of motion which these variables satisfy. While easy for linear equations, this procedure becomes quite complicated for nonlinear systems, and sometimes fails completely. The difficulties lie hidden in many places, not the least of which is the appearance of divergences which prevent the quantum theory from being definable.

An alternate approach begins with quantum variables at the outset, quantum variables defined at a lattice point. The lattice points belong to an array which covers the space dimension of interest. For the one-dimensional system, the array is a chain and the quantum variables could be any operators which satisfy a convenient algebra. The difficulties with this approach all lie buried in the procedure of constructing the continuum limit, as the spacing between the lattice points shrinks to zero.

This procedure is very general, very powerful, and so far, very vague. To be specific, it is interesting to specify the operators at each lattice point of the chain. The simplest case of all is the SU(2) algebra, the group of all unitary $2\times 2$ matrices, with determinant positive unity. The quantum structure is specified by the statement that each lattice contains just two quantum states. There are many ways to think of these two states, perhaps the most physical places a particle interpretation on them, and identifies one state with "particle present" the other with "particle absent". The one level per site problem is obviously trivial from this viewpoint, and will not be considered.

With these operators in mind, the next question to consider is the Hamiltonian, or the interactions between the operators on different lattice sites. Operating with the principle of "constructive simplicity" suggests the interaction to be between nearest neighbors on the chain, and the only Hamiltonian with this property is

$$H = -\sum_{j=1}^{N} J_\alpha S_j^\alpha S_{j+1}^\alpha \tag{12.1}$$

where the operators $S_j^\alpha$ satisfy the commutation relations

$$\left[S_j^\alpha, S_{j'}^\beta\right] = i\varepsilon^{\alpha\beta\gamma}\delta_{j,j'}S_j^\gamma \tag{12.2}$$

$\alpha$, $\beta$, and $\gamma$ are the x, y, or z components of the spin operator, j and j' run over the N sites of the chain, $J_\alpha$ is the interaction between nearest neighbors for the $\alpha^{th}$ component; $\varepsilon^{\alpha\beta\gamma}$ is the unit antisymmetric tensor, and $\delta_{j,j'}$, the Kronecker symbol. Interactions between operators with different component labels, for example $S_j^x S_{j+1}^y$, have not been included because they either violate the combined symmetries of time reversal and reflection, or can be transformed into the form of (12.1) by a unitary operator. Together with the Heisenberg equations of motion, $i\,\dot{S}_j^\alpha = [S_j^\alpha, H_s]$ the problem is now defined ($\hbar$ is taken to be one).

It is interesting that the quantum mechanics of this problem turns out to be the quantum sine-Gordon mechanics in the continuum limit, replete with fermions, bosons, fermion bound states, and a simple S matrix describing the scattering of these particles. Indeed, the relationship between the spin-½ operators and the boson field used in the conventional sine-Gordon equation provides one of the examples of "operator democracy", a principle of equivalence between different operator formulations of the same underlying physics. Although one representation may make one type of particle more explicit, and modify the manner in which the nonlinearities are presented, there is no difference in the spectrum and S matrix, and thus no preferred representation.

To proceed in the construction of the solution to (12.1), it is important to recognize the fermion content of the spin operators. This is made clear through the Jordan-Wigner transformation. Introduce fermion operators at a site "j" satisfying $\left[a_j, a_{j'}^\dagger\right]_+ = \delta_{j,j'}$ then the spin algebra is realized through the relations

$$S_j^+ = a_j^\dagger \exp\left(i\pi \sum_{k=1}^{j-1} n_k\right) \tag{12.3a}$$

---

$\hbar = h/2\pi$ (normalized Planck's constant)

$$S_j^- = (S_j^+)^\dagger \tag{12.3b}$$

$$S_j^z = a_j^\dagger a_j - \frac{1}{2} \tag{12.3c}$$

with raising and lowering operators defined by $S^\pm = S^x \pm iS^y$, and $n_k = a_k^+ a_k$. By substituting these into (12.1) one is led to a Fermi lattice gas Hamiltonian, describing particles hopping from site to site along the chain, interacting with other particles when at nearest neighbor sites, and in general behaving in a manner not intuitively associated with the underlying spin-½ Hamiltonian.

The Fermi representation of (12.1) can be written in several equivalent forms, and it turns out that the form which results from substituting (12.3) into (12.1) is not convenient. The better form is found by following this substitution with a further gauge transformation $a_j \to (i)^j a_j$ which leads to

$$H = \sum_{j=1}^{N} H_s(j) - \mu n_j \tag{12.4}$$

with

$$\begin{aligned} H_s(j) = &- \frac{i}{2} V\left[a_j^\dagger a_{j+1} - a_{j+1}^\dagger a_j\right] \\ &+ \frac{i}{2} J_\perp (-)^j \left[a_j^\dagger a_{j+1}^\dagger + a_j a_{j+1}\right] - J_z n_j n_{j+1} \end{aligned} \tag{12.5}$$

where $2V = J_x + J_y$, $2J_\perp = J_x - J_y$, and $\mu$ is a chemical potential equal to $J_z$. It is clear from the spin representation of (12.5) that the chemical potential is related to an external field. If the ground state has zero magnetization, (12.3) implies that the ground state Fermi occupation number is N/2, the situation called the half-filled band. If the ground state expectation value is subtracted from the operator $n_j$, the chemical potential drops out of the problem. This subtraction is effected by normally ordering the operators. A nonzero magnetization, as in an external field, corresponds to a shift of the occupation number from the half-filled case.

It is necessary to separate the evenly numbered lattice sites from the odd, for the Fermi fields have two components which are represented in this way. There can be large variations between operators at nearest neighbor sites, but the variations are slow within the subset of even or odd. This is demonstrated by the equations of motion for the individual lattice Fermi fields, given by computing $i\,\dot{a}_j = [a_j, H]$, namely

$$\begin{aligned} i\,\dot{a}_j = &- \frac{i}{2} V(a_{j+1} - a_{j-1}) + \frac{i}{2} J_\perp (-)^j (a_{j+1}^\dagger + a_{j-1}^\dagger) \\ &+ J_z a_j (n_{j+1} + n_{j-1}) \end{aligned} \tag{12.6}$$

with the understanding that all operators are time dependent. It is clear that the equations of motion for an operator on an even site couple to those on an odd site, and the assumption that fields vary slowly over the space of a lattice constant is, in fact, required only for the components separately.

The Fermi problem here has no length, only an index characterizing position along a chain. A continuum field theory has a length, for example in the delta function characterizing the anticommutation relations. The length is constructed from the lattice spacing, s, and is related to the length of the one-dimensional chain through L = sN. The continuum limit is then specified through the procedure of letting s tend to zero, with the length fixed. The requirement is imposed that the Hamiltonian be preserved in this limit, namely that

$$\sum_{j=1}^{N} H_s(j) = \int_0^L dx\, H_c(x) \quad ,$$

or

$$H_c(x) \to s^{-1} H_s(j) \quad , \tag{12.7}$$

where the continuum coordinate is determined by the ratio x/L = j/N. The extra factor of s introduces a length into the Hamiltonian, and it is now necessary to examine the field operators. With the two-component spinor field defined by

$$\psi(x) = s^{-1/2} \begin{bmatrix} a_j \\ a_{j+1} \end{bmatrix} \equiv \begin{bmatrix} \psi_u(x) \\ \psi_\ell(x) \end{bmatrix} \tag{12.8}$$

for j even, the anticommutator of the full spinor field is

$$[\psi(x), \psi^\dagger(x')]_+ = s^{-1}\delta_{j,j'} \to \delta(x - x') \tag{12.9}$$

which is the prescription for the continuum-limit delta function constructed from a lattice theory. The continuum limit for the operator equations of motion is made straightforward by recognizing the finite difference on the right-hand side of (12.6) as the derivative, and this interpretation leads to (we use $\partial_x$ for $\partial/\partial x$)

$$\dot{\psi}_u(x) = - V\partial_x\psi_\ell(x) - s^{-1}J_\perp\psi_\ell^\dagger(x) + 4J_z\psi_u(x)\rho_\ell(x)$$

$$\dot{\psi}_\ell(x) = - V\partial_x\psi_u(x) + s^{-1}J_\perp\psi_u(x) + 4J_z\psi_\ell(x)\rho_u(x) \tag{12.10}$$

where $\rho_u(x) = \psi_u^\dagger(x)\psi_u(x)$ and $\rho_\ell(x) = \psi_\ell^\dagger(x)\psi_\ell(x)$ are fermion densities; these can be written in simpler form, if the equations of motion are expressed in terms of new fields, defined by the linear combinations $\sqrt{2}\psi_1 = \psi_u + \psi_\ell$ and $\sqrt{2}\psi_2 = \psi_u - \psi_\ell$. The result is then

$$-\dot{\psi}_1 = v\partial_x\psi_1 + s^{-1}J_\perp\psi_2 + 4J_z\psi_1\rho_2$$

$$\dot{\psi}_2 = v\partial_x\psi_2 - s^{-1}J_\perp\psi_1 - 4J_z\psi_2\rho_1 \tag{12.11}$$

where v is a renormalized velocity, $s^{-1}J_\perp$ will turn out to be a mass, and the spatial dependence is parentheses is now to be understood.

Here are the equations of motion, which turn out to be the same as those of the continuum field theory models, to be discussed in the following section. It is only necessary to recognize in (12.11) two changes from (12.10). The first, a gauge transformation $\psi_2 \to \psi_2^\dagger$, is realizable in the continuum field theory, and has been assumed. The second concerns the renormalized velocity [which is given by $v = V - J_z(2\pi)^{-1}$]. This is not an important parameter for the continuum theory, for the velocity is fixed according to a renormalization condition.

The logic of this procedure has consisted of an analysis of the equations of motion for the lattice fermion model, and of their relation to the spin-½ Hamiltonian. The eigenvalue spectrum of the lattice problem can also be calculated. By a procedure of taking the continuum limit in the eigenvalue spectrum, it is therefore possible to calculate — and give meaning to — the continuum-limit equations themselves. After a review of the continuum models in the following section, this solution is presented.

## 12.3 Continuum Field Equations

The last section presented a discussion of the fermion field equations of motion on a lattice, for the situation with small lattice spacing. Certain parameters appear in these equations, parameters defined by exchange constants and the lattice spacing itself. It is helpful to pause in the path towards the solution to these equations, to consider the corresponding equations of motion for the fully continuum models, which will help clarify the relationship between the lattice and continuum models.

Continuum models have the virtue of locality. Local interactions and the Pauli principle combine to limit greatly the number of distinct models which can exist in one dimension. These ideas lead to a picture of soliton universality, and the appearance of certain common features in seemingly different nonlinear equations.

The occurrence of propagating solutions to nonlinear wave equations is of great importance to physics. Recognition of the profusion of possible nonlinearities leads to a rude awakening, and to further questions aimed towards classifying the types of nonlinearities which may occur. Can any nonlinear equation picked up on the street have "physically interesting" solutions? Does the restriction to "physically interesting" determine a class of permissible nonlinearities?

Questions on this grand scale usually have no answers, but in the special case of two space-time dimensions, a rather satisfying picture emerges. The framework of this picture consists of several basis postulates: locality, Lorentz invariance, a symmetry classification, and the concept of a particle. The first two are adopted resolutely as conditions on the types of theories considered. With these, it is possible to give meaning to the latter two, and so reduce the grand scale to the understandable.

Several historic models satisfying these postulates have been proposed, and recently, these have been solved in various degrees. It is not the intent here to review the foundations of the models, for several reviews exist on this subject (see for example [12.1] and the book by LIEB and MATTIS [12.2]). However, some simple properties are useful to discuss here for they help clarify the principles leading to classification of the nonlinearities, and, ultimately, towards the construction of the solution of the nonlinear equations.

The first property consists of the boson-fermion duality, or the popular concept of "operator democracy", and refers to the indistinguishability of theories constructed from these two different fields. Consider the model consisting of free massless fermions, which can propagate along a string in either direction. The Hamiltonian $H_0$ is given by

$$H_0 = -iv \int_0^L dx\left(\psi_1^\dagger \partial_x \psi_1 - \psi_2^\dagger \partial_x \psi_2\right) \tag{12.12}$$

L is the length of the string, periodic boundary conditions are used, and v is the particle velocity. The fermion field operators satisfy the anticommutation relations

$$[\psi_1(x),\psi_1^\dagger(x')]_+ = [\psi_2(x),\psi_2^\dagger(x')]_+ = \delta(x - x')$$

$$[\psi_1(x),\psi_2(x')]_+ = [\psi_1^\dagger(x),\psi_2^\dagger(x')]_+ = 0 \tag{12.13}$$

and the ground state is an infinite Fermi sea of occupied states. It is the excitations of this Fermi sea which describe physical processes, and the infinite energy associated with the sea is not a problem, for excitation energies relative to the sea energy, are finite.

The excitations are characterized by n-particle correlation functions, and it is to the calculation of these n-particle correlation functions that we now direct attention. Consider first the simplest, the single-particle functions. Introducing the Fourier transform fields by

$$a_j(k) = L^{-\frac{1}{2}} \int_0^L dx\psi_j(x)e^{ikx} \tag{12.14}$$

with i = 1, 2, and $k = 2\pi n/L$ (n integer) one finds that the Hamiltonian $H_0$ of (12.12) can be diagonalized to

$$H_0 = v \sum_k k\left[a_1^\dagger(k)a_1(k) - a_2^\dagger(k)a_2(k)\right] \tag{12.15}$$

and a typical single-particle correlation function is readily computed to be

$$\begin{aligned}\left\langle\psi(x_1,t_1)\psi^\dagger(x_2,t_2)\right\rangle &= L^{-1} \sum_{k>0} \exp\left[ik(R_{12} - v\tau_{12})\right] \\ &= i(2\pi)^{-1}(R_{12} - v\tau_{12})^{-1}\end{aligned} \tag{12.16}$$

where $R_{12} = x_1 - x_2$, $\tau_{12} = t_1 - t_2$, and an infinitesimal imaginary part has been added to $\tau_{12}$ to insure convergence of the sum over k. The sum is converted to an integral, using the prescription $\sum_k \to (2\pi)^{-1}L \int dk$.

The operator equations of motion permit another realization of both the Hamiltonian and the fermion operator. Suppose a density operator $\rho$ is introduced which the property

$$[\psi_1(x),\rho_1(-k)] = e^{-ikx}\psi_1(x) \quad , \tag{12.17}$$

implying the commutation relation of this operator with $H_0$ to be

$$[\rho_1(-k),H_0] = vk\rho_1(-k) \quad . \tag{12.18}$$

This latter equation appears to be similar to that of a harmonic oscillator problem, and it might be suspected that $\rho_1(-k)$ is a boson destruction operator. However, the commutator $[\rho_1(-k),\rho_1(k')]$ cannot be a $\delta(k - k')$, because (12.18) implies $\rho_1(-k)$ is dimensionless. Hence, the commutator must be

$$[\rho_1(-k),\rho_1(k')] = k\delta(k - k') \tag{12.19}$$

and $\rho(-k)$ differs by $\sqrt{k}$ from a canonical destruction operator. The Hamiltonian $H_0$, from the point of view of the $\rho_1$ operator, can therefore be written as a bilinear form

$$H_0(\rho_1) = v \int dk\rho_1^\dagger(-k)\rho_1(k) \quad . \tag{12.20a}$$

As seen from (12.17), the operator $\rho_1(k)$ is the Fourier transform of the density, $\psi_1^\dagger(x)\psi_1(x)$; hence it follows $\rho_1(k) = \rho_1^\dagger(-k)$. A parallel line of reasoning for the $\rho_2(k)$ operator leads to an identical conclusion, except for a minus sign from the negative velocity of the "2" fermion. Ultimately, the Hamiltonian becomes

$$H_0 = v \int_0^\infty dk[\rho_1(k)\rho_1(-k) + \rho_2(-k)\rho_2(k)] \tag{12.20b}$$

and the minus sign also appears in the equation corresponding to (12.19), for the $\rho_2$ bosons, that is $\sqrt{k}\rho_2(k)$ is the appropriate destruction operator, instead of $\sqrt{k}\rho_1(-k)$.

This bosonized Hamiltonian suggests the existence of a corresponding boson representation of the Fermi operator. Consider the commutator of $\psi_1(x)$ with the boson $H_0$ of (12.20b). The result is

$$[\psi_1(x),H_0] = v\rho_1(k)\psi_1(x) \tag{12.21}$$

which is also the commutation relation of an operator $O_1(x)$ with the same $H_0$, where

$$O_1(x) = C \exp[\phi_1(x)] \tag{12.22a}$$

and

$$\phi_1(x) = \int dk\ k^{-1}\rho_1(k)\exp(-ikx) \quad . \tag{12.22b}$$

The constant C and the behavior of the integral for large values of k are determined by requiring that

$$C^2 \left\langle \exp[\phi_1(x_1,t_1)]\exp[-\phi_1(x_2,t_2)] \right\rangle = C^2 \exp\left[ \left\langle \phi_1^2 - \phi_1(x_1,t_1)\phi_1(x_2,t_2) \right\rangle \right]$$

$$= C^2 \exp \left\{ \int_0^\infty dk\ k^{-1}\left(1 - \exp[ik(R_{12} - v\tau_{12}]\right) \exp(-\alpha k) \right\}$$

$$= C^2(i\alpha)(R_{12} - v\tau_{12} + iv\alpha)^{-1} \qquad (12.23)$$

be the same as the result of (12.16). A cutoff has been inserted in the integral to ensure convergence, and the correct answer requires $C^2 = (2\pi\alpha)^{-1}$. This cutoff corresponds to the infinitesimal imaginary part of (12.16), and can be taken equal to zero after a calculation is performed. In the steps summarized in (12.23), use of the Baker-Hausdorf formula for the evaluation of harmonic oscillator matrix elements has been used. This is discussed below.

The operator constructed in this way satisfies anticommutation relations, for at equal times, the relation

$$[\phi_1(x), \phi_1(x')] = i\ \frac{\pi}{2}\ \mathrm{sgn}(x - x') \qquad (12.24)$$

causes a sign to reverse whenever $O_1$ is interchanged with another $O_1$ at some other position. Consequently, the operator $O_1$ satisfies the correct equations of motion, and the proper anticommutation relations, and commutes with the density operators, as in (12.17). It follows that the operator $O_1$ together with the bosonized $H_0$ will give the correct equations of motion. Therefore the operators can be used interchangeably. It should be realized that the use of $O_1$ is a great computational simplification, for any expectation value is reduced to a simple Gaussian integral, which can be calculated exactly.

It is conventional to write the boson Hamiltonian in different notation, using canonical creation and destruction operators, or equivalently, canonical momentum and coordinate variables. These latter quantities are given by

$$\phi(x) = i(4\pi v)^{-\frac{1}{2}} \int dk\ k^{-1}[\rho_1(k) + \rho_2(-k)]e^{-ikx} \qquad (12.25a)$$

$$\Pi\ (x) = v^{\frac{1}{2}}(4\pi)^{-\frac{1}{2}} \int dk[\rho_2(-k) - \rho_1(k)]e^{-ikx} \qquad (12.25b)$$

and the Hamiltonian is expressed in the form

$$H_0 = \int dx\ \frac{1}{2} \left\{ \Pi^2(x) + v^2\ [\partial_x\phi(x)]^2 \right\} \qquad (12.26)$$

completing the identification of operator equivalents in the scalar one-dimensional models.

In this way, the steps from bosons to fermions are constructed, steps that will be essential for the understanding of the relationship between solitons, fermions, and the sine-Gordon equation. It is perhaps helpful to remark on the inverse process — that is going from fermions to bosons. The intermediary is the fermion density operator $\psi_1^\dagger\psi_1$ or $\psi_2^\dagger\psi_2$ which is normally ordered — that is only the matrix elements which create a particle or antiparticle (an unoccupied state, a hole in the Fermi sea) are counted, and the ground state expectation value is subtracted. These operators are bosons, and are given by

$$:\psi_1^\dagger(x)\psi_1(x): = \rho_1(x)$$

$$:\psi_2^\dagger(x)\psi_2(x): = \rho_2(x) \qquad (12.27)$$

where the colon indicates the normal ordering.

From these preliminaries, the transition to the more realistic models is short. It is only necessary to consider all possible combinations of Fermi operators consistent with Lorentz invariance and locality, the starting postulates. The operator equations of motion are simplified in a spinor notation, with $\psi_1(\psi_2)$ the upper (lower) component. They are derived from the Hamiltonian

$$H = \int dx\left[-iv\psi^\dagger\sigma_z\psi + m_0\psi^\dagger\sigma_x\psi + g\rho_1\rho_2\right] \qquad (12.28)$$

where $m_0$ represents a fermion mass, g the (local) interaction between fermions, and the $\sigma_i$ are Pauli matrices.

The first two terms in (12.28) are leading terms in an expansion of the Hamiltonian in powers of the derivatives of operators. Other higher order terms would involve more derivatives, which are dropped by the locality credo. Other local products involving more operators do not occur because our interaction of the form $\rho_1\rho_2$ is the most complicated type that can exist, since $\psi_1^2 = \psi_2^2 = 0$ due to the Pauli principle. Other interactions of the form $\rho_1\rho_1$ and the $\rho_2\rho_2$ are also possible, but from (12.21) these change the velocity—and provided this velocity is adjusted to be a constant, such terms can be dropped. The absence of more complicated products, because of the Pauli principle, requires that these fermions are spinless—a symmetry condition imposed "from outside". Within these restrictions, (12.28) is the most general type of Hamiltonian in one space dimension.

The fermion field equations follow from the Heisenberg equations of motion, and are given in space-time variables by

$$\dot{\psi}_1 = -v\partial_x\psi_1 + m_0\psi_2 + g\psi_1\rho_2$$

$$\dot{\psi}_2 = v\partial_x\psi_2 + m_0\psi_1 + g\psi_2\rho_2 \quad , \qquad (12.29)$$

where $m_0$ represents a quantity of dimension $(\text{length})^{-1}$ and is a particle mass, g is the coupling constant, and v is the velocity of (12.11). To obtain the Lorentz invariant theory, called the massive Thirring model, v is renormalized by a function of g, and the resulting velocity is chosen to be unity. This step is taken following (12.35) below. The present model, given by (12.29), is sometimes called the massive Luttinger model. The relations between the two, and to the lattice theory, are discussed below.

Perhaps a word about the confusing conventions is in place here. The conventions adopted here follow those of MATTIS and LIEB [12.2]. As suggested above the interaction term $g\rho_1\rho_2$ in fact does renormalize the Fermi velocity; this requires the coupling constant dependent renormalization of the bare velocity if the observed velocity is to be a constant. An alternative is to define the interaction also to contain appropriate combinations of $\rho_1\rho_1$ and $\rho_2\rho_2$ to compensate the velocity renormalization. Such a combination is a Lorentz invariant. In the notation here, chosen to be closest to the lattice model of Sect.12.2, such terms are absent—thus the necessity of renormalization. Different models, with different definitions of coupling constant, are related by certain algebraic equations between these coupling constants. These are listed, and the proposal for a "universal" coupling constant given, in the following section.

It is clear from dimensional considerations, that the interaction $g\rho_1\rho_2$ is not unique. An interaction involving a product of sixteen operators could be added to the Hamiltonian. Use of the symmetry principles would certainly reduce this problem to (12.28), but with different parameters $m_0$ and g (and perhaps v). These can therefore not have any fundamental significance—and the question about any underlying universality, that is model independent quantities, becomes natural.

The answer to this is not yet fully proven, but some rather deep relationships between this fermion problem and models of classical two-dimensional critical phenomena suggest an understanding. This is discussed in more detail in the subsequent section, after the solution to these nonlinear equations has been constructed.

A remaining interest concerns the boson equivalence to (12.28). With the machinery represented in (12.22), (12.15) and (12.27), it is an easy exercise to derive

$$H = \int dx\, \frac{1}{2}\left[\Pi^2 + (\partial_x\phi)^2 + m_0(2\pi\alpha)^{-1}\cos\sqrt{4\pi v}\,\phi + g\rho_1\rho_2\right] \tag{12.30}$$

It is possible to find a canonical transform which completely diagonalizes the boson problem when $m_0 = 0$, that is, diagonalizes the (massless) harmonic oscillator problem. This is effected by the canonical transformation which takes the form

$$\begin{aligned} \rho_1 &= \rho_1 \cosh\phi + \rho_2 \sinh\phi \\ \rho_2 &= \rho_2 \cosh\phi + \rho_1 \sinh\phi \end{aligned} \tag{12.31}$$

where $\tanh 2\phi = -g(2\pi v)^{-1}$. Application of this transformation to the Hamiltonian with the mass $m_0 \neq 0$, leads to the sine-Gordon equation

$$H_{sG} = \int dx\, \frac{1}{2}\left[v'^2 (\partial_x\phi)^2 + \Pi^2 + m_0(2\pi\alpha)^{-1}\cos\beta\phi\right] \tag{12.32}$$

where $\beta^2 = 4\pi v\, e^{-2\phi}$. It is significant that (12.32) results from the rather general postulates of underlying symmetries and locality; and the reasoning leading up to (12.32) suggests, indeed that this equation also must be the most general of its class, in its different but equivalent operator formulations. The transformation (12.31), also changes the velocity. As discussed below, the new, or renormalized, velocity can be chosen equal to unity without loss of generality.

It was the calculation of n-particle correlation functions which led to the discussion of the bosonized Hamiltonian and Fermi operators. These can be calculated for the free particle case, g = 0, and the massless case $m_0 = 0$, but results for the general case are not yet available. The mass correlation functions provide an interesting example for the methods discussed above. Listing the operator equivalences for the sake of completeness, we have that

$$\psi_j(x) = (2\pi\alpha)^{-\frac{1}{2}}\exp[-\phi_j(x)]$$

where

$$\phi_j(x) = (-)^j \int_{-\infty}^{\infty} dk\ k^{-1}\rho_j(k)\exp(-ikx - \frac{1}{2}\alpha|k|) \tag{12.33}$$

while the Hamiltonian is

$$H = v \int_0^{\infty} dk \Big\{ [\rho_1(k)\rho_1(-k) + \rho_2(-k)\rho_2(k)] + g[\rho_1(k)\rho_2(-k) + \rho_1(-k)\rho_2(k)] \Big\} . \tag{12.34}$$

An examination of these equations reveals that the product of any number of fermion operators always results in a complicated operator, which is still the exponential of a boson operator. One uses the identity $e^Ae^B = e^{A+B+\frac{1}{2}[A,B]}$, for $[A,B]$ a c number. The Baker-Hausdorf formula permits the calculation of expectation values of such operators, using

$$\langle e^A \rangle = e^{\frac{1}{2}\langle A^2 \rangle} .$$

This equality is true when the average is taken with any bilinear harmonic oscillator Hamiltonian. In particular it applies to the Hamiltonian (12.34). Following these steps leads one to the following result

$$C^{-M}\Big\langle O_1(x_1t_1)O_1^\dagger(x_2t_2)\ \dots\ O_1^\dagger(x_{2M},t_{2M})\Big\rangle$$

$$= \exp\Bigg[ - \cosh^2\phi \sum_{i<j}^{2M} (-)^{i-j} \ln\left(\frac{x_{ij} - v't_{ij} + i\alpha}{i\alpha}\right) - \sinh^2\phi \sum_{i<j}^{2M} (-)^{i-j} \ln\left(\frac{x_{ij} + v't_{ij} + i\alpha}{i\alpha}\right)\Bigg] . \tag{12.35}$$

Here the notation is $x_{ij} = x_i - x_j$, $t_{ij} = t_i - t_j$, and $\phi$ is defined in (12.31). For 2M odd the expectation value vanishes. For the products of $O_2$ functions, the same result as (12.35) holds, but the sign of $v'$ is changed.

The velocity appearing in these correlation functions is found from the coefficient of the $\rho_1\rho_1$ term in the Hamiltonian (12.30) after it has been diagonalized. It is given by $v'^2 = v^2 - g^2(2\pi)^{-2}$, and can be chosen to be equal to unity by adjusting v and g. Since this velocity, $v'$, is the only velocity appearing in any correlation function, this renormalization to unity is equivalent to changing the time scale. Henceforth, $v'$ is taken to be unity.

The single fermion correlation function illustrates several interacting features of the continuum theory. It is

$$\left\langle O_1(x_1t_1)O_1^{\dagger}(x_2t_2)\right\rangle = -\frac{(2\pi)^{-1}}{x_{12}-t_{12}}\left(\frac{\alpha^2}{x_{12}^2-t_{12}^2}\right)^p \qquad (12.36)$$

where $p = \sinh^2\phi$. The factor $\alpha^{2p}$ represents a wavefunction renormalization, a constant which is absorbed into the definition of the operator $O_1$. It follows from (12.35) that the 2M-point function is finite as $\alpha \to 0$ after a factor $\alpha^p$ has been absorbed into the $O_1$.

A more common correlation function is generated by the mass term, involving $O_1^+O_2$ products. This function is

$$C^{-2M}\left\langle O_1(x_1t_1)O_2^{\dagger}(x_1t_1)\ \ldots\ O_2(x_Mt_M)O_1^{\dagger}(x_Mt_M)\right\rangle$$

$$= \exp\left[2\theta\sum_{i<j}^{M}(-)^{i-j}\ln\left(\frac{x_{ij}^2-t_{ij}^2}{\alpha^2}\right)\right] \qquad (12.37)$$

where $2\theta = e^{2\phi}$. It is necessary to renormalize the operator $(O_1O_2^+ + \text{h.c.})$ with a different factor than $\alpha^{2p}$, which is $\alpha^{2\theta-1}$, and, in the sine-Gordon equation [12.5], this leads to the familiar mass renormalization (see [12.5]). Among remaining operators, $\rho_1$ and $\rho_2$ do not renormalize at all, while the combination $(O_1^+O_2^+ + \text{h.c.})$ has yet a different renormalization constant, which is $\alpha^{2/\theta-1}$.

With the operators renormalized in this way, correlation functions are finite in the limit $\alpha\to 0$. It is a consequence of this limiting procedure, that the equal time anticommutation relations are not satisfied, but the long distance and time behavior, including singularities at $x^2 = t^2$, are correctly given.

In a lattice theory, the operator algebra always refer to properties at a single site which corresponds to distances less than a lattice constant, or a cutoff length. The lattice correlation functions should then behave differently at very short distances, when compared with continuum limit functions here. These functions properly describe the behavior in the asymptotic region, the region of large separations.

With this result for the n-point correlation functions, it is possible to address the question of differences between the various models mentioned in the introductory remarks. All models have, at some stage, the appearance of these n-point functions. The form of these provides the simplest method of comparing and understanding the different models.

These functions contain several parameters, a velocity v, the cutoff factor $\alpha$, and the object giving dimension to the correlation function, $\theta$. Equivalences between the models are thus established by looking up the corresponding expressions for $\alpha$, v, and $\theta$ in the various models. The important physics is not what goes into determining these, for that is a matter of definition. The important physics is that only these parameters appear.

For the three models of interest here, the relations for the parameter $\theta$ are summarized below:

spin chain [12.4] $\theta = 1 - \pi^{-1}\cos^{-1}(-J_z/v)$

Thirring [12.5] $\theta = \frac{1}{2}(1 + g/\pi)^{-1}$

Luttinger [12.6] $\theta = \frac{1}{2}(1 + v)^{\frac{1}{2}}(1 - v)^{-\frac{1}{2}}$

where the symbols for the respective models are defined according to the conventions of the cited article. In all the continuum cases, the renormalized velocity has been chosen to be one, which is a trivial choice of unit. Finally, the parameter $\alpha$ is not specified, since it appears only as a renormalization parameter, and does not appear in the final physical expression. It may be taken to be the lattice spacing.

## 12.4 Eigenvalue Spectrum

The previous section developed the equations of motion for the continuum model of interacting massive fermions in one space dimension, while Sect.12.2 discussed the quantum spin operator formulation of a lattice problem, which has the same equations of motion in the limit of small lattice spacing. It is the interest of this section to bring together solutions of the lattice problem, perform the required limiting procedure for small lattice spacing, and construct the eigenvalue spectrum for the continuum limit theory.

The eigenvalue spectrum for the Hamiltonian of (12.1) is characterized by single-particle states, and bound states. In the spin language, the single-particle state is a spin wave, in the boson language a soliton, while it is a fermion in the remaining representation. The bound states always are constructed from these fundamental objects (or their antiparticles). The spectrum is determined by the particle mass, the binding energy, and the number of bound states. In the proper continuum limit, in which the lattice constant is taken to zero, the spectrum is Lorentz invariant, and consists of the superposition of noninteracting particles, with spectrum

$$\Delta^2(k) = \Delta_n^2 + k^2 \tag{12.38}$$

where $\Delta_n$ represents the proper mass, in general different from the bare mass $m_0$ of the last section, and n is an index representing the particle, or bound particle, in question.

The spectrum assumes this free particle form due to the existence of an infinite number of conservation laws within the spin chain problem [12.12]. These conservation laws prohibit the production of particles or the breakup of any bound-state particles. Conversely, it can be inferred from the form of the eigenvalue spectrum given in (12.38), consisting of absolutely stable particles, that these conservation laws must exist.

The determination of the masses requires an understanding of the continuum limit and the proper renormalization of the lattice parameters, $J_\alpha$, in (12.1). After the unit of length has been inserted into this equation, which leads to (12.10), it is easy to see that the limit $s \to 0$ requires $J_\perp \to 0$, that is the nearly isotropic spin problem $J_\perp \ll V$. In this limit the spin chain eigenvalue spectrum greatly simplifies. The fundamental mass in the problem is given by

$$\Delta = 8\pi V \frac{\sin\mu}{\mu} \left(\frac{J_\perp}{V}\right)^{\pi/2\mu} \tag{12.39}$$

where $\mu = \cos^{-1}(-J_{2/V})$, and the other parameters are defined in Sect.12.2. The construction of the continuum limit requires, from (12.7), that $V \to V/s$. The factors of s cancel elsewhere in the continuum limit procedure; consequently the mass behaves as

$$\Delta \sim s^{-1} J_\perp^{\pi/2\mu} \tag{12.40}$$

and to achieve a finite gap in the limit $s \to 0$, requires choosing $J_\perp = J_\perp^* s^{2\mu/\pi}$, where $J_\perp^*$ is a constant independent of s as $s \to 0$. In this manner the mass is required to be finite. Evidently this procedure is capable of determining the dependence of the gap on bare exchange parameters, $\Delta \sim J_\perp^{\pi/2\mu}$ but not the proportionality constant.

Obviously, this relation has determined an exponent, but not the ultimate mass formula. Such a formula must depend on the model, and on many parameters which might not even appear in the continuum-limit field equations, but lie hidden in conventions. It cannot be universal.

However, the ratio between the bound-state particle mass and $\Delta$ has an interesting property: the single-particle mass drops out. In the small $J_\perp$ limit, the mass formula is

$$\Delta_n = 2\Delta \sin\left(\frac{n\pi}{2} \frac{\theta}{1-\theta}\right) \tag{12.41}$$

where $\theta = 1 - \mu/\pi$. The parameter $\theta$ contains a dependence on the longitudinal exchange, $J_z$, through the parameter $\mu$. For values of this parameter such that $\theta^{-1} - 1$ is greater than one, any integer n is a solution, and the mass ratio depends on only one parameter. The spectrum of the form (12.41) was first derived for the sine-Gordon equation by TAKTADJAN and FADDEEV [12.12], and DASHEN et al. [12.14] using semi-classical methods, and is quoted in Chap.1, equation (1.105). Derivation of this spectrum from a lattice model establishes that this spectrum is not only exact, but suggests it to be universal. Universality, in this context, would imply that the ratio of bound particle to free particle mass is a universal function of the parameter $\theta$, as given in (12.41). The parameter $\theta$ can be a non-universal function of coupling constants, for it is determined from the exponent of the $\psi_1^\dagger\psi_2$ correlation function, as in (12.37).

The universal relation between exponents is a well-known property of critical exponents near a second-order phase transition. The corresponding relation here, connects the exponent of the mass gap, to an exponent of a correlation function at zero mass. The gap exponent is called $\nu$, and the relation from above establishes $\Delta \sim m_0^{\nu}$ with $2\nu = \pi/\mu = (1 - \theta)^{-1}$. The identification of $\theta$ from the correlation function of (12.37) corresponds to an order parameter exponent in the critical phenomenon problem. In addition to the scaling law relating the exponents, there is the universal function giving the bound-state mass formula.

The relationship between the exponents near the critical point of the 8-vertex model, and the exponents characterizing the correlation functions of the fermion problem, is one of the pleasing occasions where cross-fertilization has been helpful. Many further applications of this principle can be envisaged, clarifying such problems as the S matrix in the lattice theory, the thermodynamics of the quantum sine-Gordon equation, and the major remaining unsolved problem of correlation functions. Other models, with higher internal symmetries, have the quantum sine-Gordon equation as their foundation [12.15]. The complete solution of this quantum problem, will make possible substantial advances with these other models as well.

The central idea in the solution of the eight-vertex model is the extended Bethe Ansatz [12.3]. Using the equivalences discussed here, this Ansatz is clearly relevant for the construction of wave functions for the interacting Fermi problem. This application has been discussed recently [12.16], and is related to the "quantum inverse scattering method" [12.17]. Extensions to the interacting Fermi models with higher internal symmetries can be anticipated and welcomed.

## References


12.1 W. Thirring: Ann. Phys. (N.Y.) *3*, 91 (1958)
V. Glaser: Nuovo Cimento *9*, 990 (1958)
K. Johnson: Nuovo Cimento *20*, 773 (1961)
C. Sommerfield: Ann. Phys. (N.Y.) *26*, 1 (1963)
B. Klaiber: In *Lectures in Theoretical Physics*, Vol.XA, ed. by A. Barut, W. Britten (Gordon and Breach, New York 1958)

12.2 J.M. Luttinger: J. Math. Phys. *4*, 1154 (1963)
D.C. Mattis, E.H. Leib: J. Math. Phys. *6*, 304 (1965)
A. Theumann: J. Math. Phys. *8*, 2460 (1967)
E.H. Leib, D.C. Mattis: *Math. Phys. in One Dimension* (Academic Press, New York 1966)

12.3 R.J. Baxter: Phys. Rev. Lett. *26*, 834 (1971); Ann. Phys. (N.Y.) *70*, 33 (1972)

12.4 J.D. Johnson, S. Krinsky, B.M. McCoy: Phys. Rev. A*8*, 2526 (1973)

12.5 S. Coleman: Phys. Rev. D*11*, 2088 (1975)

12.6 A. Luther: Phys. Rev. B*14*, 2153 (1976)
A. Luther, I. Peschel: Phys. Rev. B*12*, 3908 (1975)

12.7 J. Frohlich: Commun. Math. Phys. (Germany) *47*, (no.3) 233 (1976)

12.8 A.B. Zamolodchikov: Commun. Math. Phys. *35*, 183 (1977); JETP Lett. *25*, 468 (1977)
M. Karowski, H.J. Thun, T.T. Truong, P. Weisz: Phys. Lett. *67*B, 321 (1977)

12.9 L.D. Faddeev, V.E. Korepin: "Quantum Theory of Solitons", in Phys. Rep. *42*C, 1 (1978)

12.10 R.K. Bullough, R.K. Dodd: In "Synergetics. A Workshop", ed. by H. Haken, Springer Series in Synergetics, Vol.2 (Springer, Berlin, Heidelberg, New York 1977) pp.92-119
12.11 D.J. Kaup: J. Math. Phys. *16*, 2036 (1975)
12.12 M. Lüscher: Nucl. Phys. B*117*, 475 (1976)
12.13 L.A. Taktadjan, L.D. Faddeev: Teor. Mat. Fiz. *25*, 147 (1975)
12.14 R.F. Dashen, B. Hasslacher, A. Neveu: Phys. Rev. D*11*, 3424 (1975)
12.15 A. Luther, V.J. Emery: Phys. Rev. Lett. *33*, 589 (1974) and A. Luther: Phys. Rev. B*15*, 403 (1977) discuss a 'backward scattering model', for example, which extends the scalar model to the case of a fermion with spin
12.16 H. Bergknoff, B. Thacker: Phys. Rev. Lett. *42*, 135 (1979)
12.17 L.D. Faddeev, E.K. Sklyanin, L.A. Taktadjan: Theor. Math. Phys. *40*, N2, 194 (1979)
L.D. Faddeev: Steklova Institute, Leningrad (1979) Preprint

# Further Remarks on John Scott Russel and on the Early History of His Solitary Wave

The first recorded observation of a solitary wave is undoubtedly that made by Russell in the month of August 1834 as described in the quotation from his paper [1.3] given in Chap.1. Russell's fascination for the solitary wave was plain then: that it continued throughout his life is also clear—see for example the quotation from Russell's book of 1865 which appears on the facing page.

Russell's career following his discovery of the solitary wave was not uneventful. We mention some points in it. In 1832-1833 he held only the temporary appointment at Edinburgh created by the death of the Professor, Dr. Leslie. He did not apply for permanent appointment to the Chair: Brewster was a candidate and the Chair went finally to J.D. Forbes. In 1838 he did apply for the vacant Chair of Mathematics, but despite a good reference from Hamilton who described Russell as 'a person of active and inventive genius' [1.5], he failed to get it. It is interesting to speculate whether soliton theory would have developed some 100 years earlier if Russell had got either of these Chairs. Forbes's interests were certainly very different from Russell's although he has left us permanent work. His reputation now rests primarily on his early work on glaciers (especially that on the Mer de Glace which he was the first to map [1.10,11]). But we believe he also introduced "Forbes's Bar" used in the measurement of the thermal conductivity of metals. The anomalous thermal behavior of one-dimensional anharmonic lattices predicted from numerical studies by FERMI, PASTA and ULAM in 1955 (the FPU problem [1.12]) stimulated the investigations which led Kruskal to the inverse scattering method for solving the KdV, whilst the soliton solutions of that equation may cause some of the difficulties in deriving from microscopic theory the Fourier law of heat conduction [1.13].[1]

Russelll had invented a steam trolley in the years before 1834; in that year the Scottish Steam Carriage Company was formed with the proposal to run a regular service between Glasgow and Edinburgh. It had a short life [1.5]. Nevertheless [1.5] it seems to have been in consequence of his association with this Company

[1]Professor Peierls has pointed out to us that even close to the melting point the magnitudes of excitations in real three-dimensional systems are so small that solitonlike contributions can surely be neglected.

that Russell received the invitation[2] from the Union Canal Company to investigate the Canal's prospects for steam navigation and so see the first soliton. Subsequently Russell made his reputation as a naval architect. His "wave-line" hulls were designed to minimise wave making resistance due to the generation of solitary wave type bow waves. In 1865 he published the first major work on naval architecture [1.14]. And in this he describes his early experiments on the production of low waves and the way they move water at a speed independent of that of the ship [1.5,14]. By 1853, however, Russell had been engaged by Brunel to construct the great iron ship the "Great Eastern" [1.5], a direct or indirect cause of many misfortunes. He entered the first of his periods of bankruptcy before that ship could be launched, became embroiled in the controversy surrounding the failure of a steam valve during the trials of that vessel in the English Channel in 1859 and the death of five seamen at which time Brunel also died, whilst in 1867 he was forced to resign (perhaps by Brunel's family [1.5]) from the Institution of Civil Engineers of which he was a founder member following a charge of unprofessional conduct associated with a second financial failure whilst acting to purchase guns for the North in the American Civil War.

In the last years of his life Russell completed the book *The Wave of Translation in the Oceans of Water, Air and Ether* published posthumously in 1882 [1.15]. That book contains again reproduced the British Association's 'Report on Waves' (1844) [1.3] where the first observation of the solitary wave is reported. It contains a number of curious speculations on the structure of matter; and it applies the formula (1.7) to compute from the velocity of sound the depth of the atmosphere (5 miles) and from the velocity of light the depth of the universe ($5\times 10^{17}$ miles)! Russell's point for the former was that, for distortionless transmission, sound must be carried by his solitary waves. This it is the velocity c of (1.7) which is to be interpreted as the sound speed [and not $c_s = \sqrt{gh}$ in (1.6) or (1.8)]. The calculated depth of 5 miles proves to be the actual equivalent depth at uniform density. However, for the size of the universe, $5\times 10^{17}$ miles is out by at least five orders of magnitude; and in any case it appears to rely on a value of g reduced arbitrarily by Russell by a factor $10^{-5}$. Nevertheless these early speculations perhaps begin already to hint at the current significance of solitons in modern physics.

Russell's 'Report on Waves' [1.3] apparently stimulated work by DE BOUSSINESQ [1.9] and RAYLEIGH [1.16]. Both authors derived the sech form of the solitary wave and the formula (1.7) for its speed. Rayleigh also gave reasons why it breaks for

---

[2]See especially Russell's Edinb. Roy. Soc. Trans. XIV (1840) paper listed below. There seems to have been no contract, but the Union Canal Company paid the expenses — according to their report of 27th January, 1835: 'Report on the practical results of experiments on Canal Navigation (Canal Office, Edinburgh).

$k \approx h$ as had been found experimentally by Russell. Rayleigh reviewed the concept of the solitary wave which Russell had introduced in his 'Report on Waves', noted that with lengths 6 or 8 times the canal depth it could apparently be treated by the theory of long waves, noted nevertheless Russell's observations of different behaviors for positive and negative waves, quoted Airy's objection, namely that his (Airy's) theory of shallow waves of great lengths admits both 'positive and negative waves' ("We are not disposed to recognise this wave as deserving of the epithets 'great' or 'primary' ..."), quoted Stoke's counter opinion, and then by the implications of his analysis came down firmly in favour of Russell.

By the time KORTEWEG and DE VRIES [1.7] derived their equation, (1.6), the $\mathrm{sech}^2$ solitary wave was "well known" and their paper was primarily concerned with refuting AIRY's opinion [1.8] that long waves in a canal must necessarily change their form. They showed that Stoke's theory of long waves [1.17] gave the first two terms of the cnoidal wave solution of their KdV equation and that whilst sinusoidal waves become steeper in the front when advancing, other waves behave differently. They showed $\eta = k\,\mathrm{sech}^2\,px$ was stationary (in a moving coordinate system) if $k = 4\sigma p^2$: for $k > 4\sigma p^2$ the waves steepen in front; for $k < 4\sigma p^2$ they steepen behind. In our notation $\sigma \equiv \frac{1}{3}h^3 - \gamma h \rho^{-1} g^{-1}$. KORTEWEG and DE VRIES [1.7] quoted DE BOUSSINESQ [1.9], RAYLEIGH [1.16] and ST. VENANT [1.18] as establishing the theory of the solitary wave but noted that (in 1895) treatises by Lamb and Basset still assert that Airy was correct in his opinion. They said that even RAYLEIGH and McCOWAN [1.19] do not directly refute Lamb's and Basset's assertions "It is the desire to settle this question definitively which has lead us into the somewhat tedious calculations which are to be found at the end of this paper." The KdV equation itself noted as 'very important' nevertheless gets rather less discussion.

In surveying the history of the solitary wave one should certainly also mention the paper actually entitled 'On the solitary wave' by J. McCOWAN [1.19] and which is quoted by KORTEWEG and DE VRIES [1.7]. McCOWAN isolated and analysed the error of STOKES [1.17] who in 1847 anyway had concluded that the degradation of a wave is an essential characteristic of its motion (and not due to friction therefore). He examined and confirmed the approximations of DE BOUSSINESQ [1.9] and RAYLEIGH [1.16] for the $\mathrm{sech}^2$ form of the solitary wave, derived (1.7) exactly, and noted the agreement with Russell's deductions and their experimental confirmation by BAZIN [1.20]. He also gave an approximate theory of the breaking of waves passing into shallower water. It is remarkable that despite the work of KORTEWEG and DE VRIES [1.7] four years later so little other work followed. Indeed the significance of the soundly based early work [1.3] of Russell's is only now being recognised as the wide range of application of the concepts of the solitary wave and soliton becomes properly appreciated.

The references quoted in this short article appear in the list of references to Chap.1. The reader may care to have other reference to the published scientific work of Russell. The following, no doubt still incomplete, list was compiled by J.C. Eilbeck from the library of the Royal Society of Edinburgh:

*Russell's Published Scientific Work*

*Russell, John Scott*: Notice of the reduction of an anomalous fact in Hydrodynamics, and of a new law of the Resistance of Fluids to the motion of floating bodies. Brit. Assoc. Rep. 1834, pp.531-534

On the motion of floating bodies. Brit. Assoc. Rep. 1835 (pt.2), p.16

On the solid of least resistance. Brit. Assoc. Rep. 1835 (pt.2), p.107-108

On the mechanism of the waves, in relation to steam navigation. Brit. Assoc. Rep. 1837 (pt.2), pp.130-131

On the fallacies of the Rotatory Steam Engine. 1837. Edinb. New Phil. Journ. XXIV., 1838, pp.35-64; Edinb. Trans. Scot. Soc. Arts, I., 1841, pp.172-202; Dingler, Polytech. Journ. LXVII., 1838, pp.332-355; LXXVIII., 1840, pp.4-18

On the economical proportion of power to tonnage in steam vessels. Brit. Assoc. Rep. 1839 (pt.2), pp.124-125

On the temperature of most effective condensation in steam vessels. Brit. Assoc. Rep. 1840 (pt.2), pp.186-187

On the most economical and effective proportion of engine power to the tonnage of the hull in steam vessels. Brit. Assoc. Rep. 1840 (pt.2), pp.188-190

Description of a Polyphotal Lamp and Reflector of single curvature employed in steam vessels, canal-boats, & c. Edinb. New Phil. Journ. XXVIII., 1840, pp.193-196

Experimental researches into the laws of certain hydrodynamical phenomena that accompany the motion of floating bodies, and have not previously been reduced into conformity with the known laws of the Resistance of Fluids. 1837. Edinb. Roy. Soc. Trans. XIV., 1840, pp.47-109

On the vibration of Suspension Bridges and other structures, and the means of preventing injury from this cause. 1839. Edinb. Trans. Scot. Soc. Arts, I., 1841, pp.304-313; Edinb. New Phil. Journ. XXVI., 1839, pp.386-396

Elementary considerations of some principles in the construction of buildings designed to accommodate spectators and auditors. 1838. Edinb. Trans. Scot. Arts, I., 1841, pp.314-318; Edinb. New Phil. Journ. XXVII., 1839, pp.131-136

Report of a Committee on the Form of Ships. Brit. Assoc. Rep. 1841, pp.325-326; 1842, pp.104-105

Supplementary Report of a Committee on Waves. Brit. Assoc. Rep. 1842 (pt.2), pp.19-21

On the indicator of speed of steam vessels. Brit. Assoc. Rep. 1842 (pt.2), p.109

On the abnormal tides of the Firth of Forth. Brit. Assoc. Rep. 1842 (pt.2), pp.115-116

Report of a series of observations on the tides of the Firth of Forth and the east coast of Scotland. Brit. Assoc. Rep. 1843, pp.110-112

Notice of a report of the Committee on the Form of Ships. Brit. Assoc. Rep. 1843, pp.112-115

On the application of our knowledge of the laws of sound to the construction of buildings. Brit. Assoc. Rep. 1843 (pt.2), pp.96-98; Majocchi, Ann. Fis. Chim. XXVIII., 1847, pp121-123

Description of a Marine Salinometer to indicate the density of brine in the boilers of marine steam-egines. Edinb. New Phil. Journ. XXXIV., 1843, pp.278-285

Report on Waves. Brit. Assoc. Rep. 1844, pp.311-390. (reference 1.3)

On the tides of the east coast of Scotland. Brit. Assoc. Rep. 1844 (pt.2), p6

On the nature of the Sound-wave. Brit. Assoc. Rep. 1844 (pt.2), p.11

On the resistance of railway trains. Brit. Assoc. Rep. 1844 (pt.2), p.96

Account of a cheap and portable self-registering Tide-Gauge, invented by John WOOD. 1844. Edinb. New Phil. Journ. XXXVIII., 1845, pp.71-76

On the terrestrial mechanism of the Tides. Edinb. Roy. Soc. Proc. I., 1845, pp.179-182

Notice of the remarkable mathematical properties of a certain parallelogram. Edinb. Roy. Soc. Proc. I., 1845, pp.187-188

On the law which connects the elastic force of vapour with its temperature. Edinb. Roy. Soc. Proc. I., 1845, pp.227-231

On the law which governs the resistance to motion of railway trains at high velocities. Brit. Assoc. Rep. 1846 (pt.2), pp.109-111

On the practical forms of breakwaters, sea walls, and other engineering works exposed to the action of the waves. Civ. Eng. Instit. Proc. VI., 1847, pp.135-143

On the practical forms of engineering works exposed to the action of the waves of the sea, and on the advantages and disadvantages of certain forms of construction for breakwaters and sea-walls. Franklin Inst. Journ. XIV., 1847, pp.13-15

On certain effects produced on sound by the rapid motion of the observer. Brit. Assoc. Rep. 1848 (pt.2), pp.37-38

On recent applications of the wave-principle to the practical construction of steam-vessels. Brit. Assoc. Rep. 1849 (pt.2), pp.30-33

On wave-line ships and yachts. Roy. Inst. Proc. I., 1851-1854, pp.115-119

On the progress of naval architecture and steam navigation, including a notice of the large ship of the Eastern Steam Navigation Company. Brit. Assoc. Rep. 1854 (pt.2), pp.160-161

Mechanical structure of the Great Eastern steamship. Brit. Assoc. Rep. 1857 (pt.2), pp.195-198

The Wave-line principle of ship construction. Naval Architects' Instit. Trans. I., 1860, pp.184-211; II., 1861, pp.230-245

Disturbing forces of locomotive engines. Civ. Eng. Instit. Proc. XXII., 1862-1863, pp.107-108

On the rolling of ships, as influenced by their forms and by the disposition of their weights. Naval Architects' Instit. Trans. IV., 1863, pp.219-231

Postscript to Mr. FROUDE's remarks on Rolling. Naval Architects' Instit. Trans. IV., 1863, pp.276-283

*Russell, John Scott* and (Sir) *John Robinson*: Report on Waves. Brit. Assoc. Rep. 1837, p.417-496; 1840, pp.441-443

*Russell, John Scott*: On gun-cotton. Quarterly Journ. Sci. I., 1864, pp.401-412

On the mechanical nature and uses of gun-cotton. 1864. Roy. Instit. Proc. IV., 1866, pp.292-299

For biography see Naval Architects Trans. 23, 1882, pp.258-261; Roy. Soc. Proc., 34, 1883, pp.xv-xvi

On the true nature of the wave of translation, and the part it plays in removing the water out of the way of a ship with least resistance. Naval Architects Trans., 20, 1879, pp.59-84

On the true nature of the resistance of armour to shot. Naval Architects Trans., 21, 1880, pp.69-92

On storm stability as distinguished from smooth-water stiffness. 1879. United Service Instit. Journ., 23, 1880, pp.821-849

The wave of translation and the work it does as the carrier wave of sound. Roy. Soc. Proc., 32, 1881, pp.382-383

For biography see also Inst. Civ. Engin. Proc. *87* (1886), pp.427-440

*Recent Biography*

John Scott Russell — A Great Victorian Engineer and Naval Architect. G.S. Emmerson, 1977 (John Murray, London) (Ref. [1.5])

*Some Other References of General Interest*

J. Roy. Soc. Arts *115* (Feb. 1967) pp.204-208 (March 1967) pp.299-302 (on John Scott Russell and Henry Cole, by G.P. Mabon)

"College Courant" Glasgow, Martinmas 1958, pp.28-37 and Whitsun 1959, pp.98-109 (on John Scott Russell and the 'Great Eastern' by A.M. Robb)

The Naval Archtect, January 1978, pp.30-35 (a review of John Scott Russell by G.S. Emmerson)

Harpers and Queen, October 1979, pp.258-259 (a poetic mention of John Scott Russell by Ian Hamilton Finlay)

## **Note Added in Proof** (Chapter 1)

For greater clarity of presentation we here expand the comment on the geometrical argument used by LUND [1.142]. The metric tensor $g_{\mu\nu}$ determines intrinsic properties of $V_n$ and fixes the 'first fundamental form' $ds^2 = g_{\mu\nu}\, dy^\mu\, dy^\nu$. The curvature tensor $L_{\mu\nu} = \partial X_\mu/\partial y^\nu \cdot \hat{X}_{n+1}$ (where $\hat{X}_{n+1}$ is the normalised vector $X_{n+1}$) describes extrinsic properties of $V_n$ relative to E and fixes the 'second fundamental form' $L_{\mu\nu}\, y^\mu\, y^\nu$. Given $g_{\mu\nu}$ and $L_{\mu\nu}$ the Gauss-Weingarten equations have a unique solution for X, $\hat{X}_{n+1}$ (for given initial values at a point) if and only if the Gauss-Codazzi system are satisfied. To reach the covariant form of (1.21) Lund chooses the first fundamental form $ds^2 \equiv Edx^2 + 2FdXdt + Gdt^2$ with $F \equiv 0$ and $E = \cos^2\theta$, $G = \sin^2\theta$. He then chooses the condition $L_{22} - L_{11} = 2 \sin\theta \cos\theta$ on the curvature tensor. The reason for this is that this together with the conditions on the metric represent the equations of motion of a relativistic string in a given external field—so this is where the physics enters the problem. The Gauss-Codazzi equations now relate $B \equiv \frac{1}{2}(L_{11} + L_{22})$ and $L_{12}$ to the two fields $\theta$ and $\lambda$. Thus the physics selects one particular class of surfaces.

At first sight these are formally surfaces of constant negative Gaussian curvature $K = -1$, since the Gauss-Weingarten system (1.122) formally takes the form (1.123) and is a generalised AKNS system, whilst it is proved from (1.123) following that all AKNS systems represent surfaces with $K = -1$. This proof, however, appeals to Gauss's fundamental theorem that the Gaussian curvature K is an intrinsic property of the surface: the AKNS system has $K = -1$ with respect to the particular metric (1.131). But Lund embeds his surface choosing an (in general different) metric, together with the condition on the extrinsic curvature tensor, $L_{\mu\nu}$. In consequence Lund's surfaces have curvature $K = (L_{11}L_{22} - L_{12}^2)/(EG - F^2)$, and, if $L_{11} - L_{22} = -2 \sin\theta \cos\theta$, $F = 0$ and $EG = \sin^2\theta \cos^2\theta$, then $K = -1$ if and only if $B^2 = L_{12}^2$. The two descriptions coincide if q is real in (1.122). For then $\lambda \equiv 0$, and $B = L_{12} (= 0)$. From (1.121) this is the sine-Gordon equation whose real surfaces represent surfaces of constant negative Gaussian curvature $K = -1$. Note that the proof that all AKNS systems represent surfaces with curvature $K = -1$ with respect to the metric (1.131) is formal in so far as the AKNS scattering problem involves two fields q and r or, for example, two complex fields q and q . The surface is therefore complex in general or more complicated, but it is a real surface with

$K = -1$ for real fields $r = \alpha q$ ($\alpha$ = constant), the case which includes the sine-Gordon equation.

Note that the work of LAMB [1.145] and LAKSHMANAN [1.146] referred to relate integrable systems like the s-G and NLS to the motion of strings. It is also possible to relate the general AKNS system to the motion of a string and then complete the connection with the geometry of the moving string with the geometry of surfaces following from (1.123) and with the vanishing curvature condition (1.125) (cf. M. Lakshmanan: Private communication and to be published).

## Additional References with Titles

In order to bring the several reference lists up to date the following references have been added in proof by some of the authors.


D. Anker, N.S. Freedman: Proc. Roy. Soc. London A*360*, 529 (1978)

A.A. Belavin, V.E. Zakharov: Multidimensional method of the inverse scattering problem and duality equations for the Yang-Mills fields. Pis'ma Zh. Eksp. Teor. Fiz. *25*(12), 603-607 (20 June 1977)

A.A. Belavin, V.E. Zakharov: Yang-Mills equations as inverse scattering problem. Phys. Lett. *73*B, 63 (1978)

V.A. Belinsky, V.E. Zakharov: Integration of the Einstein equations by the inverse scattering problem technique and the calculation of the exact soliton solutions. Zh. Eksp. Teor. Fiz. *75*, 1953-1971 (December 1978)

V.A. Belinsky, V.E. Zakharov: Stationary gravitational solitons with axial symmetry. Zh. Eksp. Teor. Fiz. *77*, 3-19 (1979)

S.V. Manakov, V.E. Zakharov, L.A. Bordag, A.R. Its, V.B. Matveev: Two-dimensional solitons of the Kadomtsev-Petviashvili equation and their interaction. Phys. Lett. *63*A, 205 (1977)

V.E. Zakharov, A.V. Mikhailov: Relativistically invariant Two-dimensional models of field theory which are integrable by means of the inverse problem method. Zh. Eksp. Teor. Fiz. *74*, 1953-1958 (June 1978)

V.E. Zakharov, L.A. Takhtajan: Equivalence of nonlinear Schrödinger equation and Heisenberg ferromagnet equation. Theor. Math. Phys. *38*(1), 26 (1979)

V.E. Zakharov, S.V. Manakov: Soliton theory. Sov. Sci. Rev. Sect.A *1*, 133 (1979)

K. Konno, M. Wadati: Simple derivation of Bäcklund transformation from Riccati form of inverse method. Prog. Theor. Phys. *53*, 1652 (1975)

M. Wadati, M. Toda: Bäcklund transformation for the exponential lattice. J. Phys. Soc. Jpn. *39*, 1196 (1975)

M. Toda, M. Wadati: Canonical transformations for the exponential lattice. J. Phys. Soc. Jpn. *39*, 1204 (1975)

M. Wadati: A remarkable transformation in nonlinear lattice problem. J. Phys. Soc. Jpn. *40*, 1517 (1976)

M. Wadati: On the exact solution of the Korteweg-de Vries equation. Sci. of Light (Tokyo) *25*, 37 (1976)

Y. Kodama, M. Wadati: Canonical transformation for sine-Gordon equation. Prog. Theor. Phys. *56*, 342 (1976)

Y. Kodama, M. Wadati: Wave propagation in nonlinear lattice. III. J. Phys. Soc. Jpn. *41*, 1499 (1976)

Y. Kodama, M. Wadati: Theory of canonical transformations for nonlinear evolution equations. I. Prog. Theor. Phys. *56*, 1740 (1976)

M. Wadati, M. Watanabe: Conservation laws of Volterra system and nonlinear self-dual network equation. Prog. Theor. Phys. *57*, 808 (1977)

M. Wadati, H. Sanuki, K. Konno, Y.H. Ichikawa: Circular polarized nonlinear Alfven waves. Rocky Mount. Math. J. *8*, 323 (1978)

Y.H. Ichikawa, M. Wadati: "Solitons in Plasmas and Other Dispersive Media—Dawn of Nonlinear Physics", in *Festschrift in honor of Professor T.Y. Wu*, ed. by S. Fujita (Gordon and Breach, London 1978) p.137

M. Wadati: Invariances and conservation laws of the Korteweg-de Vries equation. Stud. Appl. Math. *59*, 153 (1978)

M. Wadati: "Infinitesimal Transformations and Conservation Laws, Field Theoretic Approach to the Theory of Soliton", in *Research Notes in Mathematics*, Vol.26, ed. by F. Calogero (Pitman, London 1978) p.33

M. Wadati, K. Konno, Y.H. Ichikawa: A generalization of inverse scattering method. J. Phys. Soc. Jpn. *46*, 1965 (1979)

R. Hirota, M. Wadati: A functional integral representation of the soliton solution. J. Phys. Soc. Jpn. *47*, 1385 (1979)

M. Wadati, K. Konno, Y.H. Ichikawa: New integrable nonlinear equations. J. Phys. Soc. Jpn. *47*, 1698 (1979)

Y.H. Ichikawa, K. Konno, M. Wadati, H. Sanuki: Spiky soliton in circular polarized Alfven wave. J. Phys. Soc. Jpn. *48*(1) (1980)

M. Wadati, K. Sawada: New representations of the soliton solution for the Korteweg-de Vries equation. J. Phys. Soc. Jpn. *48*(1) (1980)

M. Wadati, K. Sawada: Application of the trace method to the modified Korteweg-de Vries equation. J. Phys. Soc. Jpn. *48*(1) (1980)

K. Fukushima, M. Wadati, T. Kotera, K. Sawada, Y. Narahara: Experimental and theoretical study of the recurrence phenomena in nonlinear transmission line. J. Phys. Soc. Jpn. (to be published)

T. Shimizu, M. Wadati: A new integrable nonlinear evolution equation. Prog. Theor. Phys. (to be published)

A. Degasperis: "Solitons, Boomerons and Trappons", in *Nonlinear Evolution Equations Solvable by the Spectral Transform*, Research Notes in Mathematics, Vol.26, ed. by F. Calogero (Pitman, London 1978)

A. Degasperis: "Spectral Transform and Solvability of Nonlinear Evolution Equations", in *Nonlinear Problems in Theoretical Physics*, Lectures Notes in Physics, Vol.98, ed. by A.F. Ranada (Springer, Berlin, Heidelberg, New York 1979)

F. Calogero, A. Degasperis: Reduction technique for matrix nonlinear evolution equations. J. Maths. Phys. (in press, 1980)

# Subject Index

# Cavitation and Inhomogeneities in Underwater Acoustics

Proceedings of the First International Conference Göttingen, Fed. Rep. of Germany, July 9–11, 1979
Editor: W. Lauterborn

1980. 192 figures, 6 tables. XI, 319 pages
(Springer Series in Electrophysics, Volume 4)
ISBN 3-540-09939-5

**Contents:**
Cavitation. – Sound Waves and Bubbles. – Bubble Spectrometry. – Particle Detection. – Inhomogeneities in Ocean Acoustics. – Index of Contributors.

# Ocean Acoustics

Editor: J. A. DeSanto

1979. 109 figures, 5 tables. XI, 295 pages
(Topics in Current Physics, Volume 8)
ISBN 3-540-09148-3

**Contents:**
*J. A. DeSanto:* Introduction. – *J. A. DeSanto:* Theoretical Methods in Ocean Acoustics. – *F. R. DiNapoli, R. L. Deavenport:* Numerical Models of Underwater Acoustic Propagation. – *J. G. Zornig:* Physical Modelling of Underwater Acoustics. – *J. P. Dugan:* Oceanography in Underwater Acoustics. – *N. Bleistein, J. K. Cohen:* Inverse Methods for Reflector Mapping and Sound Speed Profiling. – *R. P. Porter:* Acoustic Probing of Space-Time Scales in the Ocean. – Subject Index.

# Structural Stability in Physics

Proceedings of Two International Symposia on Applications of Catastrophe Theory and Topological Concepts in Physics Tübingen, Fed. Rep. of Germany, May 2–6 and December 11–14, 1978
Editors: W. Güttinger, H. Eikemeier

Springer Series in Synergetics
1979. 108 figures, 8 tables. VIII, 311 pages
ISBN 3-540-09463-6

**Contents:**
Introduction. – General Concepts. – Topological Aspects of Wave Motion. – Catastrophes in Infinite Dimensions. – Defects and Dislocations. – Statistical Mechanics and Phase Transitions. – Solitons. – Dynamical Systems. – Index of Contributors.

# Synergetics

A Workshop

Proceedings of the International Workshop on Synergetics at Schloss Elmau, Bavaria, May 2–7, 1977
Editor: H. Haken

Springer Series in Synergetics
1977. 136 figures. VIII, 274 pages
ISBN 3-540-08483-5

**Contents:**
General Concepts. – Bifurcation Theory. – Instabilities in Fluid Dynamics. – Instabilities in Astrophysics. – Solitons. – Nonequilibrium Phase Transitions in Chemical Reactions. – Chemical Waves and Turbulence. – Morphogenesis. – Biological Systems. – General System.